Sensors for Air Quality Monitoring

Sensors for Air Quality Monitoring

Collection Editors

Klaus Schäfer
Matthias Budde

MDPI

Basel • Beijing • Wuhan • Barcelona • Belgrade • Novi Sad • Cluj • Manchester

Collection Editors
Klaus Schäfer
Atmospheric Physics
Consultant
Garmisch-Partenkirchen
Germany

Matthias Budde
Karlsruhe Institute of
Technology
Karlsruhe
Germany

Editorial Office
MDPI AG
Grosspeteranlage 5
4052 Basel, Switzerland

This is a reprint of the Topical Collection, published open access by the journal *Sensors* (ISSN 1424-8220), freely accessible at: https://www.mdpi.com/journal/sensors/topical_collections/Sensors_Air_Quality_Monitoring.

For citation purposes, cite each article independently as indicated on the article page online and as indicated below:

Lastname, A.A.; Lastname, B.B. Article Title. *Journal Name* **Year**, *Volume Number*, Page Range.

ISBN 978-3-7258-4241-4 (Hbk)
ISBN 978-3-7258-4242-1 (PDF)
https://doi.org/10.3390/books978-3-7258-4242-1

Contents

Preface

The scientific results of national and international projects in research and development are presented, together with some overviews about the status of sensors for air quality monitoring. The authors provide outlooks for sensor development and applications of sensor network data, showing the results of sensor performance evaluations, data analyses, and proper network operation. This content is addressed not only to researchers in the field of air quality studies but also to academics, stakeholders, and technicians for air quality management. The editors would like to thank all the authors, as well as the reviewers of the submitted manuscripts, for presenting these high-quality reprints of publications in the Special Issue of the journal *Sensors*.

Klaus Schäfer and Matthias Budde
Collection Editors

sensors

Article

Comparing Airborne Particulate Matter Intake Dose Assessment Models Using Low-Cost Portable Sensor Data

Rok Novak [1,2,*], David Kocman [1], Johanna Amalia Robinson [1,2], Tjaša Kanduč [1], Dimosthenis Sarigiannis [3,4,5] and Milena Horvat [1,2]

1 Department of Environmental Sciences, Jožef Stefan Institute, 1000 Ljubljana, Slovenia; david.kocman@ijs.si (D.K.); johanna.robinson@ijs.si (J.A.R.); tjasa.kanduc@ijs.si (T.K.); milena.horvat@ijs.si (M.H.)
2 Jožef Stefan International Postgraduate School, 1000 Ljubljana, Slovenia
3 Environmental Engineering Laboratory, Department of Chemical Engineering, Aristotle University of Thessaloniki, 54124 Thessaloniki, Greece; denis@eng.auth.gr
4 HERACLES Research Centre on the Exposome and Health, Center for Interdisciplinary Research and Innovation, 54124 Thessaloniki, Greece
5 University School of Advanced Study IUSS, 27100 Pavia, Italy
* Correspondence: rok.novak@ijs.si

Received: 6 February 2020; Accepted: 2 March 2020; Published: 4 March 2020

Abstract: Low-cost sensors can be used to improve the temporal and spatial resolution of an individual's particulate matter (PM) intake dose assessment. In this work, personal activity monitors were used to measure heart rate (proxy for minute ventilation), and low-cost PM sensors were used to measure concentrations of PM. Intake dose was assessed as a product of PM concentration and minute ventilation, using four models with increasing complexity. The two models that use heart rate as a variable had the most consistent results and showed a good response to variations in PM concentrations and heart rate. On the other hand, the two models using generalized population data of minute ventilation expectably yielded more coarse information on the intake dose. Aggregated weekly intake doses did not vary significantly between the models (6–22%). Propagation of uncertainty was assessed for each model, however, differences in their underlying assumptions made them incomparable. The most complex minute ventilation model, with heart rate as a variable, has shown slightly lower uncertainty than the model using fewer variables. Similarly, among the non-heart rate models, the one using real-time activity data has less uncertainty. Minute ventilation models contribute the most to the overall intake dose model uncertainty, followed closely by the low-cost personal activity monitors. The lack of a common methodology to assess the intake dose and quantifying related uncertainties is evident and should be a subject of further research.

Keywords: dose assessment; particulate matter; minute ventilation; low-cost sensors; uncertainty assessment

1. Introduction

Application of low-cost air quality (AQ) sensors is on the rise and is being used to determine air pollution in cities [1–4], monitoring of indoor AQ [5–7], and for exposure assessment [8–10]. Traditionally exposure studies use data from monitoring stations, questionnaires, or biomarkers [11], and more recently land-use regression models [12,13] and other modelling techniques [14], while the most sought-after method is measuring intake dose on a personal level [15]. To this end, low-cost sensors that have become smaller and more energy-efficient, and now enable subjects to carry these devices with them, can significantly improve the temporal and spatial resolution of information needed [16]. However, although continuous and rapid advances in sensing technologies are resulting in improved accuracy, these devices still need extra validation and/or calibration before being put to use [9,17,18]. They could employ a wide array of options to achieve more accurate results, such as comparing with reference analysers [8] or using sophisticated artificial intelligence approaches [19]. On the other hand, assessing intake dose is not only dependent on the concentrations of pollutants, but also other factors, mainly a person's breathing rate or ventilation [20]. Several studies throughout the past three decades have shown that minute ventilation is corelated with heart rate [21–24]. Ventilation can be estimated by various approaches and models, which differ mostly by the number and type of variables used, from more generalized approaches using sex, age, and ethnicity [25] with different kinds of activities [26], to more specific and complex models with additional variables such heart rate [27], forced vital capacity and breath frequency [28], and hip circumference [29]. All of these approaches do offer some advantages, often as trade-offs to accuracy. Less complex approaches use personal information, such as age and sex, and determine minute ventilation from generalized population data [25,30], while more complex models use continuously monitored variables, such as heart rate (HR) [27,28].

The aims of this study are as follows:

- to evaluate the applicability of different intake dose models by assessing the uncertainty associated with each input variable;
- to estimate how the uncertainty propagates forward and affects the uncertainty of the model;
- to compare the results calculated with the models on two contrasting individuals;
- to evaluate the complexity of the models, time, and resource requirements and the burdens participants have in providing the data.

In this work, four different approaches to assess the PM intake dose are compared, using data obtained by two participants included in the sampling campaign conducted within the ICARUS H2020 project [31], which was separated into winter (February) and summer (May) campaigns, and took place in the first half of 2019. The participants carried a portable PM sensor and a heart rate monitor with them at all times and measured indoor and outdoor concentrations of PM during the entire seven-day period.

The uncertainty associated with each intake dose assessment model was quantified and the hypothesis was that the less complex models would provide data with more uncertainty, as they use variables that have higher uncertainties and are based on more generalizations (e.g., average minute ventilation for a 60-year-old female in a less complex model, in contrast with minute ventilation derived directly from measured heart rate in more complex models). Propagation of uncertainty from low-cost sensors and minute ventilation models to intake dose assessments was investigated. A crucial component of the overall uncertainty assessment is to determine the validity of the PM concentration data from the low-cost sensor. To this end, the performance of the low-cost PM sensor was evaluated by collocating it with a reference instrument in an office environment. Moreover, uncertainties calculated and presented for each minute ventilation model were not consistent from paper to paper, as was the nomenclature regarding exposure science and metrology. These issues are addressed and discussed.

2. Materials and Methods

2.1. Terminology and Nomenclature

Terminology and nomenclature used in this work are based on the following sources. Terms regarding the human–pollutant interaction (e.g., "personal exposure", "intake dose", "exposure assessment") were adopted from the Official International Society for Exposure Analysis glossary by Zartarian et al. [31]. Terms related to metrology and statistics (e.g., "uncertainty", "reproducibility", "validity") were adopted from the International Vocabulary of Metrology [32].

2.2. Measuring Particulate Matter Concentrations

A portable Arduino based low-cost PM measuring unit (referred to as PPM) was developed for the ICARUS project [33] by IoTech Telecommunications, Thessaloniki, Greece [34], and used in this research. Using Plantower, Beijing, China, pms5003 sensor [35], based on the laser light scattering principle, the PPM unit provides data for concentrations of PM in one-minute resolution.

PM concentration data are provided in three size classes/channels: <1 µm (PM_1), <2.5 µm ($PM_{2.5}$), and <10 µm (PM_{10}). A detailed description of the instrument with specifications is provided in the Supplementary Materials.

Weekly and daily averages of $PM_{2.5}$ and PM_{10} measurements from the collocation and personal monitors were additionally compared with values obtained from the government run AQ station, near the centre of Ljubljana (2 km from where the collocation took place), to determine if the values were close. The averages were compared to determine if the values measured by the sensors were in the same range as those measured at the AQ station.

Collocation of the PPM Unit with a Reference Instrument

The PPM unit was collocated with a GRIMM (Durag Group, Hamburg, Germany) Model 11-A (1.109) Aerosol Spectrometer (GRIMM), which was used as a research-grade reference instrument for PM measurements. The collocation lasted one week, from 5 to 12 March 2019, and was located at an office space with open windows at Jožef Stefan Institute, Ljubljana, Slovenia (LAT: 14.4879, LON: 46.0424). The PPM unit provided data with minute resolution and GRIMM with five-minute resolution. Data recovery for the measuring period was 100% for the PPM unit and 99.8% for GRIMM. A more detailed description of the instrument is available in the Supplementary Materials.

There was a four-hour period with light precipitation during the collocation, the mean (min – max) temperature throughout the week was 12.3°C (6.3°C–16.3°C) and relative humidity was 55% (36–65%).

2.3. Measuring Heart Rate and Physical Activity

Continuous HR measurements were made by using a commercial smart activity tracker (SAT), a Vívosmart 3 from Garmin International [36]. The data were in minute temporal resolution and provided values for HR, specific physical activity, steps taken, calories "burned", distance walked, and stress.

Uncertainty of SAT was estimated based on the work of Oniani et al. [37], who compared the same device (Garmin Vivosmart 3) with an ECG (electrocardiogram). In their work, four participants were selected, and equipped with four SAT devices each. An 80-min treadmill test was performed with each participant, with four different speeds of walking, while being connected to an ECG monitor. The results were presented with MAPE (mean absolute percentage error) and ICC (intraclass correlation) for each device, in comparison to the ECG.

2.4. Calculating Personal Intake Dose of Airborne Particulate Matter

Intake dose of PM was calculated as a product of PM concentration, in this case, PM_1 concentrations for one-minute average values (as the results of the collocation showed, this proved to be the measurement with the lowest uncertainty associated with it), measured with a PPM device, and minute ventilation data, which was determined by four different models described below. The intake dose model (M1–M4) calculation is presented in Equation (1):

$$intake\ dose = \dot{V}_E * PM_1, \quad (1)$$

where $\dot{V}_E$ presents minute ventilation (L min^{-1}) and PM_1 is the particulate matter concentration measured with the PPM sensor (µg m^{-3}).

Two of the described models (M1 and M2) use HR as a variable, and two (M3 and M4) do not, and in turn use average minute ventilation data for specific population groups. Comparison between the four models was performed based on the data from two participants involved in the ICARUS personal exposure assessment campaign in Ljubljana:

- P1: 60-year-old Caucasian female participant, weighing 62 kg, height 166 cm. P1 participated in the campaign between 17 and 23 February 2019, and was located in Ljubljana the entire period. P1 was employed as an office worker.
- P2: 35-year-old Caucasian male participant, weighing 66 kg, height 178 cm. P2 participated in the campaign between 14 and 21 May 2019, and was also located in Ljubljana the entire period. P2 was employed as a bike courier.

Using only one participant could skew the results, including another participant with a contrasting profile (different personal characteristics, such as sex, height, and age) enables a more thorough comparison. Two participants are enough for the purposes of this research, as the goal is to compare models and not validate them for larger groups.

2.4.1. Model 1 (M1)

Minute ventilation model in M1 ($\dot{V}_E^1$) is based on the work of Greenwald et al. [28], who modelled minute ventilation with HR, age, sex, and forced vital capacity (FVC) as variables, with the explicit goal to use these data in pollution intake dose estimates. Data from 471 subjects from 8 different studies were compiled in their research and enabled the researchers to gather a dataset of 14,550 one-minute data points. Here, their best performing model was selected, using HR data, combined with information about the subject's sex, age, height, and weight (used in determining FVC):

$$\dot{V}_E^1 = e^{-9.59} HR^{2.39} age^{0.274} sex^{-0.204} FVC^{0.520}, \tag{2}$$

where $\dot{V}_E^1$ presents minute ventilation for M1; *age* is the age of the participant in years; *sex* is the participants sex, where value 1 is male and 2 is female; and *FVC* is forced vital capacity.

FVC factor was estimated using the Global Lung Function Initiative methodology [38]. FVC for P1 was 3.32 l (lower limit at 2.61 l, upper at 4.16 l, for a 90% confidence interval) and for P2 5.37 l (lower: 4.30 l, upper: 6.45 l, for a 90% confidence interval).

M1 is calculated based on Equation (1).

2.4.2. Model 2 (M2)

$\dot{V}_E^2$ is based on the work of Zuurbier et al. [27], who used a simplified approach with only HR and sex as variables:

$$\dot{V}_E^2 = (a * HR + b)^e, \tag{3}$$

where $\dot{V}_E^2$ presents minute ventilation for M2, HR stands for heart rate, and a and b present the slope and intersect based on sex (a is 0.023 and 0.021, b is 0.57 and 1.03, for females and males, respectively). Their model is based on a study performed with 34 participants.

M2 is calculated based on Equation (1).

2.4.3. Model 3 (M3)

$\dot{V}_E^3$ follows an approach by Sarigiannis et al. [25], modelling mercury intake by combining age and ethnicity-specific data of activity patterns, inhalation rates, and body weight, with a specific type of microenvironment. Madureira et al. [30] use a similar approach for indoor intake dose of bioaerosol particles. Each observation is multiplied with a breathing rate factor corresponding to the (hourly) activity self-reported by the participant, but the model does not use any continuous variable, such as HR.

Following the described methods [25,30], in this research, activity reported by the participant in a time activity diary (TAD) was differentiated into four groups, as listed in the U.S. Environmental Protection Agency (EPA) Exposure Factors Handbook [39]: sedentary and passive activities (includes sleep, nap, resting, working behind a desk, and watching TV), light intensity activities (cleaning, cooking), moderate intensity activities (walking, working in garden), and heavy intensity activities (sports, hard manual labour). Average minute ventilation was provided by this handbook for each type of activity, differentiated by age, sex, and body weight. $\dot{V}_E^3$ uses this information to determine minute ventilation for each hourly interval.

M3 is calculated based on Equation (1).

2.4.4. Model 4 (M4)

$\dot{V}_E^4$ uses one of the most basic approaches to determine minute ventilation using only a few general data points for the subject: age, sex, and weight. The EPA Exposure Factors Handbook, which provides estimated minute ventilation according to the mean values determined for specific groups [39], also provides generalized data for time spent in micro-locations, doing specific activities for each (age and sex) group. With this information, using Equation (4), it is possible to calculate average minute ventilation, weighted for per cent of time spent doing each activity.

$$\dot{V}_E^4 = (sP * aV_P + sLi * aV_{Li} + sMi * aV_{Mi} + sHi * aV_{Hi}) * BW, \tag{4}$$

where $\dot{V}_E^4$ is minute ventilation for M4; and sP, sLi, sMi, and sHi present daily shares of time spent doing P—passive, Li—light intensity, Mi—moderate intensity, and Hi—high-intensity activities, respectively. The aV factors present average ventilations for that specific activity, according to age group and sex. Factor BW is the subjects body weight, which must be included because the $\dot{V}_E$ data in the EPA handbook are presented in "per kg of body weight" form.

M4 is calculated based on Equation (1).

2.5. *Statistical Analysis and Determining the Uncertainty*

After collocating PPM with the GRIMM, to determine the validity, a Wilks–Shapiro test was conducted for each time-averaging interval to numerically determine normality and a q–q plot was made to visually determine normality. The distribution was non-normal, which prompted the use of the Spearman rank-order correlation test. For each comparison, a scatter plot was made with a linear regression line and 95% confidence interval. RMSE (root mean square error), MAPE (mean absolute percentage error), MAE (mean absolute error), R^2, slope, and intercept values were calculated.

Summary statistics were calculated for all four models, PM_1 concentrations, and HR, and for the results iterating M1 over different heights and weights, as presented in the Supplementary Data.

The uncertainty for the PPM was estimated by collocating it with the GRIMM, which has a reproducibility of 5% for the whole range [40]. This measure is carried forward to the PPM device through the collocation process, which enabled the calculation of several statistical measures of agreement (R^2, MAPE, MAE, RMSE, MAE%). According to the GRIMM manual description, the MAPE measure is the closest, and the uncertainty from the GRIMM is carried forward through the following equation:

$$u(PPM) = \sqrt{u(Grimm)^2 + u(comparison)^2}, \tag{5}$$

where u is the uncertainty.

All models used in this research had some measure of agreement listed in their evaluation. Not all measures were the same, which makes some of the results incomparable between the models.

$\dot{V}_E^1$, from Greenwald et al. [28], had the uncertainty expressed as "per cent error", which is "the difference between predictions and observations from cross-validation", and in IQR (interquartile range). The median (IQR) per cent error was −0.664 (45.4)% [28]. With the Supplementary Data, provided by the authors of the paper, it was possible to calculate other statistical measures (R^2, MAPE, MAE, RMSE). This calculated uncertainty presupposes that all the variables used are categorical (in the cases of sex and weight, this is correct) and without uncertainty. This is not the case in this research, where the SAT device has some uncertainty associated with it, as does the FVC value, with both having different exponents in the model (2.39 for HR and 0.52 for FVC). To determine the overall uncertainty of the model, all the component uncertainties were combined.

$\dot{V}_E^2$, based on Zuurbier et al. [27], had its uncertainty presented with mean R^2 values for each sex, with SD and range [27]. In this case, the mean (SD) R^2 values were 0.89 (0.06), 0.90 (0.07), and 0.90 (0.07) for women, men, and all together, respectively. Uncertainty explained in this way is different than in $\dot{V}_E^1$ and cannot be compared. To obtain a better comparison of the $\dot{V}_E^1$ and $\dot{V}_E^2$ models, the supplementary data from Greenwald et al. [28] were used to calculate minute ventilation with the $\dot{V}_E^2$ model Equation (3) and compare it with the measured minute ventilation values in the Supplementary Data. The same statistical measures were calculated for $\dot{V}_E^2$, as they were for $\dot{V}_E^1$. The SAT uncertainty was also incorporated in the overall uncertainty of the model.

$\dot{V}_E^3$ uses data from tables provided by the EPA for certain age groups, which includes mean minute ventilation values and some specific percentiles, such as the 5th and 95th percentile. The difference between these two values provides a 90% confidence interval for the values used in this model. Table 1 shows the percentiles of minute ventilation for P1 and P2 participants involved in this research. The difference between the mean and the percentile is slightly larger at the 95th percentile than at the 5th percentile, with the average value being 34% for the 95th and 30% for the 5th percentile. The overall uncertainty estimate was determined as the mean of all the differences with SD.

Table 1. Mean, 5th, and 95th percentile minute ventilation values for P1 and P2 with calculated differences between the percentiles and the mean, provided in the Environmental Protection Agency (EPA) Exposure Handbook [39].

	Activity	Mean [L/min]	5th % [L/min]	95th % [L/min]	$\frac{\lvert 5^{th} - mean \rvert}{mean}$	$\frac{\lvert 95^{th} - mean \rvert}{mean}$
P1	Sedentary and Passive	4.1	2.9	5.6	0.30	0.36
	Light intensity	10	7.4	13	0.25	0.30
	Moderate intensity	21	14	30	0.31	0.43
	High intensity	39	24	58	0.38	0.46
P2	Sedentary and Passive	4.4	2.9	5.3	0.35	0.22
	Light intensity	11	11	13	0.05	0.22
	Moderate intensity	24	15	32	0.36	0.37
	High intensity	43	27	58	0.36	0.36

Although $\dot{V}_E^4$ uses a similar approach as $\dot{V}_E^3$, the uncertainty is different as the "share of the day" values also have 5th and 95th percentile values and are not definitive, as in $\dot{V}_E^3$. As shown in Table 2, the differences between the 5th and 95th percentiles and the mean vary quite substantially between lower and higher intensity activities, with an average of 12% (14% for P2) in the "sedentary and passive" category, and 121% (132% for P2) in the "high intensity" category. Each uncertainty interval was weighted by the percentage of the day it represents, in contrast to the percentage it "should" represent, which in this case is $\frac{1}{4}$. To assess the final uncertainty for $\dot{V}_E^4$, the uncertainty from minute ventilation, as shown in Table 2, must be added by the method used in Equation (5).

Table 2. Mean, 5th, and 95th percentile daily share of activity for P1 and P2, with calculated differences between the percentiles and the mean, provided in the EPA Exposure Handbook [39].

	Activity	Mean [hours]	5th % [hours]	95th % [hours]	$\frac{\lvert 5^{th} - mean \rvert}{mean}$	$\frac{\lvert 95^{th} - mean \rvert}{mean}$
P1	Sedentary and Passive	13	11	14	0.12	0.12
	Light intensity	6.5	4.1	9.4	0.37	0.45
	Moderate intensity	4.6	1.7	7.1	0.63	0.56
	High intensity	0.3	0.03	0.9	0.91	1.5
P2	Sedentary and Passive	12	11	14	0.13	0.14
	Light intensity	5.7	2.8	10	0.51	0.83
	Moderate intensity	5.7	1.3	8.9	0.78	0.56
	High intensity	0.4	0.03	1.0	0.92	1.71

As calculating the intake dose is a product of PM concentration values and calculated minute ventilation values, the uncertainty propagation is calculated by the method described in Equation (5).

All the calculations and visualizations were made in R v3.61 [41].

3. Results

3.1. Results of the Collocation

Figure 1 shows the correlation plots between GRIMM and PPM for the collocation with three different time intervals (5, 30, and 60 min) and all three particle size classes (1, 2.5, and 10 µm). Increasing the particle size reduces the linearity of the data points along the linear model regression line, which is also apparent in the R^2 values that start at ~0.97 for PM_1, drop to ~0.89 for $PM_{2.5}$, and further drop down to ~0.68 for PM_{10} particles. As presented in Table 3, R^2 values slightly increase with larger time intervals, as evident with PM_1 particles (from 0.97 to 0.98) and with PM_{10} particles (from 0.66 to 0.69). These increases are relatively minimal and counteract the increase in the confidence interval with larger time intervals. Similarly, as in the case of R^2, RMSE values increase as the size of the particles increases.

Mean (min–max) concentrations recorded for the PPM were 13.2 (0.4–47.4) µg/m^3, 18.4 (0.8–61.6) µg/m^3, and 20.6 (1.0–69.8) µg/m^3 for PM_1, $PM_{2.5}$, and PM_{10}, respectively, while the mean (min–max) GRIMM values were 11.7 (1.6–40) µg/m^3, 15.2 (2.4–44.6) µg/m^3, and 19.4 (2.7–103.4) µg/m^3 for PM_1, $PM_{2.5}$, and PM_{10}, respectively. These numbers generally coincide with measurements from the government-run AQ station in Ljubljana, which showed an average concentration of 11.4 µg/m^3 for $PM_{2.5}$ and 18.9 µg/m^3 for PM_{10} [42,43] for the same time period. For $PM_{2.5}$, the PPM device measured 7.0 µg/m^3 higher average values than the AQ station, and the GRIMM device 3.8 µg/m^3 higher values. For PM_{10}, these values were 1.7 µg/m^3 higher for the PPM, and 0.5 µg/m^3 higher for GRIMM. As the distance to the AQ station was 2 km, some deviation is expected and these numbers fall in this range of expectations.

Although the PPM unit was validated for 5-, 30-, and 60-min intervals, intraclass differences of R^2 and RMSE values varied less than interclass differences. The uncertainty associated with 1-min values for each size class should, therefore, be similar to the values calculated for 5-min averages. These results show that the PM_1 values with the highest possible temporal resolution (1-min) have the least uncertainty, and are used to calculate intake doses.

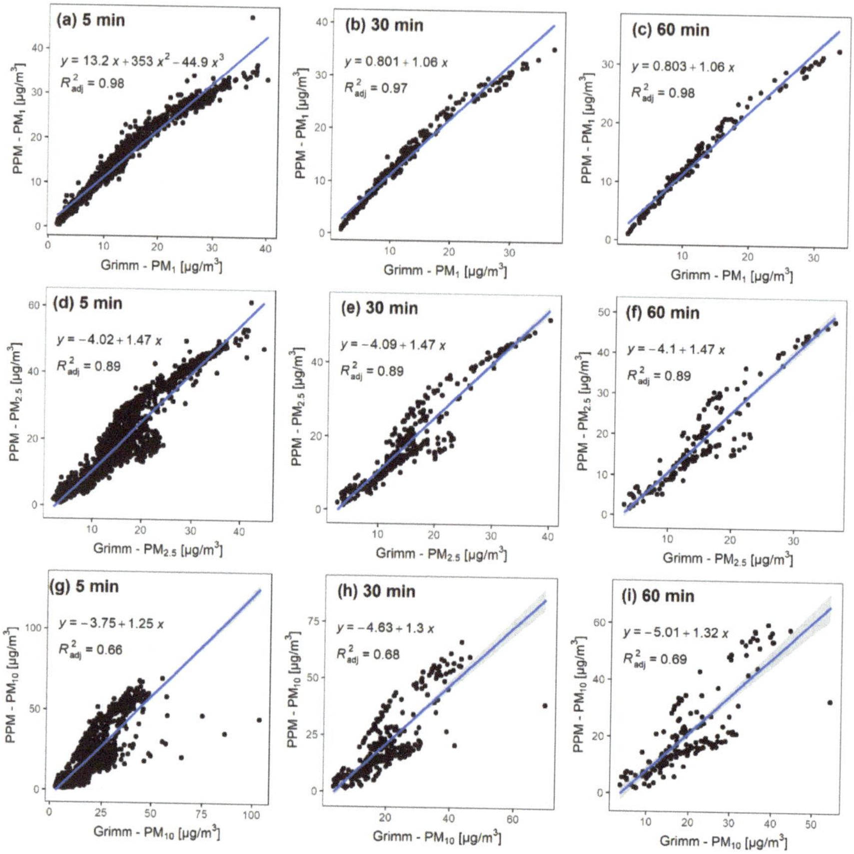

Figure 1. Correlation plots from collocating the portable Arduino based low-cost particulate matter (PM) measuring (PPM) unit with GRIMM. Rows present different sizes of particulate matter (PM_1 (**a–c**), $PM_{2.5}$ (**d–f**), PM_{10} (**g–i**)) and columns different time intervals (5 min (**a**,**d**,**g**), 30 min (**b**,**e**,**h**), 60 min (**c**,**f**,**i**)).

Table 3. Relationship between portable Arduino based low-cost particulate matter (PM) measuring (PPM) unit and GRIMM. RMSE, root mean square error.

PM class	Time	R^2	RMSE	Intercept	Slope
PM1	5 min	0.97	2.15	0.83	1.06
	30 min	0.97	2.01	0.80	1.06
	60 min	0.98	1.96	0.80	1.06
PM2.5	5 min	0.89	6.30	−4.02	1.47
	30 min	0.89	6.17	−4.09	1.47
	60 min	0.89	6.11	−4.10	1.47
PM10	5 min	0.66	9.07	−3.75	1.25
	30 min	0.68	8.76	−4.63	1.30
	60 min	0.69	8.58	−5.01	1.32

3.2. Intake Dose Results

The results of the calculations, based on all four intake dose models, are shown in Figure 2, plot (a), accompanied by plotted PM_1 values for each participant in plot (b) and HR values in plot (c).

As evident in plot (b), the PM_1 concentrations were higher for P2 than for P1. This is also evident in Table 4, where the summary statistics for PM_1 concentrations show higher numbers for P2. The mean PM_1 values for P1 and P2 were 8.1 μg/m^3 and 28.6 μg/m^3, respectively, which is more than a three-fold difference, and the maximum PM_1 values were 87.0 μg/m^3 and 338.0 μg/m^3 for P1 and P2, respectively. The measured $PM_{2.5}$ values were in the same range as values reported by the government-run AQ station in Ljubljana. Summary statistics for HR show that the maximum and standard deviation (SD) values were higher for P2, but the differences are not as pronounced as with PM concentrations.

M1 and **M2** intake dose assessments show a strong relationship. As evident in Table 4, both have similar descriptive statistics, except the maximum intake dose value, which is noticeably higher for M1 with both P1 and P2.

M3 follows a somewhat similar pattern as M1 and M2, although some deviations are evident. For P1, M3 mean value is ~15% higher than that of M1 and M2; the SD is almost double; and the maximum value is 11% and 26% higher than M1 and M2, respectively. The median and the Q1 and Q3 values are 10–30% lower. Similar ratios are found for P2, except the median value is almost half that of M1 and M2.

M4 shows noticeably different results for P1 and P2. As evident from plot (a) in Figure 2, the results of M4 for P1 mostly follow the same trend as the other models. The summary statistics for M4, shown in Table 4 (for P1), show somewhat higher values than those of M1 and M2. For P2, as shown in Figure 2, the M4 results do not follow the trend of the doses based on other models as well as for P1. Although the mean, SD, and Q3 values are lower than in the other three, the median is almost the same (262.6 ng/min for M1 and M2, and 263.3 ng/min for M4).

Table 4. Descriptive statistics for intake dose assessments based on all four models, PM_1 concentrations, and heart rate (HR) values for both participants. Recovery represents the percent of data recovered, where 100% is the entire period of observation. Sum represents the accumulated dose for the entire week of observation.

			Participant 1 (P1)			
	M1	**M2**	**M3**	**M4**	**PM1 [μg/m^3]**	**HR [bpm]**
Mean	60.2	58.8	69.6	75.9	8.1	63.8
SD	58.9	53.1	99.6	55.3	5.9	11.9
Median	43.3	44.4	29.1	65.8	7.0	62.0
Q1–Q3	25.0–73.7	26.2–74.3	16.6–63.1	37.6–104	4.0–11.0	55.0–70.0
Min–Max	0.0–729	0.0–609	0.0–820	0.0–818	0.0–87.0	38.0–148
Recovery [%]	78.2	78.2	80.7	80.7	80.7	97.2
Sum	599,520	580,128	589,983	617,486	67,352	/
			Participant 2 (P2)			
	M1	**M2**	**M3**	**M4**	**PM1 [μg/m^3]**	**HR [bpm]**
Mean	415	359	493	314	28.6	66.3
SD	476	359	698	280	25.5	18.2
Median	263	262	109	263	24.0	63.0
Q1–Q3	108–542	140–465	91.5–595	219–373	20.0–34.0	52.8–76.0
Min–Max	0.0–5110	0.0–4033	0.0–5313	0.0–3708	0.0–338	39.0–170
Recovery [%]	66.7	66.7	51.2	68.1	68.1	98.6
Sum	2,764,423	2,391,141	2,522,915	2,136,003	194,702	/

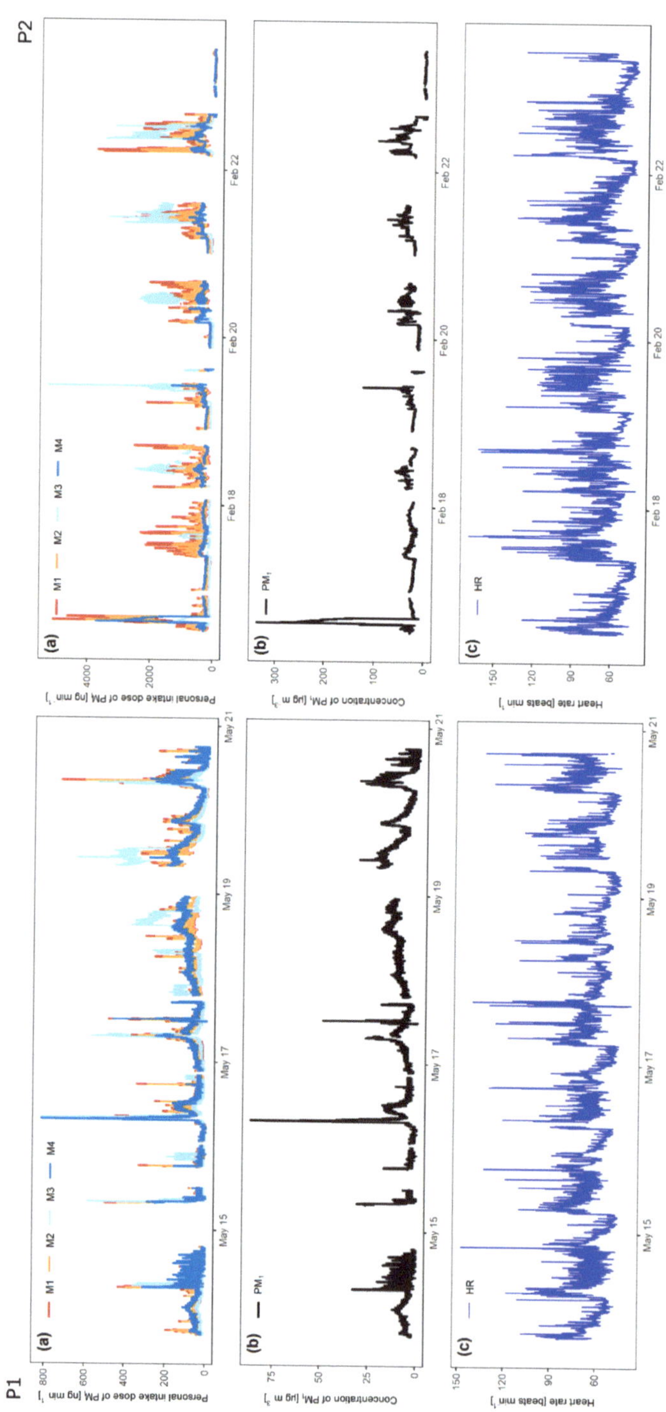

Figure 2. (**a**) Calculated intake dose of PM_1 for all four models, (**b**) measured concentrations of PM_1, and (**c**) heart rate in beats per minute. Left side for participant 1 (P1) and right side for participant 2 (P2).

3.3. Results of Quantifying Uncertainty

All the sensors used in this research have some uncertainty in their measurements, which is carried on to minute ventilation and intake dose calculations.

3.3.1. Uncertainty in PM and Heart Rate Sensors

The results for the statistical measures of agreement with the reference data calculated for the PPM were as follows: $R^2 = 0.97$, MAPE = 15.62%, MAE = 1.66, RMSE = 2.15, and MAE% = 14.17%. After including the uncertainty (listed as reproducibility) of the GRIMM, the overall uncertainty of the **PPM** is ~16%, which is estimated with the uncertainty propagation Equation (5).

The results from Oniani et al. [37] show that there is some disagreement between the **SAT** devices and the ECG. MAPE values ranged from 4.34% to 16.00% (mean: 9.82%, median: 8.85%, SD: 3.56%), and ICC from 0.91 to 0.02 (mean: 0.67, median: 0.71, SD: 0.25).

3.3.2. Uncertainties for Minute Ventilation Models

$\dot{V}_E^1$ statistical measures of agreement with the reference data were as follows: $R^2 = 0.82$, MAPE = 28%, MAE = 6.47, and RMSE = 9.41. Combining the mean value of the SAT uncertainty (9.82% ± 3.56%; weighted: 9.82% ± 3.56% *2.39 = 24% ± 9%), the mean value of the FVC 90% confidence interval (~22% ± 2%; weighted: 22% ± 2% *0.52 = 11% ± 1%) and the MAPE of the model, using the approach described in Equation (5), yields an estimate of overall uncertainty for $\dot{V}_E^1$ of ~38% ± 9%.

$\dot{V}_E^2$ calculated statistical measures were as follows: $R^2 = 0.72$, MAPE = 32%, MAE = 7.68, and RMSE = 11.68. Adding the uncertainty in the SAT (9.82% ± 3.56%, adjusted to 9.82% ± 3.56% * e = 27% ± 10%) to the 32% uncertainty in the model gives an estimate of the uncertainty of ~42% ± 10% for $\dot{V}_E^2$.

$\dot{V}_E^3$ uncertainty estimate was determined to be ~35% ± 7%, which can be considered an overall uncertainty value, in this case, for a 90% confidence interval.

$\dot{V}_E^4$ uncertainty for the "share of day" variable is ~30% ± 16% for a 90% confidence interval. The final average value with SD for overall uncertainty estimate, with minute ventilation uncertainty included, is calculated to ~46% ± 17% for a 90% confidence interval.

3.3.3. Propagation of Uncertainty

By adding the uncertainty from the PPM device, the final uncertainties for the intake dose assessment are 41% ± 9% and 45% ± 10%, for intake dose assessment models **M1** and **M2**, respectively, and 38% ± 7% and 49% ± 17% for **M3** and **M4**, respectively.

4. Discussion

Collocating the PPM with the GRIMM showed that the low-cost sensor provides valid data. These results showed that the sensor is fit for purpose, especially if the results of the smallest particles measured (PM1) are considered.

Two participants were chosen to avoid skewed results. Most of their characteristics (age, height, sex, gender, nature of their work, sampling season) were different enough to enable an indication of the model's response to the variation of respective input variables.

M1 and M2 intake dose assessment models show a strong relationship (Figure 2), deviating mostly in peak concentrations, where M1 predicts a higher intake dose than M2. This is more evident with P2, where the calculated intake dose is higher and the peaks are more pronounced. Interestingly, although M1 uses more variables and was determined based on a larger number of participants in multiple research than M2, they show similar results. M1 shows a greater response to higher concentrations of PM than M2. M1 showed the highest intake dose to be 17% and 21% higher with P1 and P2, respectively, than M2. P2 was exposed to PM_1 concentrations more than three-times higher than P1, and the difference in M1 and M2 rose. This indicates that M1 is more sensitive to elevated concentrations, which could be a crucial aspect when determining the acute intake dose.

There are more peaks, which are more pronounced in the M3 intake dose assessment model, but the median values indicate that most of the calculated values are lower than in M1 and M2. Although, the weekly intake doses do not differentiate much between the first three dose assessments and are all around 0.59 mg for P1 and between 2.39 mg and 2.76 mg for P2. Calculating the intake dose on a larger time interval would be as good with M3 as with M1 or M2. The issue, in this case, would be the exaggerated response of the model to elevated concentrations. A better realignment of the calculated values in M3 to M1 and M2 could be possible with a different interpretation of the TAD and categorizing each activity more in line with the HR associated with it.

M4 intake dose assessment model results show that M1 and M2 are influenced by the HR variable and do correspond to changes, most notably in time intervals with elevated HR and concentrations of PM, while M4 does not. As it is only an "adjusted" PM value, meaning that all the PM measurements are multiplied with the same value, it is not influenced by the higher HR differences, present with the results from P2.

Uncertainty is associated with all stages of calculating intake dose of PM. There is inherent uncertainty in the GRIMM, PPM, and Garmin devices, and uncertainty that comes from calculating minute ventilation from HR, and all the generalizations associated with it. Using different models, published in individual papers, shows that presenting uncertainty is not uniform in this field. Papers describe the relationship between modelled and measured data with different statistical measures, which makes assessing uncertainty difficult and sometimes incomparable. Although uncertainties calculated for each minute ventilation model are not entirely comparable, the uncertainties for $\dot{V}_E^1$ and $\dot{V}_E^2$ can be compared separately, as can those for $\dot{V}_E^3$ and $\dot{V}_E^4$.

Calculated statistical measures for $\dot{V}_E^1$ and $\dot{V}_E^2$ show that $\dot{V}_E^2$ has poorer agreement with the reference data (lower R^2 value) and higher errors (higher MAPE, MAE, and RMSE). The model with more variables ($\dot{V}_E^1$) can calculate data that are closer to the measured data for minute ventilation. Both models have their uses, and although $\dot{V}_E^2$ has poorer results of calculated statistical measures than $\dot{V}_E^1$, it requires less information about the participant. Relatively high standard deviations in the uncertainty ($\sim\frac{1}{4}$ of the value) show that the real uncertainty of the models is even closer than the uncertainty values themselves would suggest.

Because the uncertainties for $\dot{V}_E^3$ and $\dot{V}_E^4$ were calculated in the same manner, they can be compared. The results are as expected, where $\dot{V}_E^4$ has a higher uncertainty than $\dot{V}_E^3$ because the share of daily activity has a certain level of uncertainty, which is propagated to the minute ventilation estimate, which is also presented with an uncertainty interval.

The uncertainties calculated through a series of steps do provide some measure of the validity of each minute ventilation model. Each of the models was provided with a specific measure of agreement between the modelled and measured data, but as these measures were different, they are not entirely comparable. This was somewhat compensated with further calculations for $\dot{V}_E^1$ and $\dot{V}_E^2$. Further research is needed to validate each model directly with reference data.

5. Limitations

Collocating the PPM device took place in a room with open windows, so the environment was a mixture of outdoor and indoor, which is not the case during the deployment phase. Moreover, only one PPM device was evaluated during colocation with the reference instrument. To truly determine the validity of the sensor, more devices should be subjected to collocation for longer periods and in different seasons and conditions. The device was stationary for the entire period, which is not representative as the device is designed to be mobile. Indeed, most of the time, the device is stationary in real-life circumstances, for example, in the office, at home, in the bedroom, or in a car. The collocation of the device would be needed while it is mobile. Using the GRIMM device for the collocation was the only available option at the time, but in further validation of the PPM device, it should be compared to a certified government AQ station or similar.

Data from two participants were used for the models. There were certain differences between the participants, but they were also both Caucasian and had a similar height and weight. The latter is also discussed in the Supplementary Materials. For both participants, there were some data missing, and the TADs were somewhat incomplete or inconsistent. All of the models were also validated indirectly by validating the minute ventilation and particulate matter concentrations separately. Future research should develop methods for direct validation of the models, using real-time data with a high temporal resolution for each observed variable, with research-grade instruments.

Measures of uncertainty were provided for all minute ventilation models, but were inconsistent and not entirely comparable.

6. Conclusions

A comparison of the four different approaches to assess intake dose, using data from low-cost sensors, was presented. Collocating the PPM device with a more expensive, research-grade instrument showed that the sensor provides good data, and was reliable enough to use it to determine intake dose of PM. Agreement with the reference instrument was better with smaller-sized particles, but the differences for different time averaging intervals were only marginal (ΔR^2 = 0.01–0.03). Considering these results, PM_1 concentrations were used for modelling, with the highest temporal resolution possible (1 min).

Four different minute ventilation models with increasing levels of complexity were used to determine minute ventilation, which was then used to calculate the intake dose of PM. Intake dose assessment models M1 and M2, which used HR as a variable, showed good agreement with each other, although M1, which was more complex and used sex, age, height, weight, and FVC as variables, showed more pronounced peaks and a stronger response to elevated HR and PM concentrations than M2, which only used sex as a variable (apart from HR). Intake dose assessment M3 and M4 did not use HR as a variable, but relied on generalized population data for specific activities, differentiated by sex and age. M4 showed better agreement with M1 and M2 than M3, but this could be the result of inaccurate activity classification. With further optimization, M3 could be improved and better realigned with other models.

Comparing the uncertainties between all the minute ventilation models was not possible, owing to different measures of uncertainty being reported for each model. After some additional calculations, a direct comparison of $\dot{V}_E^1$ and $\dot{V}_E^2$ was possible and between $\dot{V}_E^3$ and $\dot{V}_E^4$. $\dot{V}_E^1$ had lower uncertainty than $\dot{V}_E^2$, which is mostly associated with the model itself and less with the SAT and other variables. The comparison of $\dot{V}_E^3$ and $\dot{V}_E^4$ showed that $\dot{V}_E^3$ had less uncertainty associated with it than $\dot{V}_E^4$, which was a direct consequence of $\dot{V}_E^4$ using another set of generalized population data to determine "share of the day" for each specific activity, for which $\dot{V}_E^3$ had data from TADs. The minute ventilation models contributed the largest share to the overall uncertainty of the intake dose assessment models, followed by the SAT and finally the PPM.

As evident in this work, there are several different approaches for calculating the intake dose of pollutants. This stems also from different goals that the developers of these models set out in their respective studies. While some validate existing models, others try to evaluate the models predicting ability. Future research can use these results to determine which model best suits their needs and resources. While more complex models provide dose calculations on a minute-by-minute basis and have less uncertainty, they also require more resources in terms of sensors used and invested time by the researchers and the participants. This paper can also provide several options for future research in PM intake dose assessment, from developing models with less uncertainty, using location data and different sensors, and using the described models on larger groups.

As low-cost sensor technology is rapidly developing, there is an ever-expanding field of possibilities of how to implement such technologies for sensing and intake dose assessments. To allow comparability between the results of these measurements and calculations, a more homogeneous approach to presenting these finding and the uncertainties associated with them is needed. This work is a contribution towards this goal—by using appropriate terms and methods, this paper will contribute to further developing a unified methodological and terminological approach in this type of research. Modelling the intake dose of PM, by determining certain variables with low-cost sensors, was shown to be possible. Although there are many advantages, there is uncertainty that comes with this kind of sampling, and researchers need to account for this aspect in reporting their data. As this technology and these approaches become more widespread and distributed in the general public, users must be made aware that these data can come with wide margins of uncertainty and should only be used as a general guideline and not a scientific fact.

Supplementary Materials: The following are available online at http://www.mdpi.com/1424-8220/20/5/1406/s1, Figure S1: Boxplots of intake dose calculations with different height and weight variations as 67 referenced in Table S3, Table S1 Basic specifications for GRIMM Model 11-A, Table S2: Excerpt from Plantower pms5003 datasheet with some relevant figures about the functionality of the sensor, Table S3: Matrix of all variations for four different weights and heights. The Supplementary Materials (SI) contain extended descriptions of the sensors used in this research, a more detailed overview of the data collection process, and a brief investigation of the influence of weight and height of the participant on the final results.

Author Contributions: R.N. and D.K. conceptualized the idea, collected the data with J.A.R. and T.K., analyzed, validated and visualized the data, and prepared the original draft. D.S. coordinated and led the design efforts for the ICARUS project. M.H. headed the project on a local level, ensured funding and contributed to the final review and editing. All authors approved the content of the manuscript. All authors have read and agreed to the published version of the manuscript.

Funding: This project has received funding from the European Union's Horizon 2020 research and innovation program under grant agreement No 690105 and the Young researchers program funded by the Slovenian Research Agency.

Acknowledgments: The authors express their gratitude to the "Jožef Stefan" Institute for allowing the use of its facilities, Janja Vaupotič and Dafina Kikaj for their assistance with the GRIMM instrument, and Janja-Snoj Tratnik for her advice in the statistical analysis. The authors gratefully acknowledge the financial support of the European Commission of the Horizon 2020 project ICARUS (grant agreement no. 690105) and the Slovenian Research Agencies "Young researchers" program. Thanks are also merited for the participants in the ICARUS campaign.

Conflicts of Interest: The authors declare no conflict of interest

Ethical Approval: Ethical approval for the ICARUS project in Slovenia was obtained from the National Medical Ethics Committee of the Republic of Slovenia (approval nr. 0120-388/2018/6 on 22 August, 2018).

References

1. Kumar, P.; Morawska, L.; Martani, C.; Biskos, G.; Neophytou, M.; Di Sabatino, S.; Bell, M.; Norford, L.; Britter, R. The rise of low-cost sensing for managing air pollution in cities. *Environ. Int.* **2015**, *75*, 199–205. [CrossRef] [PubMed]
2. Popoola, O.A.M.; Carruthers, D.; Lad, C.; Bright, V.B.; Mead, M.I.; Stettler, M.E.J.; Saffell, J.R.; Jones, R.L. Use of networks of low cost air quality sensors to quantify air quality in urban settings. *Atmos. Environ.* **2018**, *194*, 58–70. [CrossRef]
3. Feinberg, S.N.; Williams, R.; Hagler, G.; Low, J.; Smith, L.; Brown, R.; Garver, D.; Davis, M.; Morton, M.; Schaefer, J.; et al. Examining spatiotemporal variability of urban particulate matter and application of high-time resolution data from a network of low-cost air pollution sensors. *Atmos. Environ.* **2019**, *213*, 579–584. [CrossRef]
4. Johnston, S.J.; Basford, P.J.; Bulot, F.M.J.; Apetroaie-Cristea, M.; Easton, N.H.C.; Davenport, C.; Foster, G.L.; Loxham, M.; Morris, A.K.R.; Cox, S.J. City Scale Particulate Matter Monitoring Using LoRaWAN Based Air Quality IoT Devices. *Sensors* **2019**, *19*, 209. [CrossRef]
5. Rogulski, M. Indoor PM10 concentration measurements using low-cost monitors in selected locations in Warsaw. *Energy Procedia* **2018**, *147*, 137–144. [CrossRef]
6. Li, J.; Li, H.; Ma, Y.; Wang, Y.; Abokifa, A.A.; Lu, C.; Biswas, P. Spatiotemporal distribution of indoor particulate matter concentration with a low-cost sensor network. *Build. Environ.* **2018**, *127*, 138–147. [CrossRef]

7. Patel, S.; Li, J.; Pandey, A.; Pervez, S.; Chakrabarty, R.K.; Biswas, P. Spatio-temporal measurement of indoor particulate matter concentrations using a wireless network of low-cost sensors in households using solid fuels. *Environ. Res.* **2017**, *152*, 59–65. [CrossRef]
8. Castell, N.; Dauge, F.R.; Schneider, P.; Vogt, M.; Lerner, U.; Fishbain, B.; Broday, D.; Bartonova, A. Can commercial low-cost sensor platforms contribute to air quality monitoring and exposure estimates? *Environ. Int.* **2017**, *99*, 293–302. [CrossRef]
9. Morawska, L.; Thai, P.K.; Liu, X.; Asumadu-Sakyi, A.; Ayoko, G.; Bartonova, A.; Bedini, A.; Chai, F.; Christensen, B.; Dunbabin, M.; et al. Applications of low-cost sensing technologies for air quality monitoring and exposure assessment: How far have they gone? *Environ. Int.* **2018**, *116*, 286–299. [CrossRef]
10. Int Panis, L.; de Geus, B.; Vandenbulcke, G.; Willems, H.; Degraeuwe, B.; Bleux, N.; Mishra, V.; Thomas, I.; Meeusen, R. Exposure to particulate matter in traffic: A comparison of cyclists and car passengers. *Atmos. Environ.* **2010**, *44*, 2263–2270. [CrossRef]
11. Watson, A.Y.; Bates, R.R.; Kennedy, D. *Assessment of Human Exposure to Air Pollution: Methods, Measurements, and Models*; National Academies Press (US): Washington, DC, USA, 1988.
12. Mölter, A.; Simpson, A.; Berdel, D.; Brunekreef, B.; Custovic, A.; Cyrys, J.; de Jongste, J.; de Vocht, F.; Fuertes, E.; Gehring, U.; et al. A multicentre study of air pollution exposure and childhood asthma prevalence: The ESCAPE project. *Eur. Respir. J.* **2015**, *45*, 610–624. [CrossRef] [PubMed]
13. Son, Y.; Osornio-Vargas, Á.R.; O'Neill, M.S.; Hystad, P.; Texcalac-Sangrador, J.L.; Ohman-Strickland, P.; Meng, Q.; Schwander, S. Land use regression models to assess air pollution exposure in Mexico City using finer spatial and temporal input parameters. *Sci. Total Environ.* **2018**, *639*, 40–48. [CrossRef] [PubMed]
14. Brokamp, C.; Brandt, E.B.; Ryan, P.H. Assessing exposure to outdoor air pollution for epidemiological studies: Model-based and personal sampling strategies. *J. Allergy Clin. Immunol.* **2019**, *143*, 2002–2006. [CrossRef] [PubMed]
15. Koehler, K.A.; Peters, T.M. New Methods for Personal Exposure Monitoring for Airborne Particles. *Curr. Environ. Health Rep.* **2015**, *2*, 399–411. [CrossRef] [PubMed]
16. Munir, S.; Mayfield, M.; Coca, D.; Jubb, S.A.; Osammor, O. Analysing the performance of low-cost air quality sensors, their drivers, relative benefits and calibration in cities—A case study in Sheffield. *Environ. Monit. Assess* **2019**, *191*, 94. [CrossRef] [PubMed]
17. Fishbain, B.; Lerner, U.; Castell, N.; Cole-Hunter, T.; Popoola, O.; Broday, D.M.; Iñiguez, T.M.; Nieuwenhuijsen, M.; Jovasevic-Stojanovic, M.; Topalovic, D.; et al. An evaluation tool kit of air quality micro-sensing units. *Sci. Total Environ.* **2017**, *575*, 639–648. [CrossRef]
18. Clements, A.L.; Griswold, W.G.; Rs, A.; Johnston, J.E.; Herting, M.M.; Thorson, J.; Collier-Oxandale, A.; Hannigan, M. Low-Cost Air Quality Monitoring Tools: From Research to Practice (A Workshop Summary). *Sensors* **2017**, *17*, 2478. [CrossRef]
19. Arroyo, P.; Herrero, J.L.; Suárez, J.I.; Lozano, J. Wireless Sensor Network Combined with Cloud Computing for Air Quality Monitoring. *Sensors* **2019**, *19*, 691. [CrossRef]
20. Ramos, C.A.; Reis, J.F.; Almeida, T.; Alves, F.; Wolterbeek, H.T.; Almeida, S.M. Estimating the inhaled dose of pollutants during indoor physical activity. *Sci. Total Environ.* **2015**, *527–528*, 111–118. [CrossRef]
21. Vai, F.; Bonnet, J.L.; Ritter, P.; Pioger, G. Relationship between heart rate and minute ventilation, tidal volume and respiratory rate during brief and low level exercise. *Pacing Clin. Electrophysiol.* **1988**, *11*, 1860–1865. [CrossRef]
22. Onorati, P.; Marinelli, P.; Paoletti, P.; Ferrazza, A.M.; Valli, G.; Internullo, M.; Martolini, D.; Palange, P. Minute ventilation (V'E) and heart rate (HR) relationship during exercise in chronic cardiac and pulmonary diseases. *Eur. Respir. J.* **2012**, *40*, 856.
23. Valli, G.; Internullo, M.; Ferrazza, A.M.; Onorati, P.; Cogo, A.; Palange, P. Minute ventilation and heart rate relationship for estimation of the ventilatory compensation point at high altitude: A pilot study. *Extrem. Physiol. Med.* **2013**, *2*, 7. [CrossRef] [PubMed]
24. Samet, J.M.; Lambert, W.E.; James, D.S.; Mermier, C.M.; Chick, T.W. Assessment of heart rate as a predictor of ventilation. *Res. Rep. Health Eff. Inst.* **1993**, *59*, 19–55, discussion 57–59.
25. Sarigiannis, D.A.; Karakitsios, S.P.; Antonakopoulou, M.P.; Gotti, A. Exposure analysis of accidental release of mercury from compact fluorescent lamps (CFLs). *Sci. Total Environ.* **2012**, *435–436*, 306–315. [CrossRef] [PubMed]

26. Liang, L.; Gong, P.; Cong, N.; Li, Z.; Zhao, Y.; Chen, Y. Assessment of personal exposure to particulate air pollution: The first result of City Health Outlook (CHO) project. *BMC Public. Health* **2019**, *19*, 711. [CrossRef] [PubMed]
27. Zuurbier, M.; Hoek, G.; van den Hazel, P.; Brunekreef, B. Minute ventilation of cyclists, car and bus passengers: An experimental study. *Environ. Health* **2009**, *8*, 48. [CrossRef]
28. Greenwald, R.; Hayat, M.J.; Dons, E.; Giles, L.; Villar, R.; Jakovljevic, D.G.; Good, N. Estimating minute ventilation and air pollution inhaled dose using heart rate, breath frequency, age, sex and forced vital capacity: A pooled-data analysis. *PLoS ONE* **2019**, *14*, e0218673. [CrossRef]
29. Good, N.; Carpenter, T.; Anderson, G.B.; Wilson, A.; Peel, J.L.; Browning, R.C.; Volckens, J. Development and validation of models to predict personal ventilation rate for air pollution research. *J. Expo. Sci. Environ. Epidemiol.* **2019**, *29*, 568–577. [CrossRef]
30. Madureira, J.; Aguiar, L.; Pereira, C.; Mendes, A.; Querido, M.M.; Neves, P.; Teixeira, J.P. Indoor exposure to bioaerosol particles: Levels and implications for inhalation dose rates in schoolchildren. *Air Qual. Atmos. Health* **2018**, *11*, 955–964. [CrossRef]
31. Zartarian, V.; Bahadori, T.; McKone, T. Adoption of an official ISEA glossary. *J. Expo. Sci. Environ. Epidemiol.* **2005**, *15*, 1–5. [CrossRef]
32. International Vocabulary of Metrology—Basic and General Concepts and Associated Terms (VIM). 2012. Available online: https://www.bipm.org/en/publications/guides/vim.html (accessed on 4 March 2020).
33. Sarigiannis, D.; Gotti, A.; Karakitsios, S.; Chapizanis, D. 1.2_ICARUS_Report-on-conceptual-framework_final.pdf. 2017, p. 40. Available online: https://icarus2020.eu/wp-content/uploads/2017/08/D1.2_ICARUS_Report-on-conceptual-framework_final.pdf (accessed on 20 November 2019).
34. IoTECH TELECOMMUNICATIONS | LoRaWANTM NETWORK PROVIDER | LoRaWAN LPWAN CONNECTIVITY IN GREECE | INTERNET OF THINGS | SMART CITIES IoT CONNECTIVITY. Available online: https://iotech.gr/ (accessed on 13 May 2019).
35. Plantower-pms5003-manual Annotated. Available online: https://usermanual.wiki/Pdf/plantowerpms5003manualannotated.626592918/view (accessed on 3 September 2019).
36. Garmin; Subsidiaries, G.L. Or its Garmin vívosmart®3 | Fitness Activity Tracker. Available online: https://buy.garmin.com/en-US/US/p/567813 (accessed on 3 September 2019).
37. Oniani, S.; Woolley, S.I.; Pires, I.M.; Garcia, N.M.; Collins, T.; Ledger, S.; Pandyan, A. Reliability Assessment of New and Updated Consumer-Grade Activity and Heart Rate Monitors. In Proceedings of the SENSORDEVICES 2018, Venice, Italy, 16–20 September 2018; p. 6.
38. Quanjer, P.H.; Stanojevic, S.; Cole, T.J.; Baur, X.; Hall, G.L.; Culver, B.H.; Enright, P.L.; Hankinson, J.L.; Ip, M.S.M.; Zheng, J.; et al. Multi-ethnic reference values for spirometry for the 3–95-yr age range: The global lung function 2012 equations. *Eur. Respir. J.* **2012**, *40*, 1324–1343. [CrossRef] [PubMed]
39. U.S. EPA. *Exposure Factors Handbook 2011 Edition (Final Report)*; EPA/600/R-09/052F; U.S. Environmental Protection Agency: Washington, DC, USA, 2011. Available online: https://cfpub.epa.gov/ncea/risk/recordisplay.cfm?deid=236252 (accessed on 18 September 2019).
40. GRIMM Model 1.108-and-1.109.pdf. Available online: https://www.wmo-gaw-wcc-aerosol-physics.org/files/opc-grimm-model--1.108-and-1.109.pdf (accessed on 15 June 2019).
41. R: The R Project for Statistical Computing. Available online: https://www.r-project.org/ (accessed on 5 December 2019).
42. SEA—Slovenian Environment Agency. Average Daily Concentrations of PM2.5 at Ljubljana Bežigrad Station. Available online: https://www.arso.gov.si/zrak/kakovost%20zraka/podatki/PM2.5_D_dec19_slo.pdf (accessed on 1 November 2019).
43. SEA—Slovenian Environment Agency. Average Daily Concentrations of PM10 at Ljubljana Bežigrad Station. Available online: https://www.arso.gov.si/zrak/kakovost%20zraka/podatki/PM10_D_dec19_slo.pdf (accessed on 11 November 2019).

sensors

MDPI

Article

Wildfire Smoke Adjustment Factors for Low-Cost and Professional $PM_{2.5}$ Monitors with Optical Sensors

William W. Delp and Brett C. Singer *

Indoor Environment Group and Residential Building Systems Group, Lawrence Berkeley National Laboratory, Berkeley, CA 94720, USA; wwdelp@lbl.gov
* Correspondence: bcsinger@lbl.gov

Received: 27 May 2020; Accepted: 28 June 2020; Published: 30 June 2020

Abstract: Air quality monitors using low-cost optical $PM_{2.5}$ sensors can track the dispersion of wildfire smoke; but quantitative hazard assessment requires a smoke-specific adjustment factor (AF). This study determined AFs for three professional-grade devices and four monitors with low-cost sensors based on measurements inside a well-ventilated lab impacted by the 2018 Camp Fire in California (USA). Using the Thermo TEOM-FDMS as reference, AFs of professional monitors were 0.85 for Grimm mini wide-range aerosol spectrometer, 0.25 for TSI DustTrak, and 0.53 for Thermo pDR1500; AFs for low-cost monitors were 0.59 for AirVisual Pro, 0.48 for PurpleAir Indoor, 0.46 for Air Quality Egg, and 0.60 for eLichens Indoor Air Quality Pro Station. We also compared public data from 53 PurpleAir PA-II monitors to 12 nearby regulatory monitoring stations impacted by Camp Fire smoke and devices near stations impacted by the Carr and Mendocino Complex Fires in California and the Pole Creek Fire in Utah. Camp Fire AFs varied by day and location, with median (interquartile) of 0.48 (0.44–0.53). Adjusted PA-II 4-h average data were generally within ±20% of $PM_{2.5}$ reported by the monitoring stations. Adjustment improved the accuracy of Air Quality Index (AQI) hazard level reporting, e.g., from 14% to 84% correct in Sacramento during the Camp Fire.

Keywords: fine particles; air pollutant exposure; air quality monitoring; climate change impacts; health hazard assessment; respiratory health

---

1. Introduction

Throughout the Western U.S., wildland fires have increased in frequency and intensity over the past several decades due to climate change and the legacy of forest fire suppression [1–5]. Development at the wildland urban interface also has contributed to wildfire frequency and increased their cost in terms of human life and health and property damage [4,6].

Wildfire smoke contains fine particulate matter ($PM_{2.5}$), toxic particle-phase constituents, ultrafine particles, and many irritant gases including acrolein and formaldehyde. Exposure to elevated levels of wildfire smoke specifically has been linked to many adverse health outcomes [7–11]. During wildfire smoke episodes, air pollutant concentrations can increase substantially, and low-income homes, which typically have high rates of uncontrolled air leakage, are particularly vulnerable [12]. Filtration can be cost-effectively applied to reduce exposures and health impacts of wildfire smoke in buildings [13,14].

$PM_{2.5}$ is often used to track wildfire smoke because it is an established health hazard and is routinely measured at regulatory air quality monitoring stations. In the U.S., $PM_{2.5}$ is regulated under the Clean Air Act, with health-based standards of 12 $\mu g\ m^{-3}$ annual average and 35 $\mu g\ m^{-3}$ averaged over 24 h [15]. The thresholds are set by the US EPA based on a systematic review of studies that examine how deaths, strokes, and other indicators such as hospitalizations increase as ambient PM2.5 increases [16]. To provide information to the public about the hazard posed by PM2.5, both below and above the thresholds of the standard, EPA uses the air quality index, or AQI [17]. The AQI is a piecemeal linear scale that relates PM2.5 concentrations to hazard level for sensitive subgroups and the general population. The AQI is normally calculated on a daily basis; but the interval can be shortened to as little as 3 h when ambient conditions change rapidly. It is also important to note that despite the cautions communicated about the AQI, wildfire smoke may still present substantial health risk and cause harm to individuals with pre-existing health conditions and vulnerabilities beyond that presented in the PM2.5 AQI.

The U.S. Federal Reference Method (FRM) for determining fine particle concentrations is gravimetric: it requires use of certified equipment to collect particles onto a filter which is equilibrated to standard temperature and humidity conditions and weighed before and after air sampling [18,19]. Devices that use alternative measurement methods—typically with the goal of achieving hourly or more resolved data—can be approved as Federal Equivalent Methods (FEM) by producing similar results when collocated with FRM sampling at multiple sites across multiple seasons. Agreement is acceptable when linear correlation is high ($r > 0.97$) with a slope of 1.00 ± 0.05 and intercept $\leq \pm 1$ µg/m^3. In practice, FEM monitors don't always match collocated FRM measurements to these specifications [20]. Also, since FEM certification is based on seasonal statistics of daily averages, accuracy and precision may be lower for shorter averaging times.

Historically, it has been difficult to map the hazards posed by wildfire smoke because regulatory air monitoring stations are sparse in many areas. During severe events, portable monitors have been deployed to fill critical gaps in spatial coverage [21,22]; but that approach has been limited by the high cost of purchasing and maintaining equipment that meet data quality standards [23]. At least one low-cost monitor has been developed for the stated objective of filling this gap for wildfire smoke monitoring [24]; but the extent of its use to date is unclear. One study used surface measurements from regulatory air monitoring stations and low-cost particle monitor networks to translate satellite images of aerosol optical depth into $PM_{2.5}$ for mapping of wildfire spread [25]. In that study, PurpleAir II monitors were compared to co-located FEM data but the evaluation did not include high $PM_{2.5}$ concentrations (>100 µg/m^3) or PM known to be predominantly wood smoke.

For years, industrial hygienists and researchers have measured $PM_{2.5}$ inside and outside buildings using real-time monitors that quantify aerosol concentrations by light scattering. These devices are robust, have a wide dynamic range (3 or more orders of magnitude), and temporal resolution on the order of seconds. However, the intensity of scattered light depends on aerosol properties including size, shape, density, and refractive index [26,27]. Photometers are a main variant of optical monitor; they measure and translate scattering from ensembles of particles to an estimated mass concentration based on calibration with a defined aerosol [28,29]. Optical particle counters interpret scattering generated from single particles moving through a laser to estimate particle size based on assumptions about shape and optical properties; counts are aggregated over time to quantify number density by size. Aerosol mass concentrations are estimated based on assumptions or collocated measurements to determine particle density. Professional grade air quality monitors provide active flow control, sheath airflow, and component diagnostic monitoring to achieve high consistency and durability; but they cost thousands of dollars per unit and still require source-specific calibration factors or coincident gravimetric sampling.

Collections of low-cost air quality monitors that are being deployed throughout the world [30–33] could be used to track wildfire smoke events. However, the optical sensors used in the monitors vary with aerosol properties [34–39] and adjustment factors specific to wildfire smoke are needed.

The present study had the following objectives: (1) using data collected during the November 2018 Camp Fire in Northern California, determine adjustment factors (AFs) for four low-cost and three professional air quality monitors to improve their accuracy for measuring infiltrating $PM_{2.5}$ associated with wildfire smoke; (2) using publicly available data from regulatory air quality monitoring stations (AQS) and nearby PurpleAir PA-II monitors at varied distances downwind from the fire, evaluate the variability of AFs for this monitor across space and time for sites throughout the region; (3) quantify

PA-II AFs for smoke from two other recent wildfires and compare them to AFs determined for the Camp Fire; and (4) quantify the improvement in exposure estimates and AQI scores when the regional AF is applied to individual monitors throughout the region.

2. Methods

2.1. Overview

During the Camp Fire in November 2018, air quality monitors were operated at Lawrence Berkeley National Lab (LBNL) inside a small building with high infiltration air exchange and the doors to outdoors propped open at times to promote infiltration of outdoor air and particulate matter (PM). The monitors included one certified as a US EPA FEM, a wide range aerosol spectrometer designed for indoor air quality research, two professional grade photometers, and four low-cost monitors that use mass produced optical sensors. The latter group included three monitors purchased via retail distribution in the U.S. and production units of an IAQ station for networked building monitoring provided by the device maker (eLichens). $PM_{2.5}$ concentrations reported by the alternative monitors were compared to those of the FEM to quantify response and adjustment factors. The monitors are listed in Table 1.

Table 1. Descriptions of monitors used in this study.

ID	Device	Data	Particle Sensor(s) and Specifications for $PM_{2.5}$	Calibration and Quality Assurance Information Provided by Manufacturer
AQE	Air Quality Egg 2018 version	1 min	Two Plantower PMS5003 [1] Effective range: 0–500 μg/m^3 Max. range: ≥1000 μg/m^3 Max. consistency error: 0~100 μg/m^3: ±10 μg/m^3 100~500 μg/m^3: ±10%	https://airqualityegg.com/home Reports mean $PM_{2.5}$ and PM_{10} of the two sensors. Each unit checked for consistency with other devices before shipping by exposure to incense smoke in a small room.
AVP	IQAir AirVisual Pro	10 s	AirVisual AVPM25b Effective range: 0–1798 μg/m^3	https://www.airvisual.com/ Sensors calibrated through automatic process in controlled chamber, using distinct aerosols for PM_1, $PM_{2.5}$, PM_{10} using Grimm 11-A.
PAI	PurpleAir Indoor	80 s	Plantower PMS1003 Same specification as PMS5003	https://www.purpleair.com/sensors Data direct from sensor: PM_1, $PM_{2.5}$ and PM_{10} in μg/m^3, number density (#/0.1 L) of particles larger than the following optical diameters: 0.3, 0.5, 1.0, 2.5, 5.0, 10 μm.
ELI	eLichens Indoor Air Quality Pro Station	1 min	Plantower PMS7003 Same specification as PMS5003	https://www.elichens.com/elsi-indoor-air-quality-station Each station individually calibrated against regulatory AQ stations meeting EU standards. Data are adjusted in real-time for environmental conditions.

Table 1. *Cont.*

ID	Device	Data	Particle Sensor(s) and Specifications for $PM_{2.5}$	Calibration and Quality Assurance Information Provided by Manufacturer
PA-II	PurpleAir II (outdoor)	80 s	Two Plantower PMS5003	https://www.purpleair.com/sensors Same as PAI.
TEOM	Model 1045-DF Tapered Element Oscillating Microbalance with Filter Dynamic Measurement System	12 min	Range: 0 to 1,000,000 μg/m^3 Resolution: 0.1 μg/m^3, Precision: ±2.0 μg/m^3, 1-h avg	https://www.thermofisher.com/ Approved Federal Equivalent Method U.S. EPA PM-2.5 Equivalent Monitor EQPM-0609-182.
WRAS	Model 1.371 Mini Wide-Range Aerosol Spectrometer	1 min	Combined electrical mobility instrument with optical particle spectrometer. Range: 0.1 μg/m^3–100 mg/m^3 Electrical mobility sensing: 10 bins in range 10–193 nm, Optical sensing 31 bins in range 0.253–35 μm	https://www.grimm-aerosol.com Optical spectrometer calibrated using class I reference with NIST-certified, mono-disperse polystyrene latex (PSL) particles. Electrical sensor calibrated using GRIMM model 7811 with poly-disperse aerosol of particles with diameters of ~5 nm to ~300 nm generated from NaCl solution. Aerosol is dried and diffusion-neutralized. A Differential Mobility Analyzer (DMA) provides narrow size distributions simultaneously to the sensor and a reference Faraday cup electrometer.
PDR	Thermo pDR-1500	10 s	Laser optical photometer Range: 0.001–400 mg/m^3 Precision: larger of ±0.2% of reading or ±0.0005 mg/m^3. Accuracy: ±0.5% reading ±precision	https://www.thermofisher.com/ Traceable to SAE Fine Test Dust.
DT	TSI DustTrak II-8533	2 min	Laser optical photometer Range: 0.001 to 150 mg/m^3 Flow Accuracy: ±5% factory setpoint Internal flow controlled	https://tsi.com/home/ Calibrated with ISO 12103–1, A1 Ultrafine Test Dust.

[1] Plantower documentation describes the analytical method as follows: "… collect scattering light in certain angle [and] obtain the curve of scattering light change with time. [By microprocessor, calculate] equivalent particle diameter and the number of particles with different diameter per unit volume based on MIE theory. Product documentation also reports "endurance max error" after 720 h of operation: as ±15 μg/m^3 for 0~100 μg/m^3 and ±15% for 100~500 μg/m^3.

Additionally, we extracted and analyzed publicly available data collected at regulatory air quality monitoring stations (AQS) and nearby PurpleAir PA-II monitors for three Western US wildfires. For the Camp Fire, data were obtained for 12 Northern California (NorCal) AQS sites with high $PM_{2.5}$ concentrations and 53 PA-II monitors near the 12 sites. Data were identified and analyzed for a single AQS site and nearby PA-II monitor that were in an area impacted by the Carr and Mendocino Complex Fires in California and for a single AQS and PA-II monitor impacted by the Pole Creek Fire in Utah.

2.2. Wildland Fires

2.2.1. Camp Fire

At roughly 6:30 a.m. on November 8, 2018 a wildland fire started along Camp Creek Road near Poe Dam in Butte County, Northern California. The fire spread quickly and ravaged the nearby town of Paradise. It caused at least 86 fatalities and destroyed almost 19,000 buildings, many in the first few hours. It ultimately burned over 150,000 acres and was not contained until November 25. A map showing the area burned in the fire is provided in Figure 1.

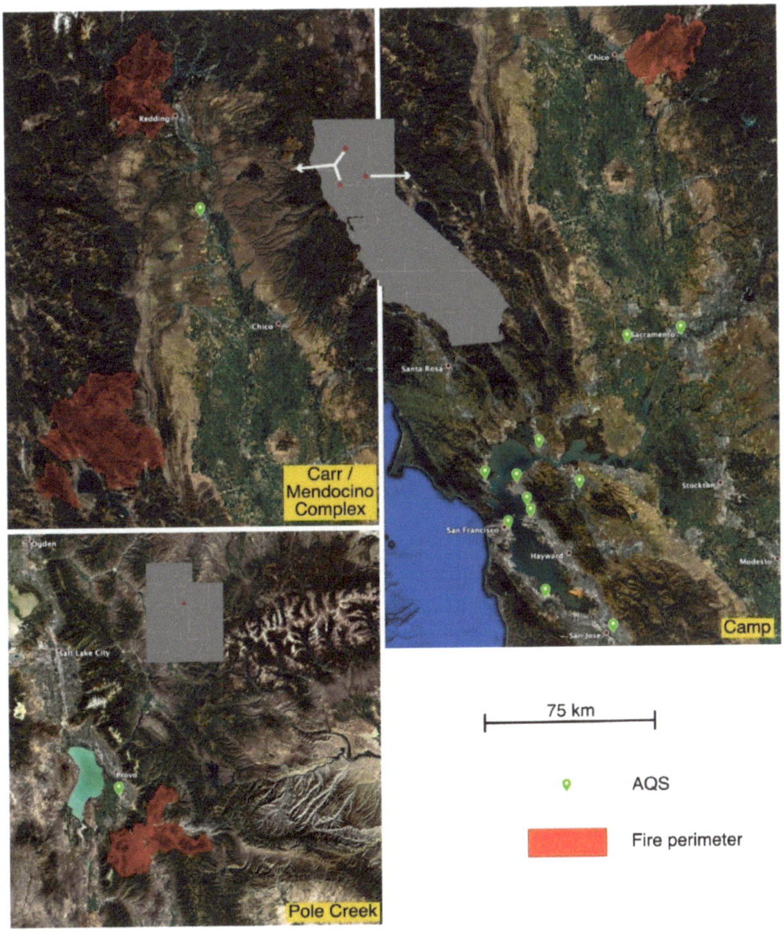

Figure 1. Areas burned (red) and locations of air quality monitoring stations (green dots) reporting hourly $PM_{2.5}$ that also had nearby PA-II monitors. Gray images are the states of California (top) and Utah (bottom) and red dots show the locations of the detailed maps with the states. The 75 km scale marker applies to all three detailed images. Satellite images were obtained via Google Earth. Fire extent data from https://fsapps.nwcg.gov/googleearth.php.

Strong off-shore katabatic winds on November 8 carried smoke throughout the most heavily populated areas of Northern California. By 11:30 a.m. the smoke plume had reached Berkeley, over 225 km away. Smokey conditions persisted in California's Central Valley and the San Francisco

Bay Area until November 21. A satellite image showing the extent of smoke coverage is provided as Figure S1 in the Supplementary Material.

2.2.2. Carr and Mendocino Complex Fires

The Carr Fire started on July 23, 2018 in the Whiskeytown district in Shasta County, California. The fire spread quickly over the next few days and burned over 28,000 acres by the evening of July 26. It ultimately consumed nearly 230,000 acres and was not fully contained until August 30. The Mendocino Complex Fire burned over 450,000 acres over the period from July 27 to November 7, 2018. The smoke from the two fires appeared to combine over large areas in the northern portion of the Sacramento Valley as shown in a satellite image provided as Figure S2. The areas burned by these two fires and the location of the impacted AQS with nearby PA-II monitor is shown in Figure 1. These two fires are henceforth described as the Carr/MC Fires.

2.2.3. Pole Creek Fire

The Pole Creek Fire was the largest fire in Utah in 2018; it consumed over 98,000 acres in a steep mountain canyon approximately 90 km SSE from Salt Lake City. The smoke from the Pole Creek Fire exhibited sharp diurnal behavior consistent with changing wind patterns in a mountain environment. The AQS at Spanish Fork was the only station that recorded significant PM from the fire and had a PurpleAir monitor nearby (4 km) with publicly available data throughout the smoke event. The area burned and locations of AQS and PA-II monitors are shown in Figure 1; a satellite image of the smoke is provided as Figure S3.

2.3. *Monitors Deployed at LBNL*

The monitors used at LBNL are described in detail below with summary notes in Table 1. Reference $PM_{2.5}$ data at LBNL were obtained using a Model 1405-DF Tapered Element Oscillating Microbalance with Filter Dynamic Measurement System (TEOM) (Thermo Scientific, Waltham, MA, USA). This device is approved as Federal Equivalent Method for 24 h measurements of $PM_{2.5}$. As a check on the TEOM, five pairs of filter-based gravimetric samples were collected and analyzed over the study period using Gillian AirCon2 pumps (Sensidyne, St. Petersburg, FL, USA) drawing 10 liters per minute (l pm) through an Personal Environmental Monitor for $PM_{2.5}$ (SKC, Eighty Four, PA, USA). The AirCon2 actively adjusts flow based on an internal sensor and reports an error if the flow deviates by more than 5% from the setting. The flow was also checked before each sample using a Gilian Gilibrator2 (Sensidyne). Gravimetric samples were collected on 37 mm diameter, 2 μm pore-size, TEFLO (Pall, Port Washington, NY, USA) PTFE filters that were equilibrated at a temperature of 19.5 ± 0.5 °C and relative humidity of 47.5 ± 1.5% for at least 24 h before weighing pre-and post-sampling. Filter weights were determined using an SE2-F ultra-microbalance (Sartorius, Goettingen, Germany).

The Model 1371 Mini Wide Range Aerosol Spectrometer (WRAS) (Grimm Aerosol Technik, Muldestausee, Germany) is designed specifically for indoor sampling and combines an optical particle sensor that quantifies particles in 31 size bins from 0.25 to 35 μm mean diameter and an electrical mobility spectrometer that quantifies particles in 10 size bins from 10 to 193 nm mean diameter. In addition to size-resolved number concentrations, the WRAS calculates PM_1, $PM_{2.5}$ and PM_{10} mass concentrations at 1-min time resolution based on the measured particle size distribution and an assumed particle density. The WRAS uses the same optical sensor as the Grimm Model EDM180, which is certified as an FEM. The EDM180 also includes a Nafion^{TM} dryer to reduce the potential for high humidity to cause significant particle growth when sampling outdoors.

Measurements were additionally made with two professional grade aerosol photometers that are used for industrial hygiene and research: a DustTrak II-8530 (DT) (TSI, Shoreview, MN, USA) that was combined with a model 801,850 heated inlet system and a Thermo pDR-1500 (PDR). These instruments have wide measurement ranges, starting at 1 μg m^{-3} and nominally extending to 150 and 400 mg m^{-3}, respectively. Both have active flow control and filtered sheath air to keep the optical path clean. The PDR features temperature and humidity compensation (via software), and the heated inlet on the DT is intended to prevent artifacts from high relative humidity (RH). The two instruments are calibrated with Arizona test dust: A1-ultrafine for the DT, and A2-fine for the PDR. Figure S4 shows the particle size distribution of the two dusts along with the average distribution of the smoke from the Camp Fire. Manuals for both monitors recommend coincident gravimetric sampling for calibrations to specific sources or environments and both offer flow-controlled internal filter collection. Based on the work of Wallace et al. [29], TSI recommends a calibration factor of 0.38 when using the DT to sample ambient air if no coincident gravimetric sample is obtained. The data reported in this paper uses the default calibration for the PDR and a calibration factor of 1.0 for the DT to avoid confusion. The PDR saved data every 10 s while the DT saved every 2 min. Notes on calibrations for the DT and PDR are found in the Supplementary Materials.

Also deployed at LBNL were four low-cost IAQ monitors: the Air Quality Egg—2018 edition (AQE) (Wicked Device LLC, Ithaca, NY, USA) the AirVisual Pro (AVP) (IQAir, Goldach, Switzerland), the PurpleAir Indoor (PAI) (PurpleAir LLC, Draper Utah, USA), and the Indoor Air Quality Pro Station from eLichens (ELI) (Gremoble, France). The AVP uses a proprietary AVPM25 b sensor and calibration procedure. The PAI uses a single PMS1003 sensor (Plantower, Beijing, China) and directly reports its output. The AQE incorporates dual Plantower PMS5003 sensors and reports the average of the two sensors. The ELI uses a single Plantower PMS5003 and reports $PM_{2.5}$ and PM_{10} after processing the data with a proprietary algorithm. The low-cost sensors appear to use a hybrid approach of optical particle sensing and photometry to estimate $PM_{2.5}$. The Plantower sensors also report PM_1 and particle number concentrations in 6 size bins (>0.3, >0.5, >1.0, >2.5, >5.0, >10 μm). The AQE and ELI saved data every minute, the AVP every 10 s, and the PAI every 80 s.

All monitors at LBNL were collocated within a 120 m^3 laboratory housed within a single-story building with two exterior doors at opposite sides. Doors were closed during most of the data collection and there were no indoor sources; all $PM_{2.5}$ in the room was thus infiltrated from outdoors. During three periods multi-hour periods totaling 26 h, the two doors were opened to increase indoor $PM_{2.5}$ to be closer to outdoor levels. The room was not thermally conditioned. Daily high temperatures outdoors at LBNL varied from 13.2 to 19.6 °C and overnight low temperatures were 6.9 to 14.7 °C during the two weeks of the Camp Fire. Room high temperatures varied from 18.1 to 24.3 Dynamic Measurement System (TEOMC and lows varied from 14.8 to 17.8 °C. The median temperature difference (inside to outside) was 3.9 °C. The outdoor air exchange rate (AER) was not measured directly during smoke monitoring; based on prior assessments we estimate that AERs were approximately 0.5 h^{-1} or lower (refer to discussion in the Supplementary Materials) when the door was closed.

2.4. Data from Regulatory Air Quality Monitoring Stations

Data from air quality monitoring stations (AQS) in California were obtained from the Air Quality and Meteorological Information System (AQMIS) maintained by the California Air Resources Board [40]. The regulatory network in California uses beta attenuation monitors (BAM) models 1020 and 1022 (Met One Instruments; Grants Pass, OR, USA) to record hourly $PM_{2.5}$. A BAM draws air through a size-selective inlet to set the PM mass fraction being measured (e.g., $PM_{2.5}$ or PM_{10}) then through a filter tape to collect sample. Collected particles change the attenuation of beta rays passing through the filter tape proportionally to the mass of particles collected. The change in mass over the measurement time interval is divided by the sample air volume to calculate $PM_{2.5}$ concentration.

Data from the Spanish Fork monitoring site that was impacted by the Pole Fire in Utah was obtained from the Utah Department of Environmental Quality website [41]. The FEM monitor operating at this site was a Thermo Scientific Model 5030i Synchronized Hybrid Ambient Real-time (SHARP) particulate monitor. The SHARP monitor combines an optical particle counter with a beta attenuation instrument. The optical portion of the instrument provides a data stream with high temporal resolution, and the beta attenuation provides a mass measurement to dynamically adjust the optical instrument and provide accurate time resolved PM mass concentration.

2.5. PurpleAir Network

The PurpleAir PA-II monitor features two Plantower PMS5003 sensors, electronics, and software to enable quick connection to the web via wifi-all packaged in a 4″ PVC cap with an outdoor power supply for weather protection. When setting up a new device the user is prompted to set the geographic location and whether it is indoors or outdoors, with outdoors as the default. PurpleAir provides a real-time, map-based data display (https://www.purpleair.com/map#1.1/0/--30) and enables downloads of data from its server. Device owners have the option of making the data publicly available or accessible only to users whom they designate. To our knowledge, PurpleAir had by far the largest distributed network of PM sensors with publicly viewable data deployed around California at the time of the Camp Fire. We also observed expansion of the network in terms of number of monitors and spatial coverage during the fire.

The default setting on the PurpleAir map presents data as the US EPA $PM_{2.5}$ AQI calculated from the $PM_{2.5}$ concentration reported by each Plantower sensor. The online map allows the user to display any of the other data streams provided by the sensors and other AQI-type values. At the time of the wildfires (and still on 07-May-2020), the site offered two "conversions" to adjust $PM_{2.5}$ concentrations and corresponding AQI values. The site attributes an "AQandU" calibration (0.778 * PA + 2.65) to a long-term University of Utah study in Salt Lake City and an "LRAPA" calibration (0.5 * PA − 0.68) to a Lane Regional Air Pollution Agency study of PA sensors. The University of Utah evaluated Plantower sensors measuring ambient air in Salt Lake City during all seasons of the year [42,43]. The LRAPA adjustment is from a winter study performed in a region of Oregon that has widespread use of wood combustion heating and ambient $PM_{2.5}$ that is predominantly composed of wood smoke when at its highest levels [44].

2.6. Identification of Paired PA-II and Regulatory AQ Monitoring Data

To conduct the analysis described herein, we manually searched the PA map to identify PA-II monitors within ~5 km of an AQS site that reported $PM_{2.5}$ on an hourly basis during any of the fires examined herein. To assess if PA-II and AQS data were appropriately paired, we considered local topography such as the presence of valleys or mountains that could result in the PA and AQS seeing different air masses and also viewed data to confirm basic synchronicity of trends. For the Camp Fire we identified 53 PA-II monitors in the vicinity of 12 NorCal AQS sites. From these 53 PA-II devices, we downloaded data from 97 sensors that reported data that appeared valid based on the review described in the next section. The median distance between the AQS and PA-II monitors was 2.7 km, the interquartile range was 1.1–4.6 km, and the full range was 0–11.6 km. For both the Carr/MC and Pole Creek Fires we found a single PA-II monitor and nearby AQS combination. For the Carr/MC Fire the AQS was approximately 50 km from the fire, and the PurpleAir was co-located with the site. The AQS selected for the Pole Creek Fire was approximately 35 km from the fire and the PA-II monitor was ~4 km from the AQS.

2.7. Analysis of Data from PA-II Monitors

The PA-II monitors report data at 80 s resolution. Many of the devices had occasional data gaps, presumably due to wifi connectivity issues. AQS data are provided at 1 h resolution. The unadjusted cf_1 data stream reported by the Plantower sensors to the PA-II monitor were used to calculate 1 h averages and both PA-II and AQS data were aggregated to 4 h averages to account for the sites not being precisely co-located. Correlations and adjustment factors were evaluated with the 4-h data streams. (The cf_1 and cf_atm data streams were switched in PA-II reporting at the time of the fire [45], but we have subsequently confirmed that the stream used in this study was the stream that is currently labeled as cf_1).

Most PA-II devices provide data for each of the two onboard sensors and occasionally the sensors read significantly different results. As a quality assurance screen, we reviewed data from all of the individual sensors of PA-II units. This was done by plotting the time series of the raw values along with the AQS data on the same plot, visually identifying the outliers. We flagged any sensor that substantially diverged from other nearby sensors, including from the same monitor. We also identified devices that appeared to be indoors but were marked as outdoors, indicated by the two sensors of the device agreeing closely and being well below the group or having occasional peaks (presumably from indoor PM emissions) that did not appear in the other outdoor devices. Faulty sensors and presumed indoor units were removed from the analysis.

3. Results

3.1. Reference Measurements at LBNL

Over the course of the fire, $PM_{2.5}$ as measured by the TEOM inside the lab at LBNL averaged 47.3 μg m^{-3} or roughly half of the event-averaged concentrations of 93.2 and 93.9 μg m^{-3} at air quality stations in Berkeley (4.5 km to the West) and Oakland-West (7.3 km to the Southwest). Figure S5 presents the time concentration profiles of hourly-averaged $PM_{2.5}$ for these sites. Duplicate filters provided consistent results and generally agreed with the TEOM (Figure S5). Some difference is expected since the TEOM sampling sequence did not perfectly align with some filter sample intervals.

3.2. Measurements with Low-Cost, Professional and Research Monitors at LBNL

Time series of $PM_{2.5}$ concentrations reported by all monitors deployed at LBNL are shown in Figure 2, which shows that all tracked with the TEOM and all but the WRAS substantially over-reported $PM_{2.5}$ throughout the event. Responses relative to TEOM varied for both professional-grade and low-cost monitors. For each monitor, we calculated the statistics of relative response (device reported $PM_{2.5}$ divided by TEOM $PM_{2.5}$) using event-integrated and 4-h average data, with results provided in Table S1. The event-integrated mean and median 4-h response factors were closest to unity for the WRAS and farthest for the DustTrak.

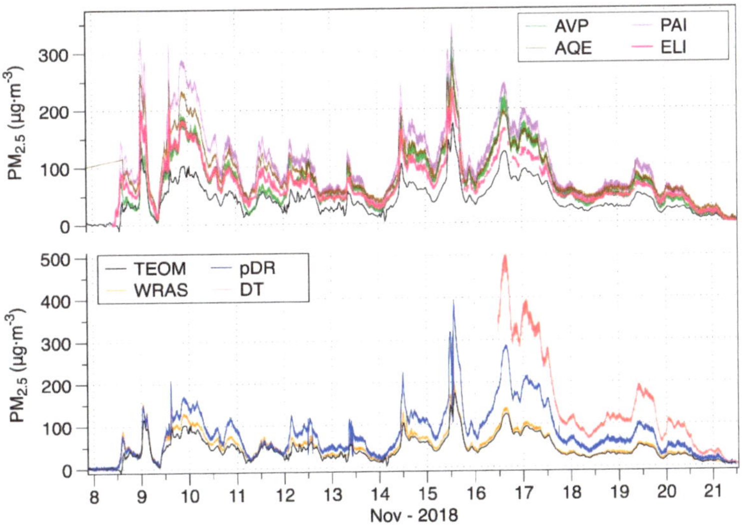

Figure 2. Time series of $PM_{2.5}$ as reported by all monitors tested at LBNL. The TEOM is a U.S. Federal Equivalent Method, and thus considered as the reference data. The top group are the monitors utilizing low-cost sensors and the bottom group are professional research grade monitors, with the TEOM reference measurement shown with both groups. Refer to Table 1 for monitor descriptions.

The DustTrak and pDR-1500 both use Arizona test dust as their calibration aerosol; and the devices nevertheless provided different responses. The DustTrak calibration is based on A1 Ultrafine dust, with mass median diameter (mmd) in the range of 3 to 5 μm. The PDR is calibrated with Arizona test dust A2 Fine, with mmd of 8 to 10 μm. These calibration aerosols have very different size distribution and optical properties than wildfire smoke, as shown in Figure S4. The two devices also use light sources of different wavelengths and measure at different scattering angles.

3.3. *Adjustment Factors Based on LBNL Measurements*

Low-cost and professional monitor measurements were related to actual $PM_{2.5}$, as measured by the TEOM, by determining the best fit parameters for linear equations with zero or non-zero intercepts. While prior studies have reported substantial non-zero intercepts when using low-cost monitors to measure ambient aerosols [46–49], we found that for the wildfire smoke, the slopes were very similar with zero or non-zero intercepts (see Figure S6). Based on this, we subsequently report adjustment factors as simple scalars with no offset.

Adjustment factors (AF) to translate the $PM_{2.5}$ reported by each instrument to the $PM_{2.5}$ reported by the TEOM (i.e., TEOM $PM_{2.5}$/device $PM_{2.5}$) were calculated for each 4-h interval of data and for the entire event. Summary results, provided in Figure 3, show that AFs varied across devices and also over time for each device. The DT had the least variability in part because it was used for only a few days. The WRAS reported concentrations closest to the TEOM with a median AF of 0.85. Median AFs for the monitors with low-cost sensors varied from 0.42 to 0.60.

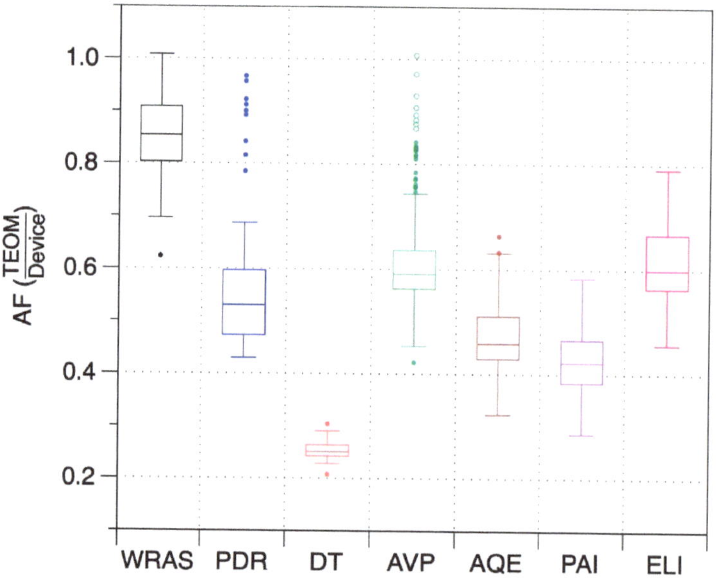

Figure 3. Distributions of 4-h average adjustment factors determined from measurements in a naturally ventilated lab over 13 days of elevated smoke from the Camp Fire.

An example of applying the AFs to adjust time-resolved data is provided in Figure 4. The bottom panel shows the AFs for 4-h average data over the course of the fire for the three AVP units. The middle panel shows that the adjusted time-series (using the median of the 4-h AFs across the event) closely match the TEOM and the top panel shows that residual errors were almost all between −30% and +20%. The same plot is provided for the PAI as Figure S7 in the Supplementary Materials. For both the AVP and PAI, three units of each device agreed closely throughout the fire.

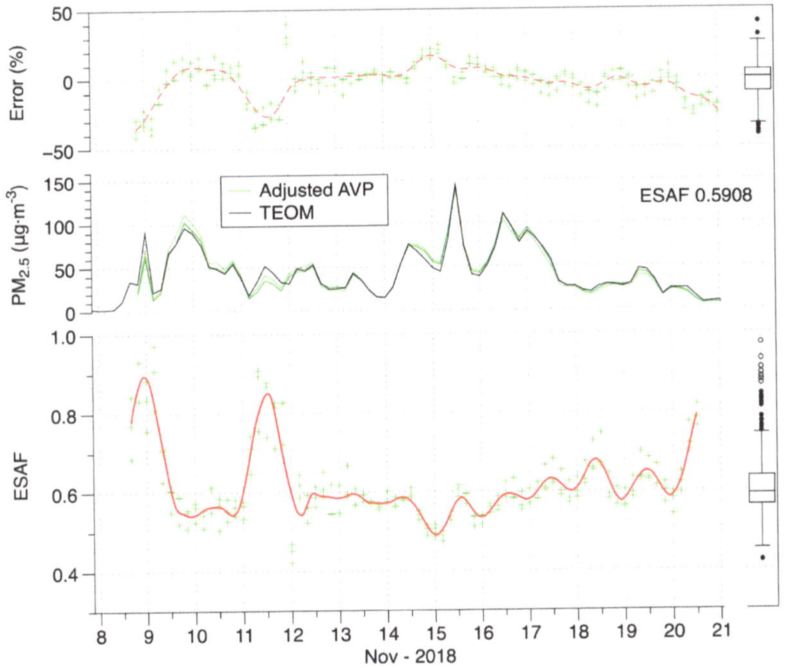

Figure 4. Event-specific adjustment factors (ESAFs), adjusted data and error of adjusted data (relative to co-located TEOM $PM_{2.5}$) for 4-h average AVP measurements in a naturally ventilated lab over 13 days of elevated smoke from the Camp Fire.

3.4. Measurements of $PM_{2.5}$ and Adjustment Factors in Northern California and Utah

Table 2 provides summary data from the 12 NorCal air quality monitoring stations that had nearby PA-II monitors at the time of the Camp Fire. The AQS sites varied in distance from the town of Paradise, which was the focal point of the fire. The positions of the AQS sites relative to the burned area are shown in Figure 1. Table S2 provides the number of PA-II monitors and reporting sensors. Given the large distances, there was relatively small variation of $PM_{2.5}$ across sites. Sacramento was closest and had the highest mean concentration. Davis was nearly the same distance from the fire but had among the lowest concentrations, possibly due to it being on the edge of the plume for much of the event (based on satellite images). Sites between Vallejo and San Francisco showed remarkable consistency with a relative standard deviation in the mean event concentration of 3.4%. Mean concentrations at the two furthest sites, San Jose and Redwood City, were about 15% lower than most of the other sites.

Table 2. Calculated linear fitting parameters with/out a zero offset and adjustment factors to quantify wildfire smoke $PM_{2.5}$ based on comparisons of 4-h means from PurpleAir PA-II monitors nearby to Northern California regulatory air quality monitoring stations during the Camp Fire in 2018.

		$PM_{2.5}$ ($\mu g\ m^{-3}$), 4-h avgs.			Linear Fits of 4-h avg Data Relating PA-II to AQS			Adjustment Factors Based on 4-h Ratios of AQS/PA-II				
AQS Site	**Distance (km)**	**Mean**	**10th**	**90th**	**Slope, Zero Intercept**	**Slope**	**Intercept**	**Mean**	**SD**	**Median**	**10th**	**90th**
Sacramento	133	134	47	239	0.498	0.510	−3.2	0.509	0.106	0.487	0.393	0.626
Davis	137	82	15	169	0.425	0.410	3.0	0.411	0.165	0.419	0.293	0.558
Vallejo	192	92	38	175	0.490	0.490	0.0	0.490	0.142	0.490	0.373	0.641
Concord	206	87	33	160	0.474	0.445	4.9	0.494	0.167	0.494	0.341	0.758
San Pablo	210	93	47	162	0.504	0.455	9.7	0.499	0.108	0.488	0.397	0.630
San Rafael	213	89	46	153	0.495	0.505	−2.0	0.631	0.255	0.635	0.439	1.092
Berkeley	219	93	54	164	0.459	0.423	6.9	0.464	0.089	0.472	0.375	0.572
Oakland-West	224	94	51	161	0.532	0.509	3.9	0.465	0.079	0.458	0.383	0.563
Oakland-Laney	226	91	53	157	0.459	0.437	4.5	0.530	0.088	0.528	0.423	0.627
San Francisco	232	93	45	156	0.511	0.498	2.5	0.504	0.160	0.520	0.318	0.692
Redwood City	258	74	33	122	0.446	0.387	8.3	0.451	0.162	0.449	0.314	0.607
San Jose	270	80	40	126	0.574	0.536	4.9	0.482	0.140	0.484	0.353	0.593

Table 2 also presents linear fits, with zero and non-zero intercepts, relating 4-h intervals of PA-II sensor and nearby AQS data. Table 2 presents AFs calculated for all AQS sites as the medians of all available 4-h AFs from sensors near the sites. Distributions of daily AFs across sites over the course of the fire are presented in Figure 5.

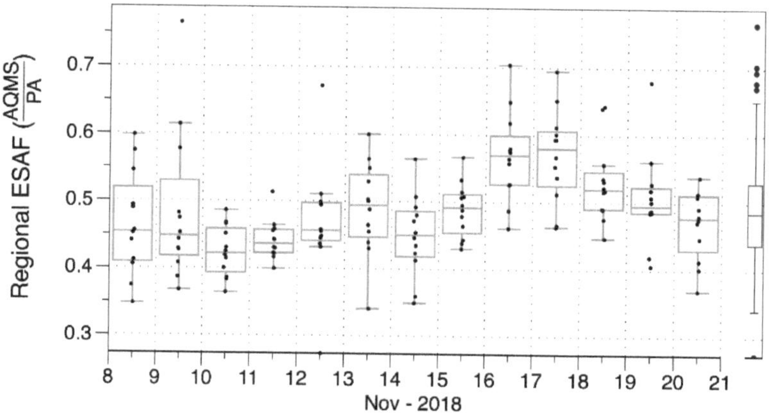

Figure 5. Variation in regional event-specific adjustment factor over time, shown as distributions of daily ESAFs required for PA-II monitors to align with measurements at 12 regulatory monitoring sites in Northern California during the Camp Fire in November 2018.

The median AF for the 12 AQS sites varied between 0.42 and 0.49 for the first 9 days, rose to 0.57–0.58 on Days 9–10, and declined over the last few days. To assess variance, we consider the relative median absolute deviation (RMAD) statistic, which is analogous to the relative standard deviation (RSD). The median RMAD for all the 4-h average AFs—including all days and all sensors —nearby to individual AQS sites was 12% with an IQR of 8–18% (Figure S8). An event specific adjustment factor (ESAF) for the Camp Fire was calculated as the median of all daily AFs for all AQS sites, providing a value of 0.485. Examples of time-series adjusted with the regional ESAF are provided for three sites in Figure 6. Whereas a prior study by Stampfer et al. [50] reported non-linear response for Plantower sensors when $PM_{2.5}$ was above 25 $\mu g\ m^{-3}$, the PA data used in this study linearly tracked with $PM_{2.5}$ to above 200 $\mu g\ m^{-3}$ as shown in Supplementary Figure S9. Since deviation from the linear relationship occurred only at very high levels, which were infrequent (Figure S9) the simple linear adjustment was applied.

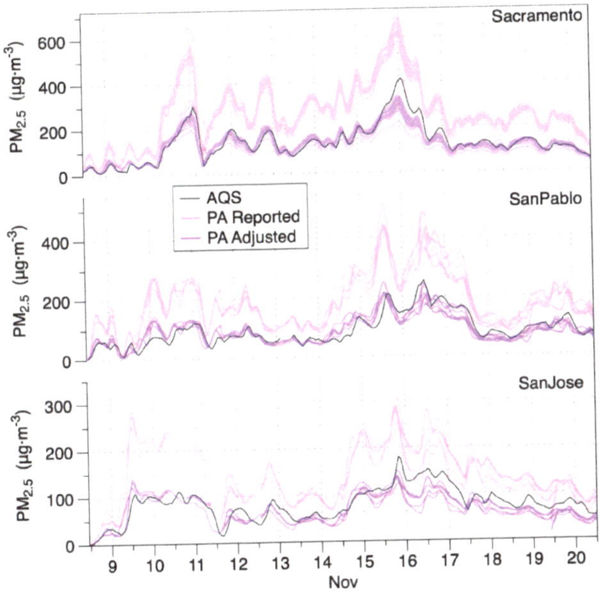

Figure 6. $PM_{2.5}$ concentrations reported by regulatory air quality monitoring stations (AQS) and all valid sensors of PurpleAir PA-II monitors nearby to three AQS sites during the Camp Fire in Northern California in November 2018. Pink data are unadjusted and purple data are adjusted using the regional event specific adjustment factor of 0.485.

Applying the regional adjustment factor to PA-II measurements throughout the area substantially reduced errors relative to default PA-II output. Figure 7 presents the summary distributions of residual errors of 4-h average data across all 12 AQS sites over all days of the smoke event using unadjusted values along with three different adjustment factors. The median unadjusted error was +102% with an IQ range of 74–133%. Using the regional ESAF from the Camp Fire produced an interquartile range of roughly ±15% in the residual error. The LRAPA adjustment available on the PurpleAir map at the time of the fire produced very similar results as using the regional AF calculated in this study. Using the AQandU correction resulted in concentration estimates that were 61% off at the median with an IQ range of 38–85%.

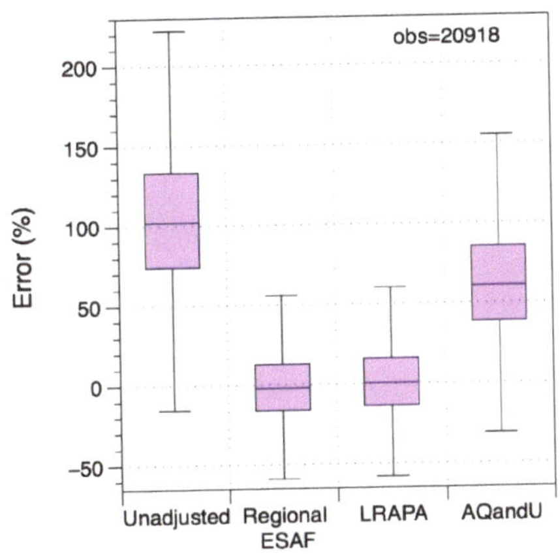

Figure 7. Distributions of errors for 4-h average $PM_{2.5}$ concentrations for Northern California PA-II monitors compared to regulatory monitors for the Camp Fire using different adjustment factors.

We applied the same analysis to the single PA-II and AQS combination that was available during the Carr/MC Fire. This site was heavily impacted by smoke from July 26 through August 12, 2018. Figure 8 shows the errors for the PA-II compared to the Red Bluff AQS using both unadjusted and adjusted data. The unadjusted PA-II monitor had a median error of +135% with an IQ range of 112–153%. When data were adjusted using the regional Camp Fire AF, median residual errors were 13% (3–22% IQR). When the Carr/MC data were adjusted with the AF determined for that fire, the IQR of the residual error was −9.5 to 7% (with median of 0%).

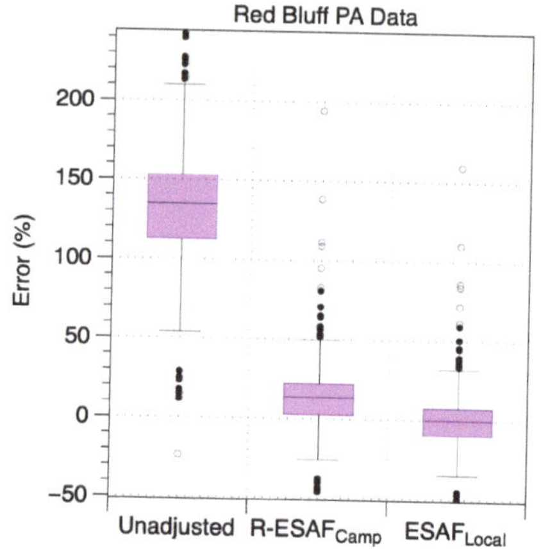

Figure 8. Distributions of errors for 4-h average $PM_{2.5}$ concentrations for Red Bluff PA-II direct readings and after applying the regional event specific adjustment factors from Camp Fire (R-ESAF) or those derived for the local Carr/Mendocino Complex Fire.

The Pole Creek Fire intermittently pushed smoke out into the Utah Valley from September 13 through September 24, 2018. Figure S8 shows the AQS and PA-II data—both unadjusted and adjusted using the Camp Fire AF, along with errors in the adjusted data. The AQS data showed a strong diurnal pattern with clean air in the evening around sunset and smoke starting to enter the valley by midnight and peaking a few hours after sunrise. When performing the analysis of PA-II and AQS data for this site, we focused only on the periods impacted by the smoke event by excluding any 4-h intervals where the AQS was <35 µg m^{-3}. When smoke was present, errors in the unadjusted PA-II $PM_{2.5}$ were +111% (66–212% IQR). Errors were reduced to 1.8% (−20 to +51% IQR) when using the Camp Fire regional ESAF. When adjusting data with the AF determined for the Pole Creek Fire, the IQR of the residual error was −21 to +48%.

Figure 9 shows the adjustment factors for the three fires. Three distributions are shown for the Camp Fire: regional ESAFs from the 12 AQS sites, the AFs determined at the Berkeley AQS, and the AFs measured inside the lab at LBNL. While AFs for the Berkeley AQS had a similar median and range as those from sites across the region, AFs for the infiltrated $PM_{2.5}$ at LBNL were lower. This could be due to several factors. First is the possibility that the particle size distribution of $PM_{2.5}$ inside the lab was different than outside around the region. Since $PM_{2.5}$ levels inside were lower by about a factor of two, there was clearly some loss of particles relative to outside. It is well established that the size distribution can change as penetration and deposition rates vary with particle size, leading to uneven losses across the range of particle sizes [51–53]. Another possible factor is that the reference instrument at LBNL was a TEOM while the air monitoring stations in Berkeley and elsewhere in California used BAM instruments. A third potential factor is the different versions of the Plantower sensor used in the PAI devices inside at LBNL (which use PMS1003 sensors) and the PA-II, which uses the PMS5003 sensor. The two sensors have the same nominal specifications and appear to use the same electronic components but they have different internal flow pathways. The Carr/MC Fire required a larger adjustment than was needed for the Camp Fire. The variability in the Pole Creek AFs may be impacted by the diurnal variability (which was still present even when analyzing only those intervals with $PM_{2.5} > 35\ \mu g/m^3$), coupled with the fact the PA-II was ~4 km away from the AQS for this site. As the plume was moving either in or out it was hitting the PA-II and AQS at different times. By contrast, for the California fires the plumes were present for days at a time reducing spatial and temporal variability.

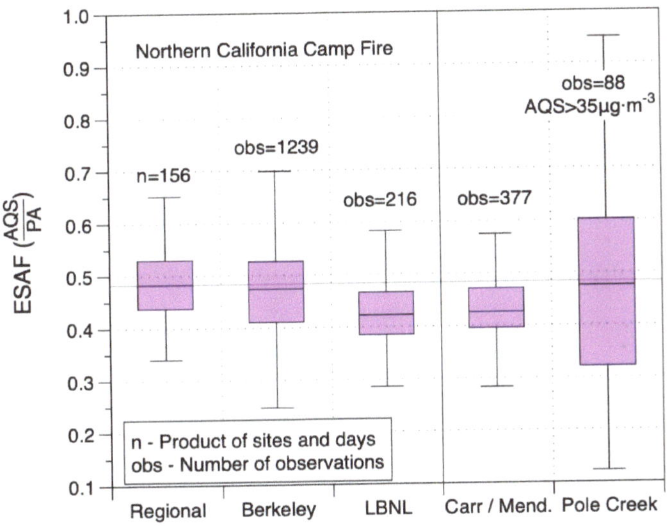

Figure 9. Event specific adjustment factors (ESAFs) calculated for PA-II monitors using data from three wildfire smoke events in 2018. The LBNL results are for indoor monitoring of infiltrated $PM_{2.5}$ in a lab with an estimated outdoor air exchange rate of 0.5 h^{-1}.

3.5. Impact of Adjustments on Air Quality Index Estimates

Adjustments to the reported concentrations translate to major changes to the associated AQIs reported for PA-II monitors. Figure 10 provides examples for PA-II monitors nearby to three monitoring sites during the Camp Fire in Northern California. For each site, three time series of 4-h AQI values are presented. The top bar is calculated from unadjusted PA-II readings. The middle is calculated from PA-II readings adjusted with the regional ESAF. And the bottom row is the 4-h AQI calculated from AQS data. At the Sacramento site, which was the closest of the three to the fire at 135 km away, unadjusted PA-II data indicated an AQI of "very unhealthy" or "hazardous" for 83% of the smoke event. Adjusted data indicated an AQI in these categories 31% of the time, which is similar to the 30% of time that the regulatory monitor reported AQI in these categories. Unadjusted sensor readings indicated the correct AQI category 14% of the time whereas the adjusted $PM_{2.5}$ provided the correct AQI category 84% of the time. At the San Pablo site (210 km from the fire), unadjusted PA-II indicated "very unhealthy" or "hazardous" for 59% of the event duration, which was much higher than the 17% of time that $PM_{2.5}$ measurements from the AQS indicated those AQI categories; adjusted PA-II data indicated these AQI categories of concern 10% of the time. Overall, the AQI category calculated from unadjusted PA-II data matched the AQS AQI only 29% of the time while the adjustment resulted in the correct AQI range 65% of the time. At the San Jose site (270 km from the fire), unadjusted PA-II indicated "very unhealthy" or "hazardous" for 34% of the event duration, whereas the 0% of time for the ESAF-adjusted data was similar to the 2% for the AQMS. Unadjusted data predicted AQI 47% of the time and adjustment led to the correct AQI category 66% of the time.

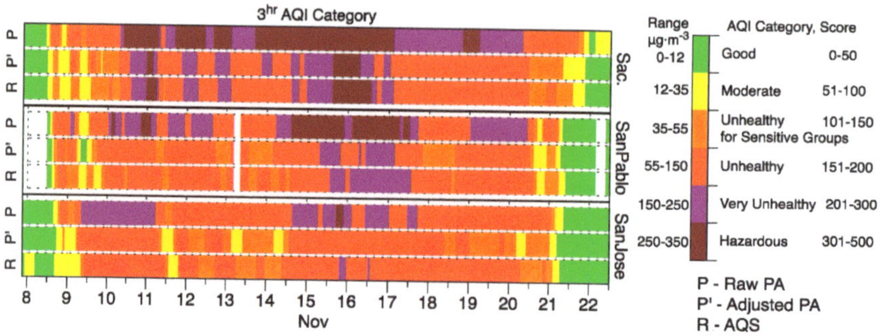

Figure 10. Comparison of Air Quality Index (AQI) values calculated using unadjusted data from PA-II monitors and the same data after correction with the regional, event specific AF to AQIs from $PM_{2.5}$ data at regulatory monitoring sites in Northern California during Camp Fire.

Despite the overall improvements in AQI predictions, it is important to note that adjusted values sometimes predicted a lower AQI hazard level than indicated by the nearby AQS.

3.6. Impact of Environmental Conditions on Adjustment Factors

PA-II adjustment factors were similar for the three fires, which occurred under varied seasonal conditions: the MC/Carr fire occurred in early summer (June), the Pole Fire occurred in late summer (September) and the Camp Fire occurred in late fall (late November). To assess the relationship between AFs for wildfire smoke and those occurring at the same sites through the year, we analyzed data from all of 2019 at measurement sites impacted by the Camp and MC/Carr fires in 2018. We first screened the data to look only at 4-h intervals with $PM_{2.5}$ above 12 μg/m^3. Figures S11 and S12 show distributions of temperature, RH, and $PM_{2.5}$ for these intervals of elevated $PM_{2.5}$ during each of the three main seasonal conditions (winter = December through February; summer = May through September; and shoulder = all other months), and for the period of each wildfire in 2018. During the Camp Fire, RH spanned a range similar to the shoulder seasons and temperatures were closer to those during winter. Consistent with this, the AF distribution was midway between the fall and winter distributions during periods of elevated $PM_{2.5}$. In Red Bluff, temperature and RH during the wildfire were similar to 2019 summer periods with elevated $PM_{2.5}$, but the distribution of AFs closely matched those during winter elevated $PM_{2.5}$. One possible reason for the similarity of wintertime and wildfire smoke AFs is that a substantial fraction of elevated winter $PM_{2.5}$ at both T-street and Red Bluff AQS sites may be associated with smoke from home heating wood combustion. The effect of environmental conditions is explored further in Figures S12 and S13, which show AFs as a function of RH, by season, at the two sites. In both cases the AFs for wildfire smoke do not change much over a very broad range of RH conditions. AFs also do not vary with RH during the winter. By contrast, AFs vary with humidity for the summer and shoulder seasons. Collectively these results provide support for the hypothesis that the AFs identified for smoke events are directly related to the characteristics of the smoke and do not vary greatly with environmental conditions.

4. Discussion

Prior research indicates varying responses of optical particle sensors to smoke generated from biomass combustion. In the most extensive and directly relevant study, McNamara et al. reported ratios of DustTrak 8520 and 8530 models to gravimetric, FRM and FEM measurements of varied instances of $PM_{2.5}$ from wood smoke [54]. Inside homes with wood stoves, which had $PM_{2.5}$ elevated from loading and stoking events, ratios of DT to gravimetric measurements were 1.60 ± 1.05 (mean AF = 0.63) across 43 sampling periods with mean (SD) gravimetric $PM_{2.5}$ = 30.7 (34.7) μg/m^3. DT to gravimetric ratios were 1.59–1.70 (AF = 0.63–0.59) for sampling in a University of Montana laboratory during three 24-h periods impacted by forest fire smoke (gravimetric $PM_{2.5}$ = 11.3, 21.2, 55.3 μg/m^3). For wintertime ambient sampling of $PM_{2.5}$ impacted by wood smoke in Libby, Montana, DT to BAM ratios were 1.43 ± 0.61 for BAM-reported $PM_{2.5}$ of 24.6 ± 8.0 μg/m^3. Dacunto et al. reported fireplace wood smoke correction factors (equivalent to AF) of 0.44–0.47 for the TSI Sidepak photometer, which is similar to the DT [55]. The previously reported AFs for wood and wildfire smoke are higher than the DT AF of 0.25 measured at LBNL for infiltrated wildfire smoke. These differences are presumed to result from variations in composition and size distributions of the measured aerosols, which had varied generation, ageing, and environmental conditions. Using an early generation of the pDR instrument, a U.S. Department of Agriculture study reported an AF of 0.53 for smoke generated in a fire laboratory [56]; the same AF was measured in our current study for the modern version of the pDR for infiltrated wildfire smoke.

There are limited published data on wood or wildfire smoke AFs for low-cost monitors. The Lane Regional Air Protection Agency (LRAPA), which is in a region of Oregon that is impacted by wood smoke from home heating, compared PA-II monitors to their network of FEMs and reported an AF equation of 0.5 * PA($PM_{2.5}$)–0.66 [44]. This is offered as a checkbox conversion on the PA map and closely aligns with the Camp Fire regional AF of 0.48. A long-term study of Plantower 1003 and 5003 sensors by the University of Utah reported equations relating hourly averaged individual sensor readings to a collocated TEOM 1405-F FEM-approved monitor by season [42]. During the months of Jun–Oct, which the authors described as "wildfire season" because of several fire events that occurred during that period, slopes of the linear fits were in the range of 1.33–1.48 (roughly corresponding to AFs of 0.68–0.75). However, since there were relatively few hours with $PM_{2.5}$ > 40 μg/m^3 over the season and the analysis did not break out the fits during the few significant wildfire events, the reported fits are not directly appropriate to adjusting Plantower sensor data during wildfire events with high $PM_{2.5}$ (e.g., >40 μg/m^3).

5. Conclusions

Low-cost air quality monitors can be used to accurately estimate hyper-local concentrations, regional dispersion, and health risk of $PM_{2.5}$ from wildfire smoke if appropriate device-specific adjustment factors are applied. Data from the existing network of outdoor PurpleAir II monitors is currently available with substantial coverage in many locations throughout the world. While the default PA-II response substantially over-reports wildfire smoke $PM_{2.5}$, the data can be scaled using the adjustment factor of 0.48 determined for the Camp Fire in Northern California, leading to substantially more accurate air quality index estimates. Based on measurements at LBNL of infiltrating smoke $PM_{2.5}$, it appears that both professional grade photometers and other monitors using Plantower low-cost optical PM sensors also substantially over-report wildfire $PM_{2.5}$ values. A simple multiplicative adjustment factor can bring the low-cost monitor response much closer to the $PM_{2.5}$ and AQI that would be reported by a regulatory monitor at the same location. Wildfire smoke AFs can vary across locations and over time during a fire event and the median AFs from one event may differ somewhat from those at other events. Yet even with these variations, application of a global AF can reduce bias from roughly a factor of two to 20–30% or less. It is also possible to apply short term adjustment factors by comparing the previous several hours of data from a deployed sensor to that form a nearby AQS if one is available.

Supplementary Materials: The following are available online at http://www.mdpi.com/1424-8220/20/13/3683/s1, Figure S1: Satellite image of the Camp fire on 11/12/2018; Figure S2: Satellite image of the Carr and Mendocino complex fires on 08/03/2018 20:43:47 GMT; Figure S3: Satellite image of the Pole Creek fire 9/13/2018 20:36:57 GMT.; Figure S4: Mass distributions of wildfie smoke and two varieties of test dust; Figure S5: Time series of $PM_{2.5}$ measured by TEOM in the laboratory at LBNL lab and at the Berkeley and Oakland-West regulatory air quality monitoring sites during the 2018 Camp Fire in northern California; Figure S6: Adjustment factors, adjusted data and error of adjusted data (relative to co-located TEOM $PM_{2.5}$) for 4-h average PAI measurements in in a naturally ventilated lab over 13 days of elevated smoke from the Camp Fire; Figure S7: Distributions of daily relative median absolute deviations; Figure S8: Running 4 hr average data for the Pole Creek Fire at Spanish Fork UT; Figure S9:

Error distributions for 4 h average data after applying Camp Fire event-specific regional adjustment factor to PA-II monitors near air quality monitoring stations in Northern California. Figure S10: Environmental conditions and PurpleAir II adjustment factors by season over running 4-h intervals with $PM_{2.5}$ above 12 μg/m^3 at the T-street air quality monitoring station in Sacramento, CA; Figure S11: Environmental conditions and PurpleAir II adjustment factors by season over running 4-h intervals with $PM_{2.5}$ above 12 μg/m^3 at the air quality monitoring station in Red Bluff, CA; Figure S12: PurpleAir II adjustment factors and relative humidity by season over running 4-h intervals with $PM_{2.5}$ above 12 μg/m^3 at the T-street air quality monitoring station in Sacramento, CA; Figure S13: PurpleAir II adjustment factors and relative humidity by season over running 4-h intervals with $PM_{2.5}$ above 12 μg/m^3 at the air quality monitoring station in Red Bluff, CA. Table S1: Median response of each monitor deployed at LBNL relative to co-located TEOM, statistics for 4 h averages and event-integrated ratio; Table S2: Sensor data used to calculate adjustment factors for quantifying infiltrating wildfire smoke $PM_{2.5}$ using 4-h means from PurpleAir PA-II monitors nearby to northern California regulatory air quality monitoring stations during the three western US wild fires.

Author Contributions: Conceptualization, W.W.D. and B.C.S.; methodology, W.W.D. and B.C.S.; formal analysis, W.W.D.; investigation, W.W.D.; data curation, W.W.D.; writing—original draft preparation, B.C.S. and W.W.D.; writing—review and editing, B.C.S. and W.W.D.; visualization, W.W.D. and B.C.S.; project administration, B.C.S.; funding acquisition, B.C.S. All authors have read and agreed to the published version of the manuscript.

Funding: This research was supported by the U.S. Environmental Protection Agency Indoor Environments Division through Interagency Agreement DW-89-9232201-7 and the U.S.–China Clean Energy Research Center –Building Energy Efficiency Program within the Building Technologies Office of the U.S. Department of Energy, and was conducted at LBNL under Contract No. DE-AC02-05CH1123.

Acknowledgments: The authors thank Zhiqiang Wang for collecting filter samples and Marion Russell and student assistants for preparing and weighing the filters. We acknowledge the use of imagery from the NASA Worldview application (https://worldview.earthdata.nasa.gov), part of the NASA Earth Observing System Data and Information System (EOSDIS).

Conflicts of Interest: The authors declare no conflict of interest.

References

1. Abatzoglou, J.T.; Williams, A.P. Impact of anthropogenic climate change on wildfire across western US forests. *Proc. Natl. Acad. Sci. USA* **2016**, *113*, 11770–11775. [CrossRef]
2. Dennison, P.E.; Brewer, S.C.; Arnold, J.D.; Moritz, M.A. Large wildfire trends in the western United States, 1984–2011. *Geophys. Res. Lett.* **2014**, *41*, 2928–2933. [CrossRef]
3. Rasker, R. Resolving the Increasing Risk from Wildfires in the American West. *Solutions* **2015**, *6*, 55–62.
4. Schoennagel, T.; Balch, J.K.; Brenkert-Smith, H.; Dennison, P.E.; Harvey, B.J.; A Krawchuk, M.; Mietkiewicz, N.; Morgan, P.; Moritz, M.A.; Rasker, R.; et al. Adapt to more wildfire in western North American forests as climate changes. *Proc. Natl. Acad. Sci. USA* **2017**, *114*, 4582–4590. [CrossRef] [PubMed]
5. Westerling, A.L.; Hidalgo, H.G.; Cayan, D.R.; Swetnam, T.W. Warming and Earlier Spring Increase Western U.S. Forest Wildfire Activity. *Science* **2006**, *313*, 940. [CrossRef] [PubMed]
6. Radeloff, V.C.; Helmers, D.P.; Kramer, H.A.; Mockrin, M.H.; Alexandre, P.M.; Bar-Massada, A.; Butsic, V.; Hawbaker, T.J.; Martinuzzi, S.; Syphard, A.; et al. Rapid growth of the US wildland-urban interface raises wildfire risk. *Proc. Natl. Acad. Sci. USA* **2018**, *115*, 3314–3319. [CrossRef] [PubMed]
7. Künzli, N.; Avol, E.; Wu, J.; Gauderman, W.J.; Rappaport, E.; Millstein, J.; Bennion, J.; McConnell, R.; Gilliland, F.D.; Berhane, K.; et al. Health effects of the 2003 Southern California wildfires on children. *Am. J. Respir. Crit. Care Med.* **2006**, *174*, 1221–1228. [CrossRef] [PubMed]
8. Liu, J.C.; Wilson, A.; Mickley, L.J.; Ebisu, K.; Sulprizio, M.P.; Wang, Y.; Peng, R.D.; Yue, X.; Dominici, F.; Bell, M. Who Among the Elderly Is Most Vulnerable to Exposure to and Health Risks of Fine Particulate Matter From Wildfire Smoke? *Am. J. Epidemiol.* **2017**, *186*, 730–735. [CrossRef]
9. Reid, C.E.; Brauer, M.; Johnston, F.H.; Jerrett, M.; Balmes, J.R.; Elliott, C.T. Critical Review of Health Impacts of Wildfire Smoke Exposure. *Environ. Health Perspect.* **2016**, *124*, 1334–1343. [CrossRef]
10. Cascio, W.E. Wildland fire smoke and human health. *Sci. Total Environ.* **2018**, *624*, 586–595. [CrossRef]
11. Black, C.; Tesfaigzi, Y.; Bassein, J.A.; Miller, L. Wildfire smoke exposure and human health: Significant gaps in research for a growing public health issue. *Environ. Toxicol. Pharmacol.* **2017**, *55*, 186–195. [CrossRef] [PubMed]

12. Shrestha, P.; Humphrey, J.L.; Carlton, E.J.; Adgate, J.L.; Barton, K.E.; Root, E.; Miller, S. Impact of Outdoor Air Pollution on Indoor Air Quality in Low-Income Homes during Wildfire Seasons. *Int. J. Environ. Res. Public Health* **2019**, *16*, 3535. [CrossRef]
13. Fisk, W.J.; Chan, W.R. Health benefits and costs of filtration interventions that reduce indoor exposure to $PM_{2.5}$ during wildfires. *Indoor Air* **2017**, *27*, 191–204. [CrossRef]
14. Barn, P.; Larson, T.; Noullett, M.; Kennedy, S.; Copes, R.; Brauer, M. Infiltration of forest fire and residential wood smoke: An evaluation of air cleaner effectiveness. *J. Expo. Sci. Environ. Epidemiol.* **2008**, *18*, 503–511. [CrossRef]
15. US EPA. *National Ambient Air Quality Standards for Particulate Matter;* Final Rule Issuing 2013; U.S. Environmental Protection Agency: Washington, DC, USA, 2013; Volume 78.
16. US EPA. *Final Report: Integrated Science Assessment for Particulate Matter;* U.S. Environmental Protection Agency: Washington, DC, USA, 2009.
17. US EPA. *Technical Assistance Document for the Reporting of Daily Air Quality—The Air Quality Index (AQI);* U.S. Environmental Protection Agency: Washington, DC, USA, 2018.
18. Chow, J.C. Measurement Methods to Determine Compliance with Ambient Air-Quality Standards for Suspended Particles. *J. Air Waste Manag. Assoc.* **1995**, *45*, 320–382. [CrossRef] [PubMed]
19. Noble, C.A.; Vanderpool, R.W.; Peters, T.M.; McElroy, F.F.; Gemmill, D.B.; Wiener, R.W. Federal reference and equivalent methods for measuring fine particulate matter. *Aerosol Sci. Technol.* **2001**, *34*, 457–464. [CrossRef]
20. US EPA. $PM_{2.5}$ Continuous Monitor Comparability Assessments. 31 July 2019. Available online: https://www.epa.gov/outdoor-air-quality-data/pm25-continuous-monitor-comparability-assessments (accessed on 30 October 2019).
21. Schweizer, D.; Cisneros, R.; Shaw, G. A comparative analysis of temporary and permanent beta attenuation monitors: The importance of understanding data and equipment limitations when creating $PM_{2.5}$ air quality health advisories. *Atmos. Pollut. Res.* **2016**, *7*, 865–875. [CrossRef]
22. CARPA. Met One E-BAM Portable $PM_{2.5}$ Monitor and AIRSIS Data Telemetry System. 2010. Available online: https://ww3.arb.ca.gov/carpa/toolkit/data/ebam-guidance-2010.pdf (accessed on 30 October 2019).
23. National Interagency Fire Center. Available online: https://www.nifc.gov/smoke/smoke_monitoring.html (accessed on 30 October 2019).
24. Kelleher, S.; Quinn, C.W.; Miller-Lionberg, D.; Volckens, J. A low-cost particulate matter ($PM_{2.5}$) monitor for wildland fire smoke. *Atmos. Meas. Tech.* **2018**, *11*, 1087–1097. [CrossRef]
25. Gupta, P.; Doraiswamy, P.; Levy, R.C.; Pikelnaya, O.; Maibach, J.; Feenstra, B.; Polidori, A.; Kiros, F.; Mills, K.C. Impact of California Fires on Local and Regional Air Quality: The Role of a Low-Cost Sensor Network and Satellite Observations. *GeoHealth* **2018**, *2*, 172–181. [CrossRef]
26. Sorensen, C.M.; Gebhart, J.; O'Hern, T.J.; Rader, D.J. Optical measurement techniques: Fundamentals and applications. In *Aerosol Measurement: Principles, Techniques, and Applications*, 3rd ed.; Kulkarni, P., Baron, P.A., Willeke, K., Eds.; Wiley: Hoboken, NJ, USA, 2011; pp. 269–312.
27. Wang, X.; Chancellor, G.; Evenstad, J.; Farnsworth, J.E.; Hase, A.; Olson, G.M.; Sreenath, A.; Agarwal, J.K. A Novel Optical Instrument for Estimating Size Segregated Aerosol Mass Concentration in Real Time. *Aerosol Sci. Technol.* **2009**, *43*, 939–950. [CrossRef]
28. Thomas, A.; Gebhart, J. Correlations between Gravimetry and Light-Scattering Photometry for Atmospheric Aerosols. *Atmos. Environ.* **1994**, *28*, 935–938. [CrossRef]
29. Wallace, L.A.; Wheeler, A.J.; Kearney, J.; Van Ryswyk, K.; You, H.; Kulka, R.H.; Rasmussen, P.; Brook, J.R.; Xu, X. Validation of continuous particle monitors for personal, indoor, and outdoor exposures. *J. Expo. Sci. Environ. Epidemiol.* **2011**, *21*, 49–64. [CrossRef] [PubMed]
30. Ahangar, F.E.; Freedman, F.R.; Venkatram, A. Using Low-Cost Air Quality Sensor Networks to Improve the Spatial and Temporal Resolution of Concentration Maps. *Int. J. Environ. Res. Public Health* **2019**, *16*, 1252. [CrossRef] [PubMed]
31. Gao, M.L.; Cao, J.J.; Seto, E. A distributed network of low-cost continuous reading sensors to measure spatiotemporal variations of $PM_{2.5}$ in Xi'an, China. *Environ. Pollut.* **2015**, *199*, 56–65. [CrossRef] [PubMed]
32. Jiao, W.; Hagler, G.; Williams, R.; Sharpe, R.; Brown, R.; Garver, D.; Judge, R.; Caudill, M.; Rickard, J.; Davis, M.; et al. Community Air Sensor Network (CAIRSENSE) project: Evaluation of low-cost sensor performance in a suburban environment in the southeastern United States. *Atmos. Meas. Tech.* **2016**, *9*, 5281–5292. [CrossRef]

33. Morawska, L.; Thai, P.K.; Liu, X.; Asumadu-Sakyi, A.; Ayoko, G.; Bartoňová, A.; Bedini, A.; Chai, F.; Christensen, B.; Dunbabin, M.; et al. Applications of low-cost sensing technologies for air quality monitoring and exposure assessment: How far have they gone? *Environ. Int.* **2018**, *116*, 286–299. [CrossRef]
34. Singer, B.C.; Delp, W.W. Response of consumer and research grade indoor air quality monitors to residential sources of fine particles. *Indoor Air* **2018**, *28*, 624–639. [CrossRef]
35. Manikonda, A.; Zíková, N.; Hopke, P.K.; Ferro, A. Laboratory assessment of low-cost PM monitors. *J. Aerosol Sci.* **2016**, *102*, 29–40. [CrossRef]
36. Zou, Y.; Young, M.; Chen, J.; Liu, J.; May, A.A.; Clark, J.D. Examining the functional range of commercially available low-cost airborne particle sensors and consequences for monitoring of indoor air quality in residences. *Indoor Air* **2019**, *30*, 213–234. [CrossRef]
37. Sousan, S.; Koehler, K.; Thomas, G.; Park, J.H.; Hillman, M.; Halterman, A.; Peters, T.M. Inter-comparison of low-cost sensors for measuring the mass concentration of occupational aerosols. *Aerosol Sci. Technol.* **2016**, *50*, 462–473. [CrossRef]
38. Dacunto, P.J.; Cheng, K.-C.; Acevedo-Bolton, V.; Jiang, R.; Repace, J.; Ott, W.R.; Hildemann, L.M.; Klepeis, N. Determining $PM_{2.5}$ calibration curves for a low-cost particle monitor: Common indoor residential aerosols. *Environ. Sci. Process. Impacts* **2015**, *17*, 1959–1966. [CrossRef] [PubMed]
39. Johnson, K.K.; Bergin, M.H.; Russell, A.G.; Hagler, G.S. Field Test of Several Low-Cost Particulate Matter Sensors in High and Low Concentration Urban Environments. *Aerosol Air Qual. Res.* **2018**, *18*, 565–578. [CrossRef]
40. California Air Resources Board, Air Quality Data (PST) Query Tool. 2019. Available online: https://www.arb.ca.gov/aqmis2/aqdselect.php (accessed on 20 June 2019).
41. Utah Department of Environmental QualityAir Quality Data Archive. 2019. Available online: http://www.airmonitoring.utah.gov/dataarchive/archpm25.htm (accessed on 20 June 2019).
42. Sayahi, T.; Butterfield, A.; Kelly, K.E. Long-term field evaluation of the Plantower PMS low-cost particulate matter sensors. *Environ. Pollut.* **2019**, *245*, 932–940. [CrossRef] [PubMed]
43. Kelly, K.; Whitaker, J.; Petty, A.; Widmer, C.; Dybwad, A.; Sleeth, D.; Martin, R.; Butterfield, A. Ambient and laboratory evaluation of a low-cost particulate matter sensor. *Environ. Pollut.* **2017**, *221*, 491–500. [CrossRef]
44. Lane Regional Air Protection Agency. Air Quality Sensors. 2019. Available online: https://www.lrapa.org/DocumentCenter/View/4147/PurpleAir-Correction-Summary (accessed on 30 October 2019).
45. Tryner, J.; L'Orange, C.; Mehaffy, J.; Miller-Lionberg, D.; Hofstetter, J.C.; Wilson, A.; Volckens, J. Laboratory evaluation of low-cost PurpleAir PM monitors and in-field correction using co-located portable filter samplers. *Atmos. Environ.* **2020**, *220*, 117067. [CrossRef]
46. Malings, C.; Tanzer, R.; Hauryliuk, A.; Saha, P.K.; Robinson, A.L.; Presto, A.A.; Subramanian, R. Fine particle mass monitoring with low-cost sensors: Corrections and long-term performance evaluation. *Aerosol Sci. Technol.* **2020**, *54*, 160–174. [CrossRef]
47. Mehadi, A.; Moosmüller, H.; Campbell, D.E.; Ham, W.; Schweizer, D.; Tarnay, L.; Hunter, J. Laboratory and field evaluation of real-time and near real-time $PM_{2.5}$ smoke monitors. *J. Air Waste Manag. Assoc.* **2020**, *70*, 158–179. [CrossRef]
48. Feenstra, B.; Papapostolou, V.; Hasheminassab, S.; Zhang, H.; der Boghossian, B.; Cocker, D.; Polidori, A. Performance evaluation of twelve low-cost $PM_{2.5}$ sensors at an ambient air monitoring site. *Atmos. Environ.* **2019**, *216*, 116946. [CrossRef]
49. Wang, Z.; WDelp, W.; Singer, B.C. Performance of low-cost indoor air quality monitors for $PM_{2.5}$ and PM_{10} from residential sources. *Build. Environ.* **2020**, *171*, 106654. [CrossRef]
50. Stampfer, O.; Austin, E.; Ganuelas, T.; Fiander, T.; Seto, E.; Karr, C.J. Use of low-cost PM monitors and a multi-wavelength aethalometer to characterize $PM_{2.5}$ in the Yakama Nation reservation. *Atmos. Environ.* **2020**, *224*, 117292. [CrossRef]
51. El Orch, Z.; Stephens, B.; Waring, M.S. Predictions and determinants of size-resolved particle infiltration factors in single-family homes in the US. *Build. Environ.* **2014**, *74*, 106–118. [CrossRef]
52. Thatcher, T.L.; Lunden, M.M.; Revzan, K.L.; Sextro, R.G.; Brown, N.J. A concentration rebound method for measuring particle penetration and deposition in the indoor environment. *Aerosol Sci. Technol.* **2003**, *37*, 847–864. [CrossRef]
53. Chen, C.; Zhao, B. Review of relationship between indoor and outdoor particles: I/O ratio, infiltration factor and penetration factor. *Atmos. Environ.* **2011**, *45*, 275–288. [CrossRef]

54. McNamara, M.L.; Noonan, C.W.; Ward, T.J. Correction Factor for Continuous Monitoring of Wood Smoke Fine Particulate Matter. *Aerosol Air Qual. Res.* **2011**, *11*, 316–323. [CrossRef]
55. Dacunto, P.J.; Cheng, K.-C.; Acevedo-Bolton, V.; Jiang, R.; Klepeis, N.; Repace, J.; Ott, W.R.; Hildemann, L.M. Real-time particle monitor calibration factors and $PM_{2.5}$ emission factors for multiple indoor sources. *Environ. Sci. Process. Impacts* **2013**, *15*, 1511–1519. [CrossRef]
56. United States Department of Agriculture. *Laboratory Evaluation of Two Optical Instruments for Real-Time Monitoring of Smoke*; Trent, P.L.A., Ed.; USDA Forest Service, Technology and Development Program: Missoula, MT, USA, 1999. Available online: https://www.fs.fed.us/t-d/pubs/pdfpubs/pdf99252806/pdf99252806.pdf (accessed on 20 June 2019).

sensors

MDPI

Article

Application of Machine Learning for the in-Field Correction of a $PM_{2.5}$ Low-Cost Sensor Network

Wen-Cheng Vincent Wang [1], Shih-Chun Candice Lung [1,2,3,*] and Chun-Hu Liu [1]

1 Research Center for Environmental Changes, Academia Sinica, Nangang, Taipei 115, Taiwan; phdzen@gate.sinica.edu.tw (W.-C.V.W.); lch0909@gate.sinica.edu.tw (C.-H.L.)
2 Department of Atmospheric Sciences, National Taiwan University, Taipei 106, Taiwan
3 Institute of Environmental Health, National Taiwan University, Taipei 106, Taiwan
* Correspondence: sclung@rcec.sinica.edu.tw

Received: 12 July 2020; Accepted: 31 August 2020; Published: 3 September 2020

Abstract: Many low-cost sensors (LCSs) are distributed for air monitoring without any rigorous calibrations. This work applies machine learning with $PM_{2.5}$ from Taiwan monitoring stations to conduct in-field corrections on a network of 39 $PM_{2.5}$ LCSs from July 2017 to December 2018. Three candidate models were evaluated: Multiple linear regression (MLR), support vector regression (SVR), and random forest regression (RFR). The model-corrected $PM_{2.5}$ levels were compared with those of GRIMM-calibrated $PM_{2.5}$. RFR was superior to MLR and SVR in its correction accuracy and computing efficiency. Compared to SVR, the root mean square errors (RMSEs) of RFR were 35% and 85% lower for the training and validation sets, respectively, and the computational speed was 35 times faster. An RFR with 300 decision trees was chosen as the optimal setting considering both the correction performance and the modeling time. An RFR with a nighttime pattern was established as the optimal correction model, and the RMSEs were 5.9 ± 2.0 μg/m^3, reduced from 18.4 ± 6.5 μg/m^3 before correction. This is the first work to correct LCSs at locations without monitoring stations, validated using laboratory-calibrated data. Similar models could be established in other countries to greatly enhance the usefulness of their $PM_{2.5}$ sensor networks.

Keywords: efficient in-field $PM_{2.5}$ correction; random forest model; particle sensing correction; in-field calibration; PM sensing device

1. Introduction

Millions of premature deaths worldwide can be attributed to particulate matter with an aerodynamic diameter less than or equal to 2.5 μm ($PM_{2.5}$) [1,2], which is one of the human carcinogens classified by the International Agency for Research on Cancer [3]. Rising $PM_{2.5}$ levels in the ambient air and their associated health impacts are important environmental health issues that concern the general public, especially in developing countries [4,5]. In eastern Asia during 1998–2000, 51% of the population lived in areas with annual mean $PM_{2.5}$ levels above the recommended guideline of the World Health Organization (35 μg/m^3). This percentage increased to 70% during 2010–2012 [6], showing the deterioration of the air quality in this region.

In resource-limited Asian countries, there are insufficient numbers of regulatory monitoring stations in urban areas with high population densities. The purpose of the $PM_{2.5}$ monitoring stations of Environmental Protection Administrations (EPAs) worldwide is to assess the well-mixed ambient pollutant levels. Therefore, such monitors are situated at a height of 10–15 m above the ground. However, the intensive emissions of community pollution sources in Asia, such as restaurants and temples, result in high $PM_{2.5}$ levels in the immediate living environments of citizens at street level [7–10]. Even living in the same airshed, residents from different communities with different emission sources are exposed to different $PM_{2.5}$ levels, demonstrating the need for community air monitoring. In Taiwan island, there are 57 regular air monitoring stations, covering an area of 35,887 km^2 [11]. Excluding the two-thirds of mountainous areas with little population, on average, every 300 km^2 of land area has only one air quality station. Even in the capital area (Taipei metropolitan), every 141 km^2 has only one air quality station on average. Thus, there are still large monitoring gaps between these stations. These gaps can be filled by the observations provided by low-cost sensors (LCSs) for ambient-level and street-level air quality monitoring.

The United Stations Environmental Protection Agency (USEPA) noted that LCSs can be used to fill the observational gaps of environmental monitoring after appropriate quality assurance and quality control procedures [12]. Currently, there are more than 4000 $PM_{2.5}$ LCSs in the citizen air quality network (CAQN [13]) in Taiwan as a result of the collaboration of citizens, private enterprises, and non-governmental organizations (NGOs) with partial government funding. The $PM_{2.5}$ sensor network on the Island of Taiwan is possibly one of the most densely distributed sensor networks in the world [14,15], with (on average) three LCSs per 1 km^2. CAQN has increased the environmental awareness of Taiwanese citizens dramatically. However, these sensors installed by citizens or NGOs are not calibrated, and data accuracy is not assured. Thus, their readings may be 2–3 times the actual concentrations [10]. Such limitations in data quality have restrained the application of these sensors in research and policy development.

LCSs need to be calibrated, since their readings usually deviate from those of research-grade instruments [10]. The actual cut-off points of particle sizes for LCSs are not consistent with the statements declared by the manufactures [16]; individual correction equations are, thus, needed for each sensor, even those from the same batch [10]. Most environmental scientists evaluate the performance of LCSs with side-by-side comparisons against research-grade instruments in the laboratory or in the field. Research-grade instruments, which are considered the golden standard for such comparisons, include instruments with measuring principles based on light scattering, oscillating tapered elements, β-attenuation, and filter-weighing, such as GRIMM instruments [10,17–20], TSI SidePake AM510 [21,22], tapered element oscillating microbalance [18,23], β-attenuation monitor [17–19,24,25], and ultrasonic personal aerosol sampler [26]. Correction equations are established to adjust the LCS readings to those of research-grade measurements. Mathematical models, such as linear regression [17–20,27] or multiple linear regression (MLR) models [23,25,28], have been used to establish these correction equations. The latter models usually consider ambient environmental factors.

To obtain accurate $PM_{2.5}$ levels for LCSs, data correction is a necessary procedure. For thousands of $PM_{2.5}$ sensors already installed by citizens in the field, an innovative method for correction is needed, since traditional laboratory or field evaluations with side-by-side comparisons require a great deal of human resources and time. Furthermore, it is impractical to correct thousands of sensors. Innovative data science techniques, such as machine learning, could also be used for correcting LCS readings [29]. Machine learning techniques have been used for air pollution issues for different purposes. Cheng et al. applied transfer learning to correct the $PM_{2.5}$ values of the citizen network in Beijing, China using the $PM_{2.5}$ levels from public environmental monitoring stations [30]. Pandey et al. predicted the levels of ultrafine particles in China through a machine learning approach [31]. Hsieh et al. used an artificial neural network and deep learning to analyze the proper position to install air pollution stations in Beijing [32]. Zheng et al. analyzed and predicted the air quality in different areas through data mining and machine learning using data from 43 cities in China [33]. Paas et al. used an artificial neural network to acquire the mass concentrations and numbers of different sizes of particles in Germany [34]. Peng et al. forecasted the levels of ozone, $PM_{2.5}$, and NO_2 in Canada using multiple linear regression and multi-layer perceptron neural networks [35]. Moreover, machine learning methods may provide a significant accuracy improvement for gaseous and particle sensors [23,36].

Therefore, we proposed a hybrid method combining laboratory evaluations and data science to ensure that the LCS networks provide accurate PM data [20]. First, the LCS data are corrected through side-by-side laboratory comparisons for "seed" LCS devices, which can be installed strategically in areas without EPA stations; secondly, statistical or machine learning methods are applied to adjust nearby uncalibrated LCS devices using data from the EPA stations or the seed LCS devices wherever available. In this way, readings from uncalibrated LCS devices in the CAQN can be corrected to nearly research-grade observations. The first part of obtaining reliable and robust correction equations is to convert the readings of LCS devices to research-grade (or FEM-comparable) measurements via side-by-side comparisons with research-grade instruments in the laboratory, as presented in [20].

The current work focuses on the second part of this process: Applying machine learning to correct data from the LCS network with the $PM_{2.5}$ values from Taiwan EPA stations. The objectives of this work are: (1) To establish data correction models based on machine learning techniques with the $PM_{2.5}$ data from the Taiwan EPA to correct the readings of the sensor network; (2) to evaluate the model performance with data from the same sensor network calibrated with laboratory evaluations; and (3) to explore the best data correction models using choices of computing efficiency and day/night periods. As can be seen in the later section, using in-field corrections with machine learning techniques, the $PM_{2.5}$ data quality of LCSs can be greatly improved. Furthermore, this is the first work to introduce the use of a nighttime dataset instead of a whole-day dataset for the establishment of a data correction model with machine learning in order to prevent interference from local emissions during the daytime. This method can be further used to conduct in-field corrections for CAQN in Taiwan, as well as for other sensor networks in other countries.

2. Materials and Methods

2.1. Sensor Network Introduction

The LCS network corrected by Taiwan EPA data in this work consists of LCS devices designed for research purposes, namely, AS-LUNG-O. LCS devices integrate LCS, power, and data transmission components. AS-LUNG-O is an LCS device integrated by our team and designed for long-term outdoor monitoring for scientific research [10]. AS stands for Academia Sinica (the research institute that supports its development), while LUNG indicates the human organ most commonly affected by air pollutants, and O indicates the "outdoor" version. AS-LUNG-O (~650 USD basic manufacturing costs) incorporates sensors for $PM_{2.5}$ (PMS3003, Plantower, Beijing, China), CO_2 (S8, Senseair AB, Sweden), temperature/humidity (SHT31, SENSIRION, Staefa ZH, Switzerland), and Global Positioning System (GPS, u-blox, Switzerland). The $PM_{2.5}$ sensor, PMS3003, has been evaluated by several research teams in laboratory environments. For example, Kelly et al. [18] obtained an R^2 of 0.73–0.97 in wind tunnels, and Sayahi et al. [37] obtained an $R^2 > 0.978$ for 242 sets of PMS3003 in a controlled chamber. These results indicated the good performance of PMS3003 compared to research-grade instruments.

The sensors for AS-LUNG-O are placed in a waterproof shelter connected to a solar panel with backup batteries for the power supply, with the option to use household electricity, where easily accessible. The size of the whole set is roughly 60 cm (W) × 50 cm (D) × 50 cm (H), with a weight of approximately 4.8 kg. Data can be transmitted wirelessly using the built-in 4G modules to a cloud database with one-min intervals. An SD card was added as a complement to avoid data loss during wireless transmission. Currently, most of the LCSs used in the CAQN (http://www.aqmd.gov/aq-spec/product/edimax) in Taiwan are PMS5003 (Plantower, Beijing, China), which is also a Plantower LCS.

AS-LUNG-O is a versatile LCS device capable of operating under the harsh weather conditions in subtropical Taiwan, which experiences high humidity (e.g., The mean relative humidity (RH) was 74% in the year of 2016 [10]) and frequent Typhoons [38]. AS-LUNG-O can be installed on the light poles in the streets and was used in a small town in a mountainous area to fill the data gaps of $PM_{2.5}$ monitoring. The incremental $PM_{2.5}$ concentration increases due to different community sources were, thus, quantified using AS-LUNG-O [10]. Therefore, other communities without EPA monitoring stations can also use AS-LUNG-O to acquire the $PM_{2.5}$ levels.

This work uses data from the AS-LUNG-O network, including 39 AS-LUNG-O sets installed in different communities since July 2017 around Taiwan (Figure 1). Twenty-eight sets were installed in urban communities in Taipei's metropolitan areas with high population densities in northern Taiwan, while two sets were placed in the suburban communities of Taipei. In addition, 6, 2, and 1 set(s) were installed in central, southern, and eastern Taiwan, respectively. Out of the 28 sets in the urban communities in Taipei, 27 were installed near certain community sources such as traffic, restaurants, temples, night markets, etc., as described in [10]. These 27 sets were set-up at street-level on light poles around 2–2.5 m above the ground. The other 12 sets were set-up at a high-level (around 10–15 m above the ground) on the rooftops of elementary schools or government buildings to assess the $PM_{2.5}$ in ambient air.

The AS-LUNG-O network is considered a research-grade sensor network, since the data of AS-LUNG-O were corrected using correction equations based on laboratory evaluations with side-by-side comparisons against a research-grade instrument for every AS-LUNG-O reading [10]. The research-grade instrument used in these laboratory evaluations was GRIMM 1.109 (GRIMM Aerosol Technik GmbH and Co. KG, Ainring, Germany). The data from GRIMM 1.109 were in excellent agreement (R^2 = 0.999, with a bias of roughly ±11%) with the data from an EDM-180 (GRIMM Aerosol Technik Ainring GmbH and Co, Ainring, German) [20], an FEM instrument designated by the USEPA for $PM_{2.5}$. The mean values of R^2 for the correction equations were 0.97, with ranges from 0.82 to 0.99 for these 39 sets. Without data correction, AS-LUNG-O overestimates $PM_{2.5}$ by about 1.5–2.9 times [10,20]. $PM_{2.5}$ observations with 1 min resolutions from AS-LUNG-O were converted to GRIMM comparable measurements according to the correction equations and saved in the cloud database for this AS-LUNG-O network.

This research-grade AS-LUNG-O network provides a great opportunity to evaluate the feasibility and performance of the correction models based on machine learning techniques. The raw $PM_{2.5}$ readings of AS-LUNG-O corrected by the laboratory correction equations are "GRIMM-calibrated $PM_{2.5}$", while those corrected by the machine learning techniques are "model-corrected $PM_{2.5}$". The performance of the machine learning correction models can be evaluated by comparing the GRIMM-calibrated $PM_{2.5}$ with model-corrected $PM_{2.5}$.

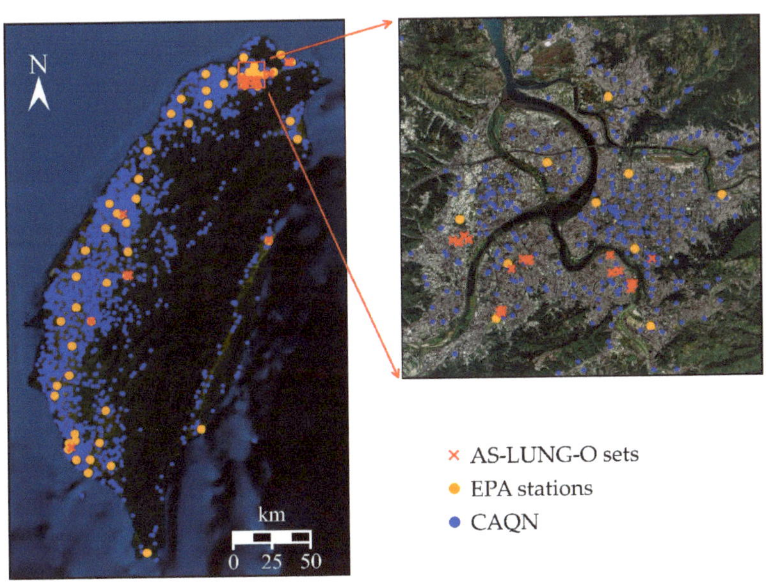

(a) (b)

Figure 1. *Cont.*

(c)

Figure 1. Distribution of AS-LUNG-O sets, Environmental Protection Administration (EPA) stations, and sensors of the citizen air quality network (CAQN) in (**a**) the whole of Taiwan island, (**b**) the Taipei metropolitan area from July 2017 to December 2018, and (**c**) an AS-LUNG-O set at street level.

2.2. The Data Correction Models

This work applied machine learning techniques to correct the raw readings of AS-LUNG-O sets with those of Taiwan EPA monitoring stations. Only EPA stations within 3 km of the AS-LUNG-O sets were selected in our work. Out of the 57 regular Taiwan EPA stations, 11 stations were selected, including 6 stations in the north, 2 in central Taiwan, 2 in the south, and 1 in the east. Hourly measurements of 24 h during July 2017 and December 2018 from these stations were used to establish the data correction models. The instruments used in Taiwan EPA [39] are Met One BAM-1020 (Met One, Inc., Grants Pass, OR, USA) and VEREWA-F701 (VEREWA, Ltd., Germany).

To establish in-field data correction models, two machine learning techniques (introduced below) were used in this work and compared with a correction model established using traditional MLR. These three models used $PM_{2.5}$ data from EPA stations as their simulation targets to adjust the raw readings of AS-LUNG-O sets within a 3 km radius. These models were constructed using a personal computer environment with an Intel® CoreTM i7-8700 and 32 GB RAM.

The inputs for these models were: (a) The raw $PM_{2.5}$ readings, (b) the temperature, RH, latitude, and longitude of the AS-LUNG-O sets, (c) the $PM_{2.5}$ levels of the nearest EPA station, and (d) the distance between AS-LUNG-O and the EPA station. These inputs (from July 2017 to December 2018) were used for 10-fold cross-validation (90% of data randomly selected for the training set; the others for the validation set) and holdout validation (50% of data randomly selected for the training set; the others for the validation set) tests to evaluate the robustness of these three models. Since there were only 1.5 years of data, we used the holdout method to generate the correction models; therefore, most of the data could be kept to validate the models, which can avoid to overvaluing the performance of the models under the situation of only using less data (10% of data) for the evaluation. Leaving more data in the validation set (50% of data in the holdout validation in this study) could increase the power of the model estimation. [40,41]. Data correction models were constructed using the training dataset, and the validation dataset was used to evaluate the correction accuracy of the models built. The MLR and machine learning models are introduced below.

The MLR model is established as follows:

$$\begin{aligned} PM2.5_{target} = {} & \beta_0 + \beta_1 \times LCS_{PM2.5_{raw}} + \beta_2 \times T + \beta_3 \times RH + \beta_4 \times Month + \beta_5 \times Day \\ & + \beta_6 \times Hr + \beta_7 \times lat + \beta_8 \times lon + \beta_9 \times D \end{aligned} \tag{1}$$

where β_0 is the intercept; β_1–β_9 are the regression coefficients; $PM2.5_{target}$ is the simulation target of the correction model, EPA $PM_{2.5}$; $LCS_{PM2.5_{raw}}$ is the raw readings from AS-LUNG-O ($\mu g/m^3$); T is the temperature (°C); RH is relative humidity (%); Month, Day, and Hr (hour) are the time values of the observations; lat and lon are the latitude and longitude of the AS-LUNG-O sets; and D is the distance of the AS-LUNG-O and the nearest EPA station (km).

Two machine learning techniques used in this work were support vector machine (SVM) and random forest (RF). SVM is based on the generalized portrait algorithm developed in the 1960s by Russian mathematicians and is a supervised learning algorithm used for classification [42,43]. The SVM algorithm is a popular machine learning tool that offers solutions for both classification and regression problems. The objective of SVM is to build an optimal hyperplane as a classifier in high dimensional space, and the data points closest to the hyperplane are called support vectors. New data are then divided by that classifier and predicted to belong to a category based on the hyperplane [44]. Our present work applies SVM to construct the support vector regression (SVR) model.

The random forest model is an ensemble learning method for classification and regression that builds a multitude of decision trees during the training process and constructs the modes of the classes or the mean predictions for classification and regression [45–47]. Using the random subspace method to build decision trees was first proposed by Ho et al. [45]. Breiman [46] further proposed to use the bagging algorithm to generate random forest to avoid over-fitting in the decision trees. The learning targets are numerical variables rather than class labels [46,47]. Our present work applies a random forest to construct a random forest regression (RFR) model.

2.3. Evaluation of the Correction Models

Figure 2 shows a flow chart of the data correction process. Raw $PM_{2.5}$ readings with a 1 min resolution were averaged to their hourly means to match the hourly observations from the nearest EPA stations within a 3 km radius. If the numbers for the raw $PM_{2.5}$ in one hour were less than 45, this hourly mean was discarded. After collecting all aforementioned input data, data correction models with three different methods can be established. A model-corrected $PM_{2.5}$ based on the optimal correction model can be obtained and then compared with the GRIMM-calibrated $PM_{2.5}$ corrected based on traditional laboratory evaluations. In this way, the performance of the $PM_{2.5}$ correction model can be evaluated accordingly.

Since the differences between the AS-LUNG-O readings and EPA observations may be affected by the community sources (of which emission activities change over time), the correction model with the best performance will be constructed based on whole-day (24 h) or nighttime (00:00–06:00) periods. The latter period was chosen because most of the community $PM_{2.5}$ sources associated with human activities were minimal during this period. The optimal correction model built with data from nighttime patterns can be used to obtain the systematic relationships of data from AS-LUNG-O and EPA instruments without interference from nearby sources around the locations of the AS-LUNG-O sets.

The indicators used for evaluating model performance are root mean square error (RMSE), Pearson correlation coefficient (r), and coefficient of determination (R^2). R^2 is used to assess the predictive or explanatory ability of the model and should be close to 1, while r shows the correlations between two variables. The equation of RMSE is as follows:

$$\mathrm{RMSE} = \sqrt{\frac{\sum_{i=1}^{n}(Y_i - M_i)^2}{n}} \tag{2}$$

The values of RMSE represent the difference between the model-corrected $PM_{2.5}$ (Mi) and referenced $PM_{2.5}$ levels (Yi) (EPA $PM_{2.5}$ used in the selection evaluation for the machine learning methods; GRIMM-corrected $PM_{2.5}$ used in the performance evaluation of the selected correction model). Thus, the closer these values are to zero, the better the model performs. Additionally, for the final model, mean absolute errors (MAEs) were also calculated for comparison with those from literature.

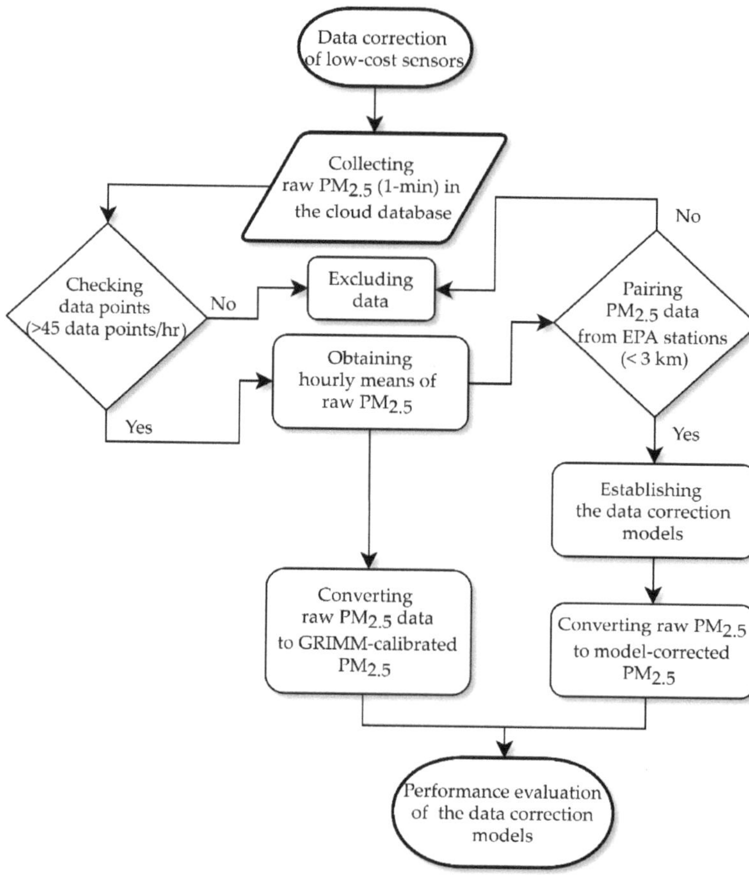

Figure 2. Flow chart of the data correction process.

3. Results

3.1. Measurements of AS-LUNG-O Sets and EPA Stations

Table 1 shows a summary of the raw $PM_{2.5}$ of the AS-LUNG-O sets and the $PM_{2.5}$ observations of the nearest EPA stations during July 2017 and December 2018, as classified by different seasons. The range of $PM_{2.5}$ for the EPA stations during this period is 2.0–135.0 μg/m^3. It can be seen that the highest $PM_{2.5}$ means and the maximum $PM_{2.5}$ occurred during winter for both the AS-LUNG-O and EPA $PM_{2.5}$ levels. The raw $PM_{2.5}$ values of the AS-LUNG-O sets were, on average, higher than those from the EPA by about 1.9–2.2 fold. These data were used to establish and evaluate the three data correction models.

Table 1. Seasonal mean with the standard deviation (SD), data range, and sample size (n) of the hourly raw $PM_{2.5}$ of AS-LUNG-O sets and EPA $PM_{2.5}$ ($\mu g/m^3$).

	Raw $PM_{2.5}$ of AS-LUNG		EPA $PM_{2.5}$		*n*
	Mean (SD)	**Range (Min, Max)**	**Mean (SD)**	**Range (Min, Max)**	
Spring	48.0 (20.6)	(3.1, 295.9)	24.3 (13.6)	(2.0, 100.0)	19,924
Summer	28.4 (21.3)	(1.0, 249.8)	12.9 (13.0)	(2.0, 75.0)	37,638
Fall	36.8 (15.3)	(1.0, 223.6)	19.3 (8.1)	(2.0, 127.0)	43,624
Winter	51.5 (29.2)	(1.0, 309.8)	27.0 (18.2)	(2.0, 135.0)	25,195

3.2. Performance Evaluation of the Correction Models

We conducted 10-fold cross-validation and holdout validation tests to evaluate the robustness of the models. The results of the 10-fold cross-validation test for MLR, SVR, and RFR were based on the same training and validation datasets. The average values of RMSE and R^2 for the results of the 10-fold cross-validation test were 6.88 ± 0.10 $\mu g/m^3$ and 0.76, 5.23 ± 0.08 $\mu g/m^3$ and 0.86, and 4.36 ± 0.06 $\mu g/m^3$ and 0.91, for MLR, SVR, and RFR, respectively. The results of the holdout evaluation test were presented in Figure 3. The differences between the RMSEs of the 10-fold cross-validation and the holdout validation tests were about averagely 0.03, 0.30, and 0.34 $\mu g/m^3$ for MLR, SVR, and RFR, respectively. Figure 3a–f show the distribution of model-corrected $PM_{2.5}$ from the AS-LUNG-O sets and EPA $PM_{2.5}$ data in the training and validation sets with three different data correction models. Based on the same training set of 63,190 data points, the computation time is 0.01, 8.16, and 0.23 minutes for building models MLR, SVR, and RFR, respectively. The R^2 of 0.76–0.99 for these three models shows these models have good explanatory abilities. In terms of RMSE, RFR is the best model (1.73) (Figure 3a,c,e). To further evaluate whether these models perform well for new datasets (Figure 3b,d,f), 63,191 data points from the validation sets were used to input these models. The RMSEs for RFR were 35% and 85% lower than those for SVR in evaluations of the training set and validation set, respectively. The results show that the R^2 values of these three models are 0.76–0.89, with the lowest RMSE value (4.7) in RFR for the validation sets. Based on the above evaluation, RFR is chosen as the best data correction model to be used for further applications.

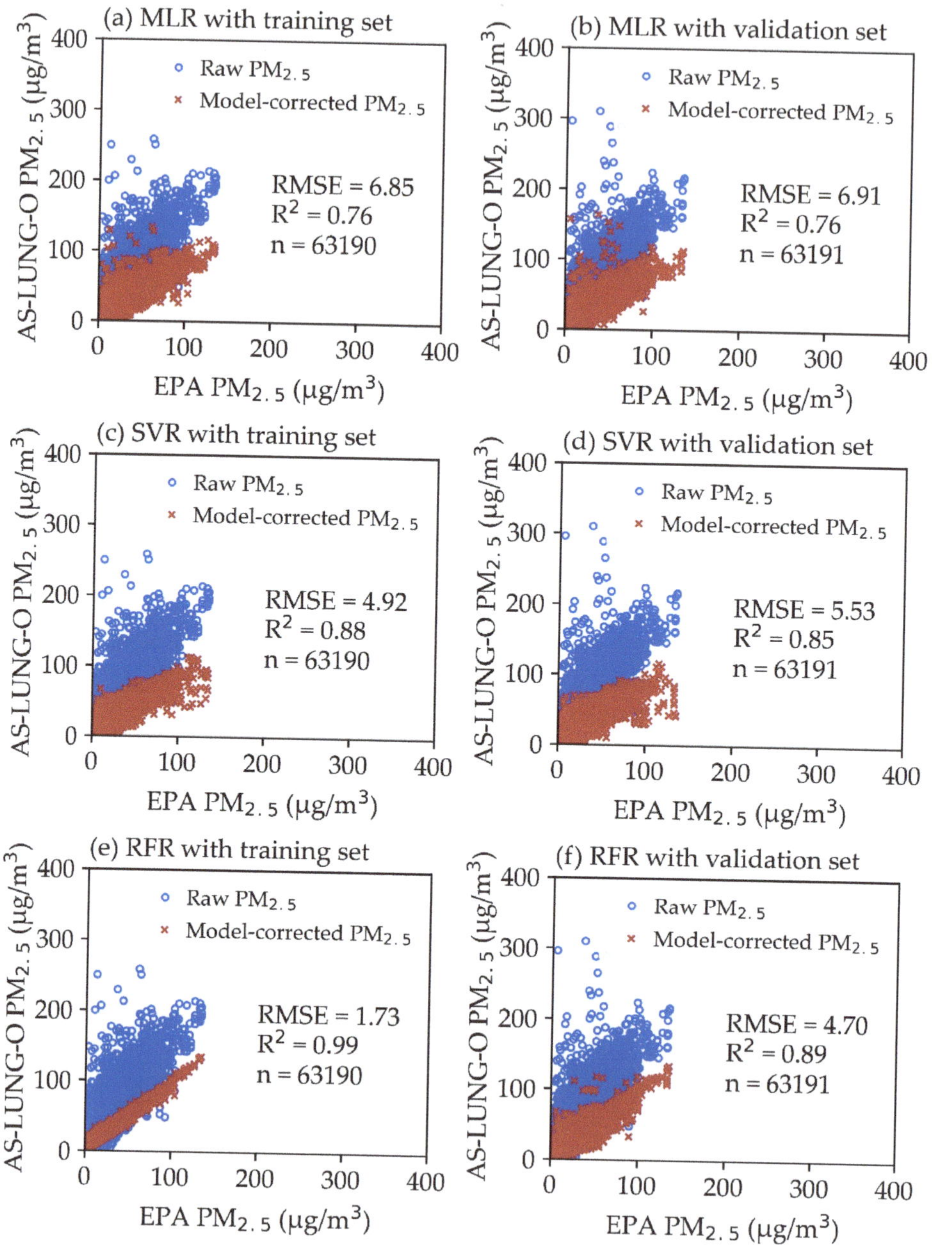

Figure 3. Model performance of (**a**) multiple linear regression (MLR) with a training set, (**b**) MLR with a validation set, (**c**) support vector regression (SVR) with a training set, (**d**) SVR with a validation set, (**e**) random forest regression (RFR) with a training set, and (**f**) RFR with a validation set. The RMSE, R^2, and n are listed in the graphs.

3.3. *Sensitivity Analysis of RFR*

To optimize the computing efficiency of the RFR model, a sensitivity analysis of RFR was conducted with 50 to 1000 decision trees (with 50-tree increases in each simulation) to assess the changes in modeling efficiency. Figure 4 shows that RFR offers good model performance (RMSE = 1.81 and R^2 = 0.9843) when there are only 50 trees. As the decision trees increase, the modeling efficiency is enhanced most significantly before the number of decision trees reaches 300. When the decision trees number is 300, the RMSE is 1.73, and the R^2 is 0.9858. Afterward, the efficiency enhancement is not significantly altered by adding more decision trees, which takes more computing time. In the overall evaluations, RFR with 300 decision trees was chosen as the model with the best efficiency.

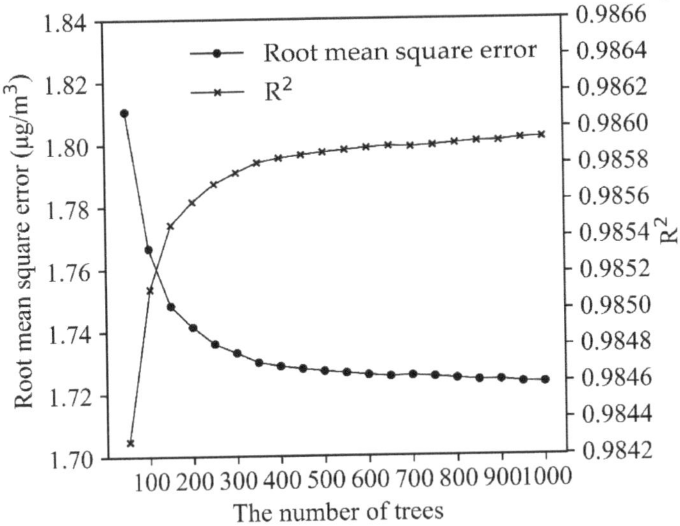

Figure 4. Model efficiency of the random forest regression model.

3.4. *Comparison of the Model-Corrected $PM_{2.5}$ and GRIMM-Calibrated $PM_{2.5}$*

3.4.1. RFR with Whole-Day and Nighttime Patterns

A performance evaluation was further conducted for the RFR with a whole-day pattern using a whole-day dataset based on four seasons. Table 2 shows that the RMSE is the lowest in the summer model (mean: 5.4 μg/m^3, ranging from 3.1 to 11.2 μg/m^3), followed by the fall model (mean: 6.1 μg/m^3, ranging from 3.7 to 10.0 μg/m^3), while the RMSE values are slightly higher in the winter and spring models (mean: 6.8 and 7.3 μg/m^3, respectively). For certain AS-LUNG-O sets at community locations, the r is as low as 0.33 between the model-corrected $PM_{2.5}$ and GRIMM-calibrated $PM_{2.5}$. This discrepancy is possibly caused by some nearby community sources that could not be detected by EPA monitoring stations. These community sources, such as traffic or restaurants, likely generate $PM_{2.5}$ in the daytime. Therefore, AS-LUNG-O may have different $PM_{2.5}$ trends from the nearby EPA station, leading to low correlations between these observations.

To focus on the systematic difference of the LCS and EPA observations, a data correction model was established for the nighttime dataset only. Since emissions from community sources resulting from human activity usually reached the lowest levels between 00:00 and 06:00, the data from this period were used to establish the model. Afterward, the established RFR with a nighttime pattern was used to correct the raw $PM_{2.5}$ for all datasets (including both daytime and nighttime). This way, the readings of AS-LUNG-O were adjusted according to the systematic differences between the AS-LUNG-O and EPA instruments, while the extra $PM_{2.5}$ increases due to community sources in the daytime could also be retained. Table 2 shows that the r-values of the model-corrected $PM_{2.5}$ and GRIMM-calibrated $PM_{2.5}$ were enhanced for the overall datasets, including both street-level and high-level AS-LUNG-O sets (including both daytime and nighttime). In the seasonal models, the r-values for RFR with a whole-day pattern were 0.83, 0.82, 0.85, and 0.90, which were improved to 0.92, 0.88, 0.88, and 0.94 for the RFR with a nighttime pattern for the spring, summer, fall, and winter models, respectively.

Table 2. Performance evaluation of the random forest regression model (RFR) with whole-day and nighttime patterns in overall AS-LUNG-O sets (street-level and high-level).

Overall	Season	RMSE [1]		Pearson Correlation		*n*
		Mean (SD [2])	Range (Min, Max)	Mean (SD)	Range (Min, Max)	
RFR with whole-day patterns	Spring	7.3 (2.6)	(4.1, 14.1)	0.83 (0.15)	(0.34, 0.96)	19,924
	Summer	5.4 (1.7)	(3.1, 11.2)	0.82 (0.11)	(0.33, 0.93)	37,638
	Fall	6.1 (1.6)	(3.7, 10.0)	0.85 (0.08)	(0.53, 0.94)	43,624
	Winter	6.8 (2.3)	(3.5, 12.9)	0.90 (0.04)	(0.79, 0.97)	25,195
RFR with nighttime patterns	Spring	6.7 (2.4)	(2.6, 10.9]	0.92 (0.05)	(0.80, 0.98)	19,924
	Summer	5.7 (1.7)	(2.8, 11.3)	0.88 (0.07)	(0.57, 0.95)	37,638
	Fall	5.7 (1.6)	(3.2, 9.9)	0.88 (0.08)	(0.68, 0.96)	43,624
	Winter	6.1 (2.3)	(2.4, 14.4)	0.94 (0.03)	(0.86, 0.98)	25,195

[1] RMSE, root mean square error; [2] SD, standard deviation.

The model improvement is most obvious in spring for certain street-level AS-LUNG-O sets. Compared with the whole-day model with an r-value of 0.34, the r-value of the nighttime model is enhanced to 0.81 (Table 3). Nevertheless, the r-values of certain AS-LUNG-O sets were not improved with the nighttime models. It is possible that the trends of $PM_{2.5}$ concentrations at these AS-LUNG-O locations were different from those in the EPA stations, regardless of whether it was during the day or night.

Table 3. Performance evaluation of the random forest regression model (RFR) with whole-day and nighttime patterns in street-level AS-LUNG-O sets.

Street-Level	Season	RMSE [1]		Pearson Correlation		*n*
		Mean (SD [2])	Range (Min, Max)	Mean (SD)	Range (Min, Max)	
RFR with whole-day patterns	Spring	7.1 (2.7)	(4.1, 14.1)	0.84 (0.15)	(0.34, 0.96)	17,255
	Summer	5.4 (1.8)	(3.1, 11.2)	0.83 (0.11)	(0.33, 0.93)	30,710
	Fall	5.8 (1.5)	(3.7, 10.0)	0.85 (0.09)	(0.53, 0.94)	32,606
	Winter	6.5 (2.2)	(3.5, 12.9)	0.91 (0.04)	(0.79, 0.97)	19,448
RFR with nighttime patterns	Spring	6.5 (2.5)	(2.6, 10.9)	0.93 (0.04)	(0.81, 0.98)	17,255
	Summer	5.6 (1.8)	(2.8, 11.3)	0.89 (0.07)	(0.57, 0.95)	30,710
	Fall	5.7 (1.7)	(3.2, 9.9)	0.89 (0.08)	(0.68, 0.96)	32,606
	Winter	5.9 (2.4)	(2.4, 14.4)	0.94 (0.02)	(0.89, 0.98)	19,448

[1] RMSE, root mean square error; [2] SD, standard deviation.

Furthermore, the results of the model evaluation were categorized as street-level and high-level (Tables 3 and 4, respectively). For data correction at a high-level, the r-values between the model-corrected $PM_{2.5}$ and GRIMM-calibrated $PM_{2.5}$ are all above 0.68 in the whole-day models, while the nighttime models between them are all above 0.75. These results indicate that the $PM_{2.5}$ levels at high-level AS-LUNG-O locations are moderately correlated with certain deviations from those of the EPA stations in a 3 km radius. This correlation was enhanced in the nighttime model. The minimum r-values of high-level AS-LUNG-O sets were improved from 0.68–0.83 to 0.75–0.86 (Table 4). This phenomenon was also observed in the $PM_{2.5}$ correction of certain street-level AS-LUNG-O sets; the minimum r-values were improved from 0.33–0.79 to 0.57–0.89 (Table 3). However, the improvement of street-level correlations was not as good as those of high-level AS-LUNG-O sets. The $PM_{2.5}$ levels sensed by street-level AS-LUNG-O locations are affected by local community sources, resulting in different $PM_{2.5}$ patterns from those of the EPA stations. It is important to keep these local features while correcting LCS data with the systematic differences between LCS and research-grade instruments in the correction procedures. Thus, based on the evaluations of RFR with whole-day patterns and nighttime patterns, the latter were selected to correct the raw $PM_{2.5}$ of AS-LUNG-Os.

Table 4. Performance evaluation of the random forest regression model (RFR) with whole-day and nighttime patterns in high-level AS-LUNG-O sets.

High-Level	Season	RMSE [1]		Pearson Correlation		n
		Mean (SD [2])	Range (Min, Max)	Mean (SD)	Range (Min, Max)	
RFR with whole-day patterns	Spring	8.8 (2.3)	(7.2, 10.4)	0.78 (0.15)	(0.68, 0.89)	2669
	Summer	5.7 (1.3)	(3.6, 7.6)	0.78 (0.07)	(0.70, 0.88)	6928
	Fall	7.3 (2.0)	(4.6, 9.9)	0.84 (0.07)	(0.75, 0.91)	11,018
	Winter	8.0 (2.9)	(4.6, 12.9)	0.88 (0.04)	(0.83, 0.94)	5747
RFR with nighttime patterns	Spring	8.2 (1.9)	(6.8, 9.5)	0.87 (0.09)	(0.80, 0.94)	2669
	Summer	5.8 (1.3)	(4.1, 7.9)	0.85 (0.06)	(0.76, 0.94)	6928
	Fall	6.2 (1.7)	(4.9, 9.3)	0.88 (0.09)	(0.75, 0.95)	11,018
	Winter	6.7 (2.2)	(4.3, 9.6)	0.92 (0.03)	(0.86, 0.94)	5747

[1] RMSE, root mean square error; [2] SD, standard deviation.

3.4.2. $PM_{2.5}$ Corrections by RFR

Figure 5a shows the RMSE values between the GRIMM-calibrated $PM_{2.5}$ and model-corrected $PM_{2.5}$ using RFR with a nighttime pattern. The RMSEs were 2.6–10.9 (mean 6.7), 2.8–11.3 (mean 5.7), 3.2–9.9 (mean 5.7), and 2.4–14.4 (mean 6.1) μg/m^3 for spring, summer, fall, and winter, respectively (Table 2). Before the model correction, the RMSE values were 15.4–32.7 (mean 23.3), 6.1–19.4 (mean 13.4), 5.3–31.2 (mean 17.4), and 5.2–40.6 (mean 21.6) μg/m^3 for spring, summer, fall, and winter, respectively. Clearly, RFR can greatly reduce the RMSE values. The significant differences in RMSEs among different seasons almost disappeared after correction with RFR (the nighttime pattern).

In addition, the MAE is used to evaluate the performance of RFR compared to the models in the literature. Figure 5b shows MAE values between the model-corrected $PM_{2.5}$ and GRIMM-calibrated $PM_{2.5}$ with RFR (nighttime pattern). The MAEs were 1.9–9.6 (mean 5.7), 2.4–9.0 (mean 4.9), 2.4–8.3 (mean 4.8), and 1.9–10.3 (mean 4.9) μg/m^3 for spring, summer, fall, and winter, respectively. Before the model correction, the MAE values were 14.6–30.1 (mean 21.5), 5.3–16.9 (mean 11.6), 4.4–27.6 (mean 15.2), and 4.1–36.6 (mean 18.8) μg/m^3 for spring, summer, fall, and winter, respectively. In summary, after correction, the RMSE was improved from 18.4 ± 6.5 to 5.9 ± 2.0 μg/m^3, and the MAE was improved from 16.2 ± 6.0 to 5.0 ± 1.8 μg/m^3. This demonstrates that the RFR model greatly reduces MAEs, lowers the overestimation of AS-LUNG-O raw data, and improves the agreements between AS-LUNG-O and EPA $PM_{2.5}$ levels.

Figure 6 shows a time series plot of the raw $PM_{2.5}$ for AS-LUNG-O, the model-corrected $PM_{2.5}$ with RFR (nighttime pattern), and the GRIMM-calibrated $PM_{2.5}$ for the whole year of 2018. After learning from the $PM_{2.5}$ observations of nearby EPA stations, the model-corrected $PM_{2.5}$ levels were close to the GRIMM-calibrated $PM_{2.5}$. In addition, after being corrected by machine learning techniques, the overestimation of the raw $PM_{2.5}$ was greatly reduced. However, with nighttime models, the peak values of model-corrected $PM_{2.5}$ were retained (as shown in the graph) to preserve the contributions of local community sources. After all, the purpose of community air quality monitoring is to assess the contributions of local sources; these important local features need to be preserved during the data correction processes.

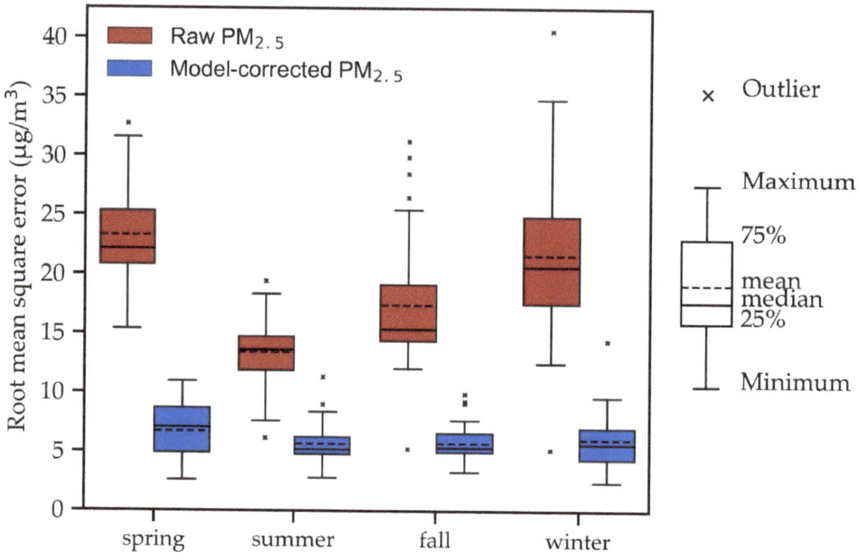

(a)

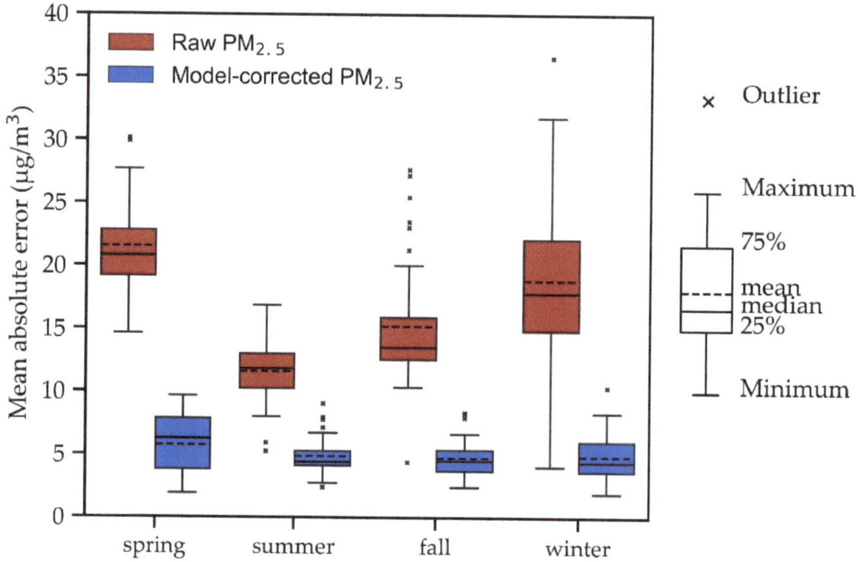

(b)

Figure 5. Correction results of RFR with a nighttime pattern for different seasons with (**a**) RMSE and (**b**) MAE as performance indicators.

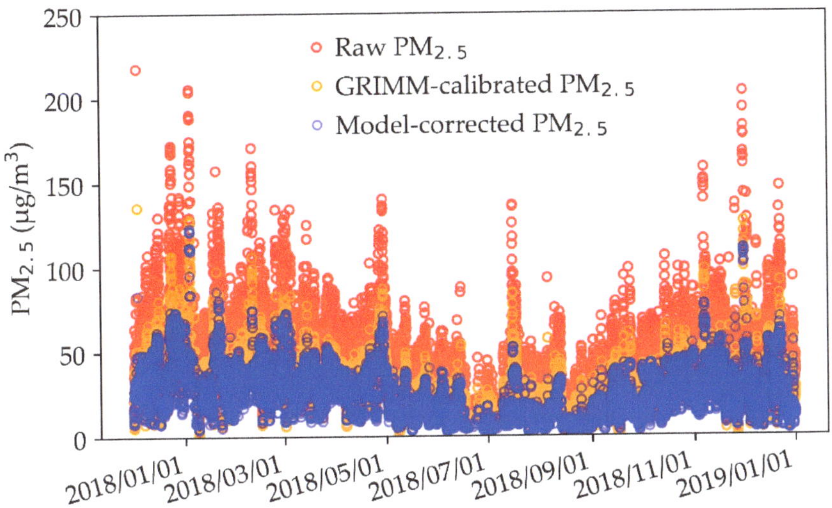

Figure 6. Time-series of raw $PM_{2.5}$, model-corrected $PM_{2.5}$, and GRIMM-calibrated $PM_{2.5}$.

4. Discussion

4.1. Comparison of in-Field $PM_{2.5}$ Correction Models

Typically, environmental research groups apply monitoring instruments calibrated by the manufacturers. For LCS sensors, Rai et al. [48] proposed a two-stage calibration process with laboratory calibration performed by the manufacturers and calibration checks performed by the end-users. This process would be ideal if the manufacturers followed the suggestions. However, demanding manufacturers to calibrate LCSs may be unrealistic, since LCSs are made in larger quantities with much lower costs than more expensive instruments. Therefore, calibration needed to be carried out by the end-users as described in the introduction; most studies use laboratory evaluations before installing LCSs to establish correction equations. Moreover, only a few studies developed in-field data correction models for $PM_{2.5}$ accuracy correction. Previously, we proposed a hybrid method combining laboratory evaluations and data science to ensure that LCS networks provide accurate PM data [20]. Statistical or machine learning methods were applied to adjust uncalibrated LCS devices with research-grade data within 3 km of the radius from either nearby EPA stations or calibrated seed LCS devices.

The focus of this work was to establish and evaluate three in-field data correction models for a $PM_{2.5}$ LCS network—namely, the MLR, SVR, and RFR models. The AS-LUNG-O network, with individual correction equations for each LCS in the laboratory, offers a great opportunity to assess the performance of the data correction models established by machine learning techniques. Based on the results of the 10-fold cross-validation and holdout validation tests, there was only a little difference between the RMSEs of the 10-fold cross-validation and holdout validation tests for MLR, SVR, and RFR. Thus, to avoid overvaluing the performance of the models, we used the holdout validation to establish our models. Among them, RFR is the best model, with an RMSE of 1.73 and an R^2 of 0.99 based on 63,190 raw hourly data of 39 AS-LUNG-O sets corrected with the $PM_{2.5}$ levels of EPA stations located within a 3 km radius. In the validation set, RFR was not overfitted, and the data corrected with the RFR model agreed well with the EPA observations. Thus, RFR was chosen as the data correction model. This work demonstrated the applicability of RFR in correcting LCS networks. Previous studies used statistical and machine learning models with data from regulatory stations to correct side-by-side LCS sets on the same locations [15,30]. This work is the first to use data from monitoring stations to correct the data of the LCS network in other locations without monitoring stations. This work is unique in providing LCS data (GRIMM-calibrated $PM_{2.5}$) based on laboratory evaluations for comparison in other locations.

Among the few in-field correction studies, one study applied the generalized additive model (GAM) to correct LCSs installed at three Taiwan EPA stations during December 2017 [15]. The RMSE values were reduced from 15.55–31.34 μg/m^3 to 4.88–9.66 μg/m^3 after correction. These results are similar to our data correction model results, showing RMSE values of 5.2–40.6 μg/m^3 that were reduced to 4.3–9.6 μg/m^3 (high-level $PM_{2.5}$ corrected by RFR with nigh-time patterns in winter, see Table 4). Our correction model used similar input data. However, while an individual GAM model needed to be established for each LCS at least once every day [15], our model, with cumulative data of more than one year, was constructed as a one-time effort, which saved much computing time. Another study applied transfer learning to correct 10 months of $PM_{2.5}$ data for LCSs located at seven public environmental monitoring stations in Beijing, China, and obtained MAEs of 7–12 μg/m^3 [30]. The RFR model in our work obtained MAEs of 1.9–10.3 μg/m^3 and performed at least comparable to or even better than those previous in-field data correction models for $PM_{2.5}$. Furthermore, our methodology is also suitable to be applied to $PM_{2.5}$ sensor networks in other countries.

One may concern that the upwind or downwind locations may also affect the relationship of AS-LUNG-O readings and EPA measurements. Since the prevailing wind in Taiwan changes with the season, the seasonal air flow variations were considered in the correction model by the variable of "season". This variable "season" also considered all other seasonal factors.

In addition, in real-world applications, the computing efficiency of a model is a key issue to determine feasibility. The impressive computer efficiency of RFR is another advantage of the model, as it can handle huge datasets for the in-field correction of sensor networks. Based on the same training dataset, RFR needed only 0.23 minutes for training a model, while SVR needed approximately 8.16 minutes. In comparison, RFR was 35 times faster. With such fast computations, this method has great potential to be expanded to CAQN in Taiwan, which includes more than 4000 uncalibrated LCSs. Decreasing run time in the real-time corrections of CAQN can be carried out by increasing the computational capacity with a larger CPU and RAM. Moreover, parallel processing can be applied in model coding to speed up the correction task. Due to its fast computing efficiency, RFR is an excellent model for big data analytics for any data applications in sensor networks.

It should be noted that the RFR can only be applied for sensor networks with $PM_{2.5}$ LCSs, which have good R^2 with research-grade instruments. If the precision of $PM_{2.5}$ LCS is not good, the RFR cannot adjust this inherent disadvantage. One previous study only enhanced the r-value between LCSs and research-grade instruments from 0.53 to 0.63 with RFR, since they used LCSs with R^2 of 0.07 to 0.27 (compared to research-grade instruments) [49,50]. On the contrary, our present work used LCS devices with good R^2 (>0.95) [10,20], so that our RFR model showed good performance in data correction for sensor networks. Thus, LCS with good precision is a prerequisite for a good data correction by RFR.

4.2. Limitations of This Work

There are some limitations to this work. RFR could greatly improve data accuracy, as shown by the reduction of RMSEs and MAEs. However, there is still an average bias of 4.8–5.7 μg/m^3 for four seasons after correction. These deviations may come from the inherent differences in ambient (high-level) $PM_{2.5}$ levels and street $PM_{2.5}$ levels. The latter is affected by various community sources, while the former is measured by EPA monitoring stations purposely assessing well-mixed ambient $PM_{2.5}$ levels free from any local emission interference. The AS-LUNG-O sets were located within 3 km radius of EPA stations. Theoretically, the $PM_{2.5}$ levels were uniformly distributed within these distances. Nevertheless, there were usually multiple localized sources (home factories, restaurants, traffic, etc.) within 100–500 m in Taiwanese communities resulting in significant spatial variations. Therefore, the $PM_{2.5}$ levels at AS-LUNG-O locations were different from those in EPA stations. This further demonstrates the necessity of establishing LCS networks to assess local ambient air.

On the other hand, the heights of AS-LUNG-O sets and EPA stations might be another variable for the correction model. Since there were only two types of height (ground level and high level) of AS-LUNG-O sets, we did not consider the altitude in the current correction model. This variable could be considered in the future.

Moreover, the performance of RFR may be improved by multiple year inputs, which may cover more environmental conditions for building the decision trees. This could be evaluated further after the accumulation of observations from the AS-LUNG-O network. Additionally, the co-localization of some AS-LUNG-O sets with EPA stations could be conducted in the future for comparison under the same environmental conditions.

5. Conclusions

Current data correction methods for $PM_{2.5}$ sensor networks are mostly established in the laboratory and in the field before LCS installation. For citizen $PM_{2.5}$ sensor networks without calibration, this work has developed an in-field data correction model with machine learning to adjust the accuracy deviations of the LCS network to enhance the data applicability of these networks. With the RFR model, the RMSEs and MAEs of the model-corrected $PM_{2.5}$ and GRIMM-calibrated $PM_{2.5}$ are greatly reduced. The contributions of local community sources to street-level $PM_{2.5}$ concentrations are also preserved by RFR with a nighttime pattern (00:00 to 06:00). This work provides a feasible method for the in-field data correction of uncalibrated $PM_{2.5}$ sensor networks with machine learning techniques. In addition, this work demonstrates the great potential of machine learning to enhance the agreement of LCSs and research-grade instruments, and thus, expands the applications of machine learning in the field of environmental monitoring.

Previously, we proposed a hybrid method combining laboratory evaluations and data science to ensure that LCS networks provide accurate PM data. Statistical or machine learning methods were applied to adjust uncalibrated LCS devices with data from nearby EPA stations or seed LCS devices that have been corrected by laboratory side-by-side comparisons and installed strategically in areas without EPA stations. The current work focuses on applying machine learning to correct the LCS network with $PM_{2.5}$ from the Taiwan EPA. Under the trend of the smart city movement, increasingly more sensors will be installed in our living surroundings for air quality monitoring. Thus, accurate data are essential to avoid false impressions of better or worse air quality leading to ineffective air pollution control strategies. With the established in-field data correction models presented in this work and the calibrated seed LCS devices, accurate $PM_{2.5}$ data from the sensor networks can be further applied to citizen science, public education, environmental research, and policymaking. Similar correction models can be established in other countries based on this example to greatly enhance the applicability and usefulness of $PM_{2.5}$ sensor networks worldwide.

Author Contributions: Conceptualization, W.-C.V.W., S.-C.C.L., and C.-H.L.; methodology, W.-C.V.W., S.-C.C.L., and C.-H.L.; validation, W.-C.V.W. and C.-H.L.; formal analysis, W.-C.V.W. and C.-H.L.; investigation, W.-C.V.W., and C.-H.L.; resources, S.-C.C.L.; data curation, W.-C.V.W. and C.-H.L.; writing—original draft preparation, W.-C.V.W. and S.-C.C.L.; writing—review and editing, W.-C.V.W., S.-C.C.L., and C.-H.L.; visualization, W.-C.V.W.; supervision, S.-C.C.L.; project administration, W.-C.V.W. and C.-H.L.; funding acquisition, S.-C.C.L. All authors have read and agreed to the published version of the manuscript.

Funding: This research was funded by ACADEMIA SINICA, Taipei, Taiwan under "Integrated Multi-source and High-resolution Heat Wave Vulnerability Assessment of Taiwan (AS-104-SS-A02)" and "Trans-disciplinary $PM_{2.5}$ Exposure Research in Urban Areas for Health-oriented Preventive Strategies (AS-SS-107-03)".

Acknowledgments: We would like to acknowledge the team in the Environmental Health Assessment Laboratory at the Research Center for Environmental Changes for their efforts to set up and maintain the AS-LUNG-O network.

Conflicts of Interest: The authors declare no conflict of interest. The funders had no role in the design of the study; in the collection, analyses, or interpretation of data; in the writing of the manuscript, or in the decision to publish the results.

References

1. Forouzanfar, M.H.; Afshin, A.; Alexander, L.T.; Anderson, H.R.; Bhutta, Z.A.; Biryukov, S.; Brauer, M.; Burnett, R.; Cercy, K.; Charlson, F.J.; et al. Global, regional, and national comparative risk assessment of 79 behavioural, environmental and occupational, and metabolic risks or clusters of risks, 1990–2015: A systematic analysis for the Global Burden of Disease Study 2015. *Lancet* **2016**, *388*, 1659–1724. [CrossRef]
2. Lelieveld, J.; Evans, J.S.; Fnais, M.; Giannadaki, D.; Pozzer, A. The contribution of outdoor air pollution sources to premature mortality on a global scale. *Nature* **2015**, *525*, 367–371. [CrossRef] [PubMed]
3. Loomis, D.; Grosse, Y.; Lauby-Secretan, B.; El Ghissassi, F.; Bouvard, V.; Benbrahim-Tallaa, L.; Guha, N.; Baan, R.; Mattock, H.; Straif, K.; et al. The carcinogenicity of outdoor air pollution. *Lancet Oncol.* **2013**, *14*, 1262–1263. [CrossRef]
4. Betha, R.; Behera, S.N.; Balasubramanian, R. 2013 Southeast Asian smoke haze: Fractionation of particulate-bound elements and associated health risk. *Environ. Sci. Technol.* **2014**, *48*, 4327–4335. [CrossRef]
5. Ji, X.; Yao, Y.X.; Long, X.L. What causes PM2.5 pollution? Cross-economy empirical analysis from socioeconomic perspective. *Energ. Policy* **2018**, *119*, 458–472. [CrossRef]
6. van Donkelaar, A.; Martin, R.V.; Brauer, M.; Boys, B.L. Use of satellite observations for long-term exposure assessment of global concentrations of fine particulate matter. *Environ. Health Persp.* **2015**, *123*, 135–143. [CrossRef]
7. Chen, C.; Zeger, S.; Breysse, P.; Katz, J.; Checkley, W.; Curriero, F.C.; Tielsch, J.M. Estimating indoor $PM_{2.5}$ and CO concentrations in households in southern Nepal: The Nepal cookstove intervention trials. *PLoS ONE* **2016**, *11*. [CrossRef]
8. Lung, S.C.C.; Hsiao, P.K.; Wen, T.Y.; Liu, C.H.; Fu, C.B.; Cheng, Y.T. Variability of intra-urban exposure to particulate matter and CO from Asian-type community pollution sources. *Atmos. Environ.* **2014**, *83*, 6–132014. [CrossRef]
9. Lung, S.C.C.; Mao, I.F.; Liu, L.J.S. Residents' particle exposures in six different communities in Taiwan. *Sci. Total Environ.* **2007**, *377*, 81–92. [CrossRef]
10. Lung, S.C.C.; Wang, W.C.V.; Wen, T.Y.J.; Liu, C.H.; Hu, S.C. A versatile low-cost sensing device for assessing PM2.5 spatiotemporal variation and quantifying source contribution. *Sci. Total Environ.* **2020**, *716*, 137145. [CrossRef]
11. Taiwan Environmental Protection Agency (Taiwan EPA), Taiwan Air Quality Monitoring Network. Available online: https://airtw.epa.gov.tw/ENG/default.aspx (accessed on 30 June 2020).
12. USEPA, Air Sensor Guidebook. United States Environmental Protection Agency (USEPA). 2014. Available online: https://www.epa.gov/air-sensor-toolbox/how-use-air-sensors-air-sensor-guidebook (accessed on 20 June 2020).
13. The Citizen Air Quality Network (CAQN). Available online: https://airbox.edimaxcloud.com (accessed on 20 June 2020).
14. Chen, L.J.; Ho, Y.H.; Lee, H.C.; Wu, H.C.; Liu, H.M.; Hsieh, H.H.; Huang, Y.T.; Lung, S.C.C. An Open Framework for Participatory $PM_{2.5}$ Monitoring in Smart Cities. *IEEE Access* **2017**, *5*, 14441–14454. [CrossRef]
15. Lee, C.H.; Wang, Y.B.; Yu, H.L. An efficient spatiotemporal data calibration approach for the low-cost PM2.5 sensing network: A case study in Taiwan. *Environ. Int.* **2019**, *130*, 104838. [CrossRef] [PubMed]
16. Kuula, J.; Makela, T.; Aurela, M.; Teinila, K.; Varjonen, S.; Gonzalez, O.; Timonen, H. Laboratory evaluation of particle-size selectivity of optical low-cost particulate matter sensors. *Atmos. Meas. Tech.* **2020**, *13*, 2413–2423. [CrossRef]
17. Holstius, D.M.; Pillarisetti, A.; Smith, K.R.; Seto, E. Field calibrations of a low-cost aerosol sensor at a regulatory monitoring site in California. *Atmos. Meas. Tech.* **2014**, *7*, 1121–1131. [CrossRef]
18. Kelly, K.E.; Whitaker, J.; Petty, A.; Widmer, C.; Dybwad, A.; Sleeth, D.; Martin, R.; Butterfield, A. Ambient and laboratory evaluation of a low-cost particulate matter sensor. *Environ. Pollut.* **2017**, *221*, 491–500. [CrossRef]
19. Mukherjee, A.; Stanton, L.G.; Graham, A.R.; Roberts, P.T. Assessing the utility of low-cost particulate matter sensors over a 12-week period in the Cuyama Valley of California. *Sensors* **2017**, *17*, 1805. [CrossRef]
20. Wang, W.C.V.; Lung, S.C.C.; Liu, C.H.; Shui, C.K. Laboratory evaluation of correction equations with multiple choices for seed low-cost particle sensing devices in sensor networks. *Sensors* **2020**, *20*, 3661. [CrossRef]

21. Dacunto, P.J.; Klepeis, N.E.; Cheng, K.C.; Acevedo-Bolton, V.; Jiang, R.T.; Repace, J.L.; Ott, W.R.; Hildemann, L.M. Determining $PM_{2.5}$ calibration curves for a low-cost particle monitor: Common indoor residential aerosols. *Environ. Sci. Process Impacts* **2015**, *17*, 1959–1966. [CrossRef]
22. Wang, Y.; Li, J.Y.; Jing, H.; Zhang, Q.; Jiang, J.K.; Biswas, P. Laboratory evaluation and calibration of three low-cost particle sensors for particulate matter measurement. *Aerosol Sci. Technol.* **2015**, *49*, 1063–1077. [CrossRef]
23. Liu, H.Y.; Schneider, P.; Haugen, R.; Vogt, M. Performance Assessment of a low-cost $PM_{2.5}$ sensor for a near four-month period in Oslo, Norway. *Atmosphere* **2019**, *10*, 41. [CrossRef]
24. Zamora, M.L.; Xiong, F.L.Z.; Gentner, D.; Kerkez, B.; Kohrman-Glaser, J.; Koehler, K. Field and laboratory evaluations of the low-cost Plantower particulate matter sensor. *Environ. Sci. Technol.* **2019**, *53*, 838–849. [CrossRef] [PubMed]
25. Zheng, T.S.; Bergin, M.H.; Johnson, K.K.; Tripathi, S.N.; Shirodkar, S.; Landis, M.S.; Sutaria, R.; Carlson, D.E. Field evaluation of low-cost particulate matter sensors in high-and low-concentration environments. *Atmos. Meas. Tech.* **2018**, *11*, 4823–4846. [CrossRef]
26. Tryner, J.; L'Orange, C.; Mehaffy, J.; Miller-Lionberg, D.; Hofstetter, J.C.; Wilson, A.; Volckens, J. Laboratory evaluation of low-cost PurpleAir PM monitors and in-field correction using co-located portable filter samplers. *Atmos. Environ.* **2020**, *220*, 117067. [CrossRef]
27. Cavaliere, A.; Carotenuto, F.; Di Gennaro, F.; Gioli, B.; Gualtieri, G.; Martelli, F.; Matese, A.; Toscano, P.; Vagnoli, C.; Zaldei, A. Development of low-cost air quality stations for next generation monitoring networks: Calibration and validation of $PM_{2.5}$ and PM_{10} sensors. *Sensors* **2018**, *18*, 2843. [CrossRef]
28. Jiao, W.; Hagler, G.; Williams, R.; Sharpe, R.; Brown, R.; Garver, D.; Judge, R.; Caudill, M.; Rickard, J.; Davis, M.; et al. Community Air Sensor Network (CAIRSENSE) project: Evaluation of low-cost sensor performance in a suburban environment in the southeastern United States. *Atmos. Meas. Tech.* **2016**, *9*, 5281–5292. [CrossRef]
29. Morawska, L.; Thai, P.K.; Liu, X.T.; Asumadu-Sakyi, A.; Ayoko, G.; Bartonova, A.; Bedini, A.; Chai, F.H.; Christensen, B.; Dunbabin, M.; et al. Applications of low-cost sensing technologies for air quality monitoring and exposure assessment: How far have they gone? *Environ. Int.* **2018**, *116*, 286–299. [CrossRef]
30. Cheng, Y.; He, X.; Zhou, Z.; Thiele, L. ICT: In-field Calibration Transfer for Air Quality Sensor Deployments. *Proc. ACM Interact. Mob. Wearable Ubiquitous Technol.* **2019**, *3*, 6. [CrossRef]
31. Pandey, G.; Zhang, B.; Jian, L. Predicting submicron air pollution indicators: A machine learning approach. *Environ. Sci. Process. Impacts* **2013**, *15*, 996–1005. [CrossRef]
32. Hsieh, H.P.; Lin, S.D.; Zheng, Y. Inferring air quality for station location recommendation based on urban big data. In Proceedings of the 21th ACM SIGKDD International Conference on Knowledge Discovery and Data Mining, Sydney, Australia, 10–13 August 2015; ACM: Sydney, Australia, 2015; pp. 437–446.
33. Zheng, Y.; Yi, X.; Li, M.; Li, R.; Shan, Z.; Chang, E.; Li, T. Forecasting fine-grained air quality based on big data. In Proceedings of the 21th ACM SIGKDD International Conference on Knowledge Discovery and Data Mining, Sydney, Australia, 10–13 August 2015; ACM: Sydney, Australia, 2015; pp. 2267–2276.
34. Paas, B.; Stienen, J.; Vorlander, M.; Schneider, C. Modelling of urban near-road atmospheric pm concentrations using an artificial neural network approach with acoustic data input. *Environments* **2017**, *4*, 26. [CrossRef]
35. Peng, H.P.; Lima, A.R.; Teakles, A.; Jin, J.; Cannon, A.J.; Hsieh, W.W. Evaluating hourly air quality forecasting in Canada with nonlinear updatable machine learning methods. *Air Qual. Atmos. Health* **2017**, *10*, 195–211. [CrossRef]
36. Zimmerman, N.; Presto, A.A.; Kumar, S.P.N.; Gu, J.; Hauryliuk, A.; Robinson, E.S.; Robinson, A.L.; Subramanian, R. A machine learning calibration model using random forests to improve sensor performance for lower-cost air quality monitoring. *Atmos. Meas. Tech.* **2018**, *11*, 291–313. [CrossRef]
37. Sayahi, T.; Kaufman, D.; Becnel, T.; Kaur, K.; Butterfield, A.E.; Collingwood, S.; Zhang, Y.; Gaillardon, P.E.; Kelly, K.E. Development of a calibration chamber to evaluate the performance of low-cost particulate matter sensors. *Environ. Pollut.* **2019**, *255*, 113131. [CrossRef] [PubMed]
38. Taiwan Central Weather Bureau (Taiwan CWB). Available online: https://www.cwb.gov.tw/eng (accessed on 28 June 2020).
39. Introduction to Air Quality Monitoring Stations of Taiwan EPA. Available online: https://airtw.epa.gov.tw/ENG/EnvMonitoring/Central/article_station.aspx (accessed on 30 June 2020).

40. Carrico, C.; Gennings, C.; Wheeler, D.C.; Factor-Litvak, P. Characterization of weighted quantile sum regression for highly correlated data in a risk analysis setting. *JABES* **2015**, *20*, 100–120. [CrossRef]
41. Tanner, E.M.; Bornehag, C.G.; Gennings, C. Repeated holdout validation for weighted quantile sum regression. *MethodsX* **2019**, *6*, 2855–2860. [CrossRef] [PubMed]
42. Vapnik, V.N.; Lerner, A. Pattern recognition using generalized portrait method. *Autom. Remote Control* **1963**, *24*, 774–780.
43. Vapnik, V.N.; Chervonenkis, A. A note on one class of perceptrons. *Autom. Remote Control* **1964**, *25*, 821–837.
44. Smola, A.J.; Scholkopf, B. A tutorial on support vector regression. *Stat. Comput.* **2004**, *14*, 199–222. [CrossRef]
45. Ho, T.K. The random subspace method for constructing decision forests. *IEEE Trans. Pattern Anal. Mach. Intell.* **1998**, *20*, 832–844.
46. Breiman, L. Random forests. *Mach. Learn.* **2001**, *45*, 5–32. [CrossRef]
47. Geurts, P.; Ernst, D.; Wehenkel, L. Extremely randomized trees. *Mach. Learn.* **2006**, *63*, 3–42. [CrossRef]
48. Rai, A.C.; Kumar, P.; Pilla, F.; Skouloudis, A.N.; Di Sabatino, S.; Ratti, C.; Yasar, A.; Rickerby, D. End-user perspective of low-cost sensors for outdoor air pollution monitoring. *Sci. Total Environ.* **2017**, *607*, 691–705. [CrossRef] [PubMed]
49. Borrego, C.; Costa, A.M.; Ginja, J.; Amorim, M.; Coutinho, M.; Karatzas, K.; Sioumis, T.; Katsifarakis, N.; Konstantinidis, K.; De Vito, S.; et al. Assessment of air quality microsensors versus reference methods: The EuNetAir joint exercise. *Atmos. Environ.* **2016**, *147*, 246–263. [CrossRef]
50. Borrego, C.; Ginja, J.; Coutinho, M.; Ribeiro, C.; Karatzas, K.; Sioumis, T.; Katsifarakis, N.; Konstantinidis, K.; De Vito, S.; Esposito, E.; et al. Assessment of air quality microsensors versus reference methods: The EuNetAir Joint Exercise—Part II. *Atmos. Environ.* **2018**, *193*, 127–142. [CrossRef]

sensors

MDPI

Article

Comparisons of Laboratory and On-Road Type-Approval Cycles with Idling Emissions. Implications for Periodical Technical Inspection (PTI) Sensors

Barouch Giechaskiel *, Tero Lähde, Ricardo Suarez-Bertoa, Victor Valverde and Michael Clairotte

European Commission, Joint Research Centre (JRC), 21027 Ispra, Italy; tero.lahde@ec.europa.eu (T.L.); ricardo.suarez-bertoa@ec.europa.eu (R.S.-B.); victor.valverde-morales@ec.europa.eu (V.V.); michael.clairotte@ec.europa.eu (M.C.)

* Correspondence: barouch.giechaskiel@ec.europa.eu; Tel.: +39-0332-78-5312

Received: 23 September 2020; Accepted: 10 October 2020; Published: 13 October 2020

Abstract: For the type approval of compression ignition (diesel) and gasoline direct injection vehicles, a particle number (PN) limit of 6×10^{11} p/km is applicable. Diesel vehicles in circulation need to pass a periodical technical inspection (PTI) test, typically every two years, after the first four years of circulation. However, often the applicable smoke tests or on-board diagnostic (OBD) fault checks cannot identify malfunctions of the diesel particulate filters (DPFs). There are also serious concerns that a few high emitters are responsible for the majority of the emissions. For these reasons, a new PTI procedure at idle run with PN systems is under investigation. The correlations between type approval cycles and idle emissions are limited, especially for positive (spark) ignition vehicles. In this study the type approval PN emissions of 32 compression ignition and 56 spark ignition vehicles were compared to their idle PN concentrations from laboratory and on-road tests. The results confirmed that the idle test is applicable for diesel vehicles. The scatter for the spark ignition vehicles was much larger. Nevertheless, the proposed limit for diesel vehicles was also shown to be applicable for these vehicles. The technical specifications of the PTI sensors based on these findings were also discussed.

Keywords: vehicle emissions; particle number; periodical technical inspection; idle; roadworthiness

1. Introduction

Air pollution, especially particulate matter (PM), has significant impacts on the health of the European population. It was estimated that in 2016 the mass of PM below 2.5 micron was responsible for about 412,000 premature deaths in Europe [1]. The road transport contributed to 11% of total PM mass primary emissions in the 28 countries of the European Union in 2017 [1]. In addition to the PM mass, there is also a concern about the contribution of traffic originated ultrafine (<0.1 μm) particles to the detriment effect on human health [2], as the road traffic is the major ultrafine particle number source in most cities [3]. Both the mass and number of traffic originated particles have shown significant reductions in the last 15–20 years [4], and it is estimated that they will further decrease in Europe until 2030 [5]. The concentration reductions can be attributed to policies (e.g., more stringent Euro emission standards), traffic management, and fleet restrictions (e.g., low emission zones) [4]. Policies focused mainly on vehicle exhaust emissions, but as the levels have decreased, the contribution of non-exhaust emissions (from brakes and tires) can contribute at similar levels [6].

In the European Union (EU) the particle number (PN) and PM mass vehicles exhaust emissions have to respect some limits defined in the regulations. The type approval of a vehicle family requires that some limits depending on the date of registration are respected (e.g., Euro 5). The procedure includes measurement of the emissions of a representative vehicle during a pre-defined driving cycle in the laboratory under well controlled ambient conditions. The type approval cycle in Europe was the NEDC (New European Driving Cycle), which was replaced by the WLTC (Worldwide harmonized Light vehicles Test Cycle) in 2017 with Euro 6c (Commission Regulation EU 2017/1151). Furthermore, in 2017 a Real-Driving Emissions (RDE) test on the road was introduced in the type-approval procedure including a Not-To-Exceed (NTE) limit for PN with Euro 6d-temp (Commission Regulation EU 2017/1154) [7]. Further provisions ensure the conformity of production (i.e., checking sample vehicles from the production line) and in-service conformity (checking vehicles already circulating in the streets). The laboratory PN instruments are based on the Particle Measurement Programme (PMP) group recommendations [8]. The on-road tests are conducted with portable emissions measurement systems (PEMS) [9]. Limits are applicable for compression ignition (diesel) vehicles since 2011 (Euro 5b) and gasoline direct injection (GDI) vehicles since 2014 (Euro 6b). The current PN limit of solid (nonvolatile) particles is 6×10^{11} p/km. However, for the first three years (2014–2017), a limit of 6×10^{12} p/km could be applied to new GDI vehicles upon request of the manufacturer [10].

The roadworthiness regulation (Directive 2014/45/EU, which repealed Directive 2009/40/EC) ensures that all circulating vehicles are kept in a safe and environmentally acceptable condition. It requires appropriate measures to prevent adverse manipulation of, or tampering with, vehicle parts and components that could have a negative bearing on required safety and environmental characteristics of the vehicle. In order to check the emissions of a vehicle, a periodical technical inspection (PTI) test is required for all circulating vehicles, typically every two years, after the first four years of circulation. Exhaust gas smoke emissions of diesel vehicles are measured with opacimeters during free acceleration (no load from idle up to cut-off speed) with gear lever in neutral and clutch engaged. Alternatively, the reading of OBD (on-board diagnostics) can be used, if available. However, in a study with 400 vehicles, 6% of them had high smoke emissions and none of them had any DPF (Diesel Particulate Filter) fault codes at the OBD reading [11], indicating that OBD systems are not always well designed to detect DPF failures. Similarly, in another study the OBD was unable to detect any DPF faults [12]. The opacity test is also obsolete for todays' vehicles because the opacity limit is quite high. A study showed that all vehicles with and without DPF could pass the current limits (1.5 m^{-1} and 0.7 m^{-1} for Euro 5 and Euro 6, respectively) [13]. Another study showed that even a 100% damaged DPF resulted in emissions well below the PTI limit, although 0.5% and 5% damage ratios resulted in values exceeding the PN and PM mass limits in type approval tests [14]. Lowering the opacity limits has the challenge that it is already close to the detection limit of the method (0.3 m^{-1}). Furthermore, a study showed that all diesel vehicles (Euro 5) with or without DPF had smoke emissions $<0.5 \text{ m}^{-1}$ [15]. Recent studies concluded that, instead of using opacimeters for the determination of smoke emissions, laser light scattering sensors could be used: they were sufficiently accurate and stable, and had the necessary dynamic response characteristics and resolution for testing modern vehicles [12,14]. However, a practicable calibration procedure needs to be defined for light scattering sensors. Furthermore, concerns have been raised for their high dependence on the particle size and the resulting low sensitivity for small nanoparticles [16].

Because current PTI procedures cannot detect such high emitters of PM, their DPFs are not repaired or replaced and the contribution of these high emitters could increase the average fleet emissions [17], even by a factor 30 [13]. Some studies showed that a small percentage of high polluting vehicles can account for the majority of the emissions [18]. Depending on the pollutant, <10% of the fleet can contribute 30–85% of the emissions [16,19–21]. Various inspections found that 5–15% of the inspected vehicles were high emitters with damaged or removed DPFs [13,22]. The fail rate increased with mileage: from 3% (<50,000 km) to 25% (>150,000 km). Identifying and removing from the road high emitters (result of damaged or tampered particulate filters) should result in an important reduction of the contribution of on-road transport to the total particle emissions.

The VERT (Verification of Emission Reduction Technologies) Association, Swiss, German, and Dutch governmental organizations, metrological institutes, scientists, and equipment manufacturers established an informal new periodical technical inspection (NPTI) technical working group. The working group is developing methodologies for both DPFs and NO_x aftertreatment systems [23]. In 2017 a White Paper summarized the proposal with a PN test at low idle for diesel vehicles [24]. A report from TNO (Netherlands Organisation for Applied Scientific Research) in the same year gave more details [25]. Low idle was chosen as it is simple: only the average of at least 15 s is needed, after a stabilization time of at least 15 s. Snap accelerations followed by low idle speed operation were excluded because they resulted in a non-defined engine behavior. The accelerations affected the activation of the EGR (exhaust gas recirculation) in a non-predictable and non-repeatable way. The Netherlands introduced a PTI regulation in November 2019, applicable to Euro 5b and later diesel vehicles, with a PN limit of 2.5×10^5 p/cm^3, to be met at low idle [26]. The foreseen implementation year is 2021, when measurement sensors will be available to the inspection centers. Belgium and Germany are considering adopting a similar regulation.

Although the work with compression ignition vehicles is at a good level, the studies on positive (spark) ignition vehicles are limited, and without clear conclusions whether the procedure used on diesel vehicles is applicable. Furthermore, there is lack of correlation data of type approval emission tests and idle concentrations. The objective of this study is to present comparisons of type approval cycles and idle concentrations for both compression ignition and positive ignition vehicles and to suggest thresholds based on the experimental results.

Section 2 describes the experimental setup and explains the analysis that was followed with an example. Emissions on the complete laboratory and on-road type-approval cycles are compared to laboratory hot and cold idle concentrations in Section 3. The impact of the DPF soot load on the measured idle concentrations is presented in Section 4 along with the implications of these results for the PTI instruments.

2. Materials and Methods

2.1. Experimental Setup

The typical experimental setup of this study is presented in Figure 1. The type-approval laboratory measurements were conducted from a tunnel where the whole exhaust gas was diluted with filtered air and using constant volume sampling (CVS).

The PN system was based on the Particle Measurement Programme (PMP) recommendations and the regulation technical requirements (2017/1151). In all cases it was the AVL (Graz, Austria) Advanced Particle Counter (APC 489) with an evaporation tube at 350 °C, and a Condensation particle Counter (CPC) with 50% detection efficiency at 23 nm [27]. The vehicle followed a pre-defined test cycle (NEDC or WLTC) and the emissions were determined in p/km (see, e.g., [28] for calculation details). In order to have comparable conditions with PTI measurements that are sampling from the tailpipe, a second PMP system identical or similar to the system at the dilution tunnel was connected to the tailpipe. The idle solid particle concentrations were determined during cold start (<300 s) or with hot engine in p/cm^3.

In many cases on road tests were conducted using a PEMS (Portable Emissions Measurement System) from AVL (MOVE). The PN-PEMS used a catalytic stripper at 300 °C and measured solid particle number concentration with a cut-off of 23 nm by means of a diffusion charger sensor. Euro 6b and older vehicles were tested with prototype PEMS, because the AVL MOVE was not available then: the Nanomet 3 (from Testo, Lenzkirch, Germany; formerly Matter Engineering) which had an evaporation tube at 300 °C and a diffusion charger to count solid particles or the modified NPET (from Horiba, Kyoto, Japan) which had a catalytic stripper at 350 °C and a CPC with 50% detection efficiency at 23 nm. Details about the PEMS instruments can be found elsewhere [9].

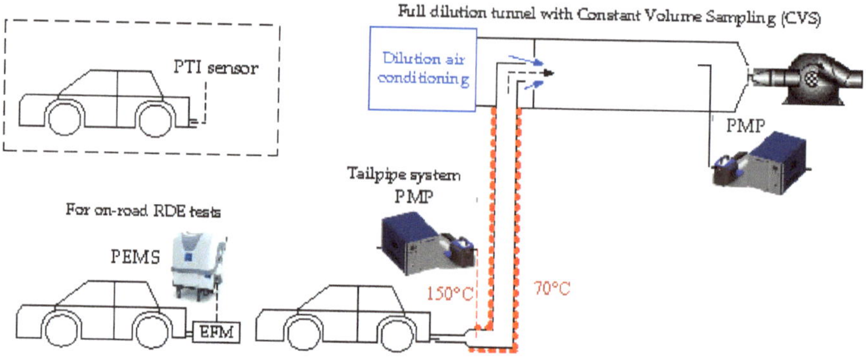

Figure 1. Experimental setup of this study in the laboratory or on the road. In the laboratory the particle number systems (PMP) where connected to the tailpipe and/or to the full dilution tunnel (CVS). On the road particle number PEMS were used. The typical new PTI setup, used in some studies of the literature, is also shown on the top left corner of the figure in dashed lines. CVS = Constant Volume Sampling; EFM = Exhaust Flow Meter; PEMS = Portable Emissions Measurement System; PMP = Particle Measurement Programme; PTI= Periodical Technical Inspection; RDE = Real-Driving Emissions.

2.2. Vehicles

Table 1 summarizes the number of vehicles and the tests that were available: WLTC or NEDC type approval laboratory emissions and hot idle concentrations were available for all cases. In total, data from 32 diesel vehicles (6 of them without DPF, and 6 from the literature), 31 GDI vehicles (4 with GPF, 2 from the literature), 18 PFI vehicles and 7 LPG or CNG vehicles were found. RDE emissions and cold idle concentrations were available only for a fraction of the vehicles. For some older and relatively high emitting vehicles, the idle concentrations were determined from the CVS because no tailpipe measurements were available. Details will be given in the next section.

Table 1. Summary of available vehicles. Number after "+" are vehicles from literature (dedicated new PTI test).

	WLTC	NEDC	RDE	Cold Idle	Hot Idle	Comment
Diesel w/o DPF	0	6 [1]	0	5	6	3 from CVS
Diesel with DPF	18 + 2	2 + 4	15	8	26	
GDI w/o GPF	17 + 1	9	13	17	27	6 from CVS
GDI with GPF	3 + 1	0	2	2	4	
PFI	14	4	13	13	18	
LPG or CNG	7	0	4	6	7	

[1] two of them L-category mini cars CNG = Compressed Natural Gas; CVS = Constant Volume Sampling; DPF = Diesel Particulate Filter; GDI = Gasoline Direct injection; GPF = Gasoline Particulate Filter; LPG = Liquefied Petroleum Gas; NEDC = New European Driving Cycle; PFI = Port Fuel Injection; PTI = Periodical Technical Inspection; WLTC = Worldwide harmonized Light Vehicles Test Cycle.

The data were taken from the following studies:

- TNO: PN emissions at low idle speed and NEDC tests of 4 different diesel vehicles with (cracked) DPF or variable bypass [22,25].
- JRC PTI study: PN emissions at low idle and WLTC of 4 different vehicles (two diesel, one GDI, one GDI with GPF), one of them (diesel) bypassing the DPF [29].
- PN-PEMS studies: Laboratory studies that compared PN-PEMS with PMP systems at the tailpipe or the dilution tunnel [30–33]. From the studies that investigated the 10 nm PEMS, only the 23 nm information was used [34,35].

- Tailpipe studies: Laboratory studies that compared PMP systems at the tailpipe and the dilution tunnel [28,36].
- Emissions monitoring: Laboratory and on-road assessment of various vehicles [37–46].
- Unpublished data: Internal data of older vehicles (non DPF diesel vehicles, or GDIs emitting $>6 \times 10^{12}$ p/km).

No other studies were identified in the literature that reported both idle and type approval emissions.

2.3. Analysis and Calculations

The data needed for the analysis were: idle emissions during cold start (within the first 300 s, but after ignition on >30 s), idling emissions with hot engine (engine on >700 s, typically around 1500 s), PN emissions of type approval cycles (NEDC or WLTC), or RDE compliant tests. Figure 2 gives an example of the first 1500 s of a cold WLTC and an RDE test, in order to explain how the idle data were estimated. The vehicle was a GDI with emissions approximately 3×10^{12} p/km.

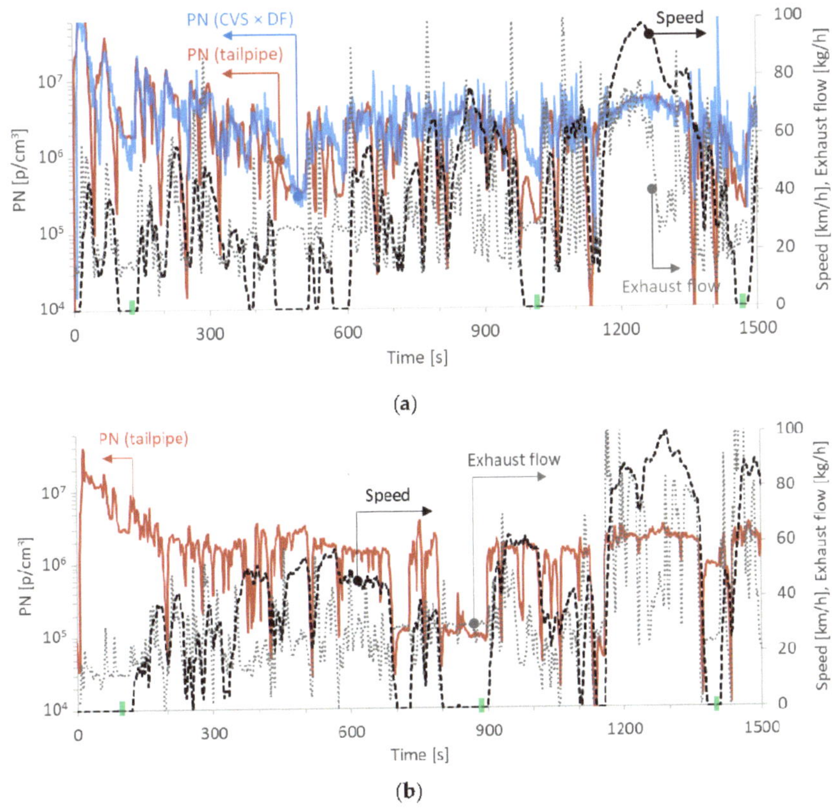

Figure 2. Real time particle number (PN) emissions, exhaust flow rate and speed trace for the first 1500 s of: (**a**) a WLTC (upper panel) and (**b**) a RDE test (lower panel) of a Gasoline Direct Injection (GDI) vehicle. Arrows indicate the appropriate y-axis. The green indexes at the time x-axis show the period that idle concentrations were calculated. CVS = Constant Volume Sampling; DF = Dilution Factor; RDE=Real-Driving Emissions; WLTC = Worldwide harmonized Light vehicles Test Cycle.

2.3.1. Idle Concentrations

Starting with the laboratory WLTC test (Figure 2a), idling periods (i.e., speed is 0 km/h and exhaust flow rate >3 kg/h) were at times around 100–120 s, 350–400 s, 1000 s and 1450 s. The "cold" idle concentration levels were considered during the first idle (around 120 s in this example), after the first ignition of the engine (>30 s) and before 300 s (defined as cold start duration in the European regulation). The exclusion of the first 30 s was decided to exclude the very high start-up emissions due to incomplete combustion [47,48]. It also ensured that the start-up emissions with gasoline fuel of CNG and LPG vehicles would not be considered. Furthermore, measuring within 30 s would be very difficult in practice with high risk of damaging the PTI systems due to condensation [49]. The "hot" idle concentrations were determined at periods where the engine was on at least 700 s (e.g., for some NEDCs at 780 s), but usually around 1000 or 1450 s. For the example of Figure 2a, the "hot" idle concentrations were determined at around 1020 and 1470 s. The concentrations were measured with PEMS or PMP systems measuring from the tailpipe (most of the cases with PMP systems). Averages of the last 10 s were used in order to minimize any influence from concentrations of previous engine modes. Data from the dilution tunnel were considered only for a few exceptional cases (7 GDIs and 3 diesel vehicles, all without particle filter) where no tailpipe measurements were available and there was lack of the specific technologies and emission levels, i.e., GDIs emitting $>6 \times 10^{12}$ p/km and diesel vehicles without DPF. Figure 2a also shows the concentrations estimated from the PMP system at the dilution tunnel, taking into account the dilution factor (DF) at the dilution tunnel. The DF was calculated by dividing the total dilution tunnel flow to the exhaust flow, typically estimated as the difference between total flow and dilution air flow. While there is a rather good agreement at most parts of the cycle, at idle the deviations are quite high (note the logarithmic scale of the y-axis). One reason is the high uncertainty of the DF determination at idling periods: the exhaust flow is the difference of two large values of similar magnitude. The other is the diffusion that takes place between the vehicle and the dilution tunnel. While at high speeds (and exhaust flow rates) the effect is negligible, at idle, due to the low flow rate and the long residence time, the effect is significant and the concentrations of the section before the idling have a big impact.

Figure 2b plots the same information as in Figure 2a for an RDE trip. The specific RDE trip was not fully compliant with the cold start provisions of the regulation (idling periods), because the test was conducted before the entry into force of the regulation. In this specific example, the "cold" idle concentrations were estimated at 100 s, while the "hot" at 900 and 1400 s. The idle emissions at 900 s were much lower than at 1400 s, nevertheless both values were used for the calculation of the average. The PN concentrations were measured with a PEMS. A similar approach was used for the other vehicles analyzed, depending on the available data.

2.3.2. Type Approval Cycle Emissions

The emission of the WLTC and the RDE tests were calculated according to the equations and the procedures described in the relative regulations (i.e., PMP system at the dilution tunnel for NEDC or WLTC, PEMS at the tailpipe for RDE tests) (Commission Regulation (EU) 2017/1151 and all amendments and corrections). In some cases, laboratory cycles with hot engine start were available. In the text all laboratory cycles (NEDC or WLTC) are with cold engine start, unless specified differently (hot NEDC or hot WLTC).

2.3.3. Measurement Uncertainty

Table 2 summarizes the results of the GDI vehicle presented in Figure 2, where all the information mentioned above was available. However, this was not always true for the rest vehicles, where only limited information was available. Although the repeatability of the emission tests is within acceptable levels (<10%, see last row of Table 2), there are significant differences at the absolute levels between the different cycles and locations. Comparing cold and hot cycles, the cold cycle emissions are almost double for this vehicle (3.6×10^{12} p/km vs 1.8×10^{12} p/km). The high differences between cold and hot cycles are well known. The enrichment of the air/fuel mixture during cold-start engine operation, in order to compensate for the reduced fuel vaporization and elevated engine components friction, leads to incomplete fuel combustion and higher emissions [50]. A difference of 23% is also noted between tailpipe and dilution tunnel (3.6×10^{12} p/km vs 2.9×10^{12} p/km) for the cold WLTC. This difference is 11% for the hot WLTC (1.8×10^{12} p/km vs 1.6×10^{12} p/km). Dedicated studies attributed these differences to exhaust flow uncertainties (for both cold and hot WLTC) and particle agglomeration in the transfer tube to the dilution tunnel especially during cold start (for cold WLTC) [28].

Table 2. Results for the vehicle presented in Figure 2. Numbers in brackets give one standard deviation or half of max-min when two repetitions were available. Number of repetitions (n): n = 4 for cold WLTC and RDE, n = 2 for hot WLTC. CVS = Constant Volume Sampling (dilution tunnel); RDE = Real-Driving Emissions test; WLTC = Worldwide harmonized Light vehicles Test Cycle.

	Cold WLTC CVS	Cold WLTC Tailpipe	Hot WLTC CVS	Hot WLTC Tailpipe	RDE [1] Tailpipe
Cold idle [p/cm^3]	8.6×10^5 (42%)	17.1×10^5 (10%)	-	-	23.4×10^5 (32%)
Hot idle [p/cm^3]	4.8×10^5 (41%)	2.8×10^5 (4%)	3.9×10^5 (5%)	1.6×10^5 (16%)	6.1×10^5 (58%)
Cycle emissions [p/km]	2.9×10^{12} (±5%)	3.6×10^{12} (±7%)	1.6×10^{12} (±7%)	1.8×10^{12} (±6%)	1.8×10^{12} (±8%)

[1] Not fully compliant with RDE regulation regarding idling cold start provisions.

The idle concentrations of a vehicle can have a wide range as well. For the vehicle of Table 2, the "cold" idle concentrations were $17–23 \times 10^5$ p/cm^3, when determined from the tailpipe, but less than half when determined from the dilution tunnel. The hot idle emissions ranged from 1.6 to 6.1×10^5 p/cm^3, the highest values measured at the RDE tests. For this specific example, the concentration that would be plotted would be the 2.8×10^5 p/cm^3. The repeatability which was calculated from the particle concentrations (p/cm^3) at different days varied from 5% to 41% at the laboratory but it was 58% on the road.

Another example is the emissions of the GDI Golden car of the inter-laboratory exercise with a Golden PN-PEMS [30]. The average laboratory (NEDC) emissions were 1×10^{12} p/km with one standard deviation of all laboratories of 24%. The on-road tests had an average of 1.1×10^{12} p/km with a similar standard deviation of 25%. The specific uncertainty includes the variability of the vehicle, and the uncertainty of the (same) Golden measurement instrument and exhaust flow meter.

The previous examples give the order of magnitude of uncertainty that the results will have. The PN emissions in one laboratory may have a variability of 10%, but the expected reproducibility levels (i.e., variability between different laboratories) are 20–40% as many studies have shown (see, e.g., reviewed studies in [31]). RDE tests could differ by a factor of two compared to the laboratory type approval cycle. The idle concentrations have a repeatability of 5–60%, but the levels can differ by a factor of 3–4, depending on when they were determined (i.e., engine conditions).

2.3.4. Start and Stop Function and Hybrids

There was one more difficulty in the collection of data. Many vehicles had the start and stop function, so for them there were no idle emissions. For these vehicles, sometimes idle emissions were available at the RDE trips and were used in the included dataset. Only cases where idle emissions could be identified were included in the dataset.

The hybrid vehicles had a similar difficulty, because the engine was almost always off during idling, regardless of the state of charge of the high voltage battery. For these vehicles, as typically the engine is working in a small operation range of revolutions per minute (rpm), the "idle concentrations" were estimated from the emissions of the vehicle at a constant low speed (below 50 km/h). For hybrid vehicles, the scatter of the WLTC emissions can also vary largely depending on the state of charge of the battery. Here, the emissions at charge sustaining mode were considered, which correspond to the maximum use of the internal combustion engine over the type-approval cycle. The number of hybrid vehicles was low (5) and the emissions also low, below 7×10^{11} p/km (except one 1.5×10^{12} p/km), so there was no reason to determine more accurately the idle levels.

3. Results

3.1. Compression Ignition Vehicles

Figure 3a illustrates the idle (p/cm^3) and cycle (p/km) emission results for the compression ignition vehicles. The 1×10^7 cm^3/km dotted line is also shown as a guide to the eye. The vehicles without DPF have emissions around 10^{14} p/km and the hot idle concentrations are around 10^7 p/cm^3. Note that one point with dark green background was a DPF equipped vehicle, with defect or removed DPF. The DPF equipped vehicles have emissions up to 7×10^{11} p/km and hot idle concentrations up to 3×10^4 p/cm^3. The dedicated tests bypassing the DPF (TNO and JRC bypass PTI) fit nicely on the 1×10^7 cm^3/km (dotted) line. The points "DPF bypass" are the JRC tests bypassing the DPF but determining the hot idle concentrations from the cycle (CVS) and a PMP system instead of a PTI instrument. The idle concentrations differ almost by a factor of two compared to the dedicated test (DPF bypass PTI). The different instruments used, in addition to the procedure itself, have contributed to this difference. The scatter of points around the 1×10^7 cm^3/km line is very high up to 1×10^5 p/cm^3, which corresponds to a detection limit of the methodology (i.e., determining high emitter from idle emissions) of 1×10^{12} p/km. The Dutch proposed idle limit of 2.5×10^5 p/cm^3 corresponds approximately to 2.5×10^{12} p/km, approximately four times the type approval laboratory limit.

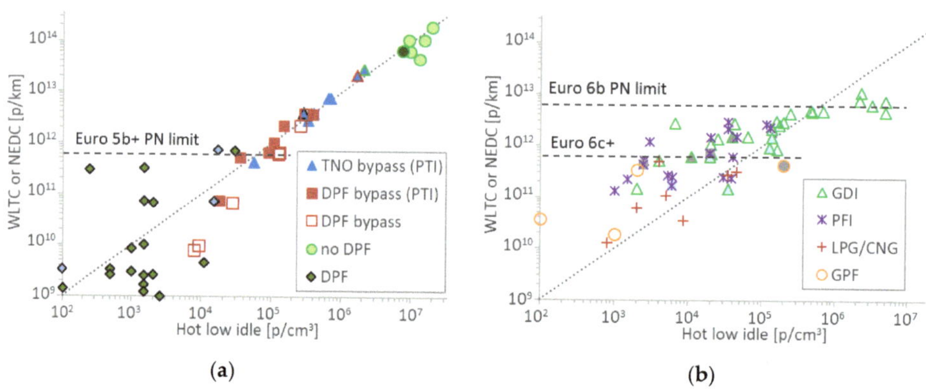

Figure 3. Scatter plots of hot idle concentrations with WLTC (Worldwide harmonized Light vehicles Test Cycle) or NEDC (New European Driving Cycle) type approval test cycles emissions at TNO or JRC (rest points): (**a**) compression ignition vehicles; (**b**) positive ignition vehicles. Each point is a vehicle. Hybrids have grey background. Vehicles with start and stop have light blue background. Horizontal lines give the respective Particle Number (PN) limits. CNG = Compressed Natural Gas; DPF=Diesel Particulate Filter; GDI = Gasoline Direct injection; GPF = Gasoline Particulate Filter; LPG = Liquefied Petroleum Gas; PFI = Port Fuel Injection; PTI = Periodical Technical Inspection; PTI = Periodical Technical Inspection.

3.2. Positive (Spark) Ignition Vehicles

Figure 3b plots the results for the spark (positive) ignition vehicles. The CNG and LPG fueled vehicles have emissions up to 5×10^{11} p/km and the hot idle concentrations up to 4.5×10^4 p/cm^3. The PFI vehicles have emissions up to 3×10^{12} p/km and hot idle concentrations up to 1.1×10^5 p/cm^3.

The GDI vehicles have emissions up to 8×10^{12} p/km and hot idle concentrations are up to 5×10^6 p/cm^3. The GPF equipped vehicles have emissions up to 4.3×10^{11} p/km and hot idle concentrations up to 2×10^5 p/cm^3. The GPF vehicle with the highest hot idle emissions was a hybrid, where the idle emissions were determined at a constant speed because the engine was always off at 0 speed. Otherwise the three vehicles propelled by the internal combustion engine fitted with a GPF had hot idle concentrations lower than 3×10^3 p/cm^3.

The scatter of points around the 1×10^7 cm^3/km line is very high. Up to 2×10^5 p/cm^3 idle concentrations there is no correlation between the type-approval emission and the hot idle concentrations, indicating that the limit of detection of the methodology for spark ignition vehicles is 3×10^{12} p/km (i.e., below 2×10^5 p/cm^3 idle concentrations it is not possible to estimate the type-approval emission, which are below 3×10^{12} p/km). Interestingly, hot idle concentrations $>2.5 \times 10^5$ p/cm^3, which is the proposed idle limit for diesel vehicles, correspond to emissions $>4 \times 10^{12}$ p/km. Vehicles with emissions $>6 \times 10^{12}$ p/km have idle concentrations $>2 \times 10^6$ p/cm^3.

3.3. Cold Idle Levels

As cold start emissions constitute a significant portion of the emissions, an effort was made to see whether "cold" idle concentrations correlate better with the type approval cycle values. Figure 4 summarizes the results. The "DPF bypass" and "no DPF" points were taken from the CVS, while the "DPF" points were taken from the tailpipe.

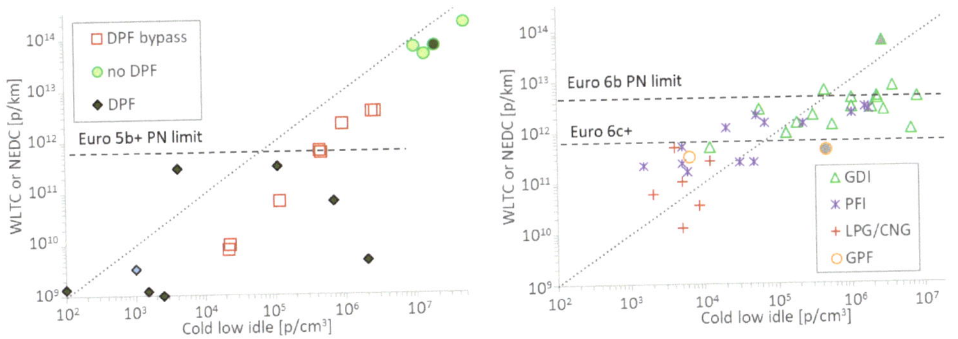

(a) (b)

Figure 4. Scatter plots of cold idle concentrations with WLTC (Worldwide harmonized Light vehicles Test Cycle) or NEDC (New European Driving Cycle) type approval test cycles emissions: (**a**) compression ignition vehicles; (**b**) positive ignition vehicles. Each point is a vehicle. Hybrids have grey background. Vehicles with start and stop have light blue background. Horizontal lines give the respective Particle Number (PN) limits. CNG = Compressed Natural Gas; DPF = Diesel Particulate Filter; GDI = Gasoline Direct injection; GPF = Gasoline Particulate Filter; LPG = Liquefied Petroleum Gas; PFI = Port Fuel Injection; PTI = Periodical Technical Inspection.

The points are shifted to the right (i.e., higher idle concentrations) compared to Figure 3 (hot idle), especially for the DPF vehicles. Furthermore, the scatter seems higher due to the high variability of cold start idle concentrations. However, the number of points are less than of Figure 3 and it is difficult to draw a solid conclusion. The cold idle concentrations of the spark ignition vehicles did not have better correlation with the type approval cycles. Only three GDI idle points were taken from CVS and all the rest from the tailpipe. No idle level was found to indicate high emitters. Actually, the correlation got worse.

3.4. Idle and RDE Correlation

Another question is how representative the idle—type-approval cycle correlation is when vehicles are driven on the road. Figure 5a compares WLTC or NEDC with RDE tests for vehicles for which both tests were available. For the vehicles tested in this study the RDE tests were between 3 times lower and two times higher than the corresponding WLTC results. This variability is higher than the maximum measurement expected uncertainty of the PEMS (sensor plus exhaust flow meter) which is 50% [9]. There was no particular trend of a specific technology or fuel for larger or smaller differences between laboratory and on-road results. Figure 5b shows hot idle concentrations versus the RDE emissions. The results are quite similar to the idle-WLTC/NEDC scatter plot (Figure 3) and they indicate that the idle and type approval cycle correlation is a good indication of the real world behavior of the vehicle. However, there are only three points at or above 2.5×10^5 p/cm^3: one having emissions around 2.4×10^{12} p/km (idle 2.5×10^5 p/cm^3), the other 9.1×10^{12} p/km (idle 5×10^5 p/cm^3), and the last one 2.9×10^{12} p/km (idle 5×10^6 p/cm^3). The last point is not in agreement with the idle and type approval cycle correlation where idle concentrations above 2×10^5 p/cm^3 corresponded to emissions $>4 \times 10^{12}$ p/km, but it's quite near (2.9×10^{12} p/km). Thus, more studies correlating RDE and idle concentrations are necessary for high emitting vehicles.

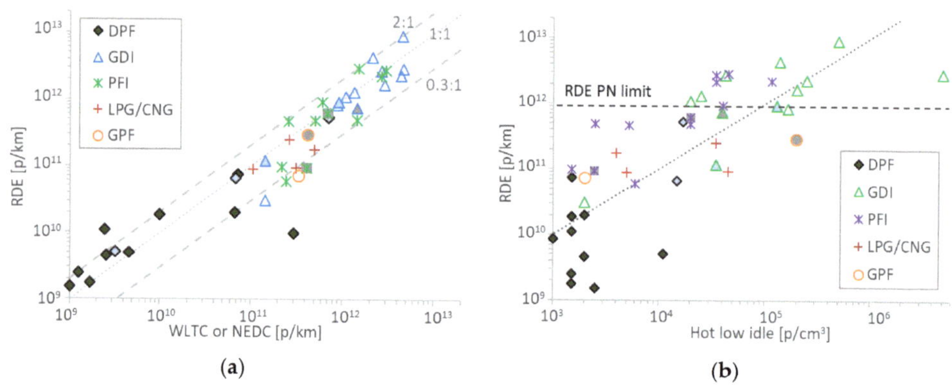

Figure 5. Scatter plots of RDE tests vs.: (**a**) cold start WLTC or NEDC; (**b**) hot idle concentrations. Hybrids have grey background. Vehicles with start and stop have light blue background. CNG=Compressed Natural Gas; DPF = Diesel Particulate Filter; GDI = Gasoline Direct injection; GPF = Gasoline Particulate Filter; LPG = Liquefied Petroleum Gas; PFI = Port Fuel Injection; PTI = Periodical Technical Inspection.

4. Discussion

The aim of this study was to compare PN concentrations at low idle with PN emissions of type approval cycles (WLTC or NEDC). The only dedicated tests (i.e., type approval test and then hot idle test) were taken from the literature. The remaining data points were based on averages of hot idle concentrations measured during the laboratory or on-road tests.

This approach had some challenges. First of all, the idle PN concentrations were not constant throughout the test. Typically, during cold start, higher concentrations were measured. But, even with hot engine the levels varied significantly. A factor of 3–4 as variability was sometimes seen, and the repeatability (i.e., the idle concentrations at the same time period over different days) was not so good (<60%). Similar behavior has been seen in other studies and has been attributed to the different exhaust gas recirculation (EGR) percentage [22]. Many vehicles utilized start and stop, so it was challenging to find the idle concentrations. Moreover, the hybrid vehicles switched the engine off when the speed was zero. The laboratory type approval tests had much better repeatability (10%), but reproducibility levels of 20–40% are common for PN measurements at different laboratories [31,51].

4.1. Limit of Detection of PTI Methodology

4.1.1. Compression Ignition Vehicles

The results of the compression ignition vehicles showed that the correlation of idle concentrations and type approval cycle emissions is quite good for idle concentrations >1 × 10^5 p/cm^3 and emission levels >1 × 10^{12} p/km. Below these levels the data had a high scatter. This finding is important because it demonstrates that the limit of detection of the PTI sensors does not need to be too low and even 2.5 × 10^4 p/cm^3 (10% of the limit), which is the minimum required accuracy in the Dutch regulation, is more than enough.

4.1.2. Spark Ignition Vehicles

The results of the positive (spark) ignition vehicles did not have a good correlation, as the scatter was very high. Nevertheless, the results indicated that the limit of detection of the methodology is around 3 × 10^{12} p/km. Hot idle concentrations of >2.5 × 10^5 p/cm^3 corresponded to >4 × 10^{12} p/km emission levels. These results are very encouraging as the same limit for diesel vehicles (2.5 × 10^5 p/cm^3) may be applicable to spark ignition vehicles. For high emitting GDIs (i.e., >6 × 10^{12} p/km), the idle emissions were >2 × 10^6 p/cm^3. However, the idle levels of the high emitting GDI vehicles were determined from the CVS, thus they could be different compared to direct tailpipe measurements, and further studies are necessary to confirm these results.

The correlation did not improve (actually got worse) when the cold idle concentrations were used. Thus, it is not necessary to define a cold test for PTI testing, which is also impractical as most of the times vehicles undergo PTI in warm conditions. As mentioned in the Dutch regulation, the test can be done with cold engine, but if it fails, it should be repeated with warm engine.

The results are in agreement and further expand the findings of a previous study on GDIs [22]: the hot idle concentrations of the tested GDI vehicles with GPF (three Euro 6d-temp) were 0.1–1.2 × 10^4 p/cm^3, while for those without GPF (three Euro 5 or Euro 6b) were 5–10 × 10^4 p/cm^3. The pre-GPF emissions of the Euro 6d-temp GDIs were high: 0.1–5.8 × 10^6 p/cm^3. Combining these results with ours, it seems that GPF equipped vehicles have similar behavior with DPF equipped diesel vehicles: the idle concentrations are lower than 2–3 × 10^4 p/cm^3 (exception the hybrid GPF vehicle). In most cases, GDI vehicles without GPF have higher idle concentrations. Thus, identifying tampering of the GPF may be possible. However, this needs more research because there are already GDI vehicles without GPF having low idle concentrations and type-approval emissions. In addition, idle levels of 10^4 p/cm^3 to 2 ×10^5 p/cm^3 correspond to relatively low emissions (6 × 10^{11} p/km to 2 × 10^{12} p/km).

4.1.3. Reasons of High Scatter

As already indicated, while for compression ignition (diesel) engines, the hot idle emissions had a relative good correlation with the type approval cycle, this was not the case for the positive (spark) ignition engines. In diesel engines fuel is injected into the engine cylinder and mixes with high temperature-pressure air [52]. As the piston moves to the top-dead-center, the mixture reaches the ignition point and the combustion of the charge starts as premixed combustion and continues as diffusion limited combustion. Although the overall air-to-fuel ratio is lean, combustion occurs when vaporized fuel mixes with air stoichiometrically. Soot emissions are generally formed at the fuel-rich side of the reaction zone in the diffusion combustion phase [53]. Diesel combustion is heterogeneous in nature, compared to spark ignition engines in which the combustible mixture is predominantly homogenous. In conventional PFI engines, fuel is injected into the intake port so that fuel and air flow simultaneously into the combustion chamber during the intake process, and a homogeneous air–fuel mixture is formed. PFI engines have low emissions at steady state conditions or light loads. Spikes of PN are seen when driving requires fuel enrichment, such as cold start, accelerations and high loads [54]. As the majority of the emissions originates from these events, the idle concentrations do not necessarily correlate with the overall cycle emissions. In GDI engines, fuel is sprayed directly into the combustion chamber. This leads to incomplete fuel evaporation due to the limited time available for fuel and air mixing, resulting in localized rich combustion and PM formation [55]. Additionally, a small amount of fuel may impinge on the piston and make direct contact with the cold cylinder walls, which may lead to diffusion combustion and subsequent PM formation [56], in particular during fuel enrichment events. Thus, it seems that the idle concentrations cannot "represent" the fuel enrichment events of the GDI engines.

The even higher scatter of vehicles equipped with particulate filters (DPF, GPF) can be explained by the different mechanism of particles' appearance. Particulate filters are generally very efficient in removing particles, with filtration efficiencies >95% [52,57]. Nonetheless, in some cases, high particle concentrations are measured during cold start [8]. The main reason for these emissions are small defects in the mat used to mount the brick in the canister, resulting in reduced filtration efficiency [58]. The defects close as the particulate filter heats up and the filtration increases. The second issue with filters is that the filtration efficiency depends on the accumulated soot and deposited ash. For example, emissions are very high immediately after a regeneration event, but drop significantly only after a few minutes of driving [8]. Similarly, over time, due to deposited ash that cannot be burnt during regenerations, the filtration efficiency on average improves. In particular for GPFs, filtration improvements of 10–15% after only 3000 km have been reported [59]. It is expected that a normal use will cover more than 3000 km in four years when the first PTI will take place.

4.2. DPF Soot Load and Idle Levels

Table 3 summarizes the WLTC idle concentrations of three DPF equipped vehicles just before and immediately after regeneration events. The cold idle concentrations were determined at 120 s, while the hot idle concentrations at 1000 s. Immediately after regeneration, the cold idle exceeded the limit of 2.5×10^5 p/cm^3 for two of the vehicles. However, the hot idle concentrations were $<3 \times 10^3$ p/cm^3 for all three vehicles. Thus, driving 10–15 min after a regeneration event should be sufficient to form a soot cake at the DPF and drop the idle concentrations at the typical levels for the specific vehicle.

Table 3. Particle number (PN) cycle emissions and cold idle (at 120 s) and hot idle (at 1000 s) concentrations for WLTCs (Worldwide harmonized Light vehicles Test Cycle) just before and immediately after regenerations.

Vehicle	PN [p/km]	Cold Idle [p/cm^3]	Hot Idle [p/cm^3]	Study
DPF #1 before	2.6×10^9	n/a	$<1.0 \times 10^3$	[34]
DPF #1 after	2.0×10^{11}	2.8×10^5	1.0×10^3	[34]
DPF #2 before	3.0×10^9	$<1.0 \times 10^3$	$<1.0 \times 10^3$	[35]
DPF #2 after	9.5×10^{11}	1.5×10^6	2.0×10^3	[35]
DPF #3 before [1]	2.0×10^9	n/a	1.0×10^3	[42]
DPF #3 after [1]	6.0×10^{10}	1.8×10^4	2.7×10^3	[42]

[1] Measured from the dilution tunnel.

4.3. Implications for PTI Sensors

The results section was based on PEMS and PMP systems. PMP systems typically weight >50 kg and are fixed in the laboratory. PN-PEMS typically weight >10 kg and can easily fit in the trunk or on the hook of a vehicle. Both instruments are handled by experienced and specialized personnel. PTI instruments should be handheld in a garage type of environment and therefore they have to be robust and easy to use. Sensors that weight only 0.4 kg have been reported, however the prototype instruments are much heavier [60]. Obviously, their technical specifications cannot be as strict as of those of PEMS or PMP systems. Table 4 summarizes the efficiency (i.e., ratio to a reference system) requirements of PEMS and PMP systems [8,9], along with those of PTI sensors based on the Dutch legislation [26], and the Swiss regulation for construction machinery [61]. The Dutch regulation can be considered representative of future PTI regulations for diesel cars of other countries. The efficiency requirements are very similar for all regulations with only minor differences (e.g., calibration size and polydispersity for the Swiss regulation). The required accuracy is around 30% at large sizes (efficiency 0.70 to 1.30). The Dutch regulation requires maximum measurement error of 25% or 2.5×10^4 p/cm^3 (whichever is larger). Although the 2.5×10^4 p/cm^3 concentration is higher than the 5×10^3 p/cm^3 zero level required for PEMS, it is sufficient, as it was shown in the results section. The maximum concentration requested for these sensors is 5×10^6 p/cm^3. This concentration is appropriate for DPF equipped vehicles, even with cracks or partial damage, but not for older diesel vehicles without DPF.

Table 4. Efficiency requirements for various systems using monodisperse (mono) or polydisperse (poly) aerosol.

Diameter	Aerosol	23 nm	30 nm	50 nm	70–100 nm	200 nm
PMP [1]	Mono	0.33–0.60	0.59–0.91	0.99–1.00	1.00–1.13	1.00–1.14
PEMS [2]	Mono	0.20–0.60	0.30–1.20	0.60–1.30	0.70–1.30	0.50–2.00
Diameter	**Aerosol**	**23 nm**	**41 nm**	**50 nm**	**80 nm**	**200 nm**
Dutch PTI [3]	Mono	0.20–0.60	-	0.60 – 1.30	0.70–1.30	-
Swiss PTI [4]	Poly	<0.50	>0.40	-	0.70–1.30	<3.00

[1] Estimated from Regulation (EU) 2017/1151 requirements for separate parts; [2] Regulation (EU) 2017/1154; [3] Dutch regulation No. IENW / BSK-2019/202498 [26]; [4] Swiss regulation for construction machinery VAMV SR 941.242 [61]. PMP = Particle Measurement Programme; PEMS = Portable Emissions Measurement Systems; PTI = Periodical Technical Inspection; VAMV = Ordinance of the FDJP on Exhaust Gas Analyzers.

At 23 nm the required efficiency is around 40% (± 20%). The requirement is quite similar to the estimated efficiency of the PMP systems (33–60%). The 23 nm size was selected by the PMP group in order to include the smaller soot particles, but at the same time exclude any nucleation mode volatile particles [8]. The PEMS and PTI requirements for the steepness of the cut-off curve are not as strict as for the PMP system. Thus, particles <23 nm may be counted. This should not be an issue for diesel vehicles, for which the majority of particles are >23 nm (e.g., [7]). The only exception is during cold start where high concentrations of nonvolatile sub-23 nm particles at idle or low speeds can be seen [42,62,63]. At hot idle the nonvolatile (solid) sub-23 nm particles are absent [63]. For spark ignition vehicles though the sub-23 nm particles can be as many as those >23 nm [7,47,64].

The efficiency requirements of PMP systems can be achieved only by condensation particle counters (CPCs) [65], while those of PEMS or PTI regulations with both CPCs and advanced diffusion charging counters [9,66,67]. Opacimeters or light scattering instruments would fail these specifications.

The volatile removal efficiency requirements of the different regulations are summarized in Table 5. The new PTI requirements are similar to the PMP requirements for systems measuring >23 nm (>99% of 30 nm tetracontane particle). The PTI requirement is much easier than the one for PEMS (mass >1 mg/m^3), yet as the lower cut-off diameter is at 23 nm the evaporation efficiency is sufficient. Diesel vehicles have high air-fuel ratio and condensation is unlikely. Volatile nucleation mode particles can be seen with high exhaust gas temperature and high sulfur content [68], but at other conditions are less likely. For spark ignition engines nucleation particles at low idle are also not probable [69]. However, volatile compounds can grow the sub-23 nm solid particles in the >23 nm range, affecting the results and the comparability between different instruments. Thus, some thermal pre-treatment is necessary. Theoretically, the tetracontane removal requirement is easily achievable by heating the aerosol at 200 °C [70]. Discussion on the thermal pre-treatment topic can be found elsewhere [57,71,72]. Commercial systems use catalytic stripper, sensors at elevated temperature or include hot dilution (see detailed discussion of available equipment elsewhere [16]).

Currently, the instruments used for PTI testing are prototypes and similar to the PEMS [17]. Thus, their measurement uncertainty is around 35% and compared to the PMP systems at the full dilution tunnel the differences are around 50% [9]. Other smaller PTI sensors had differences by a factor of two during a measurement campaign [29]. Thus, it is necessary to further characterize the instruments that will appear in the market in the future.

Table 5. Volatile removal efficiency requirements of tetracontane particles for various systems.

Diameter	Aerosol	Diameter	Number Conc.	Mass Conc.	Efficiency
PMP [1]	Mono or poly	≥30 nm	≥10^4 p/cm^3	-	≥99%
PEMS [2]	Poly	≥50 nm	-	≥1 mg/m^3	≥99%
Dutch PTI [3]	Mono	≥30	0.5–1 × 10^4 p/cm^3	-	≥95%
Swiss PTI [4]	Poly	≥30	<10^5 p/cm^3	-	≥95%

[1] Regulation (EU) 2017/1151; [2] Regulation (EU) 2017/1154; [3] Dutch regulation No. IENW / BSK-2019/202498 [26]; [4] Swiss regulation for construction machinery VAMV 941.242 [61]. PMP = Particle Measurement Programme; PEMS = Portable Emissions Measurement Systems; PTI = Periodical Technical Inspection; VAMV = Ordinance of the FDJP on Exhaust Gas Analyzers.

All results and the discussion so far focused on the current regulation which counts nonvolatile particles >23 nm. Recently, in the Global Technical Regulation (GTR 15) of the worldwide harmonized light vehicles test procedure (WLTP), the new proposal includes counting solid particles >10 nm [73]. For countries adopting this option, the new PTI procedures should be adopted accordingly. As discussed previously, for diesel vehicles the effect should be minimal (other than changing the cut-off size of the instruments), however for spark ignition vehicles further studies are needed. For example, high concentrations of sub-23 nm solid particles have been reported for CNG and PFI vehicles (see, e.g., [7,64]). For filtered equipped vehicles the effect should be small because sub-23 nm particles should be captured with high efficiency, but for faulty filters of spark ignition vehicles high concentration of both >23 nm and sub-23 nm particles will pass through the faulty filter and this needs to be examined.

5. Conclusions

The new periodical technical inspection (PTI) procedure will require measurement of vehicle exhaust particle number (PN) concentrations at idle. In this study the PN emissions of type approval cycles were compared with low idle concentrations for diesel, gasoline (GDI or PFI), CNG and LPG vehicles. For diesel vehicles the correlation was good for PN levels >1 × 10^{12} p/km (idle concentration >1 × 10^5 p/cm^3). At lower emission levels the cycle emissions depended significantly on the cold start emissions, due to the existence of DPFs. However, the correlation did not improve correlating the emissions with the cold idle concentrations.

For positive (spark) ignition vehicles, no correlation between cycle emissions and idle concentrations could be found. For this category of vehicles, the emissions are mainly produced during fuel enrichments, e.g., during cold start, accelerations, or high loads. Nevertheless, idle concentrations >2.5 × 10^5 p/cm^3 were related to emission levels >4 × 10^{12} p/km and idle concentrations >2 × 10^6 p/cm^3 were related to emission levels >6 × 10^{12} p/km. Although these levels are a first step in defining limits for this category of vehicles, more dedicated studies are necessary, especially if a correlation against RDE is also of interest. Further tests are also needed for hybrid vehicles.

The current PTI technical specifications are comparable to those of on-board and laboratory type-approval systems and of sufficient stringency. The current PTI systems are similar to the on-board systems. However, as they will become smaller in size and less complex, more tests with dedicated PTI sensors are also important in order to assess their measurement uncertainty.

Author Contributions: Conceptualization, B.G.; formal analysis, B.G.; data curation, T.L., R.S.-B., V.V. and M.C.; writing—original draft preparation, B.G.; writing—review and editing, T.L., R.S.-B., V.V. and M.C. All authors have read and agreed to the published version of the manuscript.

Funding: This research received no external funding.

Conflicts of Interest: The authors declare no conflict of interest.

Disclaimer: The opinions expressed in this manuscript are those of the authors and should in no way be considered to represent an official opinion of the European Commission. Mention of trade names or commercial products does not constitute endorsement or recommendation by the authors or the European Commission.

References

1. European Environment Agency. *Air Quality in Europe: 2019 Report*; Publications Office of the European Union: Luxembourg, 2019; ISBN 978-92-9480-088-6.
2. Li, Y.; Lane, K.; Corlin, L.; Patton, A.; Durant, J.; Thanikachalam, M.; Woodin, M.; Wang, M.; Brugge, D. Association of Long-Term Near-Highway Exposure to Ultrafine Particles with Cardiovascular Diseases, Diabetes and Hypertension. *Int. J. Environ. Res. Public Health* **2017**, *14*, 461. [CrossRef] [PubMed]
3. Rivas, I.; Beddows, D.C.S.; Amato, F.; Green, D.C.; Järvi, L.; Hueglin, C.; Reche, C.; Timonen, H.; Fuller, G.W.; Niemi, J.V.; et al. Source apportionment of particle number size distribution in urban background and traffic stations in four European cities. *Environ. Int.* **2020**, *135*, 105345. [CrossRef] [PubMed]
4. Lorelei de Jesus, A.; Thompson, H.; Knibbs, L.D.; Kowalski, M.; Cyrys, J.; Niemi, J.V.; Kousa, A.; Timonen, H.; Luoma, K.; Petäjä, T.; et al. Long-term trends in PM2.5 mass and particle number concentrations in urban air: The impacts of mitigation measures and extreme events due to changing climates. *Environ. Pollut.* **2020**, *263*, 114500. [CrossRef] [PubMed]
5. Paasonen, P.; Kupiainen, K.; Klimont, Z.; Visschedijk, A.; Denier van der Gon, H.A.C.; Amann, M. Continental anthropogenic primary particle number emissions. *Atmos. Chem. Phys.* **2016**, *16*, 6823–6840. [CrossRef]
6. Grigoratos, T.; Martini, G. Brake wear particle emissions: A review. *Environ. Sci. Pollut. Res.* **2015**, *22*, 2491–2504. [CrossRef] [PubMed]
7. Giechaskiel, B.; Lähde, T.; Suarez-Bertoa, R.; Clairotte, M.; Grigoratos, T.; Zardini, A.; Perujo, A.; Martini, G. Particle number measurements in the European legislation and future JRC activities. *Combust. Eng.* **2018**, *174*, 3–16. [CrossRef]
8. Giechaskiel, B.; Mamakos, A.; Andersson, J.; Dilara, P.; Martini, G.; Schindler, W.; Bergmann, A. Measurement of automotive nonvolatile particle number emissions within the European legislative framework: A review. *Aerosol Sci. Technol.* **2012**, *46*, 719–749. [CrossRef]
9. Giechaskiel, B.; Bonnel, P.; Perujo, A.; Dilara, P. Solid particle number (SPN) portable emissions measurement systems (PEMS) in the European legislation: A review. *Int. J. Environ. Res. Public Health* **2019**, *16*, 4819. [CrossRef]
10. Giechaskiel, B.; Joshi, A.; Ntziachristos, L.; Dilara, P. European regulatory framework and particulate matter emissions of gasoline light-duty vehicles: A review. *Catalysts* **2019**, *9*, 586. [CrossRef]
11. Kadijk, G.; Spreen, J. *Roadworthiness Test Investigations of Diesel Particulate Filters on Vehicles*; Report 10307v2; TNO: Delft, The Netherlands, 2015.
12. Boulter, P.; Buekenhoudt, P.; Stricker, P.; Mäurer, H.-J.; Nolte, C.; Ost, T.; Schulz, W.; Weitz, K.-U.; Afflerbach, G.; Förster, C.; et al. *TEDDIE: A New Roadworthiness Emission Test for Diesel Vehicles Involving NO, NO2 and PM Measurements*; CITA: Brussels, Belgium, 2011.
13. Boveroux, F.; Cassiers, S.; Buekenhoudt, P.; Chavatte, L.; De Meyer, P.; Jeanmart, H.; Verhelst, S.; Contino, F. *Feasibility Study of a New Test Procedure to Identify high Emitters of Particulate Matter during Periodic Technical Inspection*; SAE International: Warrendale, PA, USA, 2019; paper number 2019-01–1190.
14. Yamada, H. Improving methodology of particulate measurement in periodic technical inspection with high-sensitivity techniques: Laser light scattering photometry and particle number method. *Emiss. Control. Sci. Technol.* **2019**, *5*, 37–44. [CrossRef]
15. Kadijk, G. *Roadworthiness Test Investigations of Diesel Particulate Filters*; Report 10160v3; TNO: Delft, The Netherlands, 2013.
16. Burtscher, H.; Lutz, T.; Mayer, A. A new periodic technical inspection for particle emissions of vehicles. *Emiss. Control. Sci. Technol.* **2019**, *5*, 279–287. [CrossRef]
17. Boveroux, F.; Cassiers, S.; De Meyer, P.; Buekenhoudt, P.; Bergmans, B.; Idczak, F.; Jeanmart, H.; Verhelst, S.; Contino, F. Impact of mileage on particle number emission factors for Euro 5 and Euro 6 diesel passenger cars. *Atmos. Environ.* **2020**, 117975. [CrossRef]
18. Platt, S.M.; El Haddad, I.; Pieber, S.M.; Zardini, A.A.; Suarez-Bertoa, R.; Clairotte, M.; Daellenbach, K.R.; Huang, R.-J.; Slowik, J.G.; Hellebust, S.; et al. Gasoline cars produce more carbonaceous particulate matter than modern filter-equipped diesel cars. *Sci. Rep.* **2017**, *7*, 4926. [CrossRef]
19. Bishop, G.A.; Schuchmann, B.G.; Stedman, D.H.; Lawson, D.R. Multispecies remote sensing measurements of vehicle emissions on Sherman Way in Van Nuys, California. *J. Air Waste Manag. Assoc.* **2012**, *62*, 1127–1133. [CrossRef] [PubMed]

20. Huang, Y.; Organ, B.; Zhou, J.L.; Surawski, N.C.; Hong, G.; Chan, E.F.C.; Yam, Y.S. Emission measurement of diesel vehicles in Hong Kong through on-road remote sensing: Performance review and identification of high-emitters. *Environ. Pollut.* **2018**, *237*, 133–142. [CrossRef]
21. Kurniawan, A.; Schmidt-Ott, A. Monitoring the soot emissions of passing cars. *Environ. Sci. Technol.* **2006**, *40*, 1911–1915. [CrossRef]
22. Kadijk, G.; Elstgeest, M.; van der Mark, P.; Ligterink, N. *Follow-up Research into the PN Limit Value and the Measurement Method for Checking Particulate Filters with a Particle Number Counter*; Report 10006; TNO: Delft, The Netherlands, 2020.
23. Czerwinski, J.; Comte, P.; Engelmann, D.; Mayer, A.; Lutz, T.; Hensel, V. *Considerations of Periodical Technical Inspection of Vehicles with deNOx Systems*; SAE International: Warrendale, PA, USA, 2019; paper number 2019-01–0744.
24. Kadijk, G.; Mayer, A. White Paper: NPTI—The New Periodic Technical Inspection Emission Test Procedure for Vehicles with Emission Control Systems. In Proceedings of the 21st ETH conference on combustion generated nanoparticles, Zürich, Switzerland, 19–22 June 2017.
25. Kadijk, G.; Elstgeest, M.; Ligterink, N.; van der Mark, P. *Investigation into a Periodic Technical Inspection (PTI) test Method to Check for Presence and Proper Functioning of Diesel Particulate Filters in Light-Duty Diesel Vehicles—Part 2*; Report 10530; TNO: Delft, The Netherlands, 2017.
26. Ministry for Infrastructure and Public Works. *Regulation of the Minister for Environment and Housing of 21 November 2019*; No. IENW/BSK-2019/202498; Amending the Vehicle Regulations to Enable the Control of Diesel Particulate filters with a Particle Counter and a Number of Other Technical Change; Ministry for Infrastructure and Public Works: Amsterdam, The Netherlands, 2019.
27. Giechaskiel, B.; Cresnoverh, M.; Jörgl, H.; Bergmann, A. Calibration and accuracy of a particle number measurement system. *Meas. Sci. Technol.* **2010**, *21*, 045102. [CrossRef]
28. Giechaskiel, B.; Lähde, T.; Drossinos, Y. Regulating particle number measurements from the tailpipe of light-duty vehicles: The next step? *Environ. Res.* **2019**, *172*, 1–9. [CrossRef]
29. Suarez-Bertoa, R.; Lahde, T.; Giechaskiel, B. Verification of NPTI-instruments for diesel and petrol vehicles—first results. In Proceedings of the 22nd ETH Conference on Combustion Generated Nanoparticles, Zürich, Switzerland, 18–21 June 2018.
30. Riccobono, F.; Giechaskiel, B.; Mendoza Villafuerte, P. *Particle Number PEMS Inter.-Laboratory Comparison Exercise*; Publications Office of the European Union: Luxembourg, 2016.
31. Giechaskiel, B.; Casadei, S.; Mazzini, M.; Sammarco, M.; Montabone, G.; Tonelli, R.; Deana, M.; Costi, G.; Di Tanno, F.; Prati, M.; et al. Inter-laboratory correlation exercise with portable emissions measurement systems (PEMS) on chassis dynamometers. *Appl. Sci.* **2018**, *8*, 2275. [CrossRef]
32. Giechaskiel, B.; Riccobono, F.; Vlachos, T.; Mendoza-Villafuerte, P.; Suarez-Bertoa, R.; Fontaras, G.; Bonnel, P.; Weiss, M. Vehicle emission factors of solid nanoparticles in the laboratory and on the road using portable emission measurement systems (PEMS). *Front. Environ. Sci.* **2015**, *3*. [CrossRef]
33. Giechaskiel, B.; Riccobono, F.; Bonnel, P. *Feasibility Study on the Extension of the Real-Driving Emissions (RDE) Procedure to Particle Number (PN): Chassis Dynamometer Evaluation of Portable Emission Measurement Systems (PEMS) to Measure Particle Number (PN) Concentration: Phase II*; Publications Office of the European Union: Luxembourg, 2015; ISBN 978-92-79-51003-8.
34. Giechaskiel, B.; Mamakos, A.; Woodburn, J.; Szczotka, A.; Bielaczyc, P. Evaluation of a 10 nm particle number portable emissions measurement system (PEMS). *Sensors* **2019**, *19*, 5531. [CrossRef] [PubMed]
35. Giechaskiel, B.; Lähde, T.; Gandi, S.; Keller, S.; Kreutziger, P.; Mamakos, A. Assessment of 10-nm particle number (PN) portable emissions measurement systems (PEMS) for future regulations. *Int. J. Environ. Res. Public Health* **2020**, *17*, 3878. [CrossRef] [PubMed]
36. Isella, L.; Giechaskiel, B.; Drossinos, Y. Diesel-exhaust aerosol dynamics from the tailpipe to the dilution tunnel. *J. Aerosol Sci.* **2008**, *39*, 737–758. [CrossRef]
37. Clairotte, M.; Valverde, V.; Bonnel, P.; Giechaskiel, P.; Carriero, M.; Otura, M.; Fontaras, G.; Pavlovic, J.; Martini, G.; Krasenbrink, A. *Joint Research Centre 2017 Light-Duty Vehicles Emissions Testing Contribution to the EU Market Surveillance: Testing Protocols and Vehicle Emissions Performance*; Publications Office of the European Union: Luxembourg, 2018; ISBN 978-92-79-90600-8.

38. Valverde, V.; Clairotte, M.; Bonnel, P.; Giechaskiel, P.; Carriero, M.; Otura, M.; Gruening, C.; Fontaras, G.; Pavlovic, J.; Martini, G.; et al. *Joint Research Centre 2018 Light-Duty Vehicles Emissions Testing: Contribution to the EU Market Surveillance: Testing Protocols and Vehicle Emissions Performance*; Publications Office of the European Union: Luxembourg, 2019; ISBN 978-92-76-12333-0.
39. Valverde, V.; Mora, B.A.; Clairotte, M.; Pavlovic, J.; Suarez-Bertoa, R.; Giechaskiel, B.; Astorga-LLorens, C.; Fontaras, G. Emission factors derived from 13 Euro 6b light-duty vehicles based on laboratory and on-road measurements. *Atmosphere* **2019**, *10*, 243. [CrossRef]
40. Valverde-Morales, V.; Clairotte, M.; Pavlovic, J.; Giechaskiel, B.; Bonnel, P. *On-Road Emissions of Euro 6d-TEMP Vehicles: Consequences of the Entry into Force of the RDE Regulation in Europe*; SAE International: Warrendale, PA, USA, 2020; paper number 2020-01–2219.
41. Valverde, V.; Giechaskiel, B. Assessment of gaseous and particulate emissions of a Euro 6d-temp diesel vehicle driven >1300 km including six diesel particulate filter regenerations. *Atmosphere* **2020**, *11*, 645. [CrossRef]
42. Giechaskiel, B. Particle number emissions of a diesel vehicle during and between regeneration events. *Catalysts* **2020**, *10*, 587. [CrossRef]
43. Suarez-Bertoa, R.; Astorga, C. Impact of cold temperature on Euro 6 passenger car emissions. *Environ. Pollut.* **2018**, *234*, 318–329. [CrossRef]
44. Giechaskiel, B.; Suarez-Bertoa, R.; Lahde, T.; Clairotte, M.; Carriero, M.; Bonnel, P.; Maggiore, M. Emissions of a Euro 6b diesel passenger car retrofitted with a solid ammonia reduction system. *Atmosphere* **2019**, *10*, 180. [CrossRef]
45. Suarez-Bertoa, R.; Lähde, T.; Pavlovic, J.; Valverde, V.; Clairotte, M.; Giechaskiel, B. Laboratory and on-road evaluation of a GPF-equipped gasoline vehicle. *Catalysts* **2019**, *9*, 678. [CrossRef]
46. Giechaskiel, B.; Zardini, A.; Martini, G. Particle emission measurements from L-category vehicles. *SAE Int. J. Engines* **2015**, *8*, 2322–2337. [CrossRef]
47. Badshah, H.; Khalek, I.A. Solid particle emissions from vehicle exhaust during engine start-up. *SAE Int. J. Engines* **2015**, *8*, 1492–1502. [CrossRef]
48. Premnath, V.; Khalek, I. *Particle Emissions from Gasoline Direct Injection Engines during Engine Start-Up (Cranking)*; SAE International: Warrendale, PA, USA, 2019; paper number 2019-01–1182.
49. Giechaskiel, B.; Zardini, A.A.; Clairotte, M. Exhaust gas condensation during engine cold start and application of the dry-wet correction factor. *Appl. Sci.* **2019**, *9*, 2263. [CrossRef]
50. Bielaczyc, P.; Szczotka, A.; Woodburn, J. An overview of cold start emissions from direct injection spark-ignition and compression ignition engines of light duty vehicles at low ambient temperatures. *Combust. Engines* **2013**, *154*, 96–103.
51. Giechaskiel, B.; Dilara, P.; Andersson, J. Particle measurement programme (PMP) light-duty inter-laboratory exercise: Repeatability and reproducibility of the particle number method. *Aerosol Sci. Technol.* **2008**, *42*, 528–543. [CrossRef]
52. Majewski, W.A.; Khair, M.K. *Diesel Emissions and Their Control*; SAE International: Warrendale, PA, USA, 2006; ISBN 978-0-7680-0674-2.
53. Stone, R. *Introduction to Internal Combustion Engines*, 4th ed.; SAE International: Warrendale, PA, USA, 2012; ISBN 978-0-7680-2084-7.
54. Eastwood, P. *Particulate Emissions from Vehicles*; Wiley-Professional engineering publishing series; Wiley: Chichester, UK, 2008; ISBN 978-0-470-72455-2.
55. Piock, W.; Hoffmann, G.; Berndorfer, A.; Salemi, P.; Fusshoeller, B. Strategies towards meeting future particulate matter emission requirements in homogeneous gasoline direct injection engines. *SAE Int. J. Engines* **2011**, *4*, 1455–1468. [CrossRef]
56. Karlsson, R.B.; Heywood, J.B. *Piston Fuel Film Observations in an Optical Access GDI Engine*; SAE International: Warrendale, PA, USA, 2001; paper number 2001-01-2022.
57. Burtscher, H. Physical characterization of particulate emissions from diesel engines: A review. *J. Aerosol Sci.* **2005**, *36*, 896–932. [CrossRef]
58. Mamakos, A.; Martini, G.; Manfredi, U. Assessment of the legislated particle number measurement procedure for a Euro 5 and a Euro 6 compliant diesel passenger cars under regulated and unregulated conditions. *J. Aerosol Sci.* **2013**, *55*, 31–47. [CrossRef]

59. Zhang, R.; Howard, K.; Kirkman, P.; Browne, D.; Lu, Z.; He, S.; Boger, T. *A Study into the Impact of Engine Oil on Gasoline Particulate Filter Performance through a Real-World Fleet Test*; SAE International: Warrendale, PA, USA, 2019; paper number 2019-01–0299.
60. Bischof, O.F. Recent developments in the measurement of low particulate emissions from mobile sources: A review of particle number legislations. *Emiss. Control. Sci. Technol.* **2015**, *1*, 203–212. [CrossRef]
61. Swiss Federal Council. *Swiss Regulation SR 941.242 Verordnung des EJPD über Abgasmessmittel für Verbrennungsmotoren (VAMV)*; Swiss Federal Council: Bern, Switzerland, 2006.
62. Filippo, A.D.; Maricq, M.M. Diesel nucleation mode particles: Semivolatile or solid? *Environ. Sci. Technol.* **2008**, *42*, 7957–7962. [CrossRef]
63. Kirchner, U.; Scheer, V.; Vogt, R.; Kägi, R. TEM study on volatility and potential presence of solid cores in nucleation mode particles from diesel powered passenger cars. *J. Aerosol Sci.* **2009**, *40*, 55–64. [CrossRef]
64. Samaras, Z.C.; Andersson, J.; Bergmann, A.; Hausberger, S.; Toumasatos, Z.; Keskinen, J.; Haisch, C.; Kontses, A.; Ntziachristos, L.D.; Landl, L.; et al. *Measuring Automotive Exhaust Particles Down to 10 nm*; SAE International: Warrendale, PA, USA, 2020; paper number 2020-01-2209.
65. Giechaskiel, B.; Wang, X.; Horn, H.-G.; Spielvogel, J.; Gerhart, C.; Southgate, J.; Jing, L.; Kasper, M.; Drossinos, Y.; Krasenbrink, A. Calibration of condensation particle counters for legislated vehicle number emission measurements. *Aerosol Sci. Technol.* **2009**, *43*, 1164–1173. [CrossRef]
66. Schriefl, M.A.; Bergmann, A.; Fierz, M. Design principles for sensing particle number concentration and mean particle size with unipolar diffusion charging. *IEEE Sens. J.* **2019**, *19*, 1392–1399. [CrossRef]
67. Schriefl, M.A.; Nishida, R.T.; Knoll, M.; Boies, A.M.; Bergmann, A. Characterization of particle number counters based on pulsed-mode diffusion charging. *Aerosol Sci. Technol.* **2020**, *54*, 772–789. [CrossRef]
68. Giechaskiel, B.; Ntziachristos, L.; Samaras, Z.; Casati, R.; Scheer, V.; Vogt, R. *Effect of Speed and Speed-Transition on the Formation of Nucleation Mode Particles from a Light Duty Diesel Vehicle*; SAE International: Warrendale, PA, USA, 2007; paper number 2007-01-1110.
69. Badshah, H.; Kittelson, D.; Northrop, W. Particle emissions from light-duty vehicles during cold-cold start. *Sae Int. J. Engines* **2016**, *9*, 1775–1785. [CrossRef]
70. Giechaskiel, B.; Drossinos, Y. Theoretical investigation of volatile removal efficiency of particle number measurement systems. *SAE Int. J. Engines* **2010**, *3*, 1140–1151. [CrossRef]
71. Giechaskiel, B.; Maricq, M.; Ntziachristos, L.; Dardiotis, C.; Wang, X.; Axmann, H.; Bergmann, A.; Schindler, W. Review of motor vehicle particulate emissions sampling and measurement: From smoke and filter mass to particle number. *J. Aerosol Sci.* **2014**, *67*, 48–86. [CrossRef]
72. Giechaskiel, B.; Melas, A.D.; Lähde, T.; Martini, G. Non-volatile particle number emission measurements with catalytic strippers: A review. *Vehicles* **2020**, *2*, 19. [CrossRef]
73. Lahde, T.; Giechaskiel, B.; Martini, G. *Development of Measurement Methodology for Sub 23 nm Particle Number (PN) Measurements*; SAE International: Warrendale, PA, USA, 2020; paper number 2020-01–2211.

sensors

MDPI

Article

Investigating a Low-Cost Dryer Designed for Low-Cost PM Sensors Measuring Ambient Air Quality

Abdul Samad *, Freddy Ernesto Melchor Mimiaga, Bernd Laquai and Ulrich Vogt

Department of Flue Gas Cleaning and Air Quality Control, Institute of Combustion and Power Plant Technology (IFK), University of Stuttgart, Pfaffenwaldring 23, 70569 Stuttgart, Germany; freddernesto@gmail.com (F.E.M.M.); bernd.laquai@ifk.uni-stuttgart.de (B.L.); ulrich.vogt@ifk.uni-stuttgart.de (U.V.)

* Correspondence: abdul.samad@ifk.uni-stuttgart.de

Abstract: Air pollution in urban areas is a huge concern that demands an efficient air quality control to ensure health quality standards. The hotspots can be located by increasing spatial distribution of ambient air quality monitoring for which the low-cost sensors can be used. However, it is well-known that many factors influence their results. For low-cost Particulate Matter (PM) sensors, high relative humidity can have a significant impact on data quality. In order to eliminate or reduce the impact of high relative humidity on the results obtained from low-cost PM sensors, a low-cost dryer was developed and its effectiveness was investigated. For this purpose, a test chamber was designed, and low-cost PM sensors as well as professional reference devices were installed. A vaporizer regulated the humid conditions in the test chamber. The low-cost dryer heated the sample air with a manually adjustable intensity depending on the voltage. Different voltages were tested to find the optimum one with least energy consumption and maximum drying efficiency. The low-cost PM sensors with and without the low-cost dryer were compared. The experimental results verified that using the low-cost dryer reduced the influence of relative humidity on the low-cost PM sensor results.

Keywords: low-cost sensor; PM sensor; air quality sensor; low-cost dryer; humidity influence on PM; air pollutants; air quality monitoring

1. Introduction

1.1. Air Quality and Low-Cost Sensors

During the last years, the importance of good ambient air quality has strongly increased around the world. According to the World Health Organization (WHO), nine out of ten people live in places where the air quality guidelines are not fulfilled and every year poor air quality is related to 4.2 million premature deaths [1]. This is due to its relation to many negative effects on human health not only regarding respiratory diseases like deterioration of lung function, worsening of asthma symptoms, allergic reactions and airway obstruction [2], but also stroke, heart diseases and cancer [1,3–5]. Additionally, the impacts are not only related to human beings, but also to the ecosystems and earth climate system [6,7].

Considering these air pollution impacts, the air quality monitoring is nowadays of great concern, because it provides the necessary information to develop and implement suitable methods to improve ambient air quality. For this purpose, municipalities and authorities install stationary air quality monitoring stations in specific locations based on a measurement strategy and criteria for setting up the monitoring stations, e.g., height of sample inlets, distance to crossings, distance to roads, number of people impacted by air pollution, etc. [8]. These monitoring stations are usually equipped with devices, which are checked for accuracy in accordance with standards. These devices provide precise and reliable data about the air quality situation of the area. However, these instruments are in general costly and require highly trained professionals for their maintenance and operation [9]. Thus, only a small number of monitoring stations can be installed in a large area, which limits the spatial resolution. A good spatial resolution is necessary especially in areas where the pollutant distribution is not homogeneous, because of the influence of different sources as in the case of an urban environment [10].

In order to solve that issue, different citizen science groups and research institutes started investigation on using the low-cost sensors for this purpose available mainly for indoor air quality measurements and compared the results with the reference instruments [11,12]. These investigations helped to develop the low-cost sensors in a way that they could also be used for ambient air quality. Several companies have started to produce low-cost sensors, which are able to measure the air quality with lower expenses for operation and maintenance, allowing the possibility to deploy in large numbers and create a detailed air pollution map. Nevertheless, this new technology has the disadvantage that it is highly affected by different meteorological parameters, resulting in lower data quality [13]. Therefore, if a suitable solution to this problem can be found, then the low-cost sensors have the potential to work as a good support for the current conventional air quality monitoring stations [14].

1.2. Particulate Matter and Its Classification

The airborne Particulate Matter (PM) suspended in the atmosphere is formed by either natural or anthropogenic sources. PM is one of the most important pollutants when investigating air pollution due to its great impact on the environment and human health [15].

Airborne PM is a complex mixture of solid and liquid particles that can be:

(a) Primary, emitted directly into the atmosphere from either natural process such as windblown dust, smoke from forest fires, volcanic eruptions or anthropogenic processes such as automobile exhaust, smoke from power plants, etc.
(b) Secondary, formed by chemical reactions of gaseous components.

The transportation and environmental impact of PM depends on factors such as originating sources, composition and size of the PM. The classification based on size distribution can predict the residence time in the air as well as the transportation distance. In terms of health impacts, it can estimate the deposition intensity in the respiratory system. Hence, the air quality policies and emission regulations propose the PM limit values according to size fractions.

The PM size fractions are mostly represented as PM10, PM2.5 and PM1. PM10 are inhalable particles that may reach the upper part of the airways and lungs, while PM2.5 and PM1 are inhalable as well, but they can easily penetrate the lungs and perhaps might reach deeper parts of the lungs such as alveoli. The official limit values of PM are typically available for PM10 (coarse particles) and PM2.5 (fine particles), because in these fractions the PM is small enough to be inhaled and respired [16]. The impacts that PM can imply on the environment are diverse. The vegetation can be altered by the deposition of PM to the vegetated surfaces, which is mainly influenced by the PM size distribution and to a small extent on the chemical composition of the PM. Some of the effects caused by the PM can be abrasion, radiative heating and reduction of photosynthetic activity. In addition, the alkaline and acidic components may damage the surface or be absorbed through the cuticle [7].

In Figure 1, the most common PM and the diameter range in which they can be found is shown.

It is very important to understand that the size, composition and concentration of PM strongly depend on the local activities, meteorological conditions and the PM itself. The size, number and chemical compositions can be transformed by several mechanisms until the PM is removed from the atmosphere. The PM is classified into three groups depending on the path they follow once they are formed as shown in Figure 2 [15].

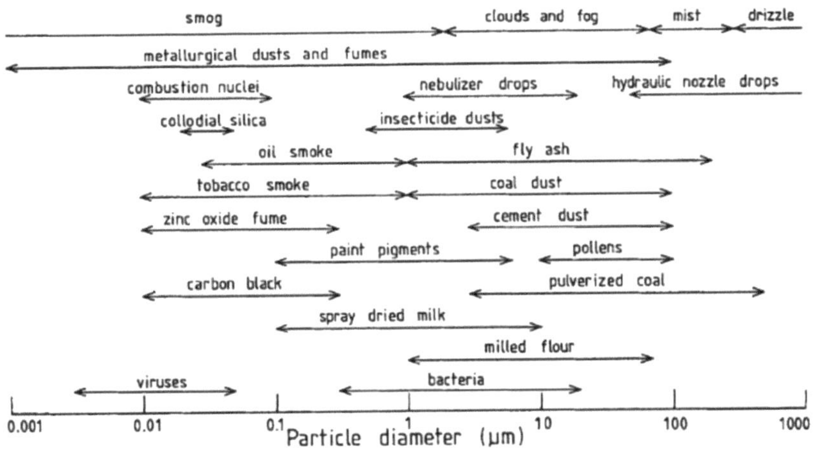

Figure 1. Particle size range for different airborne Particulate Matter (PM) [15].

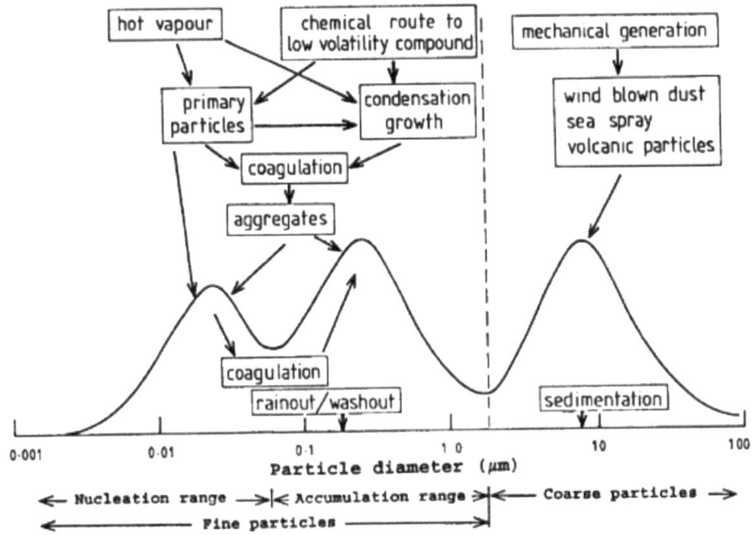

Figure 2. Schematic diagram of a typical size distribution and formation mechanisms for atmospheric particles [15].

(a) Nucleation mode (<0.2 μm diameter): Emitted from processes involving condensation of hot vapors or through gas to particle conversion in the atmosphere.
(b) Accumulation mode (0.2–2 μm diameter): These are grown from nucleation mode by coagulation or condensation of vapors.
(c) Coarse mode (>2 μm diameter): Formed by mechanical abrasion processes (soil dust, sea spray and many industrial dusts fall).

Fine particles are characterized by their etiology, their ability to remain suspended in the air and to carry material that is absorbed on their surface. The smaller the particle diameter, the longer it remains suspended in the air and the more hazardous it is [17].

1.3. Influence of Relative Humidity on Low-Cost PM Sensors

In many recent studies, it has been demonstrated that the low-cost PM sensors are highly affected when they work under humid conditions and overestimate the actual PM concentration [10,18–20]. The measurement principle for most of these low-cost PM sensors is light scattering principle with which the particles are counted. The number of particles is then converted to the PM mass concentration. The error occurs because the low-cost PM sensor counts not only the dry particles but also the wet water droplets that can happen at high levels of relative humidity and occur through condensation of the water vapor [19–21].

In 2018, Jayaratne et al. studied the behavior of the low-cost PM sensor model PMS1003 from the company Plantower. This low-cost PM sensor was tested against reference instruments in the laboratory and under field conditions. During the field experiments, there was fog formation in the early morning and the low-cost PM sensor considered these small water droplets as PM. Furthermore, the authors stated that a strong hygroscopic growth rate of PM mass was observed. The authors found that under high relative humidity levels, the low-cost PM sensors show a concentration 46% higher than the reference [19].

Akpootu and Gana presented the results obtained after the observation of hygroscopic growth on water soluble aerosols in 2013. The authors showed that at low relative humidity levels (below 50%) the aerosols do not show a significant increment in their size. However, at higher relative humidity levels, this effect is much more pronounced and can be expressed as an exponential curve [22].

Another recent study published regarding this topic was from the authors Brattich et al. that focused on two long-term measurement campaigns in order to compare the correlation between different low-cost PM sensors and reference instruments using different statistical approaches. In this publication, it was found that all low-cost PM sensors are highly affected during misty, cloudy and foggy conditions [10].

By keeping this problem in mind, the aim of this research was to develop and assess the performance of a low-cost dryer, which should be able to reduce or eliminate the negative effect of high relative humidity conditions on the ambient air quality measurements using low-cost PM sensors.

2. Measurement Technique and Methodology

2.1. Low-Cost PM Sensors and Reference Instruments

For the selection of the low-cost PM sensor for this research, previous studies performed by the authors and other researchers as well as the technical features and price of the low-cost PM sensors available in the market was taken into account. Numerous investigations were performed to compare different low-cost PM sensors with professional instruments in order to find the most suitable one for ambient air quality measurements [20,23,24]. Calibration models were also developed to improve the data quality obtained from these low-cost sensors [25]. The authors also tested well-known low-cost PM sensors such as SDS011 (Nova Fitness), PMS5003 (Plantower), OPC-N2 and OPC-N3 (Alphasense) to use them for ambient air quality measurements. OPC-N3 from the company Alphasense was chosen for this research as it showed better results compared to other low-cost PM sensors [26,27].

This low-cost PM sensor has a measurement size range from 0.35 µm to 40 µm sorting into 24 size bins. The time resolution of this low-cost PM sensor is one second and it provides a real-time histogram as well as the flowrate of the sampling air. This low-cost PM sensor also measures the temperature and relative humidity of the measurement chamber [28].

The reference instrument used to compare the results obtained from the low-cost PM sensor was an aerosol spectrometer, model EDM180 from the company Grimm Aerosol Technik Ainring GmbH & Co. KG. This device is able to measure the particle size range from 0.23 µm to 32 µm in 31 different size bins. It also measures ambient air temperature and relative humidity using a climate sensor. This device is equipped with a Nafion membrane-based dryer [29].

2.2. Experimental Setup

The experimental setup was designed to simulate the ambient air conditions on a laboratory scale. The experimental setup is shown in Figure 3. A test chamber was prepared, in which the measurements took place. The reference devices and the low-cost PM sensors along with particle generation and relative humidity control system were installed. The test chamber should be placed around 2 m above the ground as only the inlets of the reference devices and the low-cost PM sensors should be present inside the test chamber. For this reason, a metallic support that is shown in red color in Figure 3 was used to carry the test chamber. During the experiments, two low-cost PM sensors were installed in the test chamber. One of the low-cost PM sensors was equipped with a low-cost dryer (N3+dryer) and the other one without a low-cost dryer (N3). These low-cost sensors were compared with two reference instruments. One of these reference instruments was operated with a dryer (RI+dryer) and the other one without it (RI).

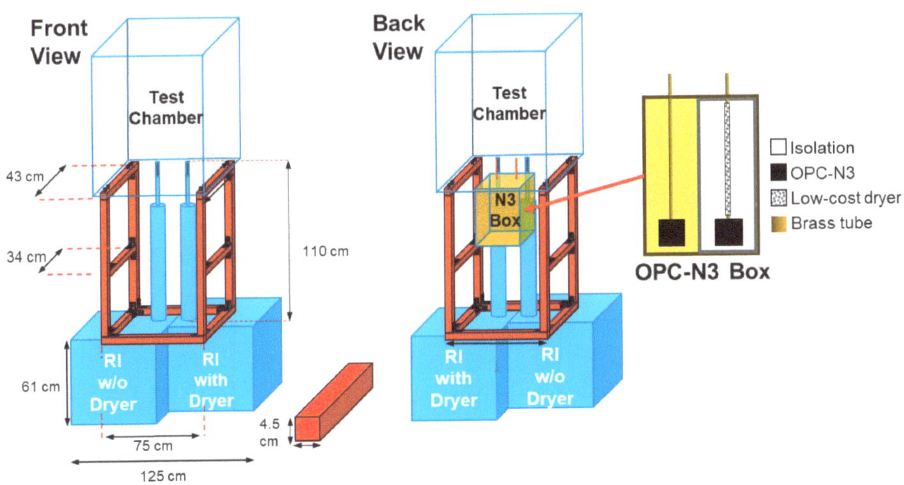

Figure 3. Experimental setup for laboratory measurements.

The following equipment were placed inside the test chamber as it can be seen in the Figure 4.

- Vaporizer to simulate the humid conditions in the test chamber.
- Ventilator to homogenize the PM distribution inside the box.
- Inlet of the reference instruments and low-cost PM sensors.
- Temperature and relative humidity sensors.

The low-cost dryer for the low-cost PM sensors was developed to reduce the relative humidity by heating the sampling air so that the influence of high relative humidity on the measured PM concentration can be reduced or eliminated. The low-cost dryer consisted of a thermally conductive brass inner tube on which winding of a metallic coil made from a resistive material such as constantan was done. A ceramic foil was put in between the inner tube and metallic coil to distribute the heat evenly throughout the inner tube. The heat is conducted through the inner tube to the sampling air. The thermally conductive properties of the ceramic foil allowed the proper heat transfer to the inner tube. The inner tube was insulated using an isolation foam to avoid any thermal loss and an outer tube was used to keep the low-cost dryer stable and to protect it from any mechanical destruction from outside. In Figure 5, the schematic diagram of the low-cost dryer developed during this research is shown.

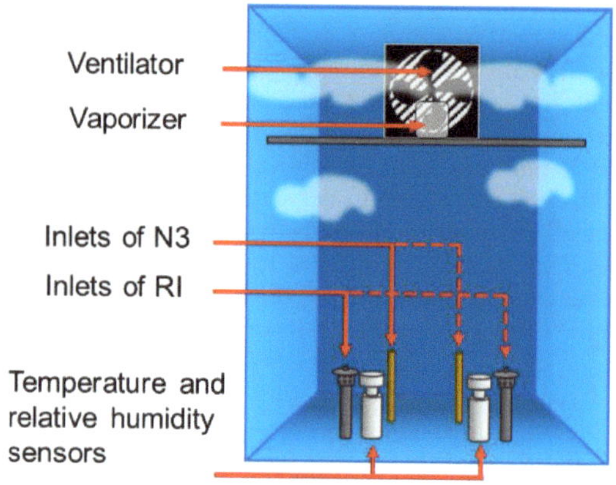

Figure 4. Test chamber configuration for experiments at different voltages.

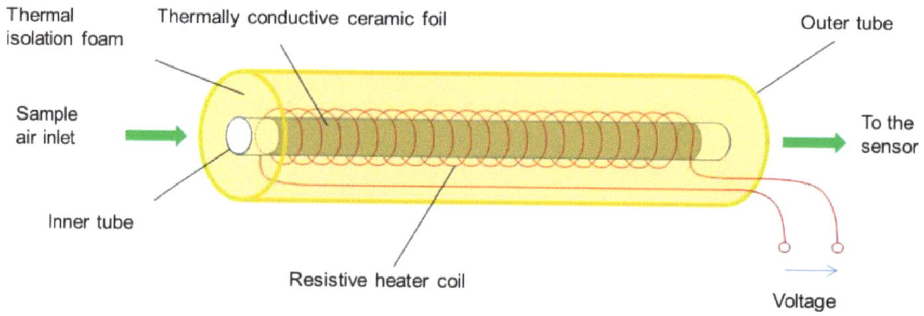

Figure 5. Schematic diagram of the low-cost dryer.

The temperature of the low-cost dryer was controlled by adjusting the voltage applied to the metallic coil. For this system, a voltage range of 5 to 9 V was tested for this setup to find the optimum voltage for the operation of low-cost dryer. Using 5 windings per cm and 5-watt power on a 45 cm active surface of the dryer, it was possible to ensure a minimum of 5 volts required for the setup. The equations below show the corresponding formulas applied with a specific resistance of 0.97 ohms per meter for the metallic coil used in this project. The calculation results are shown in Table 1.

$$R = V^2/P \tag{1}$$

$$I = V/R \tag{2}$$

Table 1. Calculations for the low-cost dryer.

Tube		Winding				Wire	Resistance		Low-Cost Dryer		
Diameter (mm)	Length (cm)	Contour (cm)	Length (cm)	Density (cm)	Total length (cm)	Length (m)	Specific (Ω/m)	Total (Ω)	Power (W)	Current (mA)	Voltage (V)
8	50	2.5	45	5	225	5.65	0.97	5.5	5	952	5.2

2.3. Methodology

This research was categorized in three steps. The first step consisted of setting up the test chamber for the experiments using the low-cost PM sensors with and without low-cost dryer and the reference instruments. Once the experimental setup was finalized, the second step was to find the optimum voltage to be used for the low-cost dryer. These experiments are further explained in Section 2.3.1. After finding the optimum voltage for the low-cost dryer, the third step was to check the performance and the efficiency of the low-cost dryer. Experiments were also performed to investigate the influence of low-cost dryer heating on the PM. In these experiments synthetic dust (Eskal14) from the company KSL Staubtechnik GmbH was used as particles and the experimental design was adjusted accordingly. These experiments are further explained in Section 2.3.2.

The experiments were carried out following a certain pattern containing different phases. This allowed to have a systematic analysis to test the performance of the low-cost dryer. The phases during the experiment were classified as following:

- Stabilizing phase: The phase in which the conditions inside the test chamber were allowed to stabilize after switching on the equipment.
- PM concentration increase phase: The phase in which the PM concentration was increased in the test chamber using a vaporizer and/or a particle distributor.
- Settling phase: The phase in which the PM concentration was allowed to settle after particle generation.
- Low-cost drying phase: The phase in which the low-cost dryer installed on one of the two low-cost PM sensors (N3+dryer) was activated.
- Final phase: The phase in which the low-cost dryer was switched off and the instruments were allowed to run for some more time.

2.3.1. Experiments to Determine the Optimum Voltage for Low-Cost Dryer Operation

The experiments to obtain the optimum voltage for dryer operation followed the pattern in which the stabilizing phase took place for 15 min. After this time, the vaporizer was switched on for three minutes, reaching a considerable PM concentration. The settling phase was seven minutes. Then the low-cost dryer installed at one of the low-cost PM sensors (N3+dryer) was switched on, while the other low-cost PM sensor was operated without the dryer (N3). The dryer was switched off after 20 to 30 min. In that way, it was possible to study and evaluate the effectivity of the low-cost dryer. The theoretical temporal variation of an ideal PM concentration curve during a standard experiment is shown graphically in Figure 6.

The above experiment was run at 6, 6.5, 7, 7.5, 8 and 9 volts. The optimum voltage was selected by comparing the ratio between the low-cost PM sensor without the low-cost dryer to the one with the dryer. This optimum voltage setting should be found to correctly dimension the low-cost dryer. A regulated thermal energy based on the sample air temperature and relative humidity could be used for the low-cost dryer operation.

2.3.2. Experiments with Synthetic Dust

These experiments were conducted using synthetic dust with a particulate size distribution between 1 μm and 10 μm. The synthetic dust used for these experiments was a temperature resistant calcium carbonate "Eskal14". This synthetic dust was chosen because of its narrow particle size distribution, excellent fluidity and suitability for wet applications [30]. The PM concentration generated during these experiments could not be calculated and hence it should be known only by means of the reference instruments. These experiments were performed using the optimum voltage applied to the low-cost dryer that was found in the previous set of experiments. The test chamber configuration for the experiments with synthetic dust is shown in Figure 7.

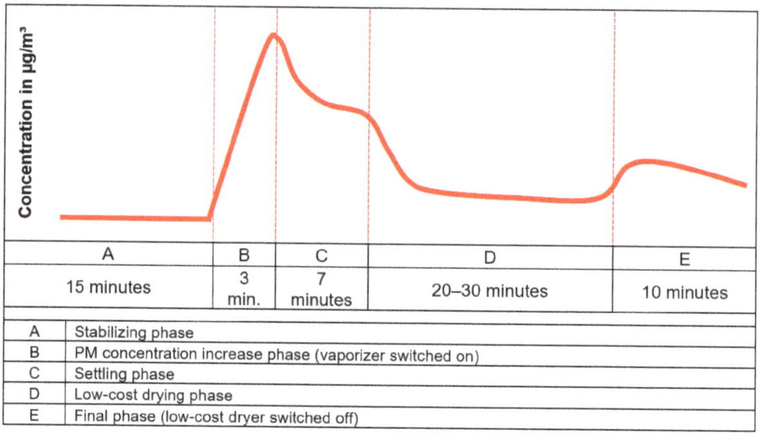

Figure 6. Temporal variation of an ideal PM concentration curve during a standard experiment.

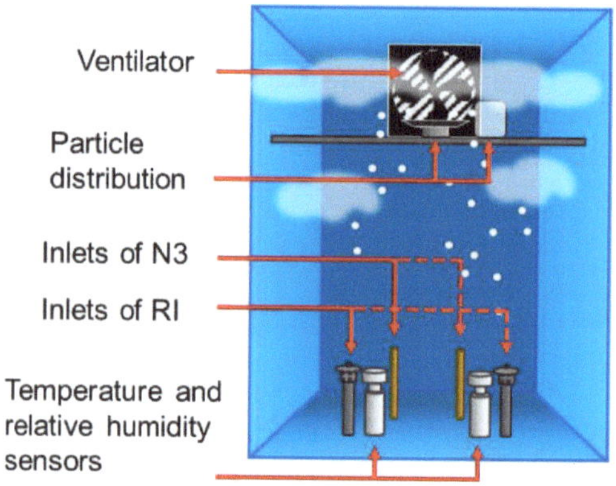

Figure 7. Test chamber configuration for the experiments with synthetic dust.

The experiments with synthetic dust were divided in two parts. In the first part, the experiments were performed using synthetic dust only. In the second part, the experiments were done using both synthetic dust and the vaporizer. These set of experiments were compared with each other to observe the behavior of PM concentration with and without the addition of water vapors through the vaporizer in the presence of synthetic dust. The experimental pattern was similar to previous experiments. However, the particle distribution technique and the experimental time were modified. The duration of these experiments was different due to different settling time of the synthetic dust particles with and without the vaporizer. The temporal variation of the PM concentrations while using only the synthetic dust is shown graphically in Figure 8 and the temporal variation of the PM concentrations while using the synthetic dust and vaporizer is shown graphically in Figure 9.

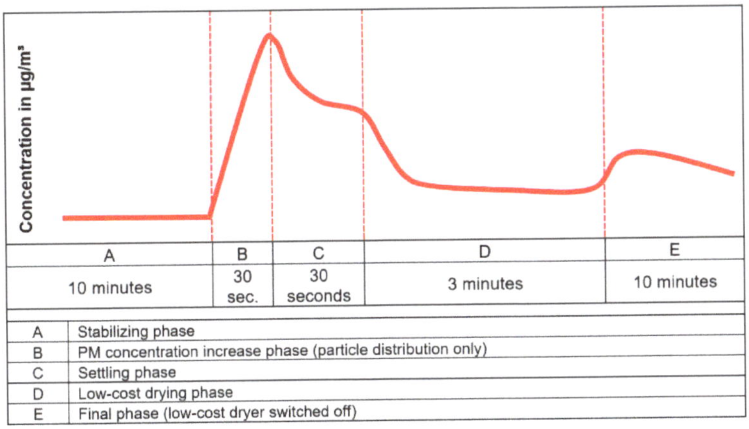

Figure 8. Temporal variation of an ideal PM concentration curve using synthetic dust only.

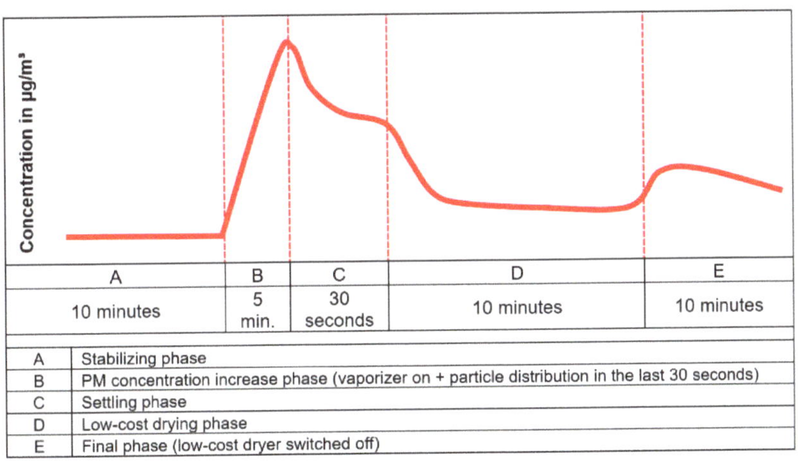

Figure 9. Temporal variation of an ideal PM concentration curve using synthetic dust and vaporizer.

3. Results and Discussion

In this section, the results obtained from the set of experiments mentioned in the previous section are shown.

3.1. Optimum Voltage for the Dryer Operation

As mentioned before, the heating of the low-cost dryer was controlled by applying different voltages. If the applied heating is less than required, then the low-cost dryer would not be able to efficiently remove the relative humidity effects on the results of the low-cost PM sensor. If the applied heating is more than required, then there is a chance that some part of the PM is evaporated that can lead to an underestimation of the PM concentration. By keeping that in mind, different voltages were tested during this research namely 6, 6.5, 7, 7.5, 8 and 9 volts. The results of 7, 8 and 9 volts are presented in Figures 10–12, respectively. These experiments were performed according to the procedure explained in Figure 6.

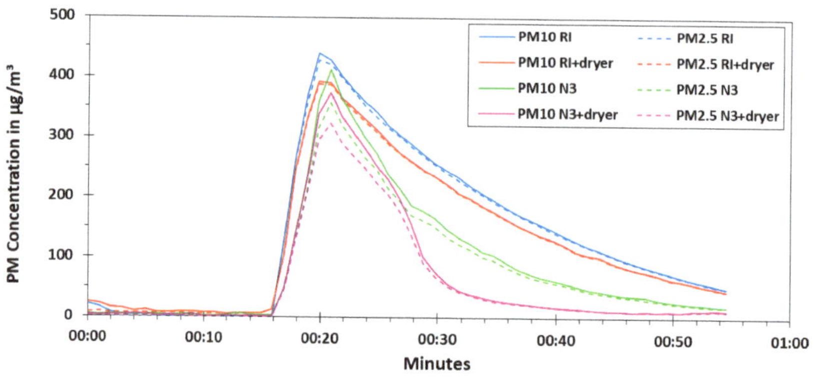

Figure 10. PM concentration comparison for testing the low-cost dryer at an applied voltage of 7 V.

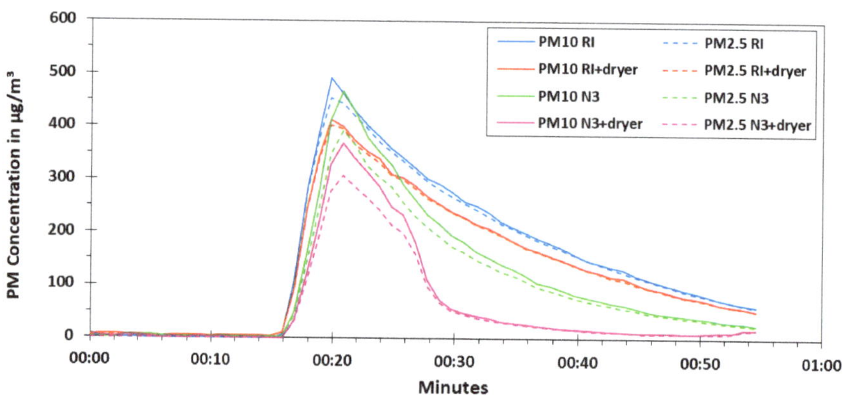

Figure 11. PM concentration comparison for testing the low-cost dryer at an applied voltage of 8 V.

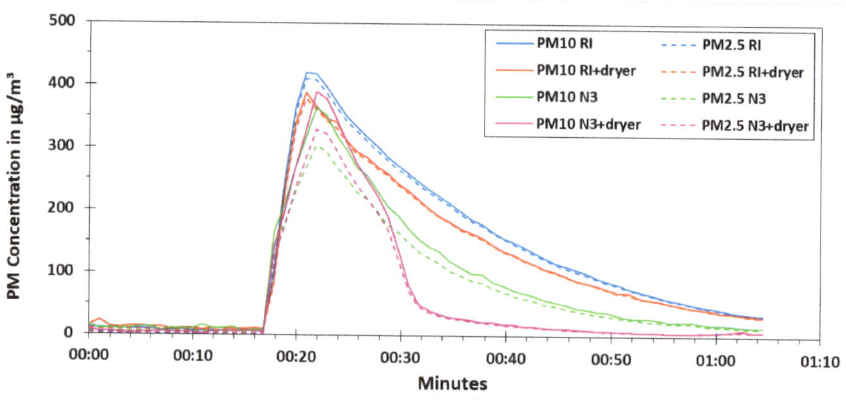

Figure 12. PM concentration comparison for testing the low-cost dryer at an applied voltage of 9 V.

In Figure 10, the experiment using the applied voltage of 7 V is shown. At the start of the experiment, the PM concentrations were allowed to stabilize in the test chamber for the first 15 min (stabilizing phase). The PM concentration measured by the low-cost PM sensors and the reference instruments at the end of this phase were below 10 $\mu g/m^3$. After switching the vaporizer on, the PM concentration measured by the low-cost PM sensors as well as the reference instruments increased. It is noticeable that the PM10 and PM2.5 concentrations measured by the low-cost PM sensors and the reference instruments were similar. This indicates that the particles (water vapors) generated by the vaporizer were fine and the majority of these were below PM2.5 fraction. The peak PM10 and PM2.5 concentrations measured by the low-cost PM sensors and the reference instruments were in the range of 350 to 450 $\mu g/m^3$ and 325 to 425 $\mu g/m^3$, respectively. Even though a ventilator was used to distribute the particles homogeneously in the test chamber, still small variation in the PM concentrations measured by the low-cost PM sensors and the reference instruments was observed, which was assumed to be due to different inlet positions of these devices. The relative humidity was also increased in the test chamber during the operation of the vaporizer. After the vaporizer was switched off, the PM concentration measured by the low-cost PM sensors and the reference instruments started to decrease. The particles were allowed to settle before the low-cost dryer was switched on for one of the low-cost PM sensors. After around 25 min from the start of the experiment, the low-cost dryer was switched on for one of the low-cost PM sensors (N3+dryer). A significant decrease in PM concentrations can be observed for N3+dryer as compared to the low-cost PM sensor without the low-cost dryer (N3). At around 30 min from the start of the experiment, the PM concentration measured by N3 was almost double than the PM concentration measured by N3+dryer. The reference instruments were operated with and without the dryer from the start of the experiment as it was not possible to change the dryer settings for reference instruments during the experiment. The PM concentration measured by the reference instrument operating with the dryer (RI+dryer) was slightly lower than the PM concentration measured by the reference instrument operating without the dryer (RI). It is interesting to see that after increasing the PM concentration using the vaporizer, the PM concentration measured by RI+dryer had a similar concentration decline curve as the one measured by RI. It is assumed that since the reference instrument dryer works on a different principle (Nafion membrane) than heating to reduce the relative humidity, it does not instantly dry out the artificially generated particles using vaporizer. In the final phase, after switching off the low-cost dryer for N3+dryer at around 50 min from the start of the experiment, the PM concentration measured by N3+dryer increased slightly and nearly matched the PM concentration measured by N3.

The results obtained from the experiments performed for testing the other voltages had the same pattern. However, by increasing the applied voltage, the decline concentration curve for N3+dryer became steeper. Figures 11 and 12 show the experiments using the voltage of 8 V and 9 V, respectively.

The behaviors of temperature and relative humidity were also observed throughout these experiments. These two parameters were measured at the inlet of the instruments using the climate sensor of the reference instruments. Apart from that, a temperature and relative humidity sensor is enclosed in the raw housing of the low-cost PM sensors. These measurements assist in understanding the performance of low-cost dryer during its operation. The experiments were performed at different temperature and relative humidity levels. As an example, the results of temperature and relative humidity for the low-cost sensor N3+dryer in the course of 8 V applied voltage experiment are shown in Figure 13. The temperature and relative humidity at the inlet of N3+dryer is shown in Figure 13 as solid orange and dashed orange lines, respectively. The temperature at the inlet of N3+dryer was constant at around 22 °C for the whole experiment. The relative humidity was marginally above 40% at the start of the experiment. A rise in relative humidity can be seen after 15 min from the start, when the vaporizer was switched on. For this experiment, the relative humidity peak was slightly above 50%. The vaporizer was switched off after three minutes of operation and the relative humidity started to decline after that. The temperature and relative humidity inside N3+dryer are shown in Figure 13 with solid pink and dashed pink lines, respectively. The temperature inside N3+dryer was somewhat below 30 °C that was moderately higher than the one at the inlet of N3+dryer while the relative humidity was lower compared to the N3+dryer inlet. This can be because of the working of electronics and mechanical parts inside N3+dryer. The increase in N3+dryer inside temperature can be observed after 30 min from the start of the experiment once the low-cost dryer is switched on. The peak temperature of approximately 37 °C was achieved at around 50 min from the start of the experiment. A slight decrease in this temperature is to be seen at the end of the experiment after switching off the low-cost dryer. The increase in N3+dryer inside relative humidity is also visible after the vaporizer is switched on. However, it is reduced considerably during the drying phase and it reached to a minimum value of below 20% which is even lower than the inside relative humidity of N3+dryer at the start of the experiment. After switching off the low-cost dryer, a minor increase in N3+dryer inside relative humidity was observed at the end of the experiment.

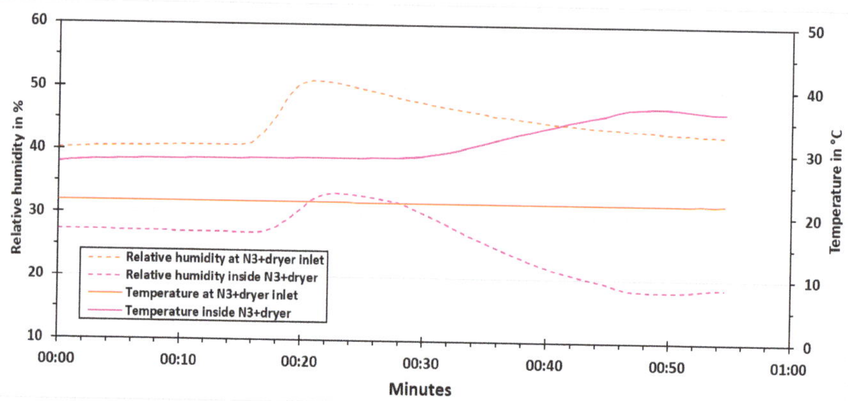

Figure 13. Temperature and relative humidity results for N3+dryer at an applied voltage of 8 V.

In order to understand the effectivity of the low-cost dryer, the ratios of PM concentrations measured by N3 and N3+dryer were calculated. Table 2 shows the results of the PM concentration ratios of N3 to N3+dryer at different voltages. These ratios were calculated when the low-cost dryer for N3+dryer was switched on during the low-cost drying phase. For a complete comparison, the ratios were obtained at three different points during the low-cost drying phase namely the start, mid and end of this phase. The results show that the average PM10 and PM2.5 concentration ratio of N3 to N3+dryer during the low-cost drying phase is in the range of 2 to 3 for the applied voltage of 6, 6.5, 7 and 7.5 V. There is a significant increase in PM concentration ratio for 8 V, where it reaches the value of around 4, which is the highest value for the tested voltages applied. Hence, the optimum voltage applied to the low-cost dryer for its operation was decided to be 8 V for further experiments using the synthetic dust.

3.2. Experiments with Synthetic Dust

These experiments were performed using the synthetic dust with the optimum voltage (8 V) for the low-cost dryer found in the previous experiments. These experiments were carried out using two different methods. The first method was to execute the experiment using only synthetic dust without using the vaporizer as mentioned before in Figure 8, while the other method included both synthetic dust and vaporizer that is explained in Figure 9. The settling time of the synthetic dust particles was much faster than particles produced by the vaporizer. Therefore, the duration of these experiments was reduced. These experiments lasted between 25 to 35 min.

Table 2. PM concentration ratios of N3 to N3+dryer at different voltages.

Low-Cost Drying Phase	6 V		6.5 V		7 V		7.5 V		8 V		9 V	
	PM10	PM2.5	PM10	PM2.5	PM10	PM2.5	PM10	PM2.5	PM10	PM2.5	PM10	PM2.5
Start	1.7	1.7	1.8	1.8	2.2	2.2	2.4	2.3	3.5	3.4	1.4	1.4
Mid	3.0	2.8	3.2	3.2	3.4	3.3	3.4	3.3	4.6	4.4	4.5	4.3
End	2.5	2.3	2.9	2.8	2.5	2.4	2.7	2.9	4.1	3.8	5.0	4.3
Average	2.4	2.3	2.6	2.6	2.7	2.6	2.8	2.8	4.1	3.9	3.6	3.3

In the experiment without vaporizer, shown in Figure 14, the stabilizing phase was for around 10 min. After that, the particles (synthetic dust) were distributed in the test chamber for around 30 s. The synthetic dust particles were coarser than PM2.5 fraction. This is the reason that PM10 concentration measured by the low-cost PM sensors was increased during the particle distribution, while no change in PM2.5 concentration was observed during this phase. Since the PM10 concentration declined rapidly after the particle distribution, therefore the settling phase was very short for around 30 s. A momentary increase in PM concentration measured by the N3+dryer as compared to the N3 during the start of particle distribution was observed, which was assumed to be due to non-homogeneous particle distribution in the start. This momentary concentration difference disappeared rapidly in the settling phase. After that, the low-cost dryer was activated for N3+dryer for around 3 min. There was no significant difference of PM10 concentration measured by the low-cost sensors during the low-cost drying phase, which was expected as the relative humidity in the test chamber was not increased using vaporizer. This experiment showed that the particles were not destroyed from the heat of the low-cost dryer. It also showed that in the absence of water vapors, the N3 and the N3+dryer measured almost the same PM concentration. Hence, the low-cost dryer does not reduce the PM concentration in dry conditions.

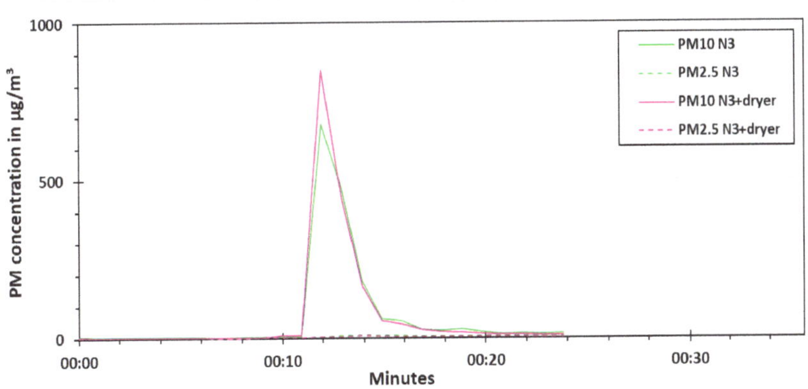

Figure 14. PM concentration for the synthetic dust experiment without vaporizer.

The same experiment was performed again with vaporizer for which the results are shown in Figure 15. The stabilizing phase was again for 10 min. In the PM concentration increase phase, the vaporizer was switched on for 5 min. The PM10 and PM2.5

concentration measured by the low-cost sensors increased to around 250 μg/m^3 due to the vaporizer. In the last 30 s of this phase, the synthetic dust was distributed in the test chamber. This caused an increase in PM10 concentration for both the low-cost PM sensors in the range of 1600 to 1800 μg/m^3. The particles were allowed to settle for 30 s. After that the low-cost dryer was switched on for N3+dryer for 10 min. The effect of low-cost dryer is evident from the PM concentration comparison for N3 and N3+dryer. The low-cost dryer swiftly dried out the particles (water vapors) from the vaporizer, while the synthetic dust particles remained that sedimented quickly due to gravity. After switching off the low-cost dryer, the PM concentration slightly increased as it was observed in the previous experiments with vaporizer.

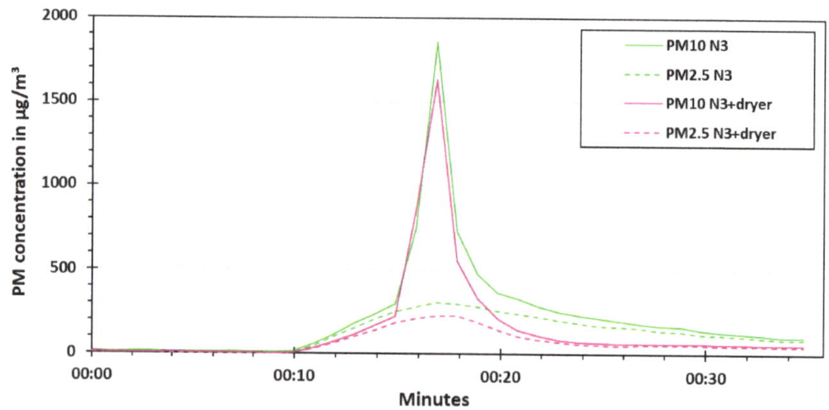

Figure 15. PM concentration for the synthetic dust experiment with vaporizer.

4. Quality Assurance

Quality assurance was done to improve the reliability of the results obtained from the experiments. A comparison of the two reference instruments as well as the two low-cost PM sensors was performed before the experiments as a quality assurance. This assured that the results obtained from the reference instruments as well as the low-cost PM sensors are comparable to each other. The reference instruments and the low-cost sensors were tested by increasing the PM concentration using the vaporizer and then letting it settle. This procedure was repeated 10 times. A linear regression correction was applied to the data, which was then used for the whole experiments to have a valid comparison of the devices.

In Figure 16, the results of PM concentrations of the reference instruments during quality assurance with the same dryer settings, i.e., dryer switched off, are shown. The particle size distribution during the quality assurance was below 2.5 μm. Therefore, the results of PM10 and PM2.5 are overlapping. It can be seen from the results that both the reference instruments follow the same pattern during the peak as well as during the fall of the PM concentration.

The low-cost PM sensors were also tested in the same way as the reference instruments. In Figure 17, the PM concentrations of the low-cost PM sensors during quality assurance with the same dryer settings, i.e., dryer switched off, are presented. The results here are similar to the ones obtained with the reference instruments. The PM10 and PM2.5 concentrations are similar as the particle size distribution lies below 2.5 μm.

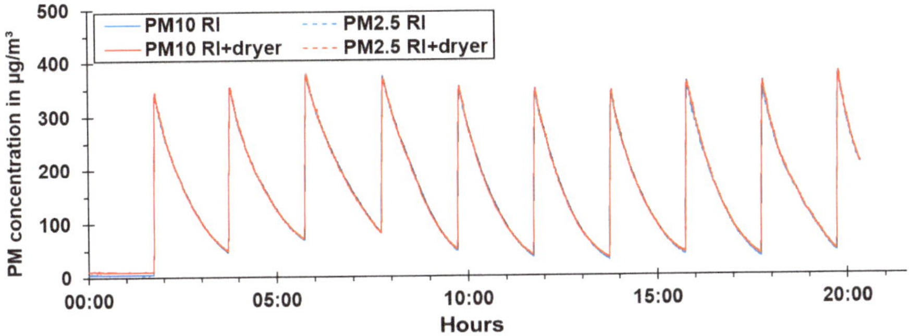

Figure 16. PM concentrations of the reference instruments during quality assurance.

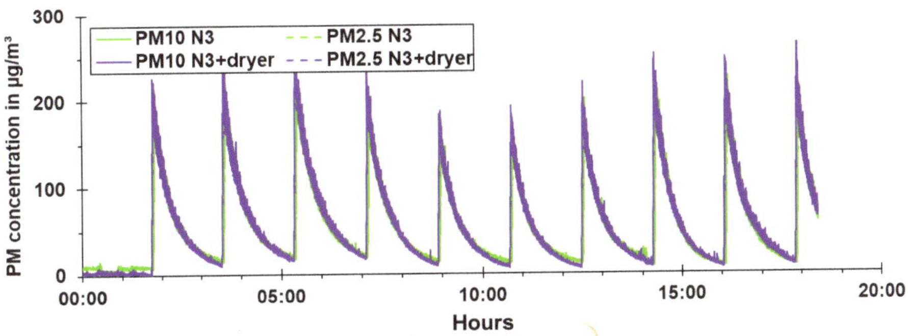

Figure 17. PM concentrations of the low-cost PM sensors during quality assurance.

In Figures 18 and 19, the correlation of the reference instruments with the low-cost PM sensors for both PM10 and PM2.5 concentrations is shown. In Figure 18a,b, the comparison of PM10 concentration of RI and RI+dryer respectively to the PM10 concentration of both low-cost PM sensors (N3 and N3+dryer) with same dryer settings are shown. It can be seen that both low-cost PM sensors showed decent correlation of above 90% with the reference instruments (RI and RI+dryer) for PM10 concentration.

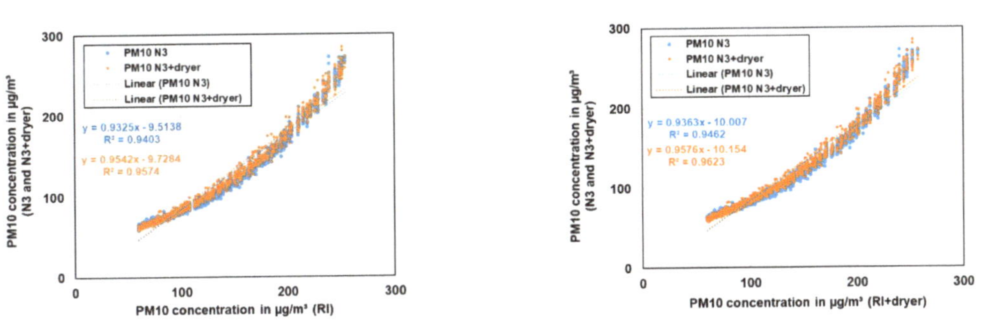

(a) (b)

Figure 18. (**a**) Comparison of PM10 concentration of the reference instrument (RI) to the PM10 concentration of both low-cost PM sensors (N3 and N3+dryer) with the same dryer settings. (**b**) Comparison of PM10 concentration of RI+dryer to the PM10 concentration of both low-cost PM sensors (N3 and N3+dryer) with the same dryer settings.

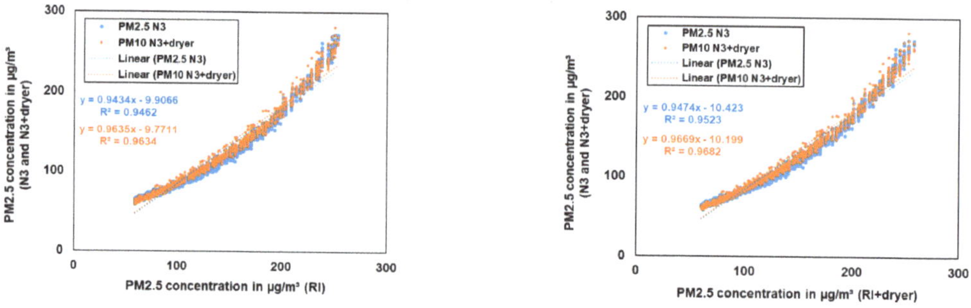

(a) (b)

Figure 19. (**a**) Comparison of PM2.5 concentration of RI to the PM2.5 concentration of both low-cost PM sensors (N3 and N3+dryer) with the same dryer settings. (**b**) Comparison of PM2.5 concentration of RI+dryer to the PM2.5 concentration of both low-cost PM sensors (N3 and N3+dryer) with the same dryer settings.

The PM2.5 concentrations were compared in a similar way. In Figure 19a, the comparison of PM2.5 concentration of RI to the PM2.5 concentration of both low-cost PM sensors (N3 and N3+dryer) is displayed, while in Figure 19b the comparison of PM2.5 concentration of RI+dryer to the PM2.5 concentration of both low-cost PM sensors (N3 and N3+dryer) with same dryer settings is shown. A correlation of around 95% was observed considering PM2.5 concentration for both low-cost PM sensors with the reference instruments (RI and RI+dryer).

5. Conclusions

It was concluded that the low-cost dryer is suitable for the application of measuring PM concentration using low-cost PM sensors. The low-cost dryer is able to eliminate the negative effects of relative humidity on the PM results measured by the low-cost PM sensors. The PM concentration comparison of the low-cost PM sensor with and without the low-cost dryer indicated that the low-cost dryer could dry out the water vapors generated from the vaporizer.

For operating the low-cost dryer with the low-cost PM sensor, the applied voltage controlled the heat applied to the low-cost dryer. The increase in applied voltage had a direct relation with the heat applied to the low-cost dryer. However, a significant increase of PM concentration ratio (PM concentrations measured by the low-cost PM sensor without the low-cost dryer to PM concentrations measured by the low-cost PM sensor with the low-cost dryer) was observed for the experiment with applied voltage of 8 V. This applied voltage helped to correctly dimension the low-cost dryer. A regulated thermal energy based on the sample air temperature and relative humidity could be used for the low-cost dryer operation.

The heat applied to the sample air through the low-cost dryer was adequate ensuring that there is no particle loss due to heating. No significant difference in the PM concentrations was observed by applying the heat and without it for the experiment with synthetic dust only, which indicated that low-cost dryer does not destroy the PM. The experiments with the synthetic dust demonstrated that the low-cost dryer should be suitable for measurements in real conditions.

The measurement technique and methodology presented in this research can be applied as it is or with some modifications to investigate the effect of the low-cost dryer on other low-cost PM sensors as a future work. The length of the low-cost dryer can be optimized for smaller platforms in order to make it more practical for field measurements.

Author Contributions: Conceptualization, A.S., B.L. and U.V.; methodology, A.S., F.E.M.M., B.L. and U.V.; software, A.S., F.E.M.M. and B.L.; validation, A.S. and F.E.M.M.; formal analysis, F.E.M.M.; investigation, A.S. and F.E.M.M.; resources, A.S., B.L. and U.V.; data curation, A.S. and F.E.M.M.; writing—original draft preparation, A.S. and F.E.M.M.; writing—review and editing, A.S., B.L. and U.V.; visualization, A.S. and F.E.M.M.; supervision, A.S., B.L. and U.V.; project administration, A.S. and U.V.; funding acquisition, U.V. All authors have read and agreed to the published version of the manuscript.

Funding: This work was performed under the project Urban Climate Under Change [UC]^2 funded by the Federal Ministry of Education and Research (BMBF) Germany.

Institutional Review Board Statement: Not applicable.

Informed Consent Statement: Not applicable.

Data Availability Statement: Not applicable.

Conflicts of Interest: The authors declare no conflict of interest.

References

1. World Health Organization. Geneva. (2018): 9 out of 10 People Worldwide Breathe Polluted Air, but More Countries Are Taking Action. 2018. Available online: https://www.who.int/news-room/detail/02-05-2018-9-out-of-10-people-worldwide-breathe-polluted-air-but-more-countries-are-taking-action (accessed on 6 November 2020).
2. Liu, H.-Y.; Dunea, D.; Iordache, S.; Pohoata, A. A Review of Airborne Particulate Matter Effects on Young Children's Respiratory Symptoms and Diseases. *Atmosphere* **2018**, *9*, 150. [CrossRef]
3. Raaschou-Nielsen, O.; Beelen, R.; Wang, M.; Hoek, G.; Andersen, Z.J.; Hoffmann, B.; Stafoggia, M.; Samoli, E.; Weinmayr, G.; Dimakopoulou, K.; et al. Particulate Matter Air Pollution Components and Risk for Lung Cancer. *Environ. Int.* **2016**, *87*, 66–73. [CrossRef] [PubMed]
4. Raaschou-Nielsen, O.; Andersen, Z.J.; Beelen, R.; Samoli, E.; Stafoggia, M.; Weinmayr, G.; Hoffmann, B.; Fischer, P.; Nieuwenhuijsen, M.; Brunekreef, B.; et al. Air Pollution and Lung Cancer Incidence in 17 European Cohorts: Prospective Analyses from the European Study of Cohorts for Air Pollution Effects (ESCAPE). *Lancet Oncol.* **2013**, *14*, 813–822. [CrossRef]
5. Bentayeb, M.; Wagner, V.; Stempfelet, M.; Zins, M.; Goldberg, M.; Pascal, M.; Larrieu, S.; Beaudeau, P.; Cassadou, S.; Eilstein, D.; et al. Association between Long-Term Exposure to Air Pollution and Mortality in France: A 25-Year Follow-Up Study. *Environ. Int.* **2015**, *85*, 5–14. [CrossRef] [PubMed]
6. Fuzzi, S.; Baltensperger, U.; Carslaw, K.; Decesari, S.; Van Der Gon, H.D.; Facchini, M.C.; Fowler, D.; Koren, I.; Langford, B.; Lohmann, U.; et al. Particulate Matter, Air Quality and Climate: Lessons Learned and Future Needs. *Atmos. Chem. Phys. Discuss.* **2015**, *15*, 8217–8299. [CrossRef]
7. Grantz, D.; Garner, J.; Johnson, D. Ecological Effects of Particulate Matter. *Environ. Int.* **2003**, *29*, 213–239. [CrossRef]
8. European Union. Sampling Points for Air Quality—Representativeness and Comparability of Measurement in Accordance with Directive 2008/50/EC on Ambient Air Quality and Cleaner Air for Europe. 2019. Available online: https://www.europarl.europa.eu/RegData/etudes/STUD/2019/631055/IPOL_STU(2019)631055_EN.pdf (accessed on 15 November 2020).
9. Karagulian, F.; Gerboles, M.; Barbiere, M.; Kotsev, A.; Lagler, F.; Borowiak, A. *Review of Sensors for Air Quality Monitoring*; EUR 29826 EN.; Publications Office of the European Uion: Luxembourg, 2019.
10. Brattich, E.; Bracci, A.; Zappi, A.; Morozzi, P.; Di Sabatino, S.; Porcù, F.; Di Nicola, F.; Tositti, L. How to Get the Best from Low-Cost Particulate Matter Sensors: Guidelines and Practical Recommendations. *Sensors* **2020**, *20*, 3073. [CrossRef] [PubMed]
11. De Vito, S.; Esposito, E.; Formisano, F.; Massera, E.; Fiore, S.; Fattoruso, G.; Salvato, M.; Buonanno, A.; Veneri, P.D.; Di Francia, G. Enabling Citizen Science with A Crowdfunded and Field Validated Smart Air Quality Monitor. *Proceedings* **2018**, *2*, 932. [CrossRef]
12. Camprodon, G.; González, Ó.; Barberán, V.; Pérez, M.; Smári, V.; de Heras, M.Á.; Bizzotto, A. Smart Citizen Kit and Station: An Open Environmental Monitoring System for Citizen Participation and Scientific Experimentation. *HardwareX* **2019**, *6*, e00070. [CrossRef]
13. Castell, N.; Dauge, F.R.; Schneider, P.; Vogt, M.; Lerner, U.; Fishbain, B.; Broday, D.; Bartonova, A. Can Commercial Low-Cost Sensor Platforms Contribute to Air Quality Monitoring and Exposure Estimates? *Environ. Int.* **2017**, *99*, 293–302. [CrossRef] [PubMed]
14. European Commission. Measuring Air Pollution with Low-Cost Sensors. 2017. Available online: https://ec.europa.eu/jrc/en/publication/brochures-leaflets/measuring-air-pollution-low-cost-sensors (accessed on 9 January 2020).
15. Kouimtzis, T.; Samara, C. *Airborne Particulate Matter*; Springer: New York, NY, USA, 1995.
16. Wiseman, C.L.S.; Zereini, F. *Urban Airborne Particulate Matter: Origin, Chemistry, Fate and Health Impacts*; Springer: Berlin/Heidelberg, Germany, 2010.
17. Rai, P.K. *Biomagnetic Monitoring of Particulate Matter: In the Indo-Burma Hotspot Region*; Elsevier: Amsterdam, The Netherlands, 2015.
18. Holstius, D.M.; Pillarisetti, A.; Smith, K.R.; Seto, E. Field Calibrations of a Low-Cost Aerosol Sensor at a Regulatory Monitoring Site in California. *Atmos. Meas. Tech.* **2014**, *7*, 1121–1131. [CrossRef]

19. Jayaratne, R.; Liu, X.; Thai, P.; Dunbabin, M.; Morawska, L. The Influence of Humidity on the Performance of a Low-Cost Air Particle Mass Sensor and the Effect of Atmospheric Fog. *Atmos. Meas. Tech.* **2018**, *11*, 4883–4890. [CrossRef]
20. Badura, M.; Batog, P.; Drzeniecka-Osiadacz, A.; Modzel, P. Evaluation of Low-Cost Sensors for Ambient PM2.5 Monitoring. *J. Sensors* **2018**, *2018*, 1–16. [CrossRef]
21. Tagle, M.; Rojas, F.; Reyes, F.; Vásquez, Y.; Hallgren, F.; Lindén, J.; Kolev, D.; Watne, Å.K.; Oyola, P. Field Performance of a Low-Cost Sensor in the Monitoring of Particulate Matter in Santiago, Chile. *Environ. Monit. Assess.* **2020**, *192*, 1–18. [CrossRef] [PubMed]
22. Akpootu, D.; Gana, N. The Effect of Relative Humidity on the Hygroscopic Growth Factor and Bulk Hygroscopicity of Water Soluble Aerosols. 2013. Available online: http://theijes.com/papers/v2-i11/Part.1/I021101048057.pdf (accessed on 7 May 2019).
23. Liu, H.Y.; Schneider, P.; Haugen, R.; Vogt, M. Performance Assessment of a Low-Cost $PM_{2.5}$ Sensor for a Near Four-Month Period in Oslo, Norway. *Atmosphere* **2019**, *10*, 41. [CrossRef]
24. Bulot, F.M.J.; Johnston, S.J.; Basford, P.J.; Easton, N.H.C.; Apetroaie-Cristea, M.; Foster, G.L.; Morris, A.K.R.; Cox, S.J.; Loxham, M. Long-Term Field Comparison of Multiple low-Cost Particulate Matter Sensors in an Outdoor Urban Environment. *Sci. Rep.* **2019**, *9*, 7497. [CrossRef] [PubMed]
25. Mahajan, S.; Kumar, P. Evaluation of Low-Cost Sensors for Quantitative Personal Exposure Monitoring. *Sustain. Cities Soc.* **2020**, *57*, 102076. [CrossRef]
26. Surgaylo, A. *Research on Comparative Measurements of Different Particulate Matter and Soot Measuring Devices and Sensors*; Research Project; University of Stuttgart: Stuttgart, Germany, 2017.
27. Samad, A.; Vogt, U.; Laquai, B. Outdoor Air Quality Measurements Using Low Cost Particulate Matter Sensors. In *Proceedings of the III International Conference on Atmospheric Dust—DUST 2018, Bari, Italy, 29–31 May 2018*; Digilabs Pub.: Bari, Italy, 2018; p. 137. ISSN 2464-9147.
28. Alphasense Ltd. *Alphasense User Manual OPC-N3 Optical Particle Counter*; Alphasense Ltd.: Braintree, UK, 2018.
29. Grimm Aerosol Technik GmbH & Co. KG. EDM 180 EDM 180 Environmental Dust Monitor for Approved PM Measurements. 2020. Available online: https://www.grimm-aerosol.com/fileadmin/files/grimm-aerosol/3%20Products/Environmental%20Dust%20Monitoring/Approved%20PM%20Monitor/EDM180_The_Proven/Product%20PDFs/Datasheet_EDM180_ENG_2020.pdf (accessed on 9 January 2020).
30. KSL Staubtechnik GmbH. Eskal Dusts. 2017. Available online: https://www.ksl-staubtechnik.de/en/glasindustrie/eskal (accessed on 7 May 2019).

sensors

MDPI

Article

Developing Relative Humidity and Temperature Corrections for Low-Cost Sensors Using Machine Learning

Ivan Vajs [1,2,*], Dejan Drajic [1,2,3], Nenad Gligoric [3,4], Ilija Radovanovic [1,2] and Ivan Popovic [2]

1 Innovation Center, School of Electrical Engineering, University of Belgrade, Bulevar Kralja Aleksandra 73, 11120 Belgrade, Serbia; ddrajic@etf.bg.ac.rs (D.D.); ilija.radovanovic@ic.etf.bg.ac.rs (I.R.)
2 School of Electrical Engineering, University of Belgrade, Bulevar Kralja Aleksandra 73, 11120 Belgrade, Serbia; popovici@etf.bg.ac.rs
3 DunavNET, DNET Labs, Trg Oslobodjenja 127, 21000 Novi Sad, Serbia; nenad.gligoric@dunavnet.eu
4 Faculty of Information Technology, Alfa BK University, Palmira Toljatija 3, 11070 Novi Beograd, Serbia
* Correspondence: ivan.vajs@ic.etf.bg.ac.rs; Tel.: +381-11-3218-455

Abstract: Existing government air quality monitoring networks consist of static measurement stations, which are highly reliable and accurately measure a wide range of air pollutants, but they are very large, expensive and require significant amounts of maintenance. As a promising solution, low-cost sensors are being introduced as complementary, air quality monitoring stations. These sensors are, however, not reliable due to the lower accuracy, short life cycle and corresponding calibration issues. Recent studies have shown that low-cost sensors are affected by relative humidity and temperature. In this paper, we explore methods to additionally improve the calibration algorithms with the aim to increase the measurement accuracy considering the impact of temperature and humidity on the readings, by using machine learning. A detailed comparative analysis of linear regression, artificial neural network and random forest algorithms are presented, analyzing their performance on the measurements of CO, NO_2 and PM10 particles, with promising results and an achieved R^2 of 0.93–0.97, 0.82–0.94 and 0.73–0.89 dependent on the observed period of the year, respectively, for each pollutant. A comprehensive analysis and recommendations on how low-cost sensors could be used as complementary monitoring stations to the reference ones, to increase spatial and temporal measurement resolution, is provided.

Keywords: air pollution measurements; low-cost sensors; calibration; machine learning; artificial neural network; temperature and relative humidity

1. Introduction

Most of the population is currently living in urban areas and a decade ago it was estimated that, at that time, the number was already higher than fifty percent [1], and the newest predictions published by WHO (World Health Organization) estimate that this number will increase up to seventy percent by the year 2050 [2]. Although the increase in population is not directly linked to the increase in pollution, a large number of people does give rise to a various number of pollution emitters. This is consequently accompanied by the increasing number of areas where the air pollution level is high above the defined ranges and could seriously affect the citizens' health [3], which is associated with a series of acute and chronic diseases and is considered as one of the major health challenges at the moment (the limits for very high air pollutions: 50 $\frac{mg}{m^3}$ for CO, 400 $\frac{\mu g}{m^3}$ for NO_2 and 180 $\frac{\mu g}{m^3}$ for PM10). In [4], it is reported that in the year 2016, in low and middle-income countries, the citizens' mortality was heavily influenced by air pollution, and air pollution was linked to more than 4.2 million deaths per year (which represents 11.6% of all deaths). To combat that problem, the WHO issued the Air Quality Guidelines [5] about the recommendation regarding the activities concerning the pollution problem. There are also EU Directives defined on the ambient air quality [6,7] and many countries developed and implemented appropriate legislation. The most recognized air pollutants are CO, NO_2, SO_2, O_3 and particulate matter (PM2.5, PM10). The EU Directive on reference methods, data validation and location of sampling points for the assessment of ambient air quality [7], contains detailed instructions and recommendations concerning the used reference methods, obtained data validation and selection of the location of sampling points for adequate air quality monitoring.

So far, in urban areas, the usual approach of the measurement of the air quality is the deployment of national networks of public monitoring stations, which are quite reliable, but, on the other hand, they are located at fixed positions, quite large and heavy [8]. Furthermore, they have a high price and annual recalibration costs, while due to the fixed and sparse positions, they provide the information only about the regional air quality while lacking the spatial resolution to provide local measurements, thus making the citizens' exposure to the pollutants untrackable.

Contemporary, new generation, low-cost, off-the-shelf sensors look like a promising solution that could be used for complementary measurements for the areas that are not, and could not be, covered by public monitoring stations. Due to their high availability, low-cost sensors have great potential to be integrated into the portable low-cost Micro Sensing Units (MSUs) that can be used for air quality measurements. MSUs are mobile, have a wireless communication module and their maintenance costs are low. By applying the Internet of Things (IoT) concept, the data are remotely and periodically in real-time sent to a server in a cloud via the appropriate communication type (2G, 3G, 4G, WiFi, LoRa, etc.) where appropriate data storage, processing and visualization are performed [9]. They could be installed across the cities utilizing the existing public infrastructure (installed on public transport vehicles, public buildings, mounted on lamp posts, etc.). Additionally, it could be carried around by individuals, i.e., pedestrians and cyclists, thus allowing crowdsourcing [10], or even attached to drones. On the other hand, their main drawbacks are a short life cycle, low accuracy and most importantly, various influential calibration factors. The collected data might not always be accurate enough (due to the nature of electrochemical processes in the sensors and the influence of relative humidity, temperature and dust on the measurements) and in-field or laboratory calibration and periodical recalibrations are necessary, while the wireless transmission, in its nature, may introduce transmission errors and in the case of a wireless network failure, could be out of use. Furthermore, every sensor should be additionally calibrated, and the measurement accuracy of every single sensor highly depends on the sensor's chemical and physical characteristics.

In the authors' previous work [11], a methodology for the calibration of off-the-shelf air quality sensors is proposed and evaluated. The calibration process is based on the use of statistical algorithms and offset values obtained from the public measurement stations. The sensors were evaluated during a nine-month campaign in order to understand the seasonal influence on their behavior and a Common Air Quality Index (CAQI) [12] was calculated and compared with the public monitoring station. Obtained results were in a high level of agreement between the compared systems. The comparison between the results has shown that low-cost sensors could be used with a relatively high reliability as a complementary network to public monitoring stations, but it was also concluded that every sensor has its own sensitivity to temperature and relative humidity that influence the measurement accuracy.

Observed CO and NO_2 sensors are electrochemical, and their performances are affected by temperature and relative humidity due to the nature of electrochemical processes ongoing during the measurements. Additionally, during the usage, the NO_2 sensor has a higher loss of sensitivity than CO, and the NO_2 gasis, by its nature, unstable at low concentrations. On the other hand, the influence of relative humidity and temperature on the PM10 sensor, which is optical, is caused by particle growth due to water absorption. The sensor sensitivity to temperature and relative humidity poses a great challenge, as it can hardly be modeled with a simple function. The linear regression (LR) model and the multi linear regression model (MLR), are the most widely used techniques to calibrate low-cost sensor data against a reference measurement. However, when modeling different dependencies is concerned, the scientific field of Artificial intelligence, more precisely machine learning (ML), has shown great promise. This field relies on different methods that have a basis in mathematical theory, and as such, have found many uses in both modern research and industry. Using the powerful tools of ML, it is possible to model a sensor's dependencies on temperature and relative humidity and thus provide a more precise and reliable, yet low-cost measurement. In recent years, different types of Artificial Neural Networks (ANN) have been used for the calibration of low-cost air quality monitoring sensors in the laboratory or field conditions. Additionally, in order to achieve better results, for some low-cost air quality sensor types, it is recommended to examine the non-linear dependencies (exponential, logarithmic, quadratics) between the influencing variables, such as Random Forest (RF) [13,14], Support Vector Machines (SVR) [14,15] and the Gradient Boosting Regression Tree (GBRT) model [16]. The aim of this paper is to compare linear, different ANN and ML algorithms for in-field calibration of a low-cost sensor platform based on the collocation method.

Related Work

The problem of field calibration methods for low-cost sensors was investigated in detail in [17,18]. The authors used the following calibration methods: LR, ANN and MLR. They have concluded that the most suitable calibration method was ANN using raw or scaled sensor inputs (higher correlation coefficient), while LR and MLR have been shown to produce lower performances, since these methods do not take into consideration all interfering factors with their weighted effect (relative humidity and temperature). For observed CO, CO_2 and NO sensors, they concluded that ambient parameters such as relative humidity and temperature are necessary as algorithm inputs for appropriate calibration.

In [19], the authors stressed that the sensors' performances are very sensitive to the environmental operating conditions, i.e., relative humidity and temperature due to the gas-sensing process that involves fairly complex reactions depending on the environment conditions, and that corresponding chemical reactions also vary from daytime to nighttime in the urban atmosphere, which additionally degrades the performance of the sensors. They did not provide measurement principles, but rather discussed in detail the sensors and measurement devices issues with the focus on calibration issues. In general, manufacturers provide some correction factors for temperature and relative humidity, but for outdoor conditions, where relative humidity and temperature could change significantly on diurnal and seasonal bases, more sophisticated corrections are required.

In the scope of the CITY-SENSE project [20], authors tried to find the optimal calibration method for low-cost gas sensors for ambient air pollutants; the LR, MLR and ANN methods were compared and it was concluded that the ANN showed the best results for CO sensors.

In [21], data were collected from devices monitoring NO_2, installed in traffic and the urban environment. A two-step calibration method was proposed; firstly, MLR was used, where the output is the value that contains the information about the error, which was then used as the input to more sophisticated algorithms: ANN, SVM and RF. The proposed method has shown that at high concentrations, NO_2 sensors could closely meet the Air Quality Directive's standards of accuracy, but they have also concluded that each individual sensor behaves differently. A very detailed analysis of the possibilities to correct the ambient PM measurement under high relative humidity (*RH*) conditions is presented in [22]. It was shown that by exploiting the measured particle size distribution, an adequate correction algorithm could be derived (using κ-Köhler theory) that highly improves measurement performance.

The authors [23] consider the problems concerning low-cost sensors calibration, having in mind the possible set of tens of thousands, or even millions, of air quality sensors deployed. They expect to use data storage and processing capability at the edge of the network [24]. For calibration, they propose the usage of a deep learning model consisting of convolution layers, fully connected neural network layers and long short-term memory (LSTM) layers that model temporal dependencies.

In [25], the authors investigated the performance for CO, NO_2 and O_3 sensors, first by using laboratory calibration, and then by conducting field experiments. They have performed the integration of ANNs with fuzzy logic, which leads to the creation of an adaptive neuro-fuzzy inference system (ANFIS) [26], thus making a single framework that uses the advantages of both techniques. The result evaluation shows that the ANFIS has high correlation coefficients in comparison to the reference system.

In [27], the authors explored the influence of relative humidity and the effect of atmospheric fog on the performance of a low-cost air particle mass PM sensor, in the laboratory and field conditions. The results have shown that there was no clear effect until relative humidity exceeded about 75%, while above this value, due to particle growth, the sensor started to show a steady increase in the measurements. The reason for this is that when the relative humidity is higher, it results in particle growth and fog that are detected by the particle monitoring equipment, that does not contain drying facilities at the sample inlets (which is the case with low-cost particle sensors). Observing this, it was concluded that this effect must be taken into account when using low-cost particle sensors in such environments.

The authors of [28] investigated the effect of relative humidity and air temperature on CO, NO, NO_2 and O_3. Tests were conducted for six relative humidity levels from 10% to 85% and four temperature levels of 10–45 °C in the laboratory. After the development of the correction algorithm, field measurements were performed (November 2019). A performance analysis showed that the developed algorithm improved the data quality of the sensors in most of the cases, as CO, NO, and NO_2 sensors showed a satisfactory improvement, while the O_3 sensor had the least improvement. When sensors were exposed to high temperatures, NO_2 and O_3 sensors mostly behaved poorly.

In [29], the authors used sensors from different manufacturers and performed a calibration by using different methods. They have concluded that for CO and NO sensors, the MLR methods were the best solution for calibration, although ANN shows the same performances as MLR for NO. For NO_2 and O_3 sensors, supervised learning models, such as SVR, RF and ANN, proved to be the best methods for calibration. For PM2.5, the best performances were obtained by using linear models, when the relative humidity measurements were less than 75%. For higher relative humidity values, the calibration using the Köhler theory is the most promising method.

In [30], an evaluation of the Aeroqual Ltd. Series 500 semiconducting metal oxide O_3 and an electrochemical NO_2 sensor was performed by comparison with UK national network reference analyzers for more than 2 months in central Edinburgh. The obtained O_3 sensor measurements were in high correlation with the reference system, while the NO_2 sensor suffered from co-sensitivity to O_3, and the measurement error correction was developed by using LR.

A developed mobile PM2.5 sensing system was presented in [31], where eight sensing nodes were mounted on different city bus lines. Sensors were calibrated by using an ANN where the inputs, relative humidity and temperature were taken into account. A Gaussian Process regression algorithm was developed and implemented, so that by using measurements obtained from multiple sensors, PM2.5 values of locations within the observed region of interest, without direct measurements, could be interpolated.

In [32], an in-field measurement was conducted for CO, NO, NO_2, O_3, PM2.5, PM10 and SO_2, and compared to the reference data. The calibration methods used were LR, ANN and RF. For the case of LR calibration, only the variable that was being calibrated was used as the input. For ANN and RF methods, all the measurements from each unit were used. In the case of CO, NO and NO_2 sensors, satisfactory performances with LR were shown, but the additional improvement was obtained after the ANN and RF calibration. For the case of O_3, ANN and especially RF calibration have shown better performances than LR. Finally, for the PM2.5, PM10 and SO_2 sensors, both the ANN and the RF improve the results in comparison to the LR, and again, as in the previous case with O_3, RF showed better performances than the ANN algorithm.

In [33], NO, NO_2 and O_3 were observed and the authors explored the performance of dynamic neural networks in comparison to the static feed-forward ANN, where relative humidity and temperature were taken into account. For all considered sensors, it was shown that the dynamic neural network architectures were superior to the classical feed-forward ANN, since its architecture considers several consecutive measurements, as opposed to the static ANN that considers only one. The design, implementation and evaluation of a novel client–cloud system are presented in [34], and two types of internet-connected particulate matter (PM2.5) monitors were created. Sensor calibration consisted of two algorithms that were combined, ANN and Gaussian Process regression. The main difference between the two algorithms was that the ANN was used for calibrating a single sensor, while the Gaussian Process regression was used to combine the data from multiple sensors with different confidence levels, which was proven in this paper to provide a significant improvement after the applied ANN calibration.

In [35], PM2.5 and PM10 were observed and three different algorithms were used for sensor calibration: LR, ANN and SVM. The algorithms were first implemented with two variants. Firstly, by using the PM concentration values, relative humidity and temperature as the inputs and the reference PM data as outputs. Secondly, the algorithms were implemented using the mentioned inputs with the addition of wind direction and wind speed. For each algorithm and particle type, the models performed better than in the first variant where wind direction and wind speed were not considered. With both input sets, the ANN was the superior algorithm.

In paper [36], authors performed a detailed study for the seasonal behavior of PM2.5, and applied different ML algorithms to perform sensor calibration, including temperature and humidity changes as factors that influence the accuracy of the sensors.

In Table 1, an overview of references used calibration methods, and commonly used metrics (correlation coefficient R and corresponding R^2 value, RMSE (Root Mean Squared Error) and NRMSE (Normalized Root Mean Squared Error) [29]) for evaluation are provided.

Table 1. Types of calibration models used in the literature.

Pollutant	Calibration Model	References	Metrics
CO	LR	Drajic [11], Spinelle [17], Spinelle [18], Topalovic [20], Samad [28], Karagulian [29], Lin [30], Borrego [32]	R, R^2, RMSE, NRMSE
CO	ANN	Spinelle [17], Spinelle [18], Topalovic [20], Motlagh [23], Alhasa [25], Karagulian [29], Borrego [32]	R, R^2, RMSE, NRMSE
CO	RF	Karagulian [29], Borrego [32]	R^2, RMSE
NO_2	LR	Drajic [11], Spinelle [17], Spinelle [18], Cordero [21], Karagulian [29], Borrego [32]	R^2, RMSE
NO_2	ANN	Spinelle [17], Spinelle [18]. Motlagh [23], Alhasa [25], Samad [28], Karagulian [29], Borrego [32], Espositi [33]	R^2, RMSE
NO_2	RF	Cordero [21], Karagulian [29], Borrego [32]	R^2, RMSE
PM10	LR	Drajic [11], Jayaratne [27], Karagulian [29], Borrego [32]	R^2, RMSE
PM10	ANN	Motlagh [23], Karagulian [29], Borrego [32]	R^2, RMSE
PM10	RF	Karagulian [29], Borrego [32]	R^2, RMSE
PM2.5	LR	Di Antonio [22], Chen [35]	R^2, RMSE
PM2.5	ANN	Gao [31], Chang [34], Chen [35]	R^2, RMSE
PM2.5	RF	Wang [36]	R^2, RMSE

It should be noted that the authors used sensors from different manufacturers, device units from different manufacturers, different measurement sampling and averaging periods, different measurement campaign periods (total period of measurements and season) and different methodology (co-location method, laboratory method in controlled environment, mobile laboratory), so it is not possible to conduct a "fair" comparison of the metrics results.

The idea of the development and deployment of a low-cost sensor network for air quality monitoring is present in modern research. In [37] authors proposed a hybrid sensor network architecture with both stationary and mobile devices. They have developed a model for predicting the pollutant level, algorithms for hybrid network deployment and deployed a sensor network in a building. In [38] the capability of a network with low-cost PM sensors to capture PM spatial and temporal variations is explored. Six devices are mounted on fences/walls in the city of Southampton. The locations were chosen to be set around a school, while one of them was placed close to the road. Promising results were obtained, and in the next step, the authors plan to improve the spatial–temporal resolution by deploying 40 air quality monitoring devices in the area of 50 km^2 around the city. The authors of [39], deployed 24 air quality devices across the city of Oslo on the kindergarten premises. The focus was on measuring the NO_2 (as one of the primary pollutants caused by traffic) to observe the proposed data fusion methodology for creating urban air quality maps. They showed that it was possible to obtain and extract valuable information from the deployed sensor network and develop urban air quality maps with high resolution by using the data fusion methodology. In [40], authors observed a network with 10 devices deployed in the city of Bari (schools, streets, port, buildings) on the fixed locations and one mobile device that was mounted on top of the public bus (CO, CO_2, NO_2, O_3, SO_2, PM1, PM2.5, PM10, T and RH). It was quite a long campaign (June 2015–December 2017) and after a detailed result analysis, it was concluded that the usage of low-cost sensor devices showed promising results that could address the data quality objective of the indicative measurements [6]. The authors of [41] developed a rapid deployment method for low-cost sensors deployment. The method has three phases: preparation, implementation and modification. In the first phase, the model is fed by basic input data (objectives, spatial data preparation, elimination rules), then the implementation phase includes information about the desired deployment density, unnecessary area elimination and algorithm settings. The proposed algorithm takes into account the geographic environment, available power supply, transmission networks, etc. The obtained result is the recommended number of sensor and deployment locations. In [42], the authors deployed 40 sensor devices (NO, NO_2, CO, CO_2) at the London Heathrow Airport and defined an analytical approach in order to distinguish long transport emissions from the airport emissions. The study was conducted during a five–week period (October–November 2012) and the implemented approach has managed to calculate ratios of the airport activities in different locations of the airport. They claim that their sensor network approach could be applied to a wide range of environmental pollution studies. A survey on existing state-of-the-art showed that the influence of RH and T on pollutant measurements is undisputable. It was also shown that different types of ML algorithms can successfully model these dependencies and improve the accuracy of various low-cost sensors. However, to the best of our knowledge, no paper has performed a comparative analysis of the calibration for low-cost sensors for CO, NO_2 and PM10, taking into account RH and T influence, while comparing the results obtained with and without the RH and T as input features to the algorithm, thus quantifying the improvements RH and T can contribute to. Furthermore, no research paper has performed the calibration of low-cost CO, NO_2 and PM10 sensors on data gathered from four different seasons, and tested the calibration of low-cost sensors using data from two consecutive years.

In this paper, the approach (LR calibration is used as a benchmark) from our earlier work [11] is taken further to additionally improve the calibration algorithms with the aim of increasing the measurement accuracy, taking into account the impact of the air *RH* and *T* on the readings by developing appropriate *RH* and *T* corrections by using ML. A detailed comparative analysis of the sensors' behavior during a long observation time is performed (2 consecutive years). The selected observed months are from four different seasons (February, April, August, October), to ensure that the analysis of the applied ML algorithms performance is conducted on various weather conditions, thus taking into account different values of relative humidity and temperature depending on the observed season.

Even though the influence of *RH* and *T* on the low-cost sensors is "well-known", and there is existing research that proves the correlation, there is no research that has quantified the differences in the performance of ML algorithms on calibration, including these two parameters (i.e., weather conditions). The calibration of a sensor was also conducted using a small sample of data from the observed month in combination with the data gathered from a preceding year.

The main contributions of the paper are the method and approach for the calibration of the low-cost sensors (CO, NO_2 and PM10) using corrective measures (impact of *RH* and *T*), evaluated on different ML algorithms for the measurements taken during four different seasons over the period of two years. It was shown that all analyzed sensors are highly operable in the observed period (in accordance with their warranty period), with acceptable performances that are significantly improved by using proposed calibration algorithms and procedures, so that they can be used reliably in MSUs to provide a better spatial resolution within air quality measurement networks.

In addition to this, the discussion section contains a detailed analysis and recommendations on how low-cost sensors could be used for complementary measurements in order to increase spatial and temporal measurement resolution in combination with existing public monitoring networks. The deployment expenses are considered; the details about one possible low-cost monitoring station are provided from a practical point of view (device weight, dimensions, data transmission technology selection, etc.). Recommendations about the selection of location and mounting of a device are given. Finally, a hybrid sensor network approach is elaborated, which consists of reference monitoring stations supported by multiple low-cost devices. In this approach, low-cost sensors are virtually co-located with the reference monitoring station, thus making the recalibration process much easier. On the other hand, reference monitoring stations are supported and are implicitly expanded with spatially distributed complementary measurements.

The paper is organized as follows: In Section 2, the calibration procedure is explained, and the used ML methods are described. In Section 3, obtained results and the evaluation of performances are presented. In Section 4, a discussion about the results and paper contribution is elaborated. Finally, Section 5 provides conclusions and directions for future work.

2. Materials and Methods

2.1. Sensors and Data Collection

The collection of the data was performed by using a single low-cost sensor station and a single public air quality Automatic Monitoring Station run by the Serbian Environmental Protection Agency as a reference. The data from the public air quality monitoring station in Belgrade (Serbia) was collected during the period February–October during 2019 and in the same period (February–October) during 2020. The low-cost sensor station sensors are used from an air quality DunavNET ekoNET device AQ10x [9] for outdoor air quality measurements. This device is equipped with CO, NO_2, SO_2, O_3 (Alphasense), temperature, air pressure, relative humidity sensors (Bosch BME 280), PM1, PM2.5 and PM10 (Plantower). The data from the device are then statistically correlated to the values captured from the official monitoring station for the exact same time intervals.

Having in mind that CO, NO_2 and PM10 are not previously evaluated in this manner and that these are the most commonly used sensors, we have selected them for further evaluation. The reference measurement stations that were used in this paper provide pollutant measurements that are averaged on an hourly basis. On the other hand, the low-cost sensors that are used provide measurements every minute are then averaged for each hour to match the reference ones. Technical specifications of sensors are given in Table 2.

Table 2. Sensor's characteristics.

Pollutant	Manufacturer	Model	Range	Unit
CO	Alphasense	CO-B4	0–50 ppm	ppm or mg/m^3
NO_2	Alphasense	NO_2-B43F	0–20 ppm	ppb or µg/m^3
PM10	Plantower	PMS7003	0~1000 µg/m^3	µg/m^3

2.2. Calibration Methods

The performance of sensor devices (MSUs) is usually assessed using the mean error and/or correlation coefficients with respect to a reference laboratory or public monitoring stations' equipment data. However, the behavior of the low-cost sensors calibrated in a laboratory can change from the laboratory to the field environment due to certain interferences (different gases, higher range of T and RH) that were not evaluated in the laboratory. In the field collocation of devices, with reference public monitoring stations or professional measuring instruments, measurements helped to compare and calibrate the low-cost sensors according to the data obtained, and in this case, the advantage is that the low-cost sensors were exposed directly to the desired environment in which they are to be deployed. Different approaches are used to increase the accuracy of the measurement and to develop correction algorithms. Although the low-cost sensors are to be tested under several established conditions and compared to reference instruments, there is a lack of uniform guidelines, protocols or standards for the application of this new technology for regulatory purposes [29].

For calibration purposes, one of the most common methods, (suitable also because of its implementation simplicity) the Least Squares Method (LSM) [43], was used. It performs line fitting based on the minimization of the sum of the squares of deviations from a straight line $S = \sum_{i=1}^{n}(y_i - a - bx_i)^2$ and calculates the line coefficients a and b. Let n be the number of experimental points, i.e., number of conducted measurements. Denoting by y_i the reference values (from the public monitoring station) and by x_i the measured values (from AQ10x device). After "calibration", i.e., calculation of parameters a and b by LSM, the next step is to calculate the correlation of the obtained "calibrated" results with the results from the public monitoring station.

In Table 3, the mean, median and standard deviation values for T and RH for observed months and years are presented:

Table 3. Averaged/Median/Standard deviation (Std) values for T and RH.

Parameter	February	April	August	October
Average T [°C] 2019	6.8	9.2	25.1	16.3
Average T [°C] 2020	7.7	11.7	23.7	18.6
Median T [°C] 2019	8.1	11.1	23.2	17.9
Median T [°C] 2020	5.9	9.7	24.9	16.1
Std T [°C] 2019	5.5	4.9	4.6	4.5
Std T [°C] 2020	3.9	5.7	4.5	3.9
Average *RH* [%] 2019	74.1	54.3	59.2	64.9
Average *RH* [%] 2020	71.3	48.9	60.1	62.1
Median *RH* [%] 2019	70.9	51.1	61.3	61.8
Median *RH* [%] 2020	72.7	52.1	59.5	64.1
Std *RH* [%] 2019	16.5	16.1	19.3	16.4
Std *RH* [%] 2020	17.4	17.1	15.1	15.8

As a benchmark for a detailed study performed in this paper, in Table 4, corresponding R^2 coefficients are given for observed gases collected during four different parts of the year 2019 (February, April, August, October). LR calibration method is applied. For all four observed periods of interest, the sample size was a 15-day period, and the reference values are obtained once per hour (averaged measurement values per one hour), yielding the sample size of $15 \times 24 = 360$ per month.

Table 4. Coefficients obtained for observed periods of 2019.

Pollutant	R^2			
	February	April	August	October
CO	0.933	0.949	0.861	0.946
NO_2	0.784	0.846	0.671	0.828
PM10	0.716	0.849	0.664	0.786

From Table 4, it can be concluded that *T* and *RH* (stated in Table 1) considerably influence the behavior of low-cost sensors, which is visible for the period of February and August when low and high *T* influence measurements (the lowest R^2 was in August when temperatures were extremely high on average and in February when the temperatures were low). *RH* also had an influence, especially in the period when these values were high. Extreme values of *T* (low and high) and *RH* (high values) could cause a "peak" in the measurements from one side, and from the other, *T* (low and high) shifts the sensitivity of measurements to the lower levels, which correspondingly produces results with lower accuracy (it is visible in February and August).

2.3. Machine Learning Algorithms

As the first step of calibration performance evaluation, several ML algorithms are selected that showed good performance in previous studies, and performed initial evaluation in order to obtain the most promising algorithms for further detailed evaluation. In this paper, a comparison between different ML algorithms using 10-fold cross-validation was performed with a 70/30 train–test split (for the data grouped together from all four observed periods). The evaluated algorithms were LR, two architectures of ANNs, RF, SVM and AdaBoost. The evaluation was performed for each measured pollutant separately, with the input for each algorithm being *RH*, *T* and the raw low-cost sensor data, and the output is the data from the reference sensor for the respective pollutant (Figure 1).

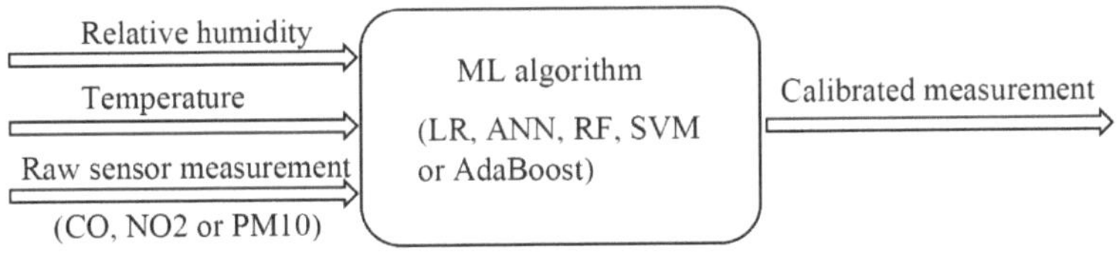

Figure 1. Measurement correction.

Each algorithm was evaluated using the metrics R^2, RMSE and NRMSE. The results of the cross-validation are shown in Table 5.

Table 5. Averaged metrics calculated on the test sets during cross-validation 2019.

Algorithm	CO		NO_2		PM10	
	R^2	RMSE	R^2	RMSE	R^2	RMSE
Linear regression	0.935	0.066	0.737	13.412	0.837	12.551
Neural network 1 (2 HL [1])	0.941	0.065	0.869	9.450	0.839	12.583
Neural network 2 (3 HL)	0.943	0.063	0.872	9.344	0.850	12.124
AdaBoost	0.924	0.074	0.843	10.360	0.846	14.560
Random forest	0.945	0.060	0.894	8.540	0.872	11.123
SVM	0.933	0.070	NC [2]	NC	0.835	12.748

[1] HL, hidden layer; [2] NC, non-convergent.

The two algorithms that have achieved the best performance (highest R^2 and the lowest RMSE) regarding all three measured pollutants are ANN [44] (with 3 HL) and RF [45]. These two algorithms were used for further calibration testing.

During the initial cross-validation, two ANN architectures were tested, one with two hidden layers, and one with three hidden layers. Each of the hidden layers had 20 neurons, and the activation function of the hidden layers was the hyperbolic tangent. The ANN with three hidden layers had achieved better results for all pollutants, so this particular architecture was used for further calibration testing in this paper. The ANN overfitting was regulated by keeping the number of neurons per layer relatively low while tracking the loss function on the validation set (25% of the training set). The RF contained 100 decision trees and each decision tree had all three features (low-cost sensor measurement, *RH* and *T*) as the input since selecting anything less than three features would make some trees lack the low-cost sensor measurement as an input, which would make them unable to create valid predictions. Both the mentioned algorithms were implemented in the Python programming language. The RF was implemented using the scikit-learn library, while the ANN was implemented in TensorFlow.

3. Results and Performance Evaluation

In this section, obtained calibration results for the selected methods (LR, ANN and RF) are presented and the performance evaluation is conducted. Firstly, we observed the behavior of the selected algorithms when data from all four months in 2019 are concatenated. In Table 6. the averaged results of the cross-validation using data from all the months are presented. In the case of LR, there is no train (calibration)/test period, rather the algorithm is applied to the whole data set. For the RF and ANN algorithms, the results on the calibration set are expected to be better than the ones on the test set, but the test results correspond to the results that the algorithm could obtain in practice. Having this in mind, the ML algorithms will be compared based on the test set results, with the benchmark results being the ones obtained by the LR performed on the entire dataset.

Table 6. All months 2019, CO, NO_2, PM10, LR, ANN, RF, calibration and test set.

Pollutant, Algorithm (Input Features)	R^2		RMSE		NRMSE
	Calibration	Test	Calibration	Test	Test
CO, LR (raw)	**0.931**		**0.068**		**0.264**
CO, ANN (raw)	0.927	0.927	0.070	0.070	
CO, ANN (raw, *RH*, *T*)	0.945	0.943	0.061	0.063	0.244
CO, RF (raw)	0.988	0.915	0.028	0.075	
CO, RF (raw, *RH*, *T*)	0.994	**0.945**	0.022	**0.060**	**0.233**
NO_2, LR (raw)	**0.793**		**11.980**		**0.455**
NO_2, ANN (raw)	0.809	0.797	11.610	11.913	
NO_2, ANN (raw, *RH*, *T*)	0.908	0.872	8.040	9.340	0.348
NO_2, RF (raw)	0.967	0.762	4.817	12.860	
NO_2, RF (raw, *RH*, *T*)	0.986	**0.894**	3.162	**8.543**	**0.325**
PM10, LR (raw)	**0.794**		**14.112**		**0.453**
PM10, ANN (raw)	0.782	0.774	14.687	14.969	
PM10, ANN (raw, *RH*, *T*)	0.910	0.850	9.482	12.121	0.389
PM10, RF (raw)	0.959	0.709	6.374	17.198	
PM10, RF (raw, *RH*, *T*)	0.982	**0.872**	4.140	**11.124**	**0.357**

It is shown that there is a clear difference between the results achieved when *RH* and *T* are included as the input to the ML algorithm calibration process. Better results were achieved regardless of which pollutant was selected, and regardless of the set type (calibration or test) in the case when *RH* and *T* are included as a calibration factor. The obtained results are to be expected since the influence of *RH* and *T* on low-cost sensors cannot be disputed. Furthermore, it is shown that both algorithms (RF and ANN) can model these influences successfully. It is also important to note that when the raw sensor data are the only input, ANN achieves superior results on the test set, regardless of the pollutant. This is most likely due to the ability of the ANN to better model non-linear functions of single variables due to the presence of activation functions. On the other hand, RF is superior if *RH* and *T* are taken into consideration.

It can be concluded that CO has the overall lowest value for NRMSE, which is expected, since CO generally shows the best R^2 value. It can also be observed that both the ANN and RF additionally lower the RMSE, and therefore the NRMSE value for each pollutant. By using the NRMSE parameters as a measure of comparison between the performances of the algorithms for different pollutants, we can see that the biggest improvement can be seen for the NO_2 with the RF algorithm. This stands in line with the biggest improvement for the R^2 factor, which is present in the same case.

In the following text, we explore the calibration results for each observed month in 2019 separately. Tables 7–10 contain the results obtained using the 10-fold cross-validation only on the data from the corresponding month in 2019 (i.e., February, April, August and October), with a 50/50 train/test split. This data split was used instead of the 70/30 one because of the size of the dataset for each individual month, to ensure testing was performed on a sufficiently large data sample. In the case of LR, there are no train/test periods, rather the algorithm is applied to the whole data set.

Table 7. February 2019, CO, NO_2, PM10, LR, ANN, RF.

Pollutant, Algorithm (Input Features)	R^2		RMSE	
	Calibration	Test	Calibration	Test
CO, LR (raw)	**0.933**		**0.053**	
CO, ANN (raw, *RH*, *T*)	0.980	**0.968**	0.031	**0.038**
CO, RF (raw, *RH*, *T*)	0.993	0.934	0.017	0.052
NO_2, LR (raw)	**0.784**		**8.940**	
NO_2, ANN (raw, *RH*, *T*)	0.857	0.832	7.986	8.625
NO_2, RF (raw, *RH*, *T*)	0.985	**0.904**	2.360	**5.976**
PM10, LR (raw)	**0.716**		**12.012**	
PM10, ANN (raw, *RH*, *T*)	0.780	0.737	11.567	12.549
PM10, RF (raw, *RH*, *T*)	0.962	**0.767**	4.436	**10.221**

The results in Table 7 show that for the CO calibration, only the ANN algorithm surpassed the reference LR results. For the other two pollutants, RF has proven to be better with a significant improvement achieved for the NO_2.

Table 8. April 2019, CO, NO_2, PM10, LR, ANN, RF.

Pollutant, Algorithm (Input Features)	R^2		RMSE	
	Calibration	Test	Calibration	Test
CO, LR (raw)	**0.949**		**0.054**	
CO, ANN (raw, *RH*, *T*)	0.982	**0.974**	0.032	**0.039**
CO, RF (raw, *RH*, *T*)	0.996	0.970	0.015	0.042
NO_2, LR (raw)	**0.846**		**9.278**	
NO_2, ANN (raw, *RH*, *T*)	0.889	0.866	9.463	10.001
NO_2, RF (raw, *RH*, *T*)	0.993	**0.943**	2.008	**5.695**
PM10, LR (raw)	**0.849**		**8.070**	
PM10, ANN (raw, *RH*, *T*)	0.888	0.867	8.111	8.680
PM10, RF (raw, *RH*, *T*)	0.984	**0.891**	2.806	**7.204**

Results for the month of April stand in line with the results from February, indicating that the ANN models the CO sensor dependencies better than RF. Furthermore, PM10 and NO_2 were better modeled by the RF, which is also in line with the results from the previous month.

Table 9. August 2019, CO, NO_2, PM10, LR, ANN, RF.

Pollutant, Algorithm (Input Features)	R^2		RMSE	
	Calibration	Test	Calibration	Test
CO, LR (raw)	**0.861**		**0.048**	
CO, ANN (raw, *RH*, *T*)	0.894	0.885	0.039	0.047
CO, RF (raw, *RH*, *T*)	0.978	**0.927**	0.019	**0.033**
NO_2, LR (raw)	**0.671**		**11.286**	
NO_2, ANN (raw, *RH*, *T*)	0.940	0.767	4.590	10.130
NO_2, RF (raw, *RH*, *T*)	0.961	**0.817**	3.620	**9.460**
PM10, LR (raw)	**0.664**		**8.740**	
PM10, ANN (raw, *RH*, *T*)	0.813	0.678	6.985	8.664
PM10, RF (raw, *RH*, *T*)	0.967	**0.731**	2.882	**7.935**

The month of August has the lowest R^2 factor for the LR, for each pollutant. The improvements of this factor, however, are still present and indicate the applicability of the ML algorithms. In this month, the RF was shown to be better than the ANN for every pollutant.

Table 10. October 2019, CO, NO_2, PM10, LR, ANN, RF.

Pollutant, Algorithm (Input Features)	R^2		RMSE	
	Calibration	Test	Calibration	Test
CO, LR (raw)	**0.946**		**0.068**	
CO, ANN (raw, *RH*, *T*)	0.969	**0.968**	0.052	**0.062**
CO, RF (raw, *RH*, *T*)	0.991	0.949	0.028	0.067
NO_2, LR (raw)	**0.828**		**13.761**	
NO_2, ANN (raw, *RH*, *T*)	0.893	0.875	10.880	11.820
NO_2, RF (raw, *RH*, *T*)	0.988	**0.914**	3.698	**9.786**
PM10, LR (raw)	**0.786**		**16.492**	
PM10, ANN (raw, *RH*, *T*)	0.910	0.819	4.550	9.570
PM10, RF (raw, *RH*, *T*)	0.977	**0.824**	5.623	**8.940**

The results from October show that the best algorithm for CO is the ANN. Regarding the NO_2 and PM10 measurements, the RF was superior to the ANN.

It is shown that for every month in 2019, the RF obtained the best results both for NO_2 and PM10 measurements. However, the results for the CO are mostly in favor of the ANN, which achieved the best results for every month except August, where the RF performed better. It is important to note that the trend of lowering the RMSE does correspond to the increase in the R^2 factor, in each observed month individually, and for every applied algorithm. The trends that the R^2 factor and RMSE follow within one month are important, but the comparison between months does have to include a careful evaluation since the lower concentrations of pollutants tend to influence the R^2 score negatively but can lower the RMSE.

As a further step of evaluation, we present the scatter plots for different pollutants and the applied algorithms, i.e., LR, ANN and RF. For ANN and RF algorithms, the values from the test set are presented. In Figure 2, the results for the case where the data from all months in 2019 is concatenated together, are presented. The axis limits were chosen to maximize the usage of the space within each graph, and as such, cause a number of outlier measurements to be on the border of some graphs.

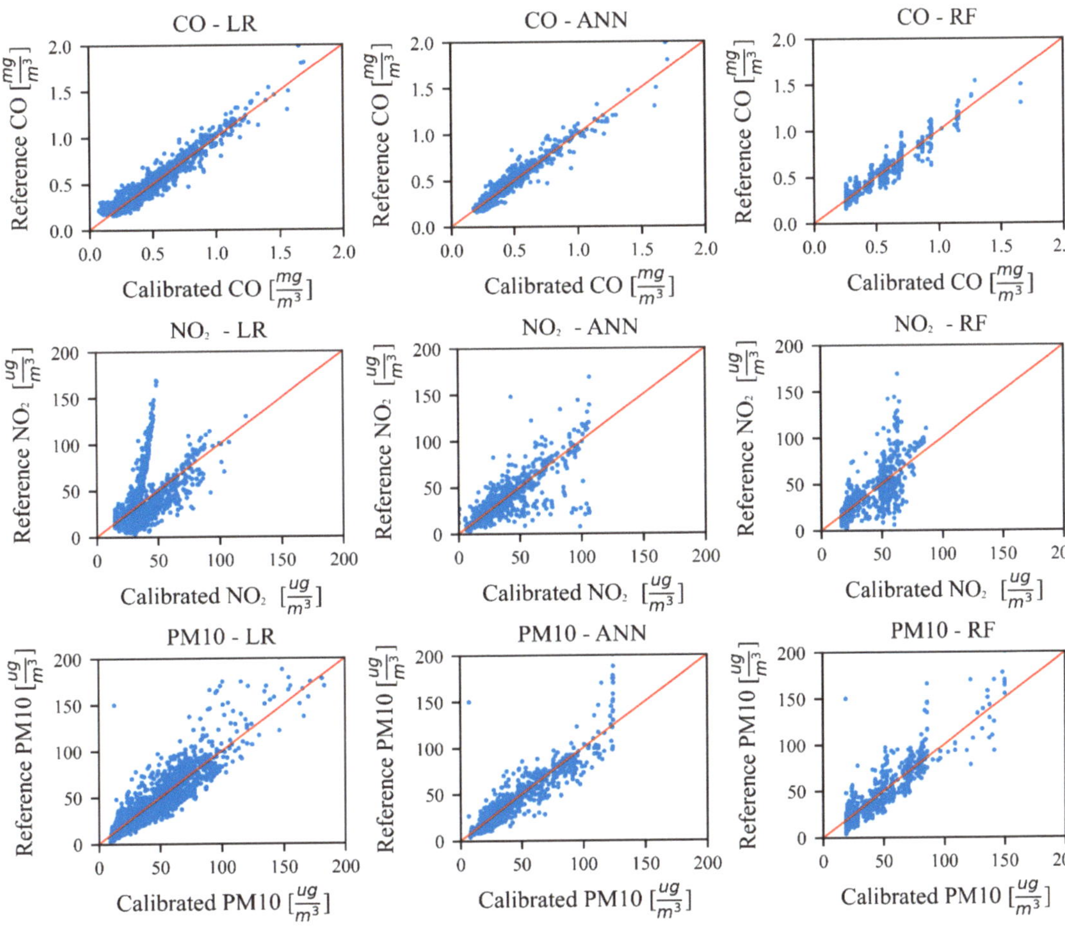

Figure 2. Test results from all observed months of 2019.

The scatter plots of the data from all months in 2019 show that if only the LR is implemented, the best correlation with the reference measurements is obtained for the CO. ANN and RF both improve the CO calibration, with the ANN having dispersed point placement and the RF having clusters. Particularly, the RF shows scatter points clustered into vertical lines. This means that for a small interval of reference measurements, the RF algorithm tends to return the same values. Although the NO_2 low-cost sensor has the same measurement principle as the CO one, the nature of these pollutants and the sensors that measure them do vary. For example, in the NO_2–LR scatter plot, it is clearly shown that by only using the raw sensor measurements as the inputs, a good linear correlation cannot be obtained, which was possible for CO. This is due to the nature of the data, as two different linear trends can be observed in the mentioned scatter plot. The ANN and RF algorithms show a clear improvement, although visibly less successful than the CO results. The PM10 scatter plots show that a single linear trend is present in the data and that both ML algorithms improve the correlation. It is interesting to note that due to the smaller number of data points (less than 50 in both the training and test set) with the higher PM10 concentration values (above 100 $\frac{ug}{m^3}$), the ANN seems to be unable to produce the higher values for PM10 concentration and maxes out at around 125 $\frac{ug}{m^3}$. The RF, on the other hand, does not seem to have this problem. The reason is due to the way both algorithms are structured, ANN has a single complex structure and adapts its weights numerically to optimize the loss function based on the data from the calibration set. Should the number of data points in a certain range be limited, their influence on the weights of the network will be insufficient to make the ANN output values in that particular range. On the other hand, the RF has many simpler structures (decision trees) where each is trained on a part of the calibration dataset, and this training process does not optimize a single model to the data, rather it fits many models on parts of the dataset.

In the following paragraphs, the results of the measurements obtained in the year 2020 are presented. The observation periods are the same as in 2019, i.e., for February, April, August and October. The methodology used for 2020 is the same as the one used for the year 2019, averaged hourly values obtained from devices were compared with measurements obtained from the reference station, for the same periods of the year on a 15-day level.

Firstly, we observe the R^2 values obtained by using LR on only the raw sensor data for the appropriate month in 2020. Secondly, we use all the data from 2019 as the training data and evaluate it on the data from a given month in 2020. Finally, we train a second RF on a sample of 4 days from the respective 2020 month and combine it with the RF trained on 2019 data. The idea is that by combining a small sample from the respective month with the data from the previous year, a significant improvement of the sensor performance could be achieved. The results obtained on the test sets (four different splits of 4/11 days of the respective 2020 month) were averaged and displayed in table format for each of the observation months of 2020. The RF algorithm was selected since it achieved the best results when using all the data from 2019 as shown in Table 5.

Observing the results obtained for the month of February 2020 (Table 11), using both the data from 2019 and 2020, the advantages of having a years' worth of measurements are clear. Regarding the CO measurements, the results obtained after the calibration on the 2019 data decrease the R^2 factor, but also lower the RMSE. A similar result, with both the R^2 and RMSE lowered, is obtained using the RF trained on the four calibration days from 2020. Finally, the CO results obtained using a linear regression on the outputs of the two RF algorithms show a merely identical R^2 to the initial data, with the lowest RMSE out of all the previously mentioned cases. The NO_2 measurements show that the linear combination of the RF algorithms shows the highest R^2 factor, followed closely by the 2019 RF algorithm. The linear combination of the RF algorithms achieves by far the lowest RMSE for the NO_2 measurements. PM10 measurements show that the linear regression based on the outputs of two RF algorithms show the highest R^2 factor alongside the lowest RMSE, which stands in line with the data from the other two pollutants. Overall, for the month of February, combining the algorithms trained on the data from 2019 and 2020 gives the best results.

Table 11. February 2020 test results, CO, NO_2, PM10.

Pollutant (Input Set)	R^2	RMSE
CO, LR (raw)	0.952	0.091
CO, RF (2019)	0.953	0.077
CO, RF (2019 + 2020)	0.957	0.065
NO_2, LR (raw)	0.830	18.564
NO_2, RF (2019)	0.853	15.667
NO_2, RF (2019 + 2020)	0.856	10.564
PM10, LR (raw)	0.833	28.356
PM10, RF (2019)	0.844	12.071
PM10, RF (2019 + 2020)	0.863	11.046

During the month of April (Table 12), there are some differences from the results obtained in February. In April, a state of emergency was declared in Serbia. This has, in turn, caused a steep decrease in the concentrations of all pollutants due to the lowered traffic. This made it more difficult for the algorithms to correctly pick up on the dependencies between the raw and reference data. The combination of two RF algorithms has a lower R^2 factor than both the raw data and the results from the 2019 RF calibration. On the other hand, the obtained RMSE for the linear combination of the RF algorithms is by far the lowest out of all the obtained results for the CO measurements. The NO_2 results are similar to the results from February with the linear combination of the two RFs having both the highest R^2 and the lowest RMSE. The PM10 results show the highest R^2 factor for the raw data measurements. The results obtained from the RF calibrated on 2019 data are acceptable but the results from the 2020 calibration data are quite poor. This is due to the high variations of PM10 values in April 2020 (measurements up to 450 $\frac{\mu g}{m^3}$, while all other months' measurements were up to 141 $\frac{\mu g}{m^3}$). This is quite interesting since the extremely high PM10 values (>200 $\frac{\mu g}{m^3}$) occurred after relaxing the state of emergency measures in Serbia. All other pollutant concentrations were also increased in the same period but not as drastically. The lowest RMSE is obtained for the linear combination of RFs but the R^2 factor is significantly decreased.

Table 12. April 2020 test results, CO, NO_2, PM10.

Pollutant (Calibration Set)	R^2	RMSE
CO, LR (raw)	0.954	0.079
CO, RF (2019)	0.955	0.064
CO, RF (2019 + 2020)	0.956	0.051
NO_2, LR (raw)	0.569	23.625
NO_2, RF (2019)	0.676	21.973
NO_2, RF (2019 + 2020)	0.689	15.316
PM10, LR (raw)	0.786	71.302
PM10, RF (2019)	0.732	49.949
PM10, RF (2019 + 2020)	0.739	48.516

The results obtained for the month of August (Table 13) show a significantly lower R^2 value on the raw data for all pollutants, compared to the previous two observed months. The CO results show that the combination of RF algorithms based on data from 2019 and 2020 has the highest R^2 value and the lowest RMSE. The NO_2 and PM10 measurements have a relatively low R^2 value on the raw data, but the RF algorithms behave differently for these two pollutants. The best results for the NO_2 are obtained for the combination of the two RF algorithms, with the R^2 value almost unchanged from the raw data, but with a significantly lower RMSE. On the other hand, the PM10 results are quite poor indicating no possibility for calibration. The lifetime of a PM sensor based on the manufacturer declaration is 1 year and at the moment of these measurements, it was already 1 year and 7 months "old", so this loss of accuracy is expected behavior. On the other hand, CO and NO_2 sensors have a warranty of 2 years, but a slight degradation of accuracy is to be expected (notable for the NO_2 sensor).

Table 13. August 2020 test results, CO, NO_2, PM10.

Pollutant (Calibration Set)	R^2	RMSE
CO, LR (raw)	0.764	0.074
CO, RF (2019)	0.787	0.054
CO, RF (2019 + 2020)	0.801	0.035
NO_2, LR (raw)	0.476	24.134
NO_2, RF (2019)	0.440	17.834
NO_2, RF (2019 + 2020)	0.477	7.917
PM10, LR (raw)	0.408	17.935
PM10, RF (2019)	0.303	8.872
PM10, RF (2019 + 2020)	0.249	8.201

The results obtained from the October data (Table 14) indicate further degradation of the PM10 sensor and an operable state of the NO_2 sensor. Although the results from August could indicate that both of the mentioned sensors suffered from significant degradation, it is clear that the NO_2 sensor was still operable in October while the PM10 sensor has lost its functionality. The calibration results for CO show that the lowest RMSE was achieved when the two RF algorithms are combined, while the highest R^2 factor is present for the raw data, but with a significantly higher RMSE. For the NO_2 results, the highest R^2 factor is obtained by combining both RF algorithms, while the lowest RMSE is obtained using only the RF trained on the data from 2019.

Table 14. October 2020 test results, CO, NO_2, PM10.

Pollutant (Calibration Set)	R^2	RMSE
CO, LR (raw)	0.901	0.081
CO, RF (2019)	0.903	0.069
CO, RF (2019 + 2020)	0.904	0.059
NO_2, LR (raw)	0.748	15.432
NO_2, RF (2019)	0.779	10.993
NO_2, RF (2019 + 2020)	0.785	10.366
PM10, LR (raw)	0.213	30.217
PM10, RF (2019)	0.134	26.418
PM10, RF (2019 + 2020)	0.219	34.650

The results obtained from the data of 2020 show that a significant improvement in the sensors' performance can be achieved by using a year's worth of data in combination with just 4 days from a respective month. The CO sensor shows a high initial correlation for each month but an increased RMSE value when compared to the measurements from 2019, although the measurement value range was similar. This does imply sensor degradation, but the degradation can be easily modeled, and the results obtained from using both 2019 and 2020 data show promising results. The NO_2 sensor does not achieve the results that are as good as the CO sensor, but it is still sufficiently accurate and shows an improvement with the implemented algorithms. The PM10 sensor has the most prominent degradation as it is practically unusable going forward from the month of August 2020 (while it is usable in February and April). Overall, apart from the limited lifetime of the PM10 sensor, the data acquired during 2019 has shown to be applicable in the calibration of the same sensor in 2020, with only 4 days from the observed month in 2020 as training data.

4. Discussion

In this paper, we have first considered data from CO, NO_2 and PM10 obtained from a 9-month measurement campaign (from February to October 2019). In order to understand the behavior of the sensors' performances, four different periods (February, April, August, October 2019) are observed, thus considering different values of RH and T. Different ML algorithms were used, that take into account RH and T in the calibration process, and the results are compared with the benchmark results obtained by the LR method. It was shown that the results from this experiment were satisfactory and that they can be further improved using the selected ML algorithms. This is important since it implies the possibility of using low-cost sensors alongside reference ones, to create better spatial and temporal measurement resolution. Generally, RF outperforms the ANN algorithm values except for the CO pollutant (although RF is better than the ANN in August). By using ML algorithms, the R^2 values are increased for all pollutants in the observed months. These improvements are summarized in Table 15.

Table 15. R^2 improvements for CO, NO2, PM10, LR, ANN, RF, by months in 2019.

Pollutant	R^2 Improvement			
	February	April	August	October
CO	0.035	0.025	0.066	0.022
NO_2	0.120	0.097	0.146	0.086
PM10	0.051	0.042	0.067	0.038

The best improvement for every pollutant out of all the months in 2019 is achieved in the month of August (and after that in February, where the influence of *RH* and *T* on sensors was the second-highest). This could seem counter-intuitive since the best achieved R^2 values for August are the lowest out of all the months. However, the measurements of the pollutants in August show the lowest R^2 score when the LR algorithm is applied, indicating the high influence of weather conditions on the measurements in that month. The highest improvement rate achieved in August is a great example of how ML algorithms can achieve much more than a simple linear calibration, as they can successfully model non-linear dependencies between features. It is also important to mention that the achieved results for every individual month are obtained using cross-validation based only on the data from that particular month. The fact that such a clear improvement can be achieved with limited data acquisition represents a significant conclusion in this field of research. Acquiring air quality data is highly time-dependent as the process cannot be sped up in order to obtain a larger dataset. By showing that ML algorithms can be used both on every individual month, and on the concatenated data from all months, it is clear that ML algorithms do not only successfully scale up with larger datasets, but also that they can be scaled down to work with rather sparse data. Regarding the improvements for the pollutants, the highest R^2 increase for every month is achieved for NO_2, followed by PM10, and finally CO. This could mean that the influence of *RH* and *T* on the low-cost sensors for NO_2 is substantial and that the ML models successfully accommodated the sensors' shortcomings. The CO correlation after LR is relatively high for each month, so a more modest improvement is expected, and PM10 particles stand somewhere in between CO and NO_2 regarding the improvement rate.

In Table 16, the improvements when using data from all of the months are summarized. Both ML algorithms show improvements, but RF shows slightly better performances than ANN in all analyzed test cases, so only the improvements for RF are presented.

Table 16. R^2 improvements for CO, NO_2, PM10, RF, all months in 2019.

Pollutant	R^2 Improvement
CO	0.014
NO_2	0.101
PM10	0.078

The improvements achieved using RF algorithms for the concatenated data from all of the months show that the ML algorithms can successfully be used on a dataset with varying weather conditions. It is also important to note that the results achieved for the concatenated data from all moths are obtained using a 70/30 train–test split, while the data for each individual month are obtained with a 50/50 train test–split. With a larger dataset and a more favorable train–test split, it would be expected that the improvements listed in Table 16 would be better than the individual improvements for each month, but that is not always the case. For example, the improvements for NO_2 for the month of April are greater than the ones achieved for all months combined. The reason for this is the wide variety of values of *RH*, *T* and NO_2 in the dataset consisting of all four months and a relatively low data count for such a feature space. If a substantial quantity of data were available, a deep learning algorithm could be implemented that would most probably successfully model all different dependencies. In this implementation, with a limited data quantity, the division of the calibration problem into monthly calibrations could be the optimal way, as is shown in the acquired results.

We have then focused on the measurement campaign conducted in the year 2020, repeating measurements with the same methodology as in the year 2019, the same four months are observed with the same measurement protocol. The observations from 2020 were used to analyze the possibility of using data from the preceding year to calibrate the same sensor in the present. It was also interesting to analyze the sensors' performance after an entire year of in-field measurements.

The obtained values for the CO sensor show that the overall performance of the sensor in 2020 is quite equivalent to the one from 2019. Considering that the R^2 values are high for this sensor, a high usability of this low-cost device for at least two years is possible. The NO_2 sensor does not have a performance as good as the CO one and the degradation is a bit more prominent. On the other hand, the R^2 factor during the 2020 months is still acceptable and shows that the NO_2 sensor is also operable after two year's worth of measurements. The PM10 sensor has shown to be the most sensitive and the results show it is operable through February and April 2020. This stands in line with the sensors' warranties, as the CO and NO_2 sensors have a 2-year warranty period and the PM10 sensor has a 1-year warranty.

The best-obtained results, using a combination of two RF algorithms, show a range of improvements. The improvements for the CO R^2 factor, ranging from 0.002 to 0.037, are overall not incredibly high. The initial R^2 for this pollutant is, however, quite high, and achieving a great improvement has shown to be unlikely. The NO_2 R^2 factor has the best improvement out of all the considered pollutants, ranging from 0.001 to 0.12. The PM10 sensor has shown an improvement of 0.03 in February, where the calibration process could be applied. The obtained results do not differ greatly from the improvements that were achieved with the 2019 data.

In this paper, a comparative analysis of ML algorithms through a span of four months during two consecutive years (2019, 2020) is performed. The months selected are from four different seasons so that the analysis of the ML algorithm performance could be performed on various weather conditions. Furthermore, a comparative analysis between different ML algorithms was performed, as well as the investigation of the influence relative humidity and temperature can have on the calibration. The difference between the performance of algorithms that are based solely on the raw pollutant measurements, and the ones that include RH and T as input features are shown. An investigation of the possibilities of calibrating a sensor from the data gathered in the preceding year is also performed. It is shown that by combining the data from 2019 and a small sample of 4 days from the observed month in 2020, the improvements could be comparable to the results obtained in 2019 when 7.5 days from the observed month were used for calibration. This opens the possibility of reducing the duration of the calibration period of a low-cost sensor in a given month by using previously acquired data. It is important to note that different low-cost devices can perform differently and that one of the limitations of this work is that the analysis was performed on a single low-cost device. It was also impossible to acquire a continual stream of data from a reference monitoring station that could cover an entire year, which would surely be beneficial for the calibration process.

Based on this comprehensive study, it is proven that the measurement accuracy of every single sensor has its own sensitivity to T, RH, etc., and that for every pollutant a different approach for increasing the reliability of measurements should be developed and applied. By applying ML algorithms on the pollutant measurements, measurement accuracy is further improved, thus allowing low-cost sensors higher reliability and capability to be used as a complementary network to public monitoring stations, which will allow much higher measurement granularity, and the ability to observe air pollution at micro-locations. Furthermore, the integration of low-cost air quality measurement sensors will enable a higher density of air pollution assessment in urban areas and the development of sophisticated location-aware services for environmental protection, intelligent traffic control, accident detection, air pollutant transport and dispersion monitoring, etc. A detailed explanation of how it could be performed is provided in the following text.

A Hybrid Sensors Network Approach

It is obvious that by increasing the number of deployed devices and providing a higher measurement frequency, one will obtain the results with better quality and accuracy, thus improving the detection of the sources of pollution and personal exposure. Low-cost devices are without a doubt more cost-effective than public monitoring stations. Based on the available vendor's information, the average ratio is between 1:20 and 1:25, i.e., the cost of one public monitoring station is comparable with the cost of 20–25 low-cost devices for the same set of pollutants observed. In order to obtain more insight into the usage of one possible low-cost device [9], we have provided more detailed device characteristics and universal recommendations about the selection of the location and mounting of the device. Device dimensions are $180 \times 180 \times 265$ mm^3, weight is 1.5 kg and power consumption is 2.5 W. Different data transmission technologies are supported: GPRS, 3G, 4G, NB-IoT, LoRa, SigFox and WiFi. Generally, a low-cost device could be mounted on a wall, pole, pillar or some other solid object. It is also important to take into account the scope of monitoring (use case), distance from the pollution source, area topography, presence of different kinds of obstructions and the availability of appropriate deployment space. The objective of urban air quality monitoring is to capture and understand pollution trends and people exposure in the observed areas (depending on the use case it could be micro (up to 0.1 km), middle (0.1–0.5 km), neighborhood (0.5–4 km) or urban scales (4–50 km) [46,47]. Urban areas usually have local microclimate areas with different pollution conditions that could be of very small scales. Finally, in order to create a more accurate estimation of pollution, which is actually the goal of this paper, it is useful to install devices with low-cost sensors as complementary measurement devices that could be installed virtually anywhere. Collecting the data from these devices allows the creation of city pollution maps that can provide a deeper understanding of pollutants spatial distributions over specific areas, and on the other hand, high temporal resolution is provided using real-time measurements conducted every minute. In order to predict air quality with a higher accuracy, ML could be applied to help identify pollution hotspots. Reference monitoring stations are accurate but placed on fixed locations and quite expensive, while low-cost devices are cheap and mobile but suffer from a problem of accuracy and calibration. The most promising solution appears to be a combination of these two kinds of monitoring stations, i.e., the creation of a hybrid sensor network that combines the best of these two monitoring approaches. In this hybrid sensor network, a reference monitoring station is supported by multiple low-cost devices. In this way, sensors are virtually co-located with the reference monitoring station and their recalibration process is much easier (thus providing higher measurement accuracy), while reference monitoring stations are enhanced by spatially distributed complementary measurements. If some of the sensors start to suffer from in-accuracy, recalibration could be performed by correlation with a reference monitoring station or cross-calibration by comparison with recently re-calibrated devices in the area.Our future work will be devoted to the development of a model for the deployment of hybrid sensor networks and recommendations for the number of nodes and their spatial distribution (density).

5. Conclusions

In this paper, different ML algorithms are applied on the low-cost sensors' measurements in order to improve the calibration algorithms taking into account the impact of the air *RH* and *T* on the readings.

The main contributions of the research described in this paper are the method and approach for the calibration of the low-cost sensors (CO, NO_2 and PM10) using corrective measures (impact of *RH* and *T*). The method was evaluated on different ML algorithms for the measurements taken during four different seasons (February, April, August, October) in a period of two consecutive years.

The CO, NO_2 and PM10, have shown satisfactory improvements after applying ML correction algorithms (the best improvements were obtained for NO_2, then for PM10 and finally for CO). RF has shown better performances for NO_2 and PM10 pollutants, while ANN was better for CO. With these corrections, the accuracy of the low-cost sensors' measurement becomes more reliable and closer to the measurements obtained from reference monitoring stations. Depending on the observed period, R^2 is in the range from 0.927–0.970 for CO, 0.817–0.943 for NO_2 and 0.731–0.891 for PM10.

After the analysis of the data from 2019, data from 2020 was taken into consideration. The 2020 data was gathered during the same months as the data from 2019 to observe sensor degradation and the possibility of calibration based on the data from the preceding year. The obtained results show that a valuable improvement on the sensors' performance can be achieved by using 2019 data in combination with just 4 days from a respective month in 2020. Regarding sensor degradation, the results are promising for the CO and NO_2 sensors, while the PM10 sensor had significant degradation in the second half of 2020.

Finally, the results of the research have shown that the low-cost sensors with adequate correction algorithms could be used as good support for the current traditional air quality monitoring stations. A detailed analysis performed on how low-cost sensors could be used for measurements in order to increase spatial and temporal measurement resolution together with public reference monitoring stations, i.e., a hybrid sensor network approach is elaborated.

For future work, the influence of weather conditions on other types of pollutant measurements using low-cost sensors (SO_2, PM2.5, O_3) will be performed. The cross-sensitivity between pollutants can also be measured, by experimenting with different pollutants as input features to the ML algorithms. The development of more complex ML models (1D convolutional neural networks and long short-term memory networks) will also be conducted, which will be trained on larger data samples. Finally, a hybrid sensor network approach will be analyzed in more detail. The possibilities of cross-calibration between low-cost sensors will be performed, by calibrating several low-cost sensors at the same measuring site and analyzing if the calibration models can be swapped between the sensors and still obtain satisfactory results.

Author Contributions: Conceptualization, I.V. and D.D.; methodology, D.D.; software, I.V.; validation, N.G., I.R. and I.P.; formal analysis, I.V.; investigation, I.V., I.R., I.P. and D.D.; resources, I.R.; data curation, I.P. and D.D.; writing—original draft preparation, I.V. and D.D.; writing—review and editing, N.G., I.R. and I.P.; visualization, I.V.; supervision, D.D. and N.G.; project administration, I.R.; funding acquisition, I.R. and I.P. All authors have read and agreed to the published version of the manuscript.

Funding: This research was funded by the Ministry of Education, Science and Technological Development of the Republic of Serbia. The APC is also covered by the Ministry of Education, Science and Technological Development of the Republic of Serbia.

Data Availability Statement: Not applicable.

Conflicts of Interest: The authors declare no conflict of interest.

References

1. State of World Population 2007. Unleashing the Potential of Urban Growth, United Nations Population Fund (UNFPA), Online Report. 2007. Available online: http://www.unfpa.org/public/publications/pid/408 (accessed on 14 March 2021).
2. World Health Organization (WHO). Global Health Observatory (GHO) data. Available online: https://www.who.int/gho/urban_health/situation_trends/urban_population_growth_text/en/ (accessed on 14 March 2021).
3. The World's Cities in 2016. Data Booklet. Available online: https://www.un.org/en/development/desa/population/publications/pdf/urbanization/the_worlds_cities_in_2016_data_booklet.pdf (accessed on 14 March 2021).

4. WHO. *Ambient Air Pollution: A Global Assessment of Exposure and Burden of Disease*; WHO Document Production Services: Geneva, Switzerland, 2016.
5. Air Quality Guidelines for Europe, 2nd ed; WHO Regional Publications, European Series, No 91. 2000. Available online: www.euro.who.int/document/e71922.pdf (accessed on 14 March 2021).
6. Directive 2008/50/EC of the European Parliament and of the Council of 21 May 2008 on Ambient Air Quality and Cleaner Air for Europe OJ L 152, 11.6.2008, p. 1–44. Available online: https://eur-lex.europa.eu/legal-content/en/ALL/?uri=CELEX%3A32008L0050 (accessed on 14 March 2021).
7. Directive 2015/1480/EC of the European Parliament and of the Council of 28 August 2015 on Ambient Air Quality and Cleaner air for Europe. O. J. Eur. Union, 2015, O. J. L 226, 29.8.2015, p. 4–11. Available online: https://eur-lex.europa.eu/legal-content/EN/TXT/?uri=celex:32015L1480 (accessed on 14 March 2021).
8. Air Quality Monitoring System. Available online: https://www.horiba.com/uk/process-environmental/products/system-engineering/air-quality-monitoring-system/ (accessed on 21 April 2021).
9. ekoNET Air Quality Device. Available online: https://ekonet.solutions/air-monitoring/ (accessed on 14 March 2021).
10. Kularatna, N.; Sudantha, B.H. An environmental air pollution monitoring system based on the IEEE 1451 standard for low cost requirements. *IEEE Sens. J.* **2008**, *8*, 415–422. [CrossRef]
11. Drajic, D.; Gligoric, N. Reliable Low-Cost Air Quality Monitoring Using Off-The-Shelf Sensors and Statistical Calibration. *Elektronika Elektrotechnika* **2020**, *26*, 32–41. [CrossRef]
12. CiteairII: Common Information to European Air. Available online: https://www.airqualitynow.eu/download/CITEAIRComparing_Urban_Air_Quality_across_Borders.pdf (accessed on 14 March 2021).
13. Zimmerman, N.; Presto, A.A.; Kumar, S.P.N.; Gu, J.; Hauryliuk, A.; Robinson, E.S.; Robinson, A.L.; Subramanian, R. A machine learning calibration model using random forests to improve sensor performance for lower-cost air quality monitoring. *Atmosph. Meas. Tech.* **2018**, *11*, 291–313. [CrossRef]
14. Bigi, A.; Mueller, M.; Grang, S.K.; Ghermandi, G.; Hueglin, C. Performance of NO, NO2 low cost sensors and three calibration approaches within a real world application. *Atmosph. Meas. Tech.* **2018**, *11*, 3717–3735. [CrossRef]
15. De Vito, S.; Esposito, E.; Salvato, M.; Popoola, O.; Formisano, F.; Jones, R.; Di Francia, G. Calibrating chemical multi-sensory devices for real world applications: An in-depth comparison of quantitative machine learning approaches. *Sens. Actuat. B Chem.* **2018**, *255*, 1191–1210. [CrossRef]
16. Johnson, N.E.; Bonczak, B.; Kontokosta, C.E. Using a gradient boosting model to improve the performance of low-cost aerosol monitors in a dense, heterogeneous urban environment. *Atmosp. Environ.* **2018**, *184*, 9–16. [CrossRef]
17. Spinelle, L.; Gerboles, M.; Villani, M.G.; Aleixandre, M.; Bonavitacola, F. Field calibration of a cluster of low-cost available sensors for air quality monitoring. Part A: Ozone and nitrogen dioxide. *Sens. Actuat. B Chem.* **2015**, *215*, 249–257. [CrossRef]
18. Spinelle, L.; Gerboles, M.; Villani, M.G.; Aleixandre, M.; Bonavitacola, F. Field calibration of a cluster of low-cost available sensors for air quality monitoring. Part B: NO, CO and CO2. *Sens. Actuat. B Chem.* **2017**, *238*, 706–715. [CrossRef]
19. Kumar, P.; Morawska, L.; Martani, C.; Biskos, G.; Neophytou, M.; Di Sabatino, S.; Bell, M.; Norford, L.; Britter, L. The rise of low-cost sensing for managing air pollution in cities. *Environ. Int.* **2015**, *75*, 199–205. [CrossRef]
20. Topalovic, D.; Davidovic, M.; Jovanovic, M.; Bartonova, A.; Ristovski, Z.; Jovasevic-Stojanovic, M. In search of an optimal calibration method of low-cost gas sensors for ambient air pollutants: Comparison of linear, multilinear and artificial neural network approaches. *Atmosp. Environ.* **2019**, *213*, 640–658. [CrossRef]
21. Cordero, J.M.; Borge, R.; Narros, A. Using statistical methods to carry out in field calibrations of low cost air quality sensors. *Sens. Actuat. B Chem.* **2018**, *267*, 245–254. [CrossRef]
22. Di Antonio, A.; Popoola, O.A.M.; Ouyang, B.; Saffell, J.; Jones, R.L. Developing a relative humidity correction for low-cost sensors measuring ambient particulate matter. *Sensors* **2018**, *18*, 2790. [CrossRef]
23. Motlagh, N.H.; Lagerspetz, E.; Nurmi, P.; Li, X.; Varjonen, S.; Mineraud, J.; Siekkinen, M.; Rebeiro-Hargrave, A.; Hussein, T.; Petäjä, T.; et al. Toward Massive Scale Air Quality Monitoring. *IEEE Commun. Mag.* **2020**, *58*, 54–59. [CrossRef]
24. Lagerspetz, E.; Hamberg, J.; Li, X.; Flores, H.; Nurmi, P.; Davies, N.; Helal, S. Pervasive Data Science on the Edge. *IEEE Perv. Comput.* **2019**, *18*, 40–49. [CrossRef]
25. Alhasa, K.M.; Mohd Nadzir, M.S.; Olalekan, P.; Latif, M.T.; Yusup, Y.; Iqbal Faruque, M.R.; Ahamad, F.; Aiyub, K.; Md Ali, S.H.; Khan, M.F.; et al. Calibration Model of a Low-Cost Air Quality Sensor Using an Adaptive Neuro-Fuzzy Inference System. *Sensors* **2018**, *18*, 4380. [CrossRef]
26. Jang, J.-S.R. ANFIS: Adaptive network-based fuzzy inference system. *IEEE Trans. Syst. Man Cybernet.* **1993**, *23*, 665–685. [CrossRef]
27. Jayaratne, R.; Liu, X.; Phong, T.K.; Dunbabin, M.; Morawska, L. The influence of humidity on the performance of a low-cost air particle mass sensor and the effect of atmospheric fog. *Atmosp. Meas. Tech.* **2018**, *11*, 4883–4890. [CrossRef]
28. Samad, A.; Obando Nuñez, D.R.; Solis Castillo, G.C.; Laquai, B.; Vogt, U. Effect of Relative Humidity and Air Temperature on the Results Obtained from Low-Cost Gas Sensors for Ambient Air Quality Measurements. *Sensors* **2020**, *20*, 5175. [CrossRef]
29. Karagulian, F.; Barbiere, M.; Kotsev, A.; Spinelle, L.; Gerboles, M.; Lagler, F.; Redon, N.; Crunaire, S.; Borowiak, A. Review of the Performance of Low-Cost Sensors for Air Quality Monitoring. *Atmosphere* **2019**, *10*, 506. [CrossRef]

30. Lin, C.; Gillespie, J.; Schuder, M.D.; Duberstein, W.; Beverland, I.J.; Heal, M.R. Evaluation and calibration of Aeroqual series 500 portable gas sensors for accurate measurement of ambient ozone and nitrogen dioxide. *Atmos. Environ.* **2015**, *100*, 111–116. [CrossRef]
31. Gao, Y.; Dong, W.; Guo, K.; Liu, X.; Chen, Y.; Liu, X.; Bu, J.; Chen, C. Mosaic: A low-cost mobile sensing system for urban air quality monitoring. In Proceedings of the 35th Annual IEEE International Conference on Computer Communications (IEEE INFOCOM 2016), San Francisco, CA, USA, 10–14 April 2016.
32. Borrego, C.; Ginja, J.; Coutinho, M.; Ribeiro, C.; Karatzas, K.; Sioumis, T.; Katsifarakis, N.; Konstantinidis, K.; De Vito, S.; Esposito, E.; et al. Assessment of air quality microsensors versus reference methods: The EuNetAir Joint Exercise—Part II. *Atmos. Environ.* **2018**, *193*, 127–142. [CrossRef]
33. Esposito, E.; De Vito, S.; Salvato, M.; Bright, V.; Jones, R.L.; Popoola, O. Dynamic neural network architectures for on field stochastic calibration of indicative low cost air quality sensing systems. *Sens. Actuat. B Chem.* **2016**, *231*, 701–713. [CrossRef]
34. Cheng, Y.; Li, X.; Li, Z.; Jiang, S.; Li, Y.; Jia, J.; Jiang, X. AirCloud: A cloud-based air-quality monitoring system for everyone. In Proceedings of the 12th ACM Conference on Embedded Network Sensor Systems (SenSys 14), New York, NY, USA, 3 November 2014; pp. 251–265.
35. Chen, C.-C.; Kuo, C.-T.; Chen, S.-Y.; Lin, C.-H.; Chue, J.-J.; Hsieh, Y.-J.; Cheng, C.-W.; Wu, C.-M.; Huang, C.-M. Calibration of low-cost particle sensors by using machine-learning method. In Proceedings of the 2018 IEEE Asia Pacific Conference on Circuits and Systems (APCCAS 2018), Chengdu, China, 26–30 October 2018; pp. 111–114.
36. Wang, W.-C.V.; Lung, S.-C.C.; Liu, C.-H. Application of Machine Learning for the in-Field Correction of a PM2.5 Low-Cost Sensor Network. *Sensors* **2020**, *20*, 5002. [CrossRef]
37. Xiang, Y.; Piedrahita, R.; Dick, R.P.; Hannigan, M.; Lv, Q.; Shang, L. A hybrid sensor system for indoor air quality monitoring. In Proceedings of the IEEE International Conference on Distributed Computing in Sensor Systems (DCoSS 2013), Cambridge, MA, USA, 20–23 May 2013; pp. 96–104.
38. Johnston, S.J.; Basford, P.J.; Bulot, F.M.J.; Apetroaie-Cristea, M.; Easton, N.H.C.; Davenport, C.; Foster, G.L.; Loxham, M.; Morris, A.K.R.; Cox, S.J. City scale particulate matter monitoring using LoRaWAN based air quality IoT devices. *Sensors* **2019**, *19*, 209. [CrossRef]
39. Schneider, P.; Castell, N.; Vogt, M.; Dauge, F.R.; Lahoz, W.A.; Bartonova, A. Mapping urban air quality in near real-time using observations from low-cost sensors and model information. *Environ. Int.* **2017**, *106*, 234–247. [CrossRef]
40. Penza, M.; Suriano, D.; Pfister, V.; Prato, M.; Cassano, G. Urban Air Quality Monitoring with Networked Low-Cost Sensor-Systems. In Proceedings of the Eurosensors, Paris, France, 3–6 September 2017; p. 573.
41. Chen, F.-L.; Liu, K.-H. Method for rapid deployment of low-cost sensors for a nationwide project in the Internet of things era: Air quality monitoring in Taiwan. *Int. J. Distrib. Sens. Netw.* **2020**, *16*, 550147720951334. [CrossRef]
42. Popoola, O.A.M.; Carruthers, D.; Lad, C.; Bright, V.B.; Mead, M.I.; Stettler, M.E.J.; Saffell, J.R.; Jones, R.L. Use of networks of low cost air quality sensors to quantify air quality in urban settings. *Atmosp. Environ.* **2018**, *194*, 58–70. [CrossRef]
43. Engelhardt, M.; Bain, L.J. *Introduction to Probability and Mathematical Statistics*; Duxbury Press: London, UK, 2000; ISBN 978-053-438-020-5.
44. Jain, A.K.; Mao, J.; Mohiuddin, K.M. Artificial neural networks: A tutorial. *Computer* **1996**, *29*, 31–44. [CrossRef]
45. Breiman, L. Random forests. *Mach. Learn.* **2001**, *45*, 5–32. [CrossRef]
46. Department of Ecology, State of Washington. Air Monitoring Site Selection and Installation Procedure. Available online: https://fortress.wa.gov/ecy/publications/documents/1602021.pdf (accessed on 14 March 2021).
47. Greater London Authority, Guide for Monitoring Air Quality in London. Available online: https://www.london.gov.uk/sites/default/files/air_quality_monitoring_guidance_january_2018.pdf (accessed on 14 March 2021).

sensors

Article

Deployment, Calibration, and Cross-Validation of Low-Cost Electrochemical Sensors for Carbon Monoxide, Nitrogen Oxides, and Ozone for an Epidemiological Study

Christopher Zuidema [1], Cooper S. Schumacher [1], Elena Austin [1], Graeme Carvlin [1], Timothy V. Larson [1,2], Elizabeth W. Spalt [1], Marina Zusman [1], Amanda J. Gassett [1], Edmund Seto [1], Joel D. Kaufman [1,3,4] and Lianne Sheppard [1,5,*]

1 Department of Environmental and Occupational Health Sciences, University of Washington, Seattle, WA 98195, USA; czuidema@uw.edu (C.Z.); coop16@uw.edu (C.S.S.); elaustin@uw.edu (E.A.); gcarvlin@uw.edu (G.C.); tlarson@uw.edu (T.V.L.); espalt@uw.edu (E.W.S.); marinaz@uw.edu (M.Z.); agassett@uw.edu (A.J.G.); eseto@uw.edu (E.S.); joelk@uw.edu (J.D.K.)
2 Department of Civil & Environmental Engineering, University of Washington, Seattle, WA 18195, USA
3 Department of Medicine, University of Washington, Seattle, WA 18195, USA
4 Department of Epidemiology, University of Washington, Seattle, WA 18195, USA
5 Department of Biostatistics, University of Washington, Seattle, WA 18795, USA
* Correspondence: sheppard@uw.edu

Abstract: We designed and built a network of monitors for ambient air pollution equipped with low-cost gas sensors to be used to supplement regulatory agency monitoring for exposure assessment within a large epidemiological study. This paper describes the development of a series of hourly and daily field calibration models for Alphasense sensors for carbon monoxide (CO; CO-B4), nitric oxide (NO; NO-B4), nitrogen dioxide (NO_2; NO2-B43F), and oxidizing gases (OX-B431)—which refers to ozone (O_3) and NO_2. The monitor network was deployed in the Puget Sound region of Washington, USA, from May 2017 to March 2019. Monitors were rotated throughout the region, including at two Puget Sound Clean Air Agency monitoring sites for calibration purposes, and over 100 residences, including the homes of epidemiological study participants, with the goal of improving long-term pollutant exposure predictions at participant locations. Calibration models improved when accounting for individual sensor performance, ambient temperature and humidity, and concentrations of co-pollutants as measured by other low-cost sensors in the monitors. Predictions from the final daily models for CO and NO performed the best considering agreement with regulatory monitors in cross-validated root-mean-square error (RMSE) and R^2 measures (CO: RMSE = 18 ppb, R^2 = 0.97; NO: RMSE = 2 ppb, R^2 = 0.97). Performance measures for NO_2 and O_3 were somewhat lower (NO_2: RMSE = 3 ppb, R^2 = 0.79; O_3: RMSE = 4 ppb, R^2 = 0.81). These high levels of calibration performance add confidence that low-cost sensor measurements collected at the homes of epidemiological study participants can be integrated into spatiotemporal models of pollutant concentrations, improving exposure assessment for epidemiological inference.

Keywords: low-cost sensors; sensor network; hazardous gases; air pollution; exposure assessment; environmental epidemiology

1. Introduction

Air pollution is a major contributor to the global burden of disease [1]. Gaseous pollutants—such as carbon monoxide (CO), oxides of nitrogen (NO_x), and ozone (O_3)—cause a range of deleterious respiratory and cardiovascular health effects [2]. Low-cost sensors and multipollutant low-cost monitors (LCMs) equipped with multiple sensors to measure air pollution are emerging tools that have the potential to change the paradigm in environmental health—one of a limited number of high-quality measurements, from regulatory agency monitors to dense networks of lower-quality sensors and monitors operated by diverse groups of users [3–9]. However, little work has been done to evaluate the application of these sensors—especially gas pollutant sensors—to exposure assessments within the context of epidemiological human health studies, which have different requirements than regulatory/non-regulatory community ambient air monitoring applications [10].

Electrochemical sensors are among the most common types of low-cost gas sensors [3,11]; they rely on a chemical reaction (oxidation or reduction) taking place between a sensor's working electrode (WE) and a target gas, producing an electrical signal proportional to the gas concentration [12,13]. Like other low-cost sensors, electrochemical sensors are small, inexpensive, portable, modular, and consume less power compared to traditional monitoring equipment, allowing for dense, networked deployment [12,14–21]. By increasing spatial coverage, these types of low-cost networks have the potential to contribute to the assessment of air pollution exposure, and can be used in epidemiological studies relying on the characterization of exposures at specific times and locations relevant to the health outcomes observed for study participants [22,23].

To overcome the lower accuracy, precision, sensitivity, and specificity of low-cost sensors, end users must rigorously calibrate them in the field/laboratory [6,12,24]. Many researchers have described procedures for calibrating electrochemical sensors in the field [6,13,17,25–29], which has generally been favored over laboratory calibration, because it is difficult to simulate ambient, real-world conditions—such as low target species concentrations, co-pollutants, and large ranges of physical parameters, such as temperature and relative humidity (RH) [23]. Additionally, recent reports advocating for standardized protocols for testing and evaluating sensor performance highlight the need for increased confidence in data quality and the demand for low-cost sensors among diverse groups [30].

Recent electrochemical sensor calibration studies have generally found that machine learning algorithms such as k-nearest neighbors, clustering, random forests, and neural network models outperform multiple linear regression models [26,31–36]. However, there is concern that unsupervised machine learning approaches treat these sensors as "black boxes", when in fact they are based on electrochemistry and designed to respond linearly to increasing concentrations of specific pollutant species when controlling for relatively few environmental covariates [3,12,13]. For this mechanistic reason, and to protect against model overfitting and a reliance on opaque machine learning algorithms, we favor a multiple linear regression approach. Additionally, multiple linear regression models offer several advantages compared to machine learning methods; these include the: (1) ease of implementation, model building, and parameter interpretation; (2) ability to generalize beyond the range of the training data; (3) provision of best estimates of offset and gain calibration terms; (4) lower data requirements; and (5) direct application to raw sensor data to obtain calibrated concentrations [37].

In this study, we used regulatory monitoring data from the Puget Sound region (encompassing the Seattle–Tacoma, WA metropolitan area) to develop and evaluate field calibration models for Alphasense carbon monoxide (CO), nitrogen monoxide (NO), nitrogen dioxide (NO_2), and ozone (O_3) B4 series gas sensors built into networked, multipollutant LCMs. We demonstrate and offer practical strategies to approach and evaluate sensor calibration, specifically for an audience of epidemiological researchers, who are familiar with multiple linear regression methods. In future works, we plan to incorporate these LCM network predictions into spatiotemporal models of air pollution that will be used in the exposure assessment of participants in two long-term epidemiological studies.

2. Materials and Methods

2.1. Study Context

This calibration study takes place within the context of two large epidemiological cohorts exploring relationships between air pollution and deleterious health effects: the "Adult Changes in Thought Air Pollution" (ACT-AP) study [38] and the "Multi-Ethnic Study of Atherosclerosis and Air Pollution" (MESA Air) study [39]. The ACT-AP study investigated the associations between chronic exposure to air pollution and the effects on brain aging and the risk of Alzheimer's disease, and was based in the Puget Sound region. The MESA Air study investigated the relationships between exposure to air pollutants and the progression of cardiovascular disease in cities in New York, Maryland, North Carolina, Minnesota, Illinois, and California. The LCMs used in both the ACT-AP and MESA Air studies shared key parts of their calibration in the Puget Sound, even though there are no MESA Air cities within the region. In both of these studies, the health outcomes are thought to be, in part, related to ambient air pollution exposure, and the goal of the exposure assessment was to obtain time-averaged air pollution concentrations incorporating data from calibrated low-cost gas sensors at the residential locations of study participants—a typical approach in air pollution epidemiological studies.

The focus of this analysis is on the Puget Sound findings, where most of our data were collected, while in Appendix A, we also provide results from one of the MESA Air cities—Baltimore, MD. Baltimore has very different environmental conditions compared to the Puget Sound, and the goals of that analysis were to (1) determine whether calibration procedures carried out in the Puget Sound region translated well to Baltimore, given their environmental differences; and (2) explore calibration options with limited co-location data, using data from both the Puget Sound and Baltimore co-location periods.

2.2. Low-Cost Monitor Deployment

From May 2017 to March 2019, we deployed 54 low-cost monitors for the ACT-AP and MESA Air studies, rotating the monitors in at least two seasons to among over 100 residential locations for the ACT-AP study (many at ACT-AP participant homes) and two regulatory agency monitoring sites measuring gas pollutants in the Puget Sound region. (Additional details about the MESA Air co-location in Baltimore are presented in Appendix A). All LCMs were periodically co-located at Puget Sound Clean Air Agency (PSCAA) sites throughout the study, and air pollutant reference data collected during periods of co-location form the basis for the sensor calibration. LCMs calibrated in this study were also rotated out of the Puget Sound region in order to collect data in other MESA Air cities.

2.3. Low-Cost Monitor and Sensor Descriptions

The LCMs were designed and constructed at the University of Washington. Each LCM was built with four electrochemical gas sensors—CO-B4, NO-B4, NO2-B43F, and OX-B431 (Alphasense Ltd., Great Notley, UK)—which detect CO, NO, NO_2, and O_3 + NO_2, respectively (Table 1). These gas sensors were selected because of their price (USD ~200), availability of sensors for gases of interest, performance, and ease of use compared to other sensor types (e.g., metal oxide sensors). The LCMs were also equipped with sensors for temperature and RH (HumidIcon HIH6130-021-001, Honeywell International Inc., Charlotte, NC). We did not include ambient air pressure sensors in the LCMs (nor did we investigate the inclusion of pressure in our calibration models), since electrochemical sensors do not meaningfully respond to changes in ambient air pressure [12,40]. The LCMs also had pairs of two different types of particulate matter sensors (Shinyei PPD42NS and Plantower PMS A003); in previous work, we have reported on the calibration and performance of these particulate matter sensors during the 2017–2018 time period [41]. Ancillary and supporting hardware included a thermostatically controlled heater, a fan, a memory card, a modem, and a microcontroller running custom firmware to sample, save, and transmit LCM data every five minutes to a secure server. Additional information about the design, specifications, and construction of the LCMs is provided in the Supplementary Materials.

Table 1. Summary of Alphasense Ltd. (Great Notley, UK) gas sensors used in the low-cost monitor network.

Model	Analyte(s)	Sensor Noise (ppb) [1]	Range (ppm) [2]	Reference
CO-B4	CO	4	1000	[42]
NO-B4	NO	15	20	[43]
NO2-B43F	NO_2	12	20	[44]
OX-B431	O_3, NO_2	4	20	[45]

[1]: Statistical uncertainty described by the manufacturer as ±2 standard deviations of measurements expressed in ppb. [2]: Limit of performance warranty.

To address the well-known issue of NO_2–O_3 cross-sensitivity, in our LCMs we implemented an industry strategy where a pair of similar oxidizing gas-type sensors is deployed—one with an O_3 filter between the sensor and the atmosphere that permits the detection of NO_2 only (NO2-B43F), and one unfiltered sensor that detects both NO_2 and O_3 (OX-B431). The filter, composed of manganese dioxide (MnO_2), acts as a catalyst in the decomposition of O_3 to O_2 [46]. By determining the NO_2 concentration via the NO2-B43F sensor, the OX-B431 sensor signal can be used to calculate the O_3 concentration [46]. The electrochemical sensors in our LCMs were also equipped with an auxiliary electrode (Aux), which provides a method of accounting for sensor drift, because it ages in the same way as the WE, but is not permitted to interact with the environment, including the target gas, temperature, and RH.

2.4. Co-Location of LCMs with Air Quality System Monitors

The US Environmental Protection Agency (EPA) collects and reports air quality and air pollution data from monitors operated by federal, state, local, and tribal air pollution control agencies through their Air Quality System (AQS). The principles of operation of AQS direct-reading instruments for gaseous pollutants vary for different gases [47], and in the Puget Sound region, instruments employ gas nondispersive infrared radiation (CO), chemiluminescence (NO, NO_2), and ultraviolet absorption (O_3) spectroscopy. Regulatory data were obtained from the EPA's AQS server and the PSCAA website [48,49]. The locations of regulatory agency monitoring sites (hereafter referred to as "agency sites") and a description of their setting are shown in Table 2. The data quality objectives (DQOs) for agency measurements require that the bias and percentage coefficient of variation be within (±) 10%, 15%, 15%, and 7% for CO, NO, NO_2, and O_3, respectively. A summary of agency DQOs for Beacon Hill for the study period is provided in Supplementary Table S1; the agency met its DQOs during all quarters of this calibration study. A schematic of the main LCM co-location site, Beacon Hill, is provided in Supplementary Figure S1. Note that 10th and Weller is a near-roadway site downwind of a major interstate highway and, thus, has higher concentrations of traffic pollution (CO, NO, and NO_2) than Beacon Hill. Furthermore, 10th and Weller does not measure O_3, because it typically forms further downwind of roadways.

Table 2. Summary of agency site characteristics, co-colocation statistics, and average gas concentrations during co-location with LCMs, temperature, and relative humidity.

Agency Site	Site Type	# LCMs Ever Co-Located	Co-Location Monitor-Days (Weeks)	CO (ppb) Mean ± SD [1]	NO (ppb) Mean ± SD [1]	NO_2 (ppb) Mean ± SD [1]	O_3 (ppb) Mean ± SD [1]	Avg Temp (°C) Mean ± SD [2]	Avg RH (%) Mean ± SD [2]
Beacon Hill	Suburban	54	204,498 (99)	223 ± 89	6 ± 10	11 ± 5	20 ± 9	11 ± 4	76 ± 12
10th and Weller	Urban	1 [3]	525 (89)	422 ± 131	27 ± 18	20 ± 7	— [4]	13 ± 5	72 ± 11

[1]: The average concentration, temperature, and RH values were averaged across daily observations at the site when both LCM and agency reference data were available, and therefore depend on co-location schedule, which differs across sites. [2]: The average temperature and RH values were estimated based on the LCM sensors, and then were calibrated with reference temperature and RH data from the Beacon Hill site in order to provide standard units. [3]: The LCM co-located at 10th and Weller was also briefly co-located at Beacon Hill. [4]: Ozone was not measured at 10th and Weller station.

2.5. Sensor Quality Assurance and Data Exclusion Criteria

Automated weekly reports were created to identify data quality issues from LCMs and allow for timely replacement of broken sensors. Sensor data were flagged for several quality criteria, including data completeness, departure from a typical range of values or daily variation, and correlation with nearby LCMs. Flags were developed with multiple levels of severity for each quality criterion, and then a weighting of flags was used to prioritize which sensors were most important to replace. Reports were developed with R markdown and CSS/HTML in a 3-panel format designed for clear and efficient communication of large amounts of information: a flag table panel clearly identified the highest priority issues; a navigation panel allowed for easy navigation to further information on any issue, and the main panel included the complete plots and tables for all sensors (Figure S2).

Throughout the study period we excluded data from malfunctioning sensors identified in our automated weekly reports and data from the first eight hours after LCMs were moved to a new location (giving LCMs time to warm up). Errors and malfunction that led to missing data included a broken sensor, data failing to be recorded, clock-related errors (e.g., no valid time recorded by the LCM), LCM power loss (e.g., LCM was unplugged), and data transmission failure. We also identified periods of high air pollution associated with the wildfire season and holiday fireworks (4 and 5 July) and excluded sensor data during model fitting to prevent high outlier concentrations from having undue influence on our calibration models, and for consistency with PM sensors in the network. In sensitivity analyses, the inclusion/exclusion of these potentially higher concentration periods had a negligible effect on LCM calibration models.

2.6. Calibration Models

Calibration models were developed using data between May 2017 and March 2019. LCMs recorded and reported data every five minutes, which were then averaged to the hourly and daily time scales. After data exclusions, we required a minimum of 75% data completeness on the five-minute timescale before averaging to the hourly or daily scales (i.e., at least 9 out of 12 5-min data points were required to include the hourly average in our analysis).

We started by estimating pollution concentrations using the manufacturer's provided calibration terms:

$$Gas\ Concentration = \frac{[(WE - V_{0,WE}) - (Aux - V_{0,\ Aux})]}{sensitivity * gain}, \tag{1}$$

The manufacturer provides both sensor-specific values of V_o and *sensitivity* upon purchase, as well as "typical" values for each type of sensor in its documentation [50]—both of which we investigated.

Next, we built a series of stepwise multiple linear regression calibration models for each gas on both the hourly and daily timescales, including WE and Aux values as separate independent terms. Additional terms included sensor ID (categorical), temperature (linear), RH (linear), interactions between the WE and temperature and WE and RH, and co-pollutant concentrations. We explored including sensor-specific slopes and sensor-specific intercepts as well as sensor-specific intercepts and common slopes, because each sensor could potentially have its own unique calibration slope and intercept. Sensor-specific intercepts were estimated by calculating baseline adjustments through an algorithm that leveraged co-location periods shared by different sensors, and assumed that the difference in baseline between sensors remained constant.

The simplest multiple linear regression model we developed (Model 1 for each gas) included terms for WE, Aux, and sensor ID; using O_3 as an example, it took the form:

$$Y_t = \beta_0 + \boldsymbol{\beta_1} \times I(ID) + \beta_2 \times WE_{ID,t}^{OX-B431} + \beta_3 \times Aux_{ID,t}^{OX-B431} + \epsilon_{ID,\ t}, \tag{2}$$

where Y_t = observation of the agency O_3 measurement (ppb) at time *t* co-located with OX-B431 sensor *ID*; β_0 and the vector $\boldsymbol{\beta_1}$ allow for sensor-specific intercepts; β_2 and β_3 = regression coefficients for WE and Aux sensor signals, respectively; I(*ID*) = unique sensor ID coded as n-1 (53) indicator (i.e., factor) variables—one for each LCM other than the reference LCM; $WE_{ID,t}^{OX-B431}$ = signal from the working electrode in mV; $Aux_{ID,t}^{OX-B431}$ = signal from the auxiliary electrode in mV; and $\epsilon_{ID,t}$ = random error. The final calibration models for each gas were more complex, and in addition to WE, Aux, and sensor ID, important terms in our model building included temperature, RH, interactions between the WE and temperature and WE and RH, and co-pollutants. For example, the final model for O_3 (Model 4) was:

$$\begin{aligned} Y_t = \beta_0 + \boldsymbol{\beta_1} \times I(ID) + \beta_2 \times WE_{ID,t}^{OX-B431} + \beta_3 \times Aux_{ID,t}^{OX-B431} \\ +\beta_4 \times NO2_{ID,t}^{cal} + \beta_5 \times Temp_{ID,t}^{spl-1} + \beta_6 \times Temp_{ID,t}^{spl-2} + \beta_7 \times Temp_{ID,t}^{spl-3} \\ +\beta_8 \times RH_{ID,t}^{spl-1} + \beta_9 \times RH_{ID,t}^{spl-2} + \beta_{10} \times Temp_{ID,t}^{spl-1} \times WE_{ID,t}^{OX-B431} \\ +\beta_{11} \times Temp_{ID,t}^{spl-2} \times WE_{ID,t}^{OX-B431} + \beta_{12} \times Temp_{ID,t}^{spl-3} \times WE_{ID,t}^{OX-B431} \\ +\beta_{13} \times RH_{ID,t}^{spl-1} \times WE_{ID,t}^{OX-B431} + \beta_{14} \times RH_{ID,t}^{spl-2} \times WE_{ID,t}^{OX-B431} + \epsilon_{ID,t} \end{aligned} \quad (3)$$

where Y_t, β_0, $\boldsymbol{\beta_1}$, I(ID), $WE_{ID,t}^{OX-B431}$, $Aux_{ID,t}^{OX-B431}$, and $\epsilon_{ID,t}$ have the same definitions as in Equation (2) above; β_2–β_{14} = regression coefficients; $NO2_{ID,t}^{cal}$ is the previously calibrated concentration of NO_2 determined by the NO2-B43F sensor in the same monitor as OX-B431 sensor *ID*; $Temp_{ID,t}^{spl-k}$ = *k*th basis functions of the temperature b-splines (knots at 4 and 21 °C), based on the temperature sensor in the same monitor; and $RH_{ID,t}^{spl-j}$ = *j*th basis functions of the relative humidity b-splines (knot at 60%), based on the RH sensor in the same monitor. Interaction terms between the temperature, RH, and working electrodes are also included for more flexible adjustment for temperature effects on the low-cost sensors. If multiple sensors $(ID_1, ID_2, \ldots, ID_m)$ are co-located at an agency site at the same time *t*, then the observed agency measurements Y_t will be the same. Final calibration models for each gas are presented in the Supplementary Materials (Equations (S1)–(S4)).

In addition to the calibration models developed for the Puget Sound region, in Appendix A, we briefly discuss models specific to Baltimore (one of the MESA Air study cities).

2.7. Cross Validation and Model Evaluation

We evaluated models with a 10-fold cross-validation (CV) technique, following prior methods used for PM sensors [41]. Model performance was evaluated with cross-validated summary measures, including the root-mean-square error (RMSE) and R^2, as well as with residual plots with reference concentration measurements, temperature, RH, and time. The 10-fold CV approach randomly partitions weeks of monitoring with co-located LCM and agency reference data into 10 folds. Typically, 10-fold CV partitions data based on individual observations. However, using data from adjacent days to both fit and evaluate models could result in artificially inflated performance measures. To minimize the effects of temporal correlation on our CV evaluation measures, we disallowed data from the same calendar week from being used to both train and test the models.

To assess sensor baseline drift over time, we modeled changes in residuals—between low-cost sensor predictions fitted with final calibration models and agency reference measurements—against deployment time. We used the slope of this best fit of residuals over time to estimate drift, focusing on sensors that were co-located with agency reference instruments over a period of at least one year, for at least 20% of the time. For the sensor of each type that had the longest duration of co-location at an agency site, we plotted the residuals between low-cost sensor predictions from the final calibration model and agency reference measurements over time.

All statistical analyses were carried out with R version 3.6.2.

3. Results

3.1. Site Descriptive Characteristics and LCM Co-Location

Measures of LCM co-location, including the number of monitor days and number of unique weeks with co-located LCMs, temperature, RH, and gas pollutants measured by reference instruments at each of the two agency sites are summarized in Table 2. Automated weekly reports identified malfunctions and led to replacement of 1, 9, 3, and 9 sensors for CO, NO, NO_2, and O_3, respectively. The Beacon Hill site had reference instruments for each of the gases under study, was co-located with each of 54 LCMs over the course of the study, and served as our primary calibration site. Beacon Hill is described as a "suburban" site by the agency, though it is located within the Seattle city limits, and is generally thought of as capturing "typical urban air quality impacts" for the region [51]. This site also generally had lower average pollutant concentrations compared to the 10th and Weller site, which had one co-located LCM for the study period (this LCM was also briefly co-located at Beacon Hill). The PSCAA considers the 10th and Weller site to be an "urban center" and a "near-road" site, located adjacent to an eight-lane highway with six additional on- and off-ramps (the distance from the station to the middle of these 14 lanes is ~60 m, and 6 m from the nearest on-ramp).

Based on the LCMs' total deployment time (i.e., the sum of co-located and non-co-located days), the percentage of time with co-located LCM-agency reference measurements was 16% (O_3), 20% (CO), and 21% (NO, and NO_2). LCM deployment for each gas is displayed in Figure S3. Co-located times were used for calibration (black points of Figure S3). Data from times when LCMs were not co-located with agency monitors but were deployed at volunteers' or study participants' houses are represented by red points of Figure S3. These LCM measurements at residential locations will be input into regional spatiotemporal pollutant models in order to improve estimates of gas pollutant exposure for participants in the ACT-AP study. One LCM remained co-located at each agency site for all or nearly all of the study period; all other LCMs were relocated throughout the study region, and included brief co-location periods at agency sites for calibration purposes. Due to QA/QC exclusions, downtime for movement or maintenance, and periods when LCMs were rotated outside of the Puget Sound region to other MESA Air cities, there were times when LCMs did not contribute to calibration or measurement data (times with neither black nor red points in Figure S3).

3.2. Evaluation of Calibration Models

Summaries of daily scale models for each gas with their performance measures are presented in Table 3 and Table S2. The NO_2 sensor showed the greatest improvement in CV performance statistics, from a basic model—which included terms for the WE, Aux, and sensor ID (Model 1: CV-RMSE = 5 ppb; CV-R^2 = 0.35)—to the final model, which included additional terms for temperature, RH, interactions between the WE and temperature splines (knots at 4 and 21 °C) and WE and RH spline (knot at RH = 60%), and CO concentration from the CO-B4 sensor (Model 4: CV-RMSE = 3 ppb; CV-R^2 = 0.79). In contrast, CO benefitted the least from the inclusion of additional terms from the basic model—which included terms for WE, Aux, and sensor ID (Model 1: CV-RMSE = 29 ppb; CV-R^2 = 0.94)—to the final model selected, which included additional terms for temperature, RH, and interaction terms between the WE and temperature and WE and RH (Model 3: CV-RMSE = 18 ppb; CV-R^2 = 0.97). To gauge sensor-specific variability, we estimated the variation of the sensor-specific intercepts across sensors for both the simplest model (Model 1) and the final model for each gas (Table S3). The final model standard deviations were 40, 24, 24, and 62 ppb for CO, NO, NO_2, and O_3, respectively.

Comparisons of daily LCM predictions using final daily calibration models and agency reference measurements are shown in Figure 1 and, overall, are in good agreement, with most data falling near and distributed evenly about the 1:1 line, as highlighted by the best fit LOESS smoother in blue. Residuals of low-cost sensor predictions calculated from final calibration models versus agency reference concentrations, temperature, and RH for CO, NO, NO_2, and O_3 are shown in Figure 2. Generally, the residuals were centered around zero, and did not exhibit trends with reference concentrations, temperature, or RH. Results from calibration models built and evaluated on the hourly scale generally followed those on the daily scale, and are presented in the Supplementary Materials (Table S2).

Table 3. Summary of daily model terms and performance measures for the manufacturer's calibration, a simple calibration model, and the final calibration model for CO, NO, NO_2, and O_3.

Gas	Terms	Model Number	CV-RMSE (ppb)	CV-R^2
CO	Manufacturer's sensor-specific slope and intercept [1]	0	150	0.49
	WE, Aux, and sensor ID	1	29	0.94
	WE, Aux, sensor ID, temperature, RH, and WE–temperature and WE–RH interactions	3	18	0.97
NO	Manufacturer's sensor-specific slope and intercept [1]	0	36	0.41
	WE, Aux, and sensor ID	1	2	0.97
	WE, Aux, Sensor ID, and temperature and RH splines [2] with WE interactions	4	2	0.97
NO_2	Manufacturer's sensor-specific slope and intercept [1]	0	24	0.08
	WE, Aux, and sensor ID	1	5	0.35
	WE, Aux, Sensor ID, temperature and RH splines [2] with WE interactions, and $[CO]_{CO\text{-}B4}$ [3]	4	3	0.79
O_3	Manufacturer's sensor-specific slope and intercept [1]	0	41	0.04
	WE, Aux, and sensor ID	1	5	0.66
	WE, Aux, Sensor ID, temperature and RH splines [2] with WE interactions, and $[NO_2]_{NO2\text{-}B43F}$ [4]	4	4	0.81

[1]: RMSE and R^2 summary measures not cross-validated. [2]: Spline knots: temperature = 40, 70 °F, RH = 60%. [3]: Previously calibrated CO concentration from the CO-B4 sensor. [4]: Previously calibrated NO_2 concentration from the NO2-B43F sensor.

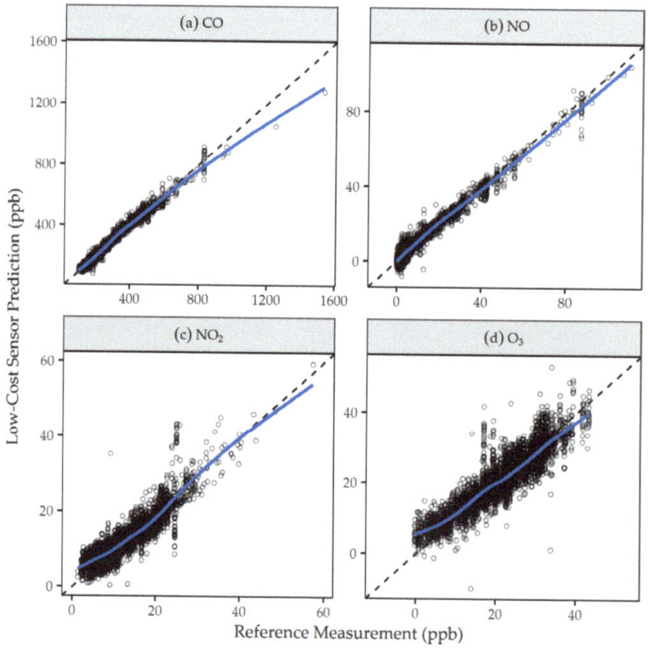

Figure 1. Comparison of daily agency reference measurement versus low-cost sensor predictions derived from the final daily models for: (**a**) CO, (**b**) NO, (**c**) NO_2, and (**d**) O_3. The dashed line is the 1:1 line; and the solid blue line is the LOESS smoother.

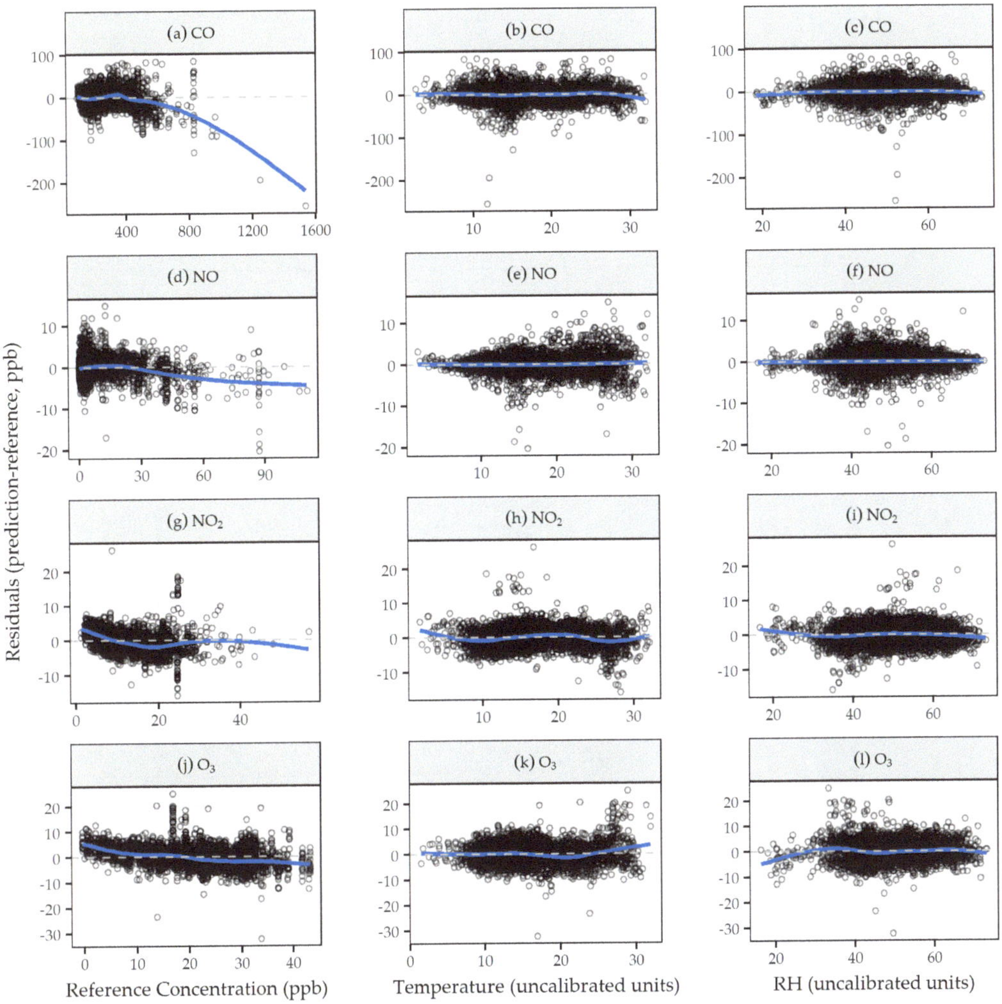

Figure 2. Residuals of low-cost sensor predictions calculated from final daily calibration models against daily agency reference measurements, temperature, and RH for: **(a–c)** CO, **(d–f)** NO, **(g–i)** NO_2, and **(j–l)** O_3. The dashed line is y = 0; and the solid blue line is the LOESS smoother.

We observed drift in each type of sensor in our network over the deployment period. The modeled changes in residuals from daily sensor predictions fitted with the final calibration models, which are an estimate of average drift, are summarized in Table S4. The mean drift (range) for each type of sensor was –11 (–21, 18); –1 (–4, 2); 1 (–3, 5); and –6 (–11, 2) ppb for CO, NO, NO_2, and O_3, respectively. Examples of this estimate of sensor drift over time are shown in Figure 3; we chose to display LCM ACT7 located at 10th and Weller for CO, NO, and NO_2, and LCM ACT2 at Beacon Hill for O_3, because these were the LCMs that spent the most amount of time co-located with an agency reference instrument (O_3 was only monitored at Beacon Hill).

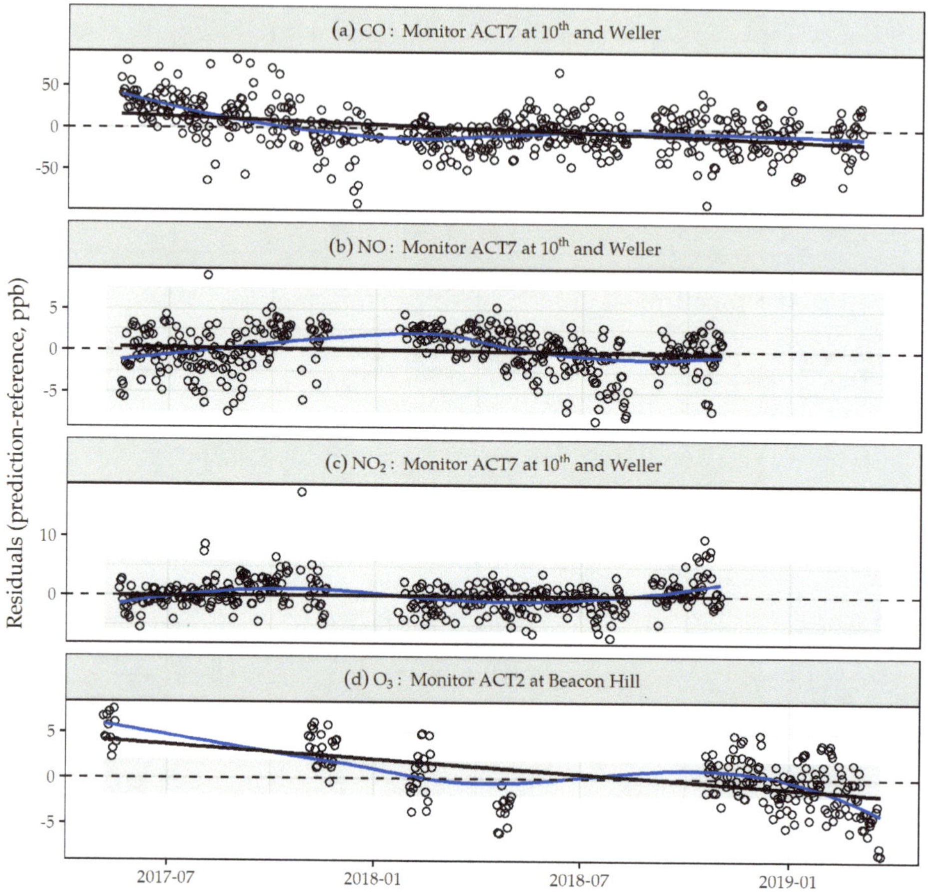

Figure 3. Examples of low-cost sensor residuals between final daily model predictions and agency reference measurement for: (**a**) CO, (**b**) NO, (**c**) NO_2, and (**d**) O_3 over the study period. Residuals over time are a proxy for drift that may also capture sources of variation not completely adjusted for in calibration. The dashed line is y = 0; the solid blue line is the LOESS smoother; and the solid black line is a least squares fit, the slope of which corresponds to the estimates provided in Table S4, while the shaded area indicates the range of the manufacturer's estimate of sensor noise provided in Table 1 (Note: axis was restricted for CO, omitting two outlying data points below −100 ppb).

4. Discussion

In this study, we demonstrated the successful deployment, field calibration, and cross-validation of a low-cost sensor network for multiple gaseous pollutants over multiple seasons and a wide range of pollutant concentrations representative of the study area. We considered multiple calibration models on the hourly and daily time scales, and showed the gains in sensor prediction performance that can be achieved by building a series of multiple linear regression models, starting with the primary variables WE, Aux, and sensor ID.

The CV-RMSE and CV-R^2 of our final daily calibration models met or exceeded the performance measures reported in other recent studies [19,28,52], providing evidence that a high level of performance compared to agency reference measurements can be attained with rigorous calibration procedures (CO: RMSE = 18 ppb, R^2 = 0.97; NO: RMSE = 2 ppb, R^2 = 0.97; NO_2: RMSE = 3 ppb, R^2 = 0.79; O_3: RMSE = 4 ppb, R^2 = 0.81). For

CO, NO, and O_3, the biggest performance gains in terms of the CV-RMSE and CV-R^2 were made between the manufacturer's calibration model and a basic multiple linear regression model that included terms for the working and auxiliary electrodes and sensor ID. For NO_2, the improvement in CV-R^2 between measurements using the manufacturer's calibration model and the basic model, and between the basic and final models, were comparable.

The results for the range of multiple linear regression models constructed exhibits the value of adding additional calibration terms; however, additional terms did not necessarily result in improved performance (Table S2). For example, for models that implemented an algorithm to calculate sensor-specific intercepts by making baseline adjustments to the WE and Aux during co-location periods shared by different sensors (Models 6 and 7), performance was not improved in the Puget Sound region. The algorithm did, however, improve model performance in another MESA Air city (Baltimore, MD), where there were more limited co-location data on which to perform a field calibration (details from Baltimore are provided in Appendix A). The models with the highest CV-RMSE and CV-R^2 were not necessarily chosen as final models, because we also considered simplicity of implementation, a trade-off of added modeling complexity for the marginal improvements observed, our desire to align model forms across pollutants for consistency, and caution of overfitting (the latter specifically relevant to Models 5–7). Our series of models provides a guide on the nature and complexity of the calibration required for a given level of performance.

Our results confirm the importance of inter-sensor differences, particularly calibration intercept terms, and the effects of temperature and RH on sensor response, consistent with previous studies [52,53], and justify their inclusion in calibration models. For two of the gases (NO_2 and O_3), we observed that sensor performance was dependent on inclusion of other gases in the calibration model, although the reasons differed. For example, we found that including the low-cost CO sensor predictions in the NO_2 sensor calibration model may have improved calibration performance because the two gases share a common traffic-related source, and the concentration of CO can provide information on the calibration of NO_2. In contrast, creating the best O_3 model depends on the inclusion of NO_2 concentration due to the function of the OX-B431 sensor, since its output is the combination of the signal from NO_2 and O_3, and therefore requires the concentration of NO_2, which is determined using the previously calibrated NO2-B43F sensor. In other words, the order in which sensors are calibrated matters.

Even though our low-cost sensors were equipped with auxiliary electrodes to counter the effects of aging, we still observed changes in sensor drift over time. The potential effects of this drift differed by gas, given the noise of the sensors' signals and the low mean pollutant concentrations in the study region. For example, the observed mean drift (range) among 10 CO sensors was −11 (−21, 18) ppb, highly variable, greater than the sensor noise, and between 3 and 5% of the mean pollutant concentrations measured by agency monitors. In contrast, the observed mean drift (range) among 12 NO_2 sensors was 1 (−3, 5) ppb, more uniform, less than the sensor noise, and approximately 10% of the typical concentrations. While the range of calibration models we built addressed several of the well-documented challenges of these low-cost gas sensors (including sensor-specific calibration slopes and intercepts, physical parameters such as temperature and RH, and cross-sensitivity with co-pollutants), we chose not to account for the effects of baseline drift that were not captured in other variables. Instead, we characterized the drift using the residuals of predictions from our final models. While an imperfect proxy for drift (for example, because our final models may not perfectly capture seasonal fluctuations or other unaccounted for factors), the results are easily converted to and interpreted as changes in gas concentration.

We faced several logistical and methodological challenges in calibrating and deploying these gas sensors for epidemiology. The LCMs in our network were generally limited in their co-location with agency reference instruments, because extended periods of co-location prevented an LCM from being deployed elsewhere in the study region at the homes of ACT-AP study participants for pollutant exposure predictions. These competing interests forced a compromise between duration of co-location in order to achieve better calibration and deployment for epidemiological purposes. Because of sensor-specific responses, each low-cost sensor would have ideally been repeatedly co-located with agency reference instruments at the same time, in order to avoid differences in calibration conditions, and for enough time to be exposed to the full ranges of pollutant concentration, temperature, and RH. Multiple simultaneous co-location periods would also assist in quantifying sensor drift. In practice, this ideal scenario was not possible due to space and logistical constraints at agency sites; however, a compromise involving groups of sensors with shared schedules may have been better than our less rigorously designed timing.

A compromise design may have allowed for more convenient adjustment of sensor-specific differences, thus improving our ability to address other calibration challenges. In contrast with our previous experience with low-cost PM sensors, which did not exhibit such prominent sensor-specific differences, the same sensor co-location design was not as problematic because the PM sensors did not require sensor-specific adjustments [41]. In hindsight, our study design was better suited for low-cost PM sensor calibration rather than gas sensors, because it allowed for both long and continuous periods with agency reference instruments for calibration and deployment at many other sites in and beyond the Puget Sound region. Another challenge we encountered in the Puget Sound region using these low-cost sensors was that typical pollution levels were often lower than the noise of the sensors' signals, which is often used in the estimation of the limit of detection. For example, the sensor noise for NO reported by the manufacturer is 15 ppb, and 66% of all agency NO measurements were below 15 ppb (92% at Beacon Hill and 25% at 10th and Weller). With typical NO concentrations less than 15 ppb (Table 2), it is not surprising that 12% of NO sensor predictions were below zero.

In this study all of our calibration procedures to produce low-cost sensor predictions were completed post-deployment, and only retrospectively did we predict gas concentrations with LCMs. While this procedure suits our ultimate epidemiological objectives, where long-term average pollutant concentrations are required for exposure assessment, this may not be practicable for end users who require more immediate or "real-time" predictions from low-cost sensors. The potential for sensors to serve as real-time direct-reading instruments is compelling; however, the error associated with those predictions may be higher if the sensors undergo a less rigorous or extensive calibration procedure.

5. Conclusions

This paper demonstrates the field calibration of low-cost electrochemical gas sensors in an LCM network with regulatory agency monitoring data. Models using manufacturer-provided calibration terms performed poorly. However, the performance of the sensors improved substantially with rigorous multiple linear regression calibration procedures. We found that the inclusion of environmental factors—such as temperature and RH, co-pollutants, and terms for sensor ID—was important, contributing to performance gains. Increasing the duration of sensor co-location with regulatory agency instruments to improve calibration models is at odds with deployment for measurement purposes, and these competing interests must be managed. Calibrated low-cost electrochemical gas sensor data can provide measurements of ambient air pollution that have the potential to improve exposure assessment in environmental epidemiology studies.

Supplementary Materials: The following are available online at https://www.mdpi.com/article/10.3390/s21124214/s1: Materials and Methods: Low-cost monitor and sensor descriptions; Equations (S1–S5): Final calibration models for CO, NO, NO_2, and O_3; Table S1: Summary of quarterly agency data quality indicators for the study period at the Beacon Hill site. Target data quality objectives are provided for each gas; Table 2: Descriptions of calibration models with summary performance statistics of sensor predictions. Models were fitted and predictions were generated on the same timescales (hourly or daily); Table S3: Estimates of intercept variability across sensors for simple and final daily scale calibration models (in ppb); Table S4: Estimated sensor drift for monitors co-located with agency reference instruments over at least one year, estimated in ppb by estimating the slope of a best fit least squares regression of residuals over time; Figure S1. Schematic of the main low-cost monitor calibration site, Beacon Hill, in Seattle, WA; Figure S2: Example of automated weekly QA/QC reports to identify sensor errors and exclude data; Figure S3: Deployment of low-cost monitors in the Puget Sound region for CO, NO, NO_2, and O_3. Black indicates days LCMs were co-located with an agency reference instrument, and red indicates days they were not co-located. Monitors at the top of each panel were MESA Air monitors and located outside of the Puget Sound region for much of the study period, and during those times contributed neither calibration data nor data characterizing pollutant concentrations in the Puget Sound; Data S1: Daily calibration dataset.

Author Contributions: Conceptualization, E.S., J.D.K., and L.S.; methodology, C.S.S. and L.S.; software, C.Z., C.S.S., E.A., and G.C.; validation, C.S.S.; formal analysis, C.Z. and C.S.S.; investigation, C.Z., C.S.S., and A.J.G.; resources, E.S, J.D.K., and L.S.; data curation, C.S.S., E.W.S., and A.J.G.; writing—original draft preparation, C.Z. and C.S.S.; writing—review and editing, C.Z, C.S.S., E.A., G.C., T.V.L., E.W.S., M.Z., A.J.G., E.S., and L.S.; visualization, C.Z. and C.S.S.; supervision, E.W.S., A.J.G., E.S., and L.S.; project administration, E.W.S., A.J.G., E.S., J.D.K., and L.S.; funding acquisition, J.D.K. and L.S. All authors have read and agreed to the published version of the manuscript.

Funding: This research was funded by National Institute for Environmental Health Science (NIEHS), grant numbers R56ES026528 and P30ES007033, and NIEHS and National Institute on Aging (NIA) grant number R01ES026187. C.Z. was supported by the University of Washington's Biostatistics, Epidemiology, and Bioinformatics Training in Environmental Health (BEBTEH) grant number T32ES015459, from the NIEHS.

Institutional Review Board Statement: Not applicable.

Informed Consent Statement: Not applicable.

Data Availability Statement: The data presented in this study are available in the Supplementary Materials.

Conflicts of Interest: The authors declare no conflict of interest.

Appendix A

This appendix presents background, methods, and results for the MESA Air city Baltimore, MD.

Appendix A.1. Background

As part of the MESA Air study, LCMs were deployed in six metropolitan areas between spring 2017 and winter 2019: New York, NY; Baltimore, MD; Chicago, IL; Los Angeles, CA; Minneapolis and Saint Paul, MN; and Winston Salem, NC. Within each city except Baltimore, five to seven LCMs were deployed, with half co-located at regulatory agency monitoring sites. In Baltimore, we deployed 30 LCMs, providing the most data and the best opportunity to explore various calibration approaches for the CO, NO_2, and NO low-cost gas sensors. The goals of this analysis were to (1) determine whether calibration procedures carried out in the Puget Sound region translated well to Baltimore, a city with very different environmental conditions compared to the Puget Sound region; and (2) explore calibration options with limited co-location data using data from both the Puget Sound and Baltimore co-location periods.

Appendix A.2. Methods

Compared to the Puget Sound, where there were 205,023 monitor-days of co-location for calibration and evaluation, the number of monitor-days was much more limited in Baltimore, with 498 (CO), 1604 (NO), and 2092 (NO_2). In addition, as shown in Figure A1, only a subset of the 30 Baltimore LCMs was co-located in Baltimore (4 for CO and 13 for NO and NO_2), limiting our ability to conduct calibration co-location in Baltimore and account for inter-sensor differences. We therefore explored two options for calibration: (1) fitting models in the Puget Sound region, then evaluating those models with data from the limited amount of co-location data with agency reference instruments in Baltimore; and (2) pre-adjusting the WE and Aux values based on co-location in Puget Sound, then fitting and evaluating models without sensor-specific intercepts using co-location data in Baltimore.

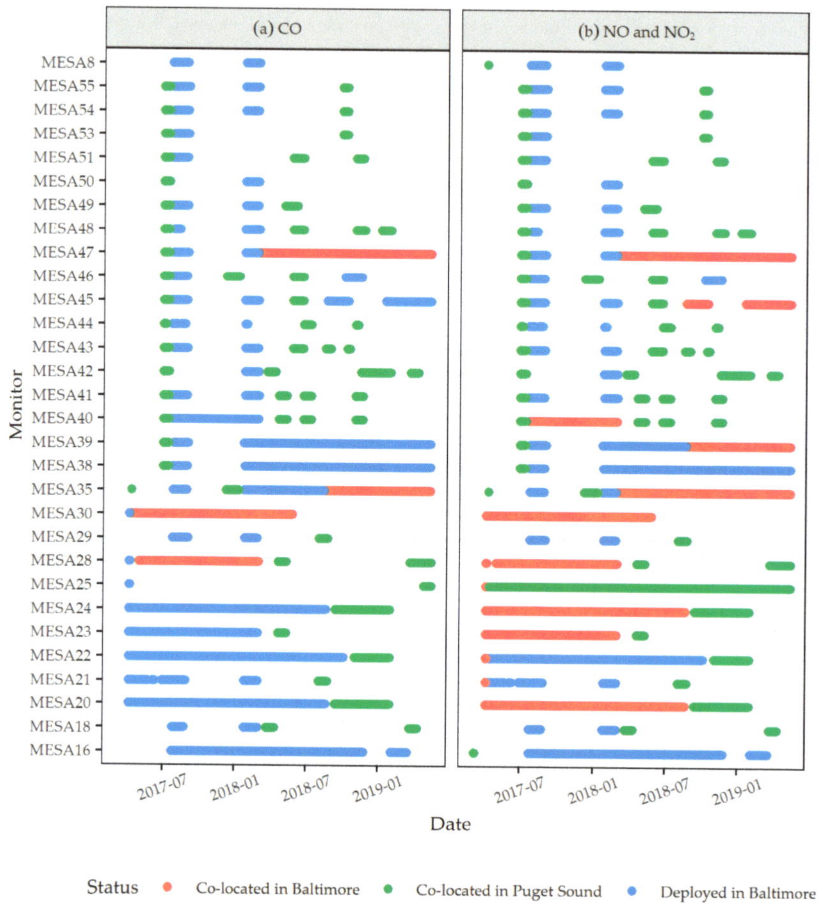

Figure A1. MESA Air study LCM co-location in Baltimore, MD for: (**a**) CO, and (**b**) NO and NO_2.

The first strategy took advantage of the larger number of monitors that were co-located with agency reference instruments (in theory providing an opportunity to adjust for inter-sensor variability), but suffered from a low number of co-location monitor days, and ignored important environmental/climactic differences between the Puget Sound and Baltimore that could affect calibration. The second strategy offered the advantage of using Baltimore co-location data while still accounting for inter-sensor differences as an alternative to fitting sensor intercepts. In this Appendix, we present the results for both of these approaches.

The first model, B1, was comparable to Model 1 for each gas presented in the main text of this paper for the Puget Sound, including terms for WE, Aux, and sensor ID, and was fit in the Puget Sound and evaluated in Baltimore. For the rest of the models developed for Baltimore, we pre-adjusted the WE and Aux of each sensor based on co-location in the Puget Sound, then created a series of multiple linear regression models with different covariates (B2–B8, approximately following the progression of Models 1–7 in the Puget Sound). The pre-adjustment algorithm we developed to address inter-sensor differences had the following steps:

1. Consider all pairwise comparisons for sensors that were ever co-located, and create a matrix for both WE and Aux that records these pairwise average differences.
2. Fill in missing data using a weighting scheme based on the time of co-location and relying on multiple degrees of separation.
3. After several iterations, the sensor differences relative to a single reference sensor are obtained, which can be used to adjust the sensor signal (mV).

We calculated the CV RMSE and R^2 with respect to agency reference measurements to assess the performance of the Baltimore models, similar to our methods in the Puget Sound.

Appendix A.3. Results and Discussion

The calibration models developed in the Puget Sound performed worse than the LCMs that remained in the region for the duration of the study (discussed in the main text of this paper), and did not translate well to Baltimore (Table A1). For CO and NO, the CV-RMSE increased and the CV-R^2 decreased when evaluating models fit in the Puget Sound. For NO_2, the performance was poor in the Puget Sound, and remained poor when applied in Baltimore.

Table A1. Summary of model B1 with terms for WE, Aux, and Sensor ID. Model B1 was fit in the Puget Sound and evaluated in Baltimore, MD on the daily timescale.

	Fit in Puget Sound				Evaluated in Baltimore			
	# Co-Location Sites	# Monitor Days Co-Location	CV-RMSE (ppb)	CV-R^2	# Co-Location Sites	# Monitor Days Co-location	CV-RMSE (ppb)	CV-R^2
CO	1	494	19	0.97	2	498	56	0.51
NO	1	520	5	0.89	4	1604	8	0.45
NO_2	1	507	6	0.22	4	2029	6	0.20

The Baltimore models with pre-adjusted terms for WE and Aux (based on co-location in the Puget Sound), then fit and evaluated on Baltimore co-location, are presented in Table A2. The models that included pre-adjusted WE and Aux terms had the advantage of using Baltimore-specific data, while still adjusting for sensor differences calculated during co-location periods in the Puget Sound. In effect, we approximated sensor calibration intercepts based on co-location in the Puget Sound, where each sensor had co-location data, then generalized calibration coefficients for the remaining model terms based on the limited number of LCMs with Baltimore co-location data.

While the pre-adjustment improved performance slightly in this study, this algorithm may not translate well to other regions, and reproduction of the technique should be approached cautiously, because it is not a well-established method. Baltimore calibration models had worse CV performance measures compared to those developed in the Puget Sound region, and we attribute this to the more limited co-location data available for calibration. Generally, the Baltimore CV performance measures were poor, and did not meet our acceptance criteria.

Table A2. Descriptions and prediction performance statistics of calibration models with each sensor pre-adjusted in the Puget Sound, then fit and evaluated in Baltimore, MD. Models were fit and predictions were generated on the daily timescale.

Model	Terms	CO		NO		NO_2	
		CV-RMSE (ppb)	CV-R^2	CV-RMSE (ppb)	CV-R^2	CV-RMSE (ppb)	CV-R^2
B2	Pre-adjusted WE, pre-adjusted Aux	51	0.53	7	0.61	5	0.33
B3	Model B2 with temperature and RH	39	0.73	7	0.58	5	0.40
B4	Model B3 with WE–temperature and WE–RH interactions	36	0.76	7	0.62	5	0.36
B5	Model B3 with WE– and Aux–temperature and WE– and Aux–RH interactions	37	0.75	7	0.64	5	0.29
B6	Model B2 with WE–temperature spline and WE–RH spline interactions	41	0.70	7	0.63	5	0.41
B7	Model B4 with WE–Aux interaction	37	0.74	7	0.62	5	0.45
B8	Model B4 with WE spline	38	0.74	7	0.61	5	0.36

References

1. Lim, S.S.; Vos, T.; Flaxman, A.D.; Danaei, G.; Shibuya, K.; Adair-Rohani, H.; AlMazroa, M.A.; Amann, M.; Anderson, H.R.; Andrews, K.G.; et al. A Comparative Risk Assessment of Burden of Disease and Injury Attributable to 67 Risk Factors and Risk Factor Clusters in 21 Regions, 1990–2010: A Systematic Analysis for the Global Burden of Disease Study 2010. *Lancet* **2012**, *380*, 2224–2260. [CrossRef]
2. Kampa, M.; Castanas, E. Human Health Effects of Air Pollution. *Environ. Pollut.* **2008**, *151*, 362–367. [CrossRef]
3. Baron, R.; Saffell, J. Amperometric Gas Sensors as a Low Cost Emerging Technology Platform for Air Quality Monitoring Applications: A Review. *ACS Sens.* **2017**, *2*, 1553–1566. [CrossRef]
4. Kotsev, A.; Schade, S.; Craglia, M.; Gerboles, M.; Spinelle, L.; Signorini, M. Next Generation Air Quality Platform: Openness and Interoperability for the Internet of Things. *Sensors* **2016**, *16*, 403. [CrossRef]
5. Kumar, P.; Morawska, L.; Martani, C.; Biskos, G.; Neophytou, M.; Di Sabatino, S.; Bell, M.; Norford, L.; Britter, R. The Rise of Low-Cost Sensing for Managing Air Pollution in Cities. *Environ. Int.* **2015**, *75*, 199–205. [CrossRef]
6. Lewis, A.C.; Lee, J.D.; Edwards, P.M.; Shaw, M.D.; Evans, M.J.; Moller, S.J.; Smith, K.R.; Buckley, J.W.; Ellis, M.; Gillot, S.R.; et al. Evaluating the Performance of Low Cost Chemical Sensors for Air Pollution Research. *Faraday Discuss.* **2016**, *189*, 85–103. [CrossRef]
7. Piedrahita, R.; Xiang, Y.; Masson, N.; Ortega, J.; Collier, A.; Jiang, Y.; Li, K.; Dick, R.P.; Lv, Q.; Hannigan, M.; et al. The next Generation of Low-Cost Personal Air Quality Sensors for Quantitative Exposure Monitoring. *Atmos. Meas. Tech.* **2014**, *7*, 3325–3336. [CrossRef]
8. Pigliautile, I.; Marseglia, G.; Pisello, A.L. Investigation of CO2 Variation and Mapping Through Wearable Sensing Techniques for Measuring Pedestrians' Exposure in Urban Areas. *Sustainability* **2020**, *12*, 3936. [CrossRef]
9. Snyder, E.G.; Watkins, T.H.; Solomon, P.A.; Thoma, E.D.; Williams, R.W.; Hagler, G.S.W.; Shelow, D.; Hindin, D.A.; Kilaru, V.J.; Preuss, P.W. The Changing Paradigm of Air Pollution Monitoring. *Environ. Sci. Technol.* **2013**, *47*, 11369–11377. [CrossRef]
10. Cromar, K.R.; Duncan, B.N.; Bartonova, A.; Benedict, K.; Brauer, M.; Habre, R.; Hagler, G.S.W.; Haynes, J.A.; Khan, S.; Kilaru, V.; et al. Air Pollution Monitoring for Health Research and Patient Care. An Official American Thoracic Society Workshop Report. *Annals ATS* **2019**, *16*, 1207–1214. [CrossRef]
11. Xiong, L.; Compton, R.G. Amperometric Gas Detection: A Review. *Int. J. Electrochem. Sci.* **2014**, *9*, 30.
12. Mead, M.I.; Popoola, O.A.M.; Stewart, G.B.; Landshoff, P.; Calleja, M.; Hayes, M.; Baldovi, J.J.; McLeod, M.W.; Hodgson, T.F.; Dicks, J.; et al. The Use of Electrochemical Sensors for Monitoring Urban Air Quality in Low-Cost, High-Density Networks. *Atmos. Environ.* **2013**, *70*, 186–203. [CrossRef]
13. Spinelle, L.; Gerboles, M.; Villani, M.G.; Aleixandre, M.; Bonavitacola, F. Field Calibration of a Cluster of Low-Cost Available Sensors for Air Quality Monitoring. Part A: Ozone and Nitrogen Dioxide. *Sens. Actuators B Chem.* **2015**, *215*, 249–257. [CrossRef]
14. Heimann, I.; Bright, V.B.; McLeod, M.W.; Mead, M.I.; Popoola, O.A.M.; Stewart, G.B.; Jones, R.L. Source Attribution of Air Pollution by Spatial Scale Separation Using High Spatial Density Networks of Low Cost Air Quality Sensors. *Atmos. Environ.* **2015**, *113*, 10–19. [CrossRef]
15. Ikram, J.; Tahir, A.; Kazmi, H.; Khan, Z.; Javed, R.; Masood, U. View: Implementing Low Cost Air Quality Monitoring Solution for Urban Areas. *Environ. Syst. Res.* **2012**, *1*, 10. [CrossRef]
16. Jiang, Q.; Kresin, F.; Bregt, A.K.; Kooistra, L.; Pareschi, E.; van Putten, E.; Volten, H.; Wesseling, J. Citizen Sensing for Improved Urban Environmental Monitoring. *J. Sens.* **2016**, *2016*, 1–9. [CrossRef]
17. Jiao, W.; Hagler, G.; Williams, R.; Sharpe, R.; Brown, R.; Garver, D.; Judge, R.; Caudill, M.; Rickard, J.; Davis, M.; et al. Community Air Sensor Network (CAIRSENSE) Project: Evaluation of Low-Costsensor Performance in a Suburban Environment in the Southeastern UnitedStates. *Atmos. Meas. Tech.* **2016**, *9*, 5281–5292. [CrossRef] [PubMed]

18. Moltchanov, S.; Levy, I.; Etzion, Y.; Lerner, U.; Broday, D.M.; Fishbain, B. On the Feasibility of Measuring Urban Air Pollution by Wireless Distributed Sensor Networks. *Sci. Total Environ.* **2015**, *502*, 537–547. [CrossRef]
19. Pang, X.; Shaw, M.D.; Lewis, A.C.; Carpenter, L.J.; Batchellier, T. Electrochemical Ozone Sensors: A Miniaturised Alternative for Ozone Measurements in Laboratory Experiments and Air-Quality Monitoring. *Sens. Actuators B Chem.* **2017**, *240*, 829–837. [CrossRef]
20. Sorte, S.; Arunachalam, S.; Naess, B.; Seppanen, C.; Rodrigues, V.; Valencia, A.; Borrego, C.; Monteiro, A. Assessment of Source Contribution to Air Quality in an Urban Area Close to a Harbor: Case-Study in Porto, Portugal. *Sci. Total Environ.* **2019**, *662*, 347–360. [CrossRef]
21. Sun, L.; Wong, K.; Wei, P.; Ye, S.; Huang, H.; Yang, F.; Westerdahl, D.; Louie, P.; Luk, C.; Ning, Z. Development and Application of a Next Generation Air Sensor Network for the Hong Kong Marathon 2015 Air Quality Monitoring. *Sensors* **2016**, *16*, 211. [CrossRef]
22. Jerrett, M.; Donaire-Gonzalez, D.; Popoola, O.; Jones, R.; Cohen, R.C.; Almanza, E.; de Nazelle, A.; Mead, I.; Carrasco-Turigas, G.; Cole-Hunter, T.; et al. Validating Novel Air Pollution Sensors to Improve Exposure Estimates for Epidemiological Analyses and Citizen Science. *Environ. Res.* **2017**, *158*, 286–294. [CrossRef]
23. Morawska, L.; Thai, P.K.; Liu, X.; Asumadu-Sakyi, A.; Ayoko, G.; Bartonova, A.; Bedini, A.; Chai, F.; Christensen, B.; Dunbabin, M.; et al. Applications of Low-Cost Sensing Technologies for Air Quality Monitoring and Exposure Assessment: How Far Have They Gone? *Environ. Int.* **2018**, *116*, 286–299. [CrossRef] [PubMed]
24. Lewis, A.; Edwards, P. Validate Personal Air-Pollution Sensors. *Nature* **2016**, *535*, 29–31. [CrossRef]
25. Cross, E.S.; Williams, L.R.; Lewis, D.K.; Magoon, G.R.; Onasch, T.B.; Kaminsky, M.L.; Worsnop, D.R.; Jayne, J.T. Use of Electrochemical Sensors for Measurement of Air Pollution: Correcting Interference Response and Validating Measurements. *Atmos. Meas. Tech.* **2017**, *10*, 3575–3588. [CrossRef]
26. Hagan, D.H.; Isaacman-VanWertz, G.; Franklin, J.P.; Wallace, L.M.M.; Kocar, B.D.; Heald, C.L.; Kroll, J.H. Calibration and Assessment of Electrochemical Air Quality Sensors by Co-Location with Regulatory-Grade Instruments. *Atmos. Meas. Tech. Katlenburg-Lindau* **2018**, *11*, 315–328. [CrossRef]
27. Masson, N.; Piedrahita, R.; Hannigan, M. Quantification Method for Electrolytic Sensors in Long-Term Monitoring of Ambient Air Quality. *Sensors* **2015**, *15*, 27283–27302. [CrossRef]
28. Popoola, O.A.M.; Stewart, G.B.; Mead, M.I.; Jones, R.L. Development of a Baseline-Temperature Correction Methodology for Electrochemical Sensors and Its Implications for Long-Term Stability. *Atmos. Environ.* **2016**, *147*, 330–343. [CrossRef]
29. Spinelle, L.; Gerboles, M.; Villani, M.G.; Aleixandre, M.; Bonavitacola, F. Field Calibration of a Cluster of Low-Cost Commercially Available Sensors for Air Quality Monitoring. Part B: NO, CO and CO2. *Sens. Actuators B Chem.* **2017**, *238*, 706–715. [CrossRef]
30. Duvall, R.M.; Clements, A.L.; Hagler, G.; Kamal, A.; Kilaru, V.; Goodman, L.; Frederick, S.; Johnson Barkjohn, K.K.; VonWald, I.; Greene, D.; et al. *Performance Testing Protocols, Metrics, and Target Values for Ozone Air Sensors: Use in Ambient, Outdoor, Fixed Site, Non-Regulatory and Informational Monitoring Applications*; US Environmental Protection Agency: Washington, DC, USA, 2021.
31. Casey, J.G.; Collier-Oxandale, A.; Hannigan, M. Performance of Artificial Neural Networks and Linear Models to Quantify 4 Trace Gas Species in an Oil and Gas Production Region with Low-Cost Sensors. *Sens. Actuators B Chem.* **2019**, *283*, 504–514. [CrossRef]
32. Casey, J.G.; Hannigan, M.P. Testing the Performance of Field Calibration Techniques for Low-Cost Gas Sensors in New Deployment Locations: Across a County Line and across Colorado. *Atmos. Meas. Tech.* **2018**, *11*, 6351–6378. [CrossRef]
33. Han, P.; Mei, H.; Liu, D.; Zeng, N.; Tang, X.; Wang, Y.; Pan, Y. Calibrations of Low-Cost Air Pollution Monitoring Sensors for CO, NO2, O3, and SO2. *Sensors* **2021**, *21*, 256. [CrossRef] [PubMed]
34. Malings, C.; Tanzer, R.; Hauryliuk, A.; Kumar, S.P.N.; Zimmerman, N.; Kara, L.B.; Presto, A.A. R. Subramanian Development of a General Calibration Model and Long-Term Performance Evaluation of Low-Cost Sensors for Air Pollutant Gas Monitoring. *Atmos. Meas. Tech.* **2019**, *12*, 903–920. [CrossRef]
35. Topalović, D.B.; Davidović, M.D.; Jovanović, M.; Bartonova, A.; Ristovski, Z.; Jovašević-Stojanović, M. In Search of an Optimal In-Field Calibration Method of Low-Cost Gas Sensors for Ambient Air Pollutants: Comparison of Linear, Multilinear and Artificial Neural Network Approaches. *Atmos. Environ.* **2019**, *213*, 640–658. [CrossRef]
36. Zimmerman, N.; Presto, A.A.; Kumar, S.P.N.; Gu, J.; Hauryliuk, A.; Robinson, E.S.; Robinson, A.L. R. Subramanian A Machine Learning Calibration Model Using Random Forests to Improve Sensor Performance for Lower-Cost Air Quality Monitoring. *Atmos. Meas. Tech.* **2018**, *11*, 291–313. [CrossRef]
37. Maag, B.; Zhou, Z.; Thiele, L. A Survey on Sensor Calibration in Air Pollution Monitoring Deployments. *IEEE Internet Things J.* **2018**, *5*, 4857–4870. [CrossRef]
38. ACT-AP Air Pollution, the Aging Brain and Alzheimer's Disease | Environmental & Occupational Health Sciences. Available online: https://deohs.washington.edu/air-pollution-aging-brain-and-alzheimers-disease (accessed on 8 November 2019).
39. MESA MESA Air Study | Environmental & Occupational Health Sciences. Available online: https://deohs.washington.edu/mesaair/mesa-air-study (accessed on 11 March 2020).
40. Chou, J. *Hazardous Gas Monitors: A Practical Guide to Selection, Operation and Applications*; McGraw-Hill Professional Publishing: New York, NY, USA, 2000; ISBN 0-07-135876-5.
41. Zusman, M.; Schumacher, C.S.; Gassett, A.J.; Spalt, E.W.; Austin, E.; Larson, T.V.; Carvlin, G.; Seto, E.; Kaufman, J.D.; Sheppard, L. Calibration of Low-Cost Particulate Matter Sensors: Model Development for a Multi-City Epidemiological Study. *Environ. Int.* **2020**, *134*, 105329. [CrossRef]

42. Alphasense Ltd. CO-B4 Carbon Monoxide Sensor. Available online: http://www.alphasense.com/WEB1213/wp-content/uploads/2019/09/CO-B4.pdf (accessed on 8 November 2019).
43. Alphasense Ltd. NO-B4 Nitric Oxide Sensor. Available online: http://www.alphasense.com/WEB1213/wp-content/uploads/2019/09/NO-B4.pdf (accessed on 8 November 2019).
44. Alphasense Ltd. NO2-B43F Nitrogen Dioxide Sensor. Available online: http://www.alphasense.com/WEB1213/wp-content/uploads/2019/09/NO2-B43F.pdf (accessed on 8 November 2019).
45. Alphasense Ltd. OX-B431 Oxidising Gas Sensor. Available online: http://www.alphasense.com/WEB1213/wp-content/uploads/2019/09/OX-B431.pdf (accessed on 8 November 2019).
46. Hossain, M.; Saffell, J.; Baron, R. Differentiating NO_2 and O_3 at Low Cost Air Quality Amperometric Gas Sensors. *ACS Sens.* **2016**, *1*, 1291–1294. [CrossRef]
47. EPA List of Designated Reference and Equivalent Methods. Available online: https://www.epa.gov/sites/production/files/2019-08/documents/designated_reference_and-equivalent_methods.pdf (accessed on 1 May 2020).
48. EPA Air Quality System (AQS). Available online: https://www.epa.gov/aqs (accessed on 20 March 2020).
49. PSCAA Puget Sound Clean Air Agency, WA | Official Website. Available online: https://pscleanair.gov/ (accessed on 20 March 2020).
50. Alphasense Ltd. Alphasense 4-Electrode Individual Sensor Board (ISB); User Manual 085-2217. 2019. Available online: http://www.apollounion.com/en/p-Alphasense-4-electrode-Individual-Sensor-Board-486.html (accessed on 19 June 2021).
51. PSCAA. 2019 Air Quality Data Summary; Puget Sound Clean Air Agency. 2020. Available online: https://pscleanair.gov/DocumentCenter/View/4164/Air-Quality-Data-Summary-2019 (accessed on 19 June 2021).
52. Castell, N.; Dauge, F.R.; Schneider, P.; Vogt, M.; Lerner, U.; Fishbain, B.; Broday, D.; Bartonova, A. Can Commercial Low-Cost Sensor Platforms Contribute to Air Quality Monitoring and Exposure Estimates? *Environ. Int.* **2017**, *99*, 293–302. [CrossRef]
53. Spinelle, L.; Aleixandre, M.; Gerboles, M.; European Commission; Joint Research Centre; Institute for Environment and Sustainability. *Protocol of Evaluation and Calibration of Low-Cost Gas Sensors for the Monitoring of Air Pollution*; Publications Office: Luxembourg, 2013; ISBN 978-92-79-32691-2.

sensors

MDPI

Article

An Approximation for Metal-Oxide Sensor Calibration for Air Quality Monitoring Using Multivariable Statistical Analysis

Diego Sales-Lérida [1,*], Alfonso J. Bello [2], Alberto Sánchez-Alzola [2] and Pedro Manuel Martínez-Jiménez [1]

1 Department of Automation Engineering, Electronics and Computer Architecture and Networks, University of Cádiz, 11519 Cádiz, Spain; pedromanuel.martinez@uca.es

2 Department of Statistic and Operations Research, University of Cádiz, 11510 Cádiz, Spain; alfonsojose.bello@uca.es (A.J.B.); alberto.sanchez@uca.es (A.S.-A.)

* Correspondence: diego.lerida@uca.es

Abstract: Good air quality is essential for both human beings and the environment in general. The three most harmful air pollutants are nitrogen dioxide (NO_2), ozone (O_3) and particulate matter. Due to the high cost of monitoring stations, few examples of this type of infrastructure exist, and the use of low-cost sensors could help in air quality monitoring. The cost of metal-oxide sensors (MOS) is usually below EUR 10 and they maintain small dimensions, but their use in air quality monitoring is only valid through an exhaustive calibration process and subsequent precision analysis. We present an on-field calibration technique, based on the least squares method, to fit regression models for low-cost MOS sensors, one that has two main advantages: it can be easily applied by non-expert operators, and it can be used even with only a small amount of calibration data. In addition, the proposed method is adaptive, and the calibration can be refined as more data becomes available. We apply and evaluate the technique with a real dataset from a particular area in the south of Spain (Granada city). The evaluation results show that, despite the simplicity of the technique and the low quantity of data, the accuracy obtained with the low-cost MOS sensors is high enough to be used for air quality monitoring.

Keywords: air quality; metal-oxide sensor; monitoring; multivariable regression models; model calibration

1. Introduction

Good air quality is essential for both humanity and the natural environment. Economic activities such as energy production, industry and agriculture, as well as the dramatic rise in traffic, release air pollutants into the environment that can lead to serious problems for our health [1]. In fact, the poor quality of air is the cause of more than 400,000 premature deaths in Europe each year, as well as a decrease in quality of life by causing or exacerbating asthma and respiratory problems [2,3].

There are several pollutants involved in air quality characterization, such as SOx, CO, NOx, O_3, or particulate matter pollution [4,5]. From all of them, three of the most harmful air pollutants, in terms of damage to ecosystems, are nitrogen dioxide (NO_2), ozone (O_3) and particulate matter (specifically PM2.5, which is directly related to traffic) [6–9]. Thus, it is very important to monitor and analyze these elements in the air, especially in towns and cities, in order to detect dangerously high levels and take actions to reduce pollution [10,11].

In this regard, several agencies around the world are responsible for the air quality monitoring of their corresponding regions, such as the European Environment Agency (EEA) in Europe, or the Environmental Protection Agency (EPA) in the United States. Their data give relevant and reliable information to policymaking agents [12]. In particular, the use of air quality models to assess the potential changes in urban air quality concentrations is a fundamental element of air quality management. In this type of modeling, the input data require high spatio-temporal resolution to capture the variability in the urban environment. However, one of the main technical difficulties nowadays is the lack, or low quality, of input data on concentrations [13]. Due to the high cost of monitoring stations, only a few examples of this type of infrastructure have been deployed in cities, providing limited spatial coverage [14].

In order to address this problem, recent environmental agencies' reports suggest that cities should participate in the input data acquisition, complementing official monitoring data with additional measurements of local air quality [13]. In this sense, the cities are increasingly aware of the potential for low-cost 'citizen science' sensors to help support the results of their air quality modeling [15,16]. These sensors offer air pollution monitoring at a lower cost and smaller size than conventional methods, making it possible for them to be installed in many more locations [17–19]. However, the accuracy of input data in air quality modeling is as important as the quantity of measures. Thus, the use of citizen science and citizen participation in air quality monitoring by means of these low-cost sensors is only feasible if they can provide accurate information [20,21].

Currently, the three most popular types of low-cost air quality sensors are electrochemical sensors (EC), metal-oxide sensors (MOS) and photoionization detectors (PID) [22,23]. Since the objective is to achieve the widest possible distribution of air monitoring sensors in cities, their price is an essential factor. In this sense, the cost of EC and PID sensors is prohibitive for most consumers (they can cost more than EUR 100). On the contrary, the cost of MOS sensors, which are usually below EUR 10, as well as their small dimensions, make them an excellent option for use by citizens [24,25]. However, it should be noticed that, in air quality monitoring, the pollutant concentration that sensors should capture is usually very small: in the order of parts per billion (ppb) or "$\mu g/m^3$". In this sense, the World Health Organization (WHO) casts some doubts on the reliability of low-cost sensors when the calibration methods provided by manufacturers are employed, because these methods may be questionable regarding very low concentrations [26]. Thus, the WHO, as well as the EEA [10], only recommends the use of these devices for air quality monitoring through an exhaustive calibration process and subsequent precision analysis [27,28].

In most of the works in the literature, sensor calibration is performed under laboratory conditions [29–31]. In this type of approach, controlled environments are created by injecting known concentrations of the specific pollutants to be measured. However, in these ideal laboratory conditions, other variables that are present in real environments are not taken into account. On the one hand, there could be particles of other components that are different from the pollutants to be measured in the specific region where the sensors should be used which are not considered in a laboratory. On the other hand, although other environmental factors in the specific region can be simulated in a laboratory, such as the temperature and relative humidity of the air, they may differ from the actual conditions [32].

In order to face these problems, several on-field calibration techniques have been proposed in the literature [33–36], which are based on the data obtained from the monitoring stations of the regional government agencies. This way, sensors are calibrated using the specific environmental conditions of the region where they will be used, and are, therefore, adapted to its temperature, humidity and air composition. In most of these works, the proposed calibration techniques are complex and not very intuitive, and they are applied by experts in the field. In addition, in those studies, a large amount of calibration data is available from sensors, since they have been placed close to the reference monitoring stations for long periods of time. However, we should remember that the objective of these low-cost sensors is the use of citizen science and citizen participation in air quality monitoring. Thus, in real situations, the sensors will be calibrated by field workers who are usually not so expert in applying complex techniques, and the available data for calibration may be limited, since locations close to monitoring stations cannot be used for long periods of time.

In this work, we present an on-field calibration technique for low-cost MOS sensors that tries to solve both problems commented on above: it can be easily applied by non-expert operators, and it can be used even with only a small amount of calibration data. The proposed technique is based on the well-known regression analysis tool [37–39], which is widely used for data modeling in a great variety of fields. In our approach, we have studied the different kinds of regression techniques in the literature, and we have selected the most appropriate one, taking into account the number of independent variables, the type of dependent variables and the shape of the regression curve. We apply and evaluate this technique with a real dataset from a particular area in the south of Spain (Granada city). The training and test data were used to fit and validate the model, respectively, using the R software [40]. The evaluation results show that, despite the simplicity of the technique and the low quantity of data, the accuracy obtained with the low-cost MOS sensors is high enough to be used for air quality monitoring. In addition, the proposed method is adaptive, in the sense that the calibration can be refined as more data become available.

The rest of the paper is organized as follows. In Section 2, we briefly present the sensors that are usually employed to measure the air pollutant concentrations, giving more details to the low-cost MOS used in this work, we describe and analyze the dataset used to validate the calibration technique, and we explain the calibration methodology. In Section 3, we apply this methodology to fit the pollutant concentrations corresponding to ozone (O_3), nitrogen dioxide (NO_2) and carbon monoxide (CO). The obtained results are statistically studied and discussed in Section 4, while Section 5 contains the main conclusions of this paper.

2. Material and Methods

2.1. Sensors

Before going into details about the sensors used in this work to measure air pollutant concentrations, we should clarify that the unit selected to express these concentrations will be "µg/m^3" because this is the form used by the European Commission for regulation in the European framework.

The European air quality standards set by the Ambient Air Quality Directive (EU, 2008) for the protection of human health [41], the air quality guidelines (AQGs) set by the World Health Organization (WHO) [42], and their subsequent revisions, define several aspects of values for the different pollutants, like typical qualitative levels, the averaging period, the time by which limit values can be overcome in a year, or alert values. In Spain, there are certain laws that refer to these standards; the most recent of their revisions were passed on 28 January 2011 in the form of the directive RD102/2011. Table 1 shows some of its aspects.

Table 1. Qualitative levels as referred to for the quantitative levels of each pollutant, and the averaging period used in each of them, following the European EEA standards.

Qualitative Index	SO2 µg/m^3 (24 h Average Value)	O3 µg/m^3 (8 h Average Value)	NO2 µg/m^3 (1 h Average Value)	CO µg/m^3 (8 h Measured Value)	PM10 µg/m^3 (24 h Measured Value)
Good	0–63	0–60	0–100	0–5000	0–25
Moderate	63–125	60–120	100–200	5000–10,000	25–50
Poor	125–187	120–180	200–300	10,000–15,000	50–75
Very Poor	>187	>180	>300	>15,000	>75

As mentioned in Section 1, in this study we have proposed the use of MOS sensors, since they are the most accessible to users from an economic point of view. These sensors are composed of a semiconductor layer, generally, tin dioxide (SnO_2), which makes them especially sensitive to other oxides, and, by controlling the doping of the semiconductor, it is possible to make the material more sensitive to certain parameters. Therefore, when there are higher concentrations of these parameters in sampled air, the conductivity of this layer changes its values. It is worth mentioning that this conductivity keeps a direct relation with temperature, and, in general terms, they change in a proportional form. In addition, it should be noticed that, after a certain temperature, the sensibility to target gases can decrease, negatively affecting the quality of sensor detection. To take advantage of this property, electrodes are inserted into the detection layer of the sensor in order to increase its temperature in a controlled way (by using a heating circuit, such as a voltage divider with resistors) [43–45].

In particular, the MOS sensors used in this work are the ones incorporated in the devices developed in the "EcoBici (Kers bike)" research project (file number G-GI3002/IDIC) which resulted in a patented invention, application number P201600319 and publication number ES2638715 [46]. These devices were designed to take air quality values, accumulating the data and being able to configure the time in which the averages are sent to a web server, in real time, through the deployment of a sensor network using XBee technology (protocol ZigBee). The parameters measured by these devices are CO, O_3 and NO_2. It should be noticed that these sensors are non-specific sensors since they can measure other gases apart from the main gas [43], but these secondary gases are not those considered in this paper. It is worth mentioning that O_3 and NO_2 are linked by the Leighton relationship. Nevertheless, the proposed methodology is not affected by this relationship since it is already considered in the parameter estimation.

For the calibration tests, the devices were adapted to send the temporal average of the three parameters every 10 min in order to be synchronized with the calibration equipment. Figure 1 shows the three sensors incorporated in EcoBici end devices, which include an MQ-7 sensor for CO measuring [43], an MQ-131 sensor for O_3 [44] and an MiCS-2714 sensor for NO_2 [45].

Figure 1. Telemetering devices from the EcoBici project.

The concentration values given by the curves in datasheets [43–45] are much higher than the values that should be measured in terms of air quality. Although some of the sensor manufacturers guarantee that the device is able to detect the presence of gas at tens of ppb, our own experience can confirm the information from the WMO, cited in Section 1, and discourage the use of these curves for low concentrations.

In order to carry out the measurement campaign for field calibration, we used the highly sophisticated equipment located in the sampling stations belonging to the Environment Council of the Andalusian government. In these sampling stations, which are mostly composed of measurement analyzers, the pollutant concentrations are taken continuously, 24 h/day, 365 days per year, except for breakdowns. The cost of this type of equipment generally exceeds the barrier of EUR 10,000, and it is used to analyze a single parameter. It should be noted that each autonomous community or region has its own criteria to collect the data. In the case of Andalusia, the analyzers used in their stations take a sample of the ambient air, previously conditioned and homogenized, and analyze it in periods ranging from 10 s to 10 min, depending on the pollutant to be analyzed. This information is averaged in 10 min periods, stored and published by the Spanish Ministry of Air Quality [47], and on the Andalusian Council website (available from the following day) [48].

In order to select the most suitable sampling stations for calibration campaigns, several factors should be taken into account, such as the latest calibration reviews of the station, accessibility, and measurements range obtained of the different parameters in the station in several days. Regarding the data range, it is highly important to choose a station that can provide a wide range of values in the different parameters to be calibrated. For example, if a station where quantitative O_3 values do not exceed 50 ppb after several days is selected, the sensor may not be properly calibrated for higher concentration values. According to this criterion, a station localized in Granada city was selected from more than 100 Andalusian Council monitoring stations. Figure 2 shows a photo of the Granada sampling station, where it is possible to identify the EcoBici devices on it, next to the station analyzers.

Figure 2. Field calibration of 10 EcoBici devices on the front of an Andalusian Council sampling station in Seville City.

Finally, it is important to take into account the particular conditions of temperature and humidity in Granada city, since both parameters affect the best adjustment of sensors, as will be seen in the data section. In fact, both parameters were requested by the agency in charge of the sampling station after the measurement campaign. In any case, if these data could not be obtained from the corresponding agency, another option would be to place temperature and humidity sensors in the devices.

2.2. Description of Dataset

The real dataset of the work in the present paper involves measurements, taken by both analyzers and sensors, of three particular gaseous pollutants: ozone, nitrogen dioxide and carbon monoxide, in addition to temperature and humidity measurements by the agency. The observations are collected in 490 registers which were taken from midnight, 00:00 h, 08/05/2016 until 09:30 h, 11/05/2016, at a ten-minute frequency. The respective pollutant variables corresponding to the analyzers, from now on also called patterns, have been denoted as "O_3", "NO_2" and "CO", the respective pollutant variables corresponding to the sensors as "O_3s", "NO_2s" and "COs", the temperature variable as "temp", and the humidity variable as "hum". To obtain a better fit of the models, we have added a new variable, called "COsR", which is a version of COs without trend. The rectified COsR time series has been obtained by the ratio of the sensor values and its adjusted least squares regression line. Moreover, we have translated the time series to the sensor range modifying the scale. Therefore, finally, we count 9 variables of work in the dataset: temp, hum, O_3, NO_2, CO, O_3s, NO_2s, COs and COsR.

The following sections show how to predict the pattern values for the gaseous pollutants O_3, NO_2 and CO, applying multivariable regression models and selecting the best fit by using the measurements of the sensors, O_3s, NO_2s and COs, and the values of temperature and humidity. That is, a general expression of the model would be:

$$Y = f(X_1, X_2, \ldots, X_5), \tag{1}$$

where Y represents the pattern values, $(X_1, X_2, \ldots, X_5)$ represent the measurements of the sensors and the temperature and humidity values, and f represents the convenient functional form of the model.

2.3. Methodology

The prediction and model assessment (or validation) are closely related to each other. Particularly, in our task, several models have been considered, of which, those that we have observed to best fit in each case will be analyzed and presented. It is important to mention that, although we have considered different more complex functional forms for the regression models, they have not managed to significantly improve the fits obtained by simple multilinear regression models in all cases. Therefore, the expression of the model used for the fit takes the form:

$$Y = \alpha_0 + \alpha_1 X_1 + \alpha_2 X_2 + \alpha_3 X_3 + \alpha_4 X_4 + \alpha_5 X_5, \tag{2}$$

where $\alpha_i \in \mathbb{R}$, for i = 0,1, … ,5, are the independent term and the contribution of the variables X_i in the model. Both fitting to a dataset and choosing the best multilinear regression model can be easily done using the *lm* and *step* functions from the R stats package (there are many works on the internet that show how to do it, such as [49,50]).

In order to evaluate the best fitting model, we have performed the following method. We have split the sample into two disjoint subsets to estimate the prediction error, treating one subset as the training set and the other as the test set (split by vertical lines in Figures 3 and 4). We used the training set to regress each gaseous pollutant on the rest of the variables. Afterward, we predicted a new gaseous pollutant value by applying the fitted model to the new values of the test set. The prediction was compared with the real response value and the prediction ability of the regression model. This provided a measure for the quality of the prediction, which was evaluated by its mean squared prediction error.

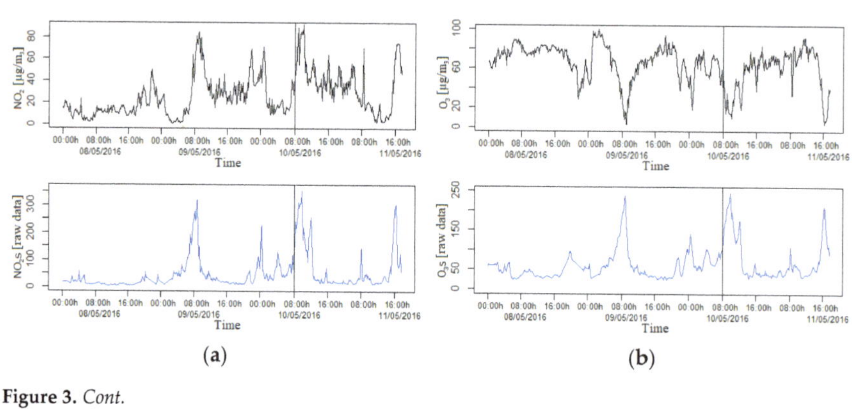

Figure 3. *Cont.*

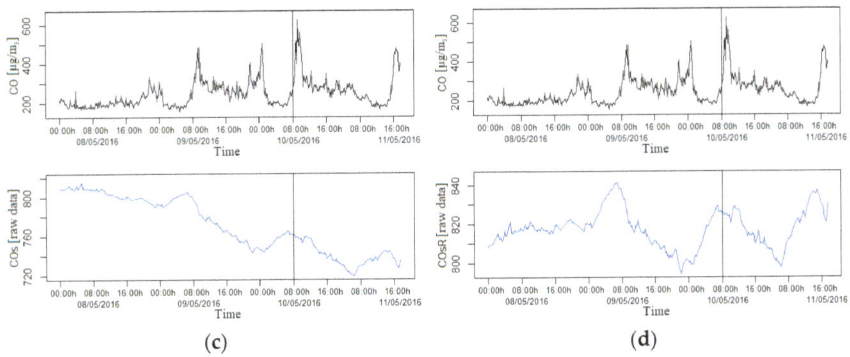

Figure 3. (**a**–**d**) Plots of the evolution over time of the indicated gaseous pollutants in black lines for the patterns and in blue lines for the sensors (and for the transformation of the variable COs, COsR). Each vertical line separates the training dataset (on the left) from the test dataset (on the right).

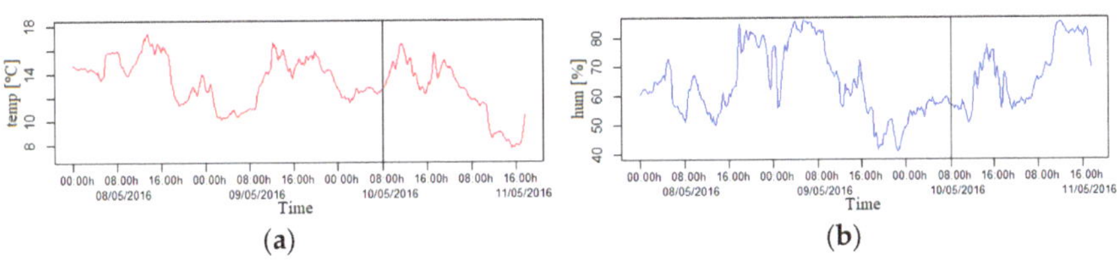

Figure 4. (**a**) The red line corresponds with the evolution over time of the temperature; (**b**) the blue line corresponds with the evolution in time of the humidity. Each vertical line separates the training dataset (on the left) from the test dataset (on the right).

Training and Test Sets

The methodology applied for each pollutant is similar. Firstly, we evaluate the different regression models using the dataset with all records and choose the one that best fits. Secondly, in order to perform a prediction test, we divide the whole dataset into two subsets: the training dataset and the test dataset.

The training dataset contained the measurements corresponding to the period from 00:00 h on 08/05/2016 until 08:00 h on 10/05/2016. Thirdly, using this subset, we fit the regression model chosen by fixing the coefficients of the model using the least squares method. The test dataset contained the measurements corresponding to the period from 08:10 h on 10/05/2016 until 09:30 h on 11/05/2016. It is important to mention that the test dataset contained an entire daily cycle, which let us include the possible daily periodicities. Fourthly, with the regression model fitted in the previous phase, we obtain the predictions for the test dataset and compare the results with respect to the pattern values of the test dataset.

3. Results

3.1. Analysis of Dataset

We can observe in Figure 3a that, in a different proportion, the evolution over time of the measurements taken by the sensor for nitrogen dioxide is closely related and also directly to the pattern values. In addition, in the same sense, we can observe in Figure 3b that there is a high association between ozone measurements, but in this case with an inverse relationship. The previous observations are supported by the correlation coefficients: $\rho(O_3,O_3s) = -0.8227$, $\rho(NO_2,NO_2s) = 0.6118$.

In Figure 3c, we do not observe the existence of an evident relationship between the carbon monoxide measurements captured by the sensor and its corresponding pattern values. In addition, $\rho(CO,COs) = -0.3735$, which is a low correlation. In line with

COs, it is possible to appreciate the existence of a decreasing trend in concentration over time that does not exist in the pattern values curve. In order to better visualize any relationship, we have decided to eliminate the slope of the curve, creating the new variable COsR. However, as we can see in Figure 3d, there is still no evidence of any relationship after removing the slope, and, in this case, an even lower correlation is obtained (ρ(CO,COsR) = −0.1467). We kept the variable COsR in the dataset because the results in the model-fitting work improved.

3.2. Fitting Ozone

3.2.1. Selection of the Model

In the case of ozone, first, we considered a multilinear regression model with different combinations among the measurements of the sensors for O_3s, NO_2s and COs, in addition to the temperature and humidity measurements. Afterward, we chose the measures of COs instead of its version without a decreasing trend, COsR, obtaining a better adjustment and results. In particular, the model that best fits is:

$$O_3 = \alpha_0 + \alpha_1 \text{ COsR} + \alpha_2 \text{ NO}_2\text{s} + \alpha_3 \text{ O}_3\text{s} + \alpha_4 \text{ temp} + \alpha_5 \text{ hum}, \tag{3}$$

where $\alpha_i \in \mathbb{R}$, for i = 0,1, … ,5.

Adjusting the model by the least squares method to the dataset with all records, we obtain the α_i values contained in Table 2. We observed that all variables considered were significant for the model. In addition, we know that the model manages to explain 75.08% of the total variability of O_3, and the predictions of the model have a correlation of 0.8665 with the measures of the O_3 pattern. In the left plot of Figure 5, we compared the values predicted by the model with the measurements of the O_3 pattern. We can see in the histogram of the net prediction errors of the model that visually these do not differ too much from adjusting to a normal distribution (although the normality hypothesis was rejected when the Shapiro–Wilk test was applied).

Table 2. The table contains the summary of the model described in Equation (3), adjusted to the dataset with all records. We can see the estimated values and the standard errors of the coefficients, the test statistics and *p*-values of their significance tests (for i = 0,1, … ,5, the null hypothesis is $\alpha_i = 0$), the statistics of the residuals and the goodness of fit.

Coefficients	Estimate	Std. Error	*t* Value	*p*-Value
α_0	−406.43899	54.43049	−7.467	3.85×10^{-13}
α_1	0.66569	0.07036	9.461	$<2 \times 10^{-16}$
α_2	0.09424	0.02109	4.468	9.82×10^{-6}
α_3	−0.56357	0.03175	−17.752	$<2 \times 10^{-16}$
α_4	−1.01488	0.31333	−3.239	0.00128
α_5	−0.44478	0.07294	−6.098	2.20×10^{-9}
Residuals:				
Min	1Q	Median	3Q	Max
−31.110	−5.171	1.232	6.224	30.671
R-squared: 0.7508				

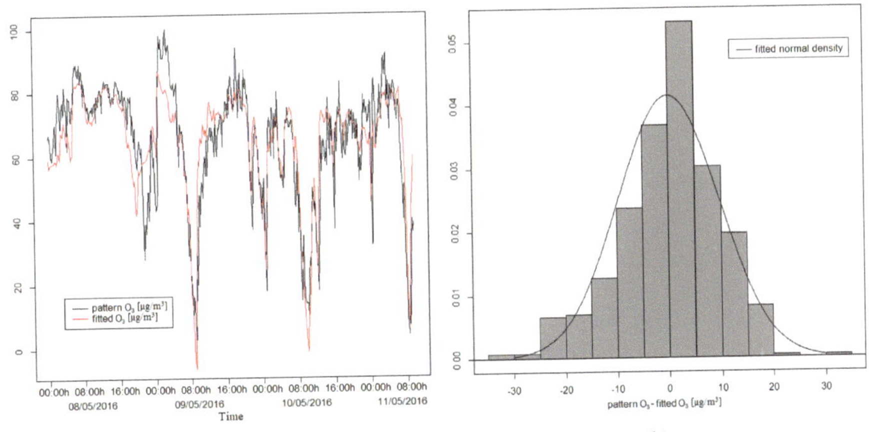

Figure 5. (**a**) The fitted values by the model and the pattern values for ozone; (**b**) the histogram of the net prediction errors of the model.

3.2.2. Evaluation of the Selected Model

Now, adjusting the model by the least squares method to the training dataset, we obtain the α_i values contained in Table 3. We can see that all variables considered in the model are significant and that it manages to explain 71.27% of the total variability of O_3 for the training dataset. In Figure 6, for the test dataset, we can compare the values predicted by the model with the measurements of the O_3 pattern and, in the histogram of the net prediction errors of the model, we can observe that these do not differ from a normal distribution. In addition, applying the Shapiro–Wilk test, we obtain a p-value of 0.4424, being able to consider the net prediction errors as normal, with mean $\mu = -4.2807$ and standard deviation $\sigma = 10.8789$. The predictions of the model have a correlation of 0.8824 with the measures of the O_3 pattern for the test dataset.

Table 3. The table contains the summary of the model described in Equation (3), adjusted to the training dataset.

Coefficients	Estimate	Std. Error	t Value	p-Value
α_0	−413.68158	69.18787	−5.979	5.81×10^{-9}
α_1	0.69410	0.08865	7.830	6.68×10^{-14}
α_2	0.10644	0.02793	3.812	0.000165
α_3	−0.61144	0.04218	−14.497	$<2 \times 10^{-16}$
α_4	−1.99947	0.42228	−4.735	3.25×10^{-6}
α_5	−0.43273	0.08206	−5.274	2.42×10^{-7}
Residuals:				
Min	1Q	Median	3Q	Max
−32.143	−4.406	1.603	5.626	19.206
R-squared: 0.7127				

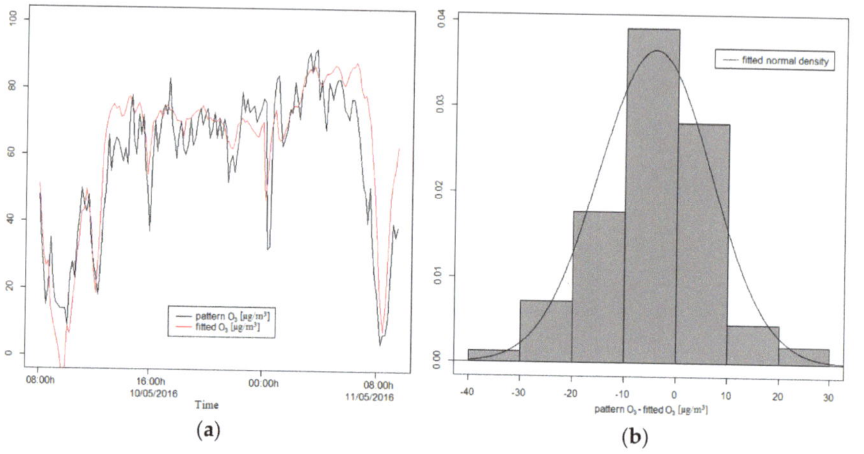

Figure 6. Both graphs with the test dataset. (**a**) The fitted values by the model and the pattern values for ozone; (**b**) the histogram of the net prediction errors of the model.

3.3. Fitting Nitrogen Dioxide

As in the ozone case, firstly we have selected the best fit for nitrogen dioxide, which corresponds to the following multilinear regression model:

$$NO_2 = \alpha_0 + \alpha_1\ COs + \alpha_2\ NO_2s + \alpha_3\ O_3s + \alpha_4\ temp + \alpha_5\ hum, \quad (4)$$

where $\alpha_i \in \mathbb{R}$, for i = 0,1, … ,5.

In Table A1 we can see the α_i values when we fit the model to the dataset with all records. It can also be seen that all variables are significant for the model, and that the model manages to explain 68.10% of the total variability of NO_2. The model predictions have a correlation of 0.8252 with the NO_2 pattern. In Figure A1, it is possible to compare the NO_2 values predicted by the model with those of the pattern and the histogram of the net prediction errors of the model, which do not differ too much from adjusting to a normal distribution.

To evaluate the chosen model, it was adjusted to the training dataset, obtaining the α_i values contained in Table A2. In this case, all variables considered are also significant and it managed to explain 65.55% of the total variability of NO_2. In Figure A2, for the test dataset, we can see the NO_2 values predicted by the model, and in the histogram of the net prediction errors of the model, we can see that they also did not differ from a normal distribution. The predictions of the model had a correlation of 0.8301 with the measures of the NO_2 pattern for the test dataset.

3.4. Fitting Carbon Monoxide

The selected model for carbon monoxide has the following expression:

$$CO = \alpha_0 + \alpha_1\ COsR + \alpha_2\ NO_2s + \alpha_3\ O_3s + \alpha_4\ temp + \alpha_5\ hum, \quad (5)$$

where $\alpha_i \in \mathbb{R}$, for i = 0,1, … ,5.

In this case, adjusting the model to the dataset with all records, the model explains 57.93% of the total variability of CO, and the corresponding predictions have a correlation of 0.7611 with the CO pattern. In Table A3 we can see the α_i values that were obtained, all variables being significant. We can also see the predictions of the model and the histogram of its net prediction errors in Figure A3.

Regarding the evaluation of the selected model, once it was adjusted to the training dataset, it explained 47.49% of the total variability and its predictions had a correlation of

0.7769 with the CO pattern. All variables considered are significant for the model, as we can see in Table A4, in addition to the values of the coefficients. In Figure A4, for the test dataset, we can observe the CO values predicted by the model and the histogram of the net prediction errors of the model, which do not differ from a normal distribution.

4. Discussion

We observed that all the models generated overcame the global significance contrasts (p-values < 0.01) and almost all the individual significance contrasts. In particular, the p-values of NO_2s from Tables A3 and A4 show that the null hypothesis cannot be rejected by 10% of the significance level (0.09886 and 0.15552, respectively), the reason why the coefficient of NO_2s in the model of CO is statistically equal to 0. Nevertheless, when this variable is removed from the model, although it simplifies it, neither the adjustment nor the prediction improves. Furthermore, an extension of the dataset will induce the NO_2 sensor to have a higher influence in the model, providing a better fit for them, as happens with the other pollutants. For this reason, we decided to keep this variable in the model.

Focusing on ozone measurements, and considering all the datasets, the model obtained explains 75.08% of the variability of the data (R-squared), leaving less than 25% to the residuals. We also observed a high direct correlation (0.8665) with the measures of this pollutant pattern. This coefficient indicated a good correspondence between the observations and the predictions of this sensor. Moreover, the histogram of the prediction errors was not normally distributed (Shapiro–Wilk test rejected), although we observed a rough 0 symmetry distribution (Figure 5b).

Nevertheless, when we consider the training dataset for this pollutant, the least square model when adjusted has a lower value of R-squared (71.27%), although it is close to the previous goodness of fit. In this case, the model overcomes the Shapiro–Wilk test on the prediction errors (they follow a normal distribution $n(-4.28;10.88)$, p-value = 0.4424). It can be seen that 95% of the central prediction errors are between −23 and 12.2 $\mu g/m^3$ with a median of −4.2 $\mu g/m^3$. The interquartile range is 13.1 $\mu g/m^3$. In the boxplot shown in Figure 7a, we observe only 3 outlier values from 337 data points. In Figure 7b, the theoretical normal quantiles, compared to prediction errors, display a good agreement in the central quantiles (points near the straight blue line).

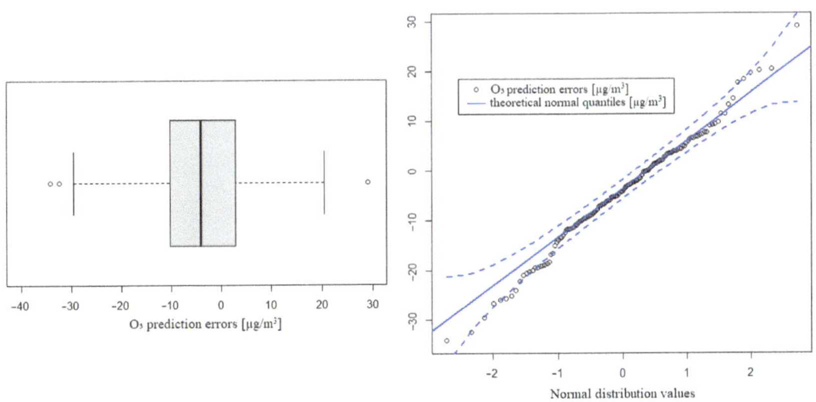

(a) (b)

Figure 7. (**a**) Boxplot of O_3 prediction errors; (**b**) theoretical normal quantiles compared to prediction errors.

In the case of nitrogen dioxide, we observed that the model obtained explained 68.10% of the variability with a 0.8252 correlation with the NO_2 pattern. The prediction errors histogram has a slight right asymmetry, although it does not differ excessively from a normal distribution (Figure A1b). Focusing on the training data, we observed a similar value of R-squared (65.55%), and the prediction errors were distributed with the same right asymmetry as before. The mean value of the prediction errors was negative (-11.5 µg/m^3), with a standard deviation of 12.14 µg/m^3. These values indicated that the prediction values were greater than the real values, so the model overestimated the NO_2 values. The asymmetric coefficient is 0.4292, so the distribution shows a right asymmetry with more concentration of negative values of the prediction errors. This bias also shows that the model was overestimating the pattern value measures. We found that 95% of the central prediction errors are between -28.3 and 9.1 µg/m^3 with a median of -12.6 µg/m^3. The interquartile range is 16.2 µg/m^3. Figure 8a shows the rough symmetry of the NO_2 distribution and Figure 8b presents the deviation of the theoretical normal quantities and the NO_2 prediction errors.

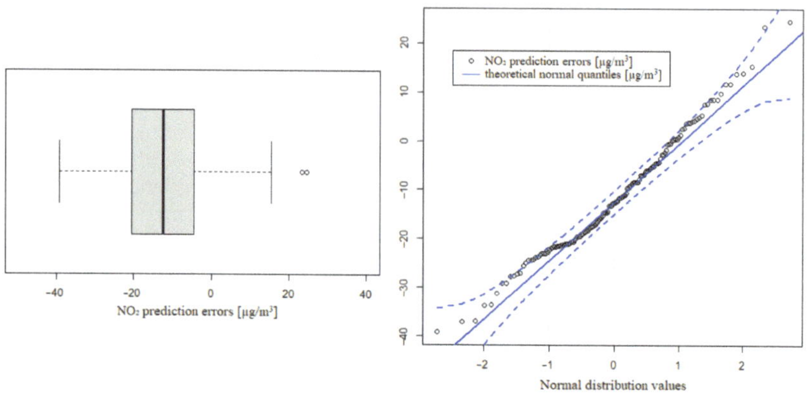

(**a**) (**b**)

Figure 8. (**a**) Boxplot of NO_2 prediction errors; (**b**) theoretical normal quantiles compared to prediction errors.

Regarding the carbon monoxide values, we needed to use the detrended measures of the CO sensor (COs) because the fit is better than COsR. In this case, the variability explained is lower (57.93%) regarding all the data, dropping to 47.49% of the total variability considering the training dataset. Clearly, the distribution of the prediction errors did not follow a normal distribution, with a strong right asymmetry with some values well over 100 µg/m^3 (asymmetric coefficient: 1.4045). The mean value is 29.84 µg/m^3 with a standard deviation of 55.51 µg/m^3. These values indicate that the prediction values were lower than the real ones, so the model underestimated the CO values. Clearly, the adjustment of CO was not as good as the fit of the other pollutants, even after the detrending process. We found that 95% of the central prediction errors were between -34.3 and 164.5 µg/m^3, with a median of -17.5 µg/m^3. The interquartile range is also the highest one, with 53 µg/m^3. Figure 9a shows the clear right asymmetry of the CO distribution and Figure 9b presents the deviation of the theoretical normal quantities and the CO prediction errors.

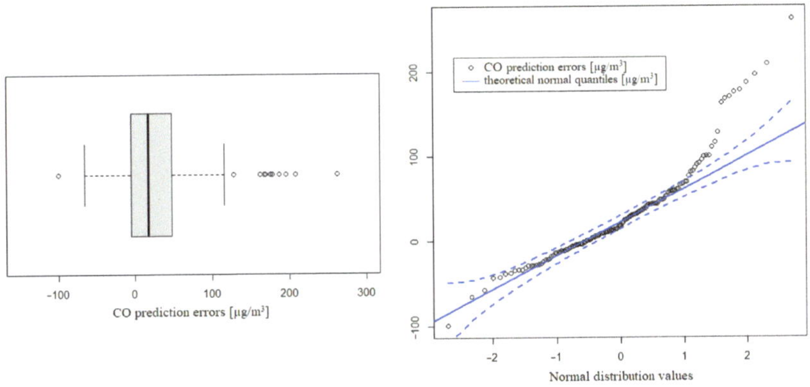

(a) (b)

Figure 9. (**a**) Boxplot of CO prediction errors; (**b**) theoretical normal quantiles compared to prediction errors.

5. Conclusions

In this paper, we present an on-field calibration technique for low-cost MOS sensors, using an adaptive method based on multivariate regression and rigorous statistical analysis. The results show a good adjustment with, at worst, almost 50% of the variability explained by the model. In particular, we found 71.27%, 65.55% and 47.49% of the variability explained for O_3, NO_2 and CO, respectively. Considering the short time interval used to estimate the model (less than 2.5 days), and achieving these adjustment values, it is expected that expanding the time series would improve the results.

In the case of O_3, we obtained the best fit. Ozone prediction errors followed a symmetrical distribution with no bias (the Shapiro–Wilks normality test passed 95% confidence). On the other hand, the NO_2 and CO prediction errors distribution had a right symmetry, which indicates a greater tail of the distribution in positive values. In these pollutants, the prediction values are generally overestimated with respect to the pattern ones. Overall, we observed a better quality on the fit with higher data.

We observed that the values of CO have the worst fit, which affected the R-squared with the variables considered. To model it, we needed to detrend the sensor measures of monoxide to include them in the calculus. Despite that, the prediction errors were greater than the others, with an average of 29 $\mu g/m^3$ and a marked right bias. We consider that this lack of adjustment in CO was caused by the high time of response of the sensor, the daily variability of this pollutant and the short time interval. Although its calibration may be improved using other more complex models, we consider that for a first approach, the linear multivariate regression is the best-balanced model.

Despite the limitations of the sensors and the dataset used, we obtained a good fit of these gaseous pollutants with respect to the values of the analyzers, while using measurements obtained with low-cost MOS sensors. After the application of our methodology, we observed that the O_3 and NO_2 adjusted parameters can be used to give reliable information to citizens and could be used by government agencies for policymaking.

In future works, we will explore other and more complex statistical modeling to enhance the results. We will also verify the possibility of calibrating other MOS sensors through the use of sensors calibrated with the proposed methodology, instead of using control stations. In addition, two of the main disadvantages of the MQ-7 sensor are the delay in the response of the measure and the discontinuous operation mode. In relation to the delay, this is due to the fact that it was designed to measure in ranges 100 times greater than those measured in air quality. Nowadays, new CO sensors working in continuous mode, with the capacity for measuring lower concentrations, has emerged, and these will be considered to replace MQ-7 sensors in future experiments.

Author Contributions: D.S.-L., A.J.B., A.S.-A. and P.M.M.-J. contributed to conceptualization, organization, and performance analysis of this paper, including writing and review. All authors have read and agreed to the published version of the manuscript.

Funding: This research received no external funding.

Institutional Review Board Statement: Not applicable.

Informed Consent Statement: Not applicable.

Data Availability Statement: The data presented in this study are available on request from the corresponding author.

Acknowledgments: The researchers would like to thank the University of Cadiz for the grant obtained through its "Programa de Fomento e Impulso de la actividad de Investigación y Transferencia". The authors would also like to thank to the Environmental Technology researching group and Acoustic Engineering Laboratory researching group, TEP-181 and TEP-195, respectively, for the access to the devices and data of the EcoBici Project (number G-GI3002/IDIC). Alfonso J. Bello acknowledges the support received from the 2014–2020 ERDF Operational Program and by the Department of Economy, Knowledge, and Business and the University of the Regional Government of Andalusia, Spain, under grant: FEDER-UCA18-107519.

Conflicts of Interest: The authors declare no potential conflict of interest.

Appendix A

This section contains tables and figures that complement Sections 3.3 and 3.4.

Table A1. The table contains the summary of the model described in Equation (4) adjusted to the dataset with all records.

Coefficients	Estimate	Std. Error	*t* Value	*p*-Value
α_0	205.52408	13.85254	14.837	$<2 \times 10^{-16}$
α_1	−0.38428	0.02208	−17.402	$<2 \times 10^{-16}$
α_2	−0.10342	0.02386	−4.335	0.000017738
α_3	0.48487	0.03688	13.146	$<2 \times 10^{-16}$
α_4	4.89923	0.43472	11.270	$<2 \times 10^{-16}$
α_5	0.41505	0.07775	5.338	0.000000144
Residuals:				
Min	1Q	Median	3Q	Max
−29.964	−7.158	−1.064	5.946	45.364
R-squared: 0.681				

Table A2. The table contains the summary of the model described in Equation (4) adjusted to the training dataset.

Coefficients	Estimate	Std. Error	*t* Value	*p*-Value
α_0	306.28198	23.25716	13.169	$<2 \times 10^{-16}$
α_1	−0.56296	0.03834	−14.685	$<2 \times 10^{-16}$
α_2	−0.12499	0.02869	−4.357	1.76×10^{-5}
α_3	0.53380	0.04452	11.989	$<2 \times 10^{-16}$
α_4	6.29591	0.55686	11.306	$<2 \times 10^{-16}$
α_5	0.72325	0.09612	7.525	5.04×10^{-13}
Residuals:				
Min	1Q	Median	3Q	Max
−25.390	−6.507	−0.237	5.456	40.765
R-squared: 0.6555				

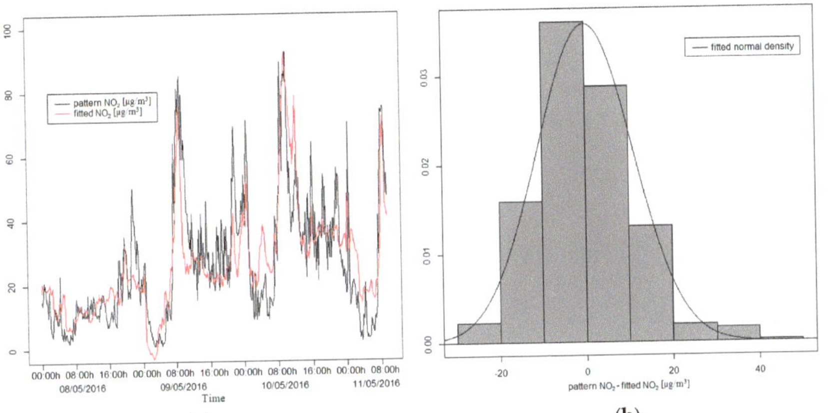

(a) (b)

Figure A1. (**a**) The fitted values by the model and the pattern values for nitrogen dioxide; (**b**) the histogram of the net prediction errors of the model.

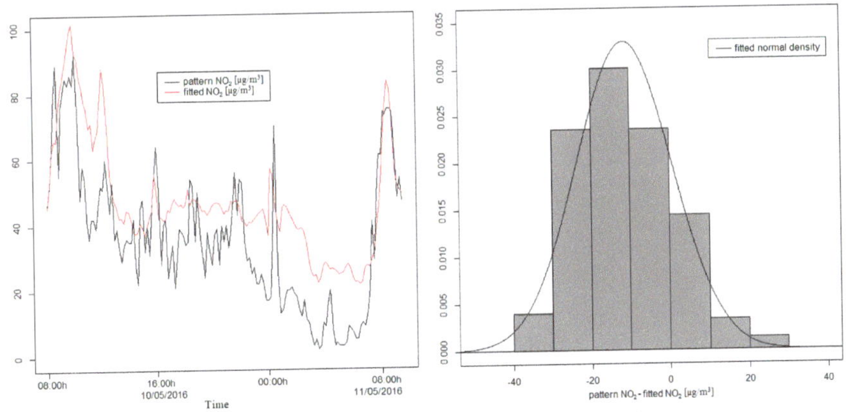

(a) (b)

Figure A2. Both graphs with the test dataset. (**a**) The fitted values by the model and the pattern values for nitrogen dioxide; (**b**) the histogram of the net prediction errors of the model.

Table A3. The table contains the summary of the model described in Equation (5) adjusted to the dataset with all records.

Coefficients	Estimate	Std. Error	*t* Value	*p*-Value
α_0	3604.9455	261.9849	13.760	$<2 \times 10^{-16}$
α_1	−4.4098	0.3387	−13.021	$<2 \times 10^{-16}$
α_2	0.1679	0.1015	1.654	0.09886
α_3	1.1791	0.1528	7.716	6.90×10^{-14}
α_4	4.5623	1.5081	3.025	0.00262
α_5	1.6188	0.3511	4.611	5.13×10^{-6}
Residuals:				
Min	1Q	Median	3Q	Max
−132.41	−26.88	−3.60	25.60	210.30
R-squared: 0.5793				

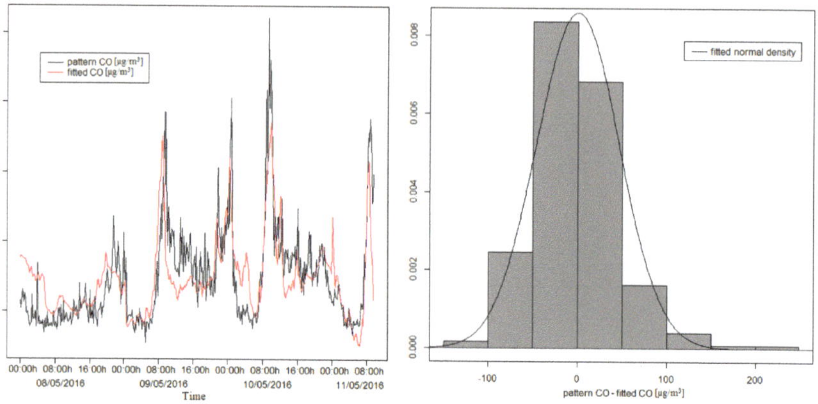

(a) (b)

Figure A3. (**a**) The fitted values by the model and the pattern values for carbon monoxide; (**b**) the histogram of the net prediction errors of the model.

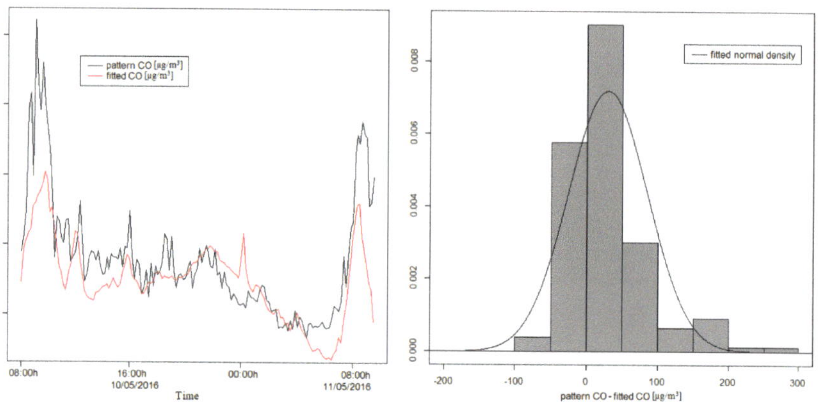

(a) (b)

Figure A4. Both graphs with the test dataset. (**a**) The fitted values by the model and the pattern values for carbon monoxide; (**b**) the histogram of the net prediction errors of the model.

Table A4. The table contains the summary of the model described in Equation (5) adjusted to the training dataset.

Coefficients	Estimate	Std. Error	*t* Value	*p*-Value
α_0	4077.0760	321.4378	12.684	$<2 \times 10^{-16}$
α_1	−5.0366	0.4119	−12.229	$<2 \times 10^{-16}$
α_2	0.1847	0.1297	1.424	0.15552
α_3	0.8636	0.1959	4.407	0.00001417
α_4	6.0151	1.9619	3.066	0.00235
α_5	2.0935	0.3812	5.492	0.00000008
Residuals:				
Min	1Q	Median	3Q	Max
−72.537	−27.893	−3.568	22.071	224.530
R-squared: 0.4749				

References

1. Brunekreef, B.; Holgate, S.T. Air pollution and health. *Lancet* **2002**, *360*, 1233–1242. [CrossRef]
2. Stanaway, G.; Afshin, A.; Gakidou, E.; Lim, S.; Abate, K.; Cristiana, A.; Abbasi, N.; Abbastabar, H.; Abd-Allah, F.; Abdela, J.; et al. Global, regional, and national comparative risk assessment of 84 behavioural, environmental and occupational, and metabolic risks or clusters of risks for 195 countries and territories, 1990–2017: A systematic analysis for the Global Burden of Disease Study 2017. *Lancet* **2018**, *392*, 1923–1994.
3. Jaimini, U.; Banerjee, T.; Romine, W.; Thirunarayan, K.; Sheth, A.; Kalra, M. Investigation of an indoor air quality sensor for asthma management in children. *IEEE Sens. Lett.* **2017**, *1*, 1–4. [CrossRef]
4. Marinello, S.; Butturi, M.A.; Gamberini, R. How changes in human activities during the lockdown impacted air quality parameters: A review. *Environ. Prog. Sustain. Energy* **2021**, *40*, e13672. [CrossRef] [PubMed]
5. Losacco, C.; Perillo, A. Particulate matter air pollution and respiratory impact on humans and animals. *Environ. Sci. Pollut. Res.* **2018**, *25*, 33901–33910. [CrossRef]
6. Orellano, P.; Reynoso, J.; Quaranta, N.; Bardach, A.; Ciapponi, A. Short-term exposure to particulate matter (PM10 and PM2.5), nitrogen dioxide (NO_2), and ozone (O_3) and all-cause and cause-specific mortality: Systematic review and meta-analysis. *Environ. Int.* **2020**, *142*, 105876. [CrossRef]
7. Liu, J.C.; Peng, R.D. Health effect of mixtures of ozone, nitrogen dioxide, and fine particulates in 85 US counties. *Air Qual. Atmos. Health* **2018**, *11*, 311–324. [CrossRef]
8. Olstrup, H.; Johansson, C.; Forsberg, B.; Åström, C. Association between mortality and short-term exposure to particles, ozone and nitrogen dioxide in Stockholm, Sweden. *Int. J. Environ. Res. Public Health* **2019**, *16*, 1028. [CrossRef]
9. Ritz, B.; Hoffmann, B.; Peters, A. The effects of fine dust, ozone, and nitrogen dioxide on health. *Dtsch. Ärzteblatt Int.* **2019**, *116*, 881. [CrossRef]
10. European Environment Agency (EEA). Air Quality in Europe-2019 Report. No 10/2019. 2019. Available online: https://www.eea.europa.eu/publications/air-quality-in-europe-2019 (accessed on 12 March 2021).
11. United States Environmental Protection Agency (EPA). Report: EPA's FYs 2020–2021 Top Management Challenges. No 20-N-0231. 2020. Available online: https://www.epa.gov/sites/production/files/2020-07/documents/_epaoig_20200721-20-n-0231_0.pdf (accessed on 12 March 2021).
12. Kuklinska, K.; Wolska, L.; Namieśnik, J. Air quality policy in the U.S. and the EU—A review. *Atmos. Pollut. Res.* **2015**, *6*, 129–137. [CrossRef]
13. European Environment Agency (EEA). Europe's Urban Air Quality-Re-Assessing Implementation Challenges in Cities. N°24-2018. 2018. Available online: https://www.eea.europa.eu/publications/europes-urban-air-quality (accessed on 12 March 2021).
14. Castell, N.; Dauge, F.R.; Schneider, P.; Vogt, M.; Lerner, U.; Fishbain, B.; Broday, D.; Bartonova, A. Can commercial low-cost sensor platforms contribute to air quality monitoring and exposure estimates? *Environ. Int.* **2017**, *99*, 293–302. [CrossRef]
15. Motlagh, N.H.; Lagerspetz, E.; Nurmi, P.; Li, X.; Varjonen, S.; Mineraud, J.; Siekkinen, M.; Rebeiro-Hargrave, A.; Hussein, T.; Petaja, T.; et al. Toward massive scale air quality monitoring. *IEEE Commun. Mag.* **2020**, *58*, 54–59. [CrossRef]
16. Kaivonen, S.; Ngai, E.C.H. Real-time air pollution monitoring with sensors on city bus. *Digit. Commun. Netw.* **2020**, *6*, 23–30. [CrossRef]
17. Munir, S.; Mayfield, M.; Coca, D.; Jubb, S.A.; Osammor, O. Analysing the performance of low-cost air quality sensors, their drivers, relative benefits and calibration in cities—A case study in Sheffield. *Environ. Monit. Assess.* **2019**, *191*, 94. [CrossRef] [PubMed]
18. Gryech, I.; Ben-Aboud, Y.; Guermah, B.; Sbihi, N.; Ghogho, M.; Kobbane, A. MoreAir: A Low-Cost Urban Air Pollution Monitoring System. *Sensors* **2020**, *20*, 998. [CrossRef]
19. Piedrahita, R.; Xiang, Y.; Masson, N.; Ortega, J.; Collier-Oxandale, A.; Jiang, Y.; Li, K.; Dick, R.; Lv, Q.; Hannigan, M.; et al. The next generation of low-cost personal air quality sensors for quantitative exposure monitoring. *Atmos. Meas. Technol.* **2014**, *7*, 3325–3336. [CrossRef]
20. Che, W.; Frey, H.C.; Fung, J.C.; Ning, Z.; Qu, H.; Lo, H.K.; Lau, A.K. PRAISE-HK: A personalized real-time air quality informatics system for citizen participation in exposure and health risk management. *Sustain. Cities Soc.* **2020**, *54*, 101986. [CrossRef]
21. Wesseling, J.; de Ruiter, H.; Blokhuis, C.; Drukker, D.; Weijers, E.; Volten, H.; Tielemans, E. Development and implementation of a platform for public information on air quality, sensor measurements, and citizen science. *Atmosphere* **2019**, *10*, 445. [CrossRef]
22. Thorson, J.; Collier-Oxandale, A.; Hannigan, M. Using A Low-Cost Sensor Array and Machine Learning Techniques to Detect Complex Pollutant Mixtures and Identify Likely Sources. *Sensors* **2019**, *19*, 3723. [CrossRef]
23. Rai, A.C.; Kumar, P.; Pilla, F.; Skouloudis, A.N.; Di Sabatino, S.; Ratti, C.; Yasar, A.; Rickerby, D. End-user perspective of low-cost sensors for outdoor air pollution monitoring. *Sci. Total Environ.* **2017**, *607–608*, 691–705. [CrossRef]
24. Burgués, J.; Marco, S. Low Power Operation of Temperature-Modulated Metal Oxide Semiconductor Gas Sensors. *Sensors* **2018**, *18*, 339. [CrossRef]
25. Martinez, D.; Burgués, J.; Marco, S. Fast Measurements with MOX Sensors: A Least-Squares Approach to Blind Deconvolution. *Sensors* **2019**, *19*, 4029. [CrossRef]
26. World Meteorological Organization. Low-Cost Sensors for the Measurement of Atmospheric Composition: Overview of Topic and Future Applications Europe's Urban Air Quality. Available online: https://library.wmo.int/doc_num.php?explnum_id=9881 (accessed on 12 March 2021).

27. Masson, N.; Piedrahita, R.; Hannigan, M. Approach for quantification of metal oxide type semiconductor gas sensors used for ambient air quality monitoring. *Sens. Actuators B Chem.* **2014**, *208*, 339–345. [CrossRef]
28. Maag, B.; Zhou, Z.; Thiele, L. A Survey on Sensor Calibration in Air Pollution Monitoring Deployments. *IEEE Internet Things J.* **2018**, *5*, 4857–4870. [CrossRef]
29. Afshar-Mohajer, N.; Zuidema, C.; Sousan, S.; Hallett, L.; Tatum, M.; Rule, A.M.; Thomas, G.; Peters, T.M.; Koehler, K. Evaluation of low-cost electro-chemical sensors for environmental monitoring of ozone, nitrogen dioxide, and carbon monoxide. *J. Occup. Environ. Hyg.* **2018**, *15*, 87–98. [CrossRef] [PubMed]
30. Schultealbert, C.; Baur, T.; Schütze, A.; Böttcher, S.; Sauerwald, T. A novel approach towards calibrated measurement of trace gases using metal oxide semiconductor sensors. *Sens. Actuators B Chem.* **2017**, *239*, 390–396. [CrossRef]
31. Leidinger, M.; Schultealbert, C.; Neu, J.; Schütze, A.; Sauerwald, T. Characterization and calibration of gas sensor systems at ppb level—A versatile test gas generation system. *Meas. Sci. Technol.* **2017**, *29*, 015901. [CrossRef]
32. Wei, P.; Ning, Z.; Ye, S.; Sun, L.; Yang, F.; Wong, K.C.; Westerdahl, D.; Louie, P.K.K. Impact Analysis of Temperature and Humidity Conditions on Electrochemical Sensor Response in Ambient Air Quality Monitoring. *Sensors* **2018**, *18*, 59. [CrossRef]
33. Jiang, Y.; Zhu, X.; Chen, C.; Ge, Y.; Wang, W.; Zhao, Z.; Cai, J.; Kan, H. On-field test and data calibration of a low-cost sensor for fine particles exposure assessment. *Ecotoxicol. Environ. Saf.* **2021**, *211*, 111958. [CrossRef]
34. Spinelle, L.; Gerboles, M.; Villani, M.G.; Aleixandre, M.; Bonavitacola, F. Field calibration of a cluster of low-cost available sensors for air quality monitoring. Part A: Ozone and nitrogen dioxide. *Sens. Actuators B Chem.* **2015**, *215*, 249–257. [CrossRef]
35. De Vito, S.; Massera, E.; Piga, M.; Martinotto, L.; Di Francia, G. On field calibration of an electronic nose for benzene estimation in an urban pollution monitoring scenario. *Sens. Actuators B Chem.* **2008**, *129*, 750–757. [CrossRef]
36. Peterson, P.J.D.; Aujla, A.; Grant, K.H.; Brundle, A.G.; Thompson, M.R.; Vande Hey, J.; Leigh, R.J. Practical Use of Metal Oxide Semiconductor Gas Sensors for Measuring Nitrogen Dioxide and Ozone in Urban Environments. *Sensors* **2017**, *17*, 1653. [CrossRef] [PubMed]
37. Berry, W.D.; Feldman, S. *Multiple Regression in Practice (Sage University Paper Series, Quantitative Applications in the Social Sciences)*; Sage: Beverly Hills, CA, USA, 1985.
38. Montgomery, D.C.; Peck, E.A.; Vining, G.G. *Introduction to Linear Regression Analysis*, 5th ed.; John Wiley & Sons, Inc.: Hoboken, NJ, USA, 2012; ISBN 978-0-470-54281-1.
39. Yan, X.; Su, X.G. *Linear Regression Analysis: Theory and Computing*; World Scientific Publishing Co. Pte. Ltd.: Singapore, 2009.
40. R Core Team. *R: A Language and Environment for Statistical Computing*; R Foundation for Statistical Computing: Vienna, Austria, 2020; Available online: https://www.R-project.org/ (accessed on 28 May 2021).
41. Official Journal of the European Union. Directive 2008/50/EC of the European Parliament and the Council of 21 May 2008 on Ambient Air Quality and Cleaner Air for Europe. June 2008. Available online: https://eur-lex.europa.eu/LexUriServ/LexUriServ.do?uri=OJ:L:2008:152:0001:0044:en:PDF (accessed on 28 May 2021).
42. World Meteorological Organization. Global Air Quality Guidelines. Available online: https://www.euro.who.int/en/health-topics/environment-and-health/air-quality/activities/update-of-who-global-air-quality-guidelines (accessed on 28 May 2021).
43. Technical Data MQ-7 Gas Sensor. Available online: https://www.sparkfun.com/datasheets/Sensors/Biometric/MQ-7.pdf (accessed on 12 March 2021).
44. MQ-131 Ozone Gas Sensor. Available online: https://aqicn.org/air/view/sensor/spec/o3.winsen-mq131.pdf (accessed on 12 March 2021).
45. Data Sheet MiCS-2714. Available online: https://www.sgxsensortech.com/content/uploads/2014/08/1107_Datasheet-MiCS-2714.pdf (accessed on 12 March 2021).
46. Sales-Lérida, D.; Sales-Márquez, D.; Hernandez-Molina, R.; Cueto-Ancela, J.L. Sistema de Telemedición de Calidad Del Aire Para la Visualización en Tiempo Real de una Red de Dispositivos Compactos. Spain Patent Number ES2638715, 16 August 2018.
47. Visor de Calidad del Aire. Available online: https://www.miteco.gob.es/es/calidad-y-evaluacion-ambiental/temas/atmosfera-y-calidad-del-aire/calidad-del-aire/visor/default.aspx (accessed on 12 March 2021).
48. Informes Diarios de Calidad del Aire. Available online: http://www.juntadeandalucia.es/medioambiente/site/portalweb/menuitem.7e1cf46ddf59bb227a9ebe205510e1ca/?vgnextoid=7e612e07c3dc4010VgnVCM1000000624e50aRCRD&vgnextchannel=910f230af77e4310VgnVCM1000001325e50aRCRD (accessed on 12 March 2021).
49. How to Perform Multiple Linear Regression in R. Available online: https://www.statology.org/multiple-linear-regression-r/ (accessed on 3 July 2021).
50. A Complete Guide to Stepwise Regression in R. Available online: https://www.statology.org/stepwise-regression-r/ (accessed on 3 July 2021).

sensors MDPI

Article

The Use of Public Data from Low-Cost Sensors for the Geospatial Analysis of Air Pollution from Solid Fuel Heating during the COVID-19 Pandemic Spring Period in Krakow, Poland

Tomasz Danek and Mateusz Zaręba *

Department of Geoinformatics and Applied Computer Science, Faculty of Geology, Geophysics and Environmental Protection, AGH University of Science and Technology, 30-059 Krakow, Poland; tdanek@agh.edu.pl

* Correspondence: zareba@agh.edu.pl

Abstract: In this paper, we present a detailed analysis of the public data provided by low-cost sensors (LCS), which were used for spatial and temporal studies of air quality in Krakow. A PM (particulate matter) dataset was obtained in spring in 2021, during which a fairly strict lockdown was in force as a result of COVID-19. Therefore, we were able to separate the effect of solid fuel heating from other sources of background pollution, mainly caused by urban transport. Moreover, we analyzed the historical data of PM2.5 from 2010 to 2019 to show the effect of grassroots efforts and pro-clean-air legislation changes in Krakow. We designed a unique workflow with a time-spatial analysis of PM1, PM2.5, and PM10, and temperature data from Airly(c) sensors located in Krakow and its surroundings. Using geostatistical methods, we showed that Krakow's neighboring cities are the main sources of air pollution from solid fuel heating in the city. Additionally, we showed that the changes in the law in Krakow significantly reduced the PM concentration as compared to neighboring municipalities without a fossil fuel prohibition law. Moreover, our research demonstrates that informative campaigns and education are important initiating factors in order to bring about cleaner air in the future.

Keywords: air pollution measurements; air quality monitoring; LCS; particulate matter; air quality in Krakow; anthropogenic emission; spatio-temporal geostatistics; fossil fuels

1. Introduction

Air pollution is a major problem for modern society. Research shows that PM exposure is responsible for over 7% of deaths and more than 4% of disabilities globally [1]. Health problems can be related to short-term and long-term exposure [2]. Short-term exposure can cause asthma [3], high blood pressure, myocardial infarction, and even death, as a result of damage to the respiratory or cardiovascular systems [4,5]. Long-term exposure may contribute to the development of diseases such as lung cancer [6], pneumonia [7], crescendo angina (a type of acute coronary syndrome) [8], chronic obstructive pulmonary disease [9], and it can be even responsible for low birth weight [10]. It was observed that PM exposure is also a factor for certain neurological diseases such as Parkinson's and Alzheimer's disease [11].

Suburbanization and the fast-growing urban population have caused an increased demand for heating. This is particularly apparent from the beginning of autumn until the end of spring in Central Europe. The geographical location of the city of Krakow (Poland), which is situated in the Vistula River valley, and the local weather conditions favor the accumulation of pollutants in the city [12]. For years, Krakow has had a bad reputation for significantly exceeding the norms for the concentration of particulate matter in the air (the 8th worst city in the European Union (EU) according to the World Health

Organization (WHO) report [13]). In 2012, the organization *Krakow Smog Alert* initiated an informative campaign focused on the bad air quality in Krakow [14]. A year later, this initiative turned into a mass social movement. The citizens of Krakow organized a public protest in which they demanded legal changes and better air quality. By the end of the same year, the Małopolska Voivodeship (the highest administrative region in Poland, which is akin to a province in other countries) assembly passed a law prohibiting the use of solid fuels for heating households. As a result of an appeal against the law, it did not enter into force as planned in 2018. In 2015, the Małopolska Voivodeship assembly passed another law prohibiting the use of solid fuels in Krakow that successfully entered into force in 2019 [15]. Presently, it is prohibited to use coal, wood, or biomass in Krakow city for central heating, any systems that emit hot air, or liquid heating installations, including fireplaces, space heaters, stoves, etc. The regulations also include a ban on the use of coal, wood, and biomass for food preparation. According to the new rules, gaseous fuels such as liquefied natural gas or light heating oil are allowed.

Air monitoring in the European Union member states is regulated by the directive 2008/50/EC of the European Parliament and of the Council of 21 May 2008 on ambient air quality and cleaner air for Europe (AAQD) [16]. In the EU, air quality stations are divided into three groups: urban traffic sites (located in urbanized areas or near high-traffic zones), urban background sites (for general population measurements), and regional background sites (general rural observations). According to the European Union Commission Staff working document *SWD(2019) 427 final*, in 2017, there were 4332 monitoring stations in all member states (278 in Poland), including 3130 PM10 total sampling points per pollutant (288 in Poland) and 1543 PM2.5 total sampling points per pollutant (111 in Poland) [17]. Krakow is located in zone PL1201 and has a total population of 756,183 and an area of 327 km^2. In zone PL1201, three automatic measurements of PM10 and PM2.5, two manual measurements for PM10, and one manual measurement for PM2.5 are available [18]. Some of Krakow's surrounding areas are in zone PL1203, which is the zone for the whole Małopolska area (over 14,700 km^2). In this zone, six automatic measurements and 10 manual measurements for PM10 are available, and four manual measurements for PM2.5 are available [19]. The reference measurements and instruments used in Poland are described in the following documents: PN-EN 12341 (for gravimetric measurements) and PN-EN 16450 (for automatic measurements) [20]. The current EU and Polish standards are 25 μg/m^3 (1-year averaging period) for PM2.5 concentrations, and 50 μg/m^3 (24-h averaging period) and 40 μg/m^3 (1-year averaging period) for PM10 concentrations [21].

In general, the low-cost sensors (LCS) can be categorized into electrochemical sensors, photoionization detectors, optical particle counters, and optical sensors. Currently, the European Union regulation does not allow for the use of LCS for official air quality reporting. This is related to the questionable data quality from these sensors as compared with official gravimetric measurements. LCS measurements can be affected by many weather-related factors; however, in well-prepared environments and stations, they can provide a similar data quality to official government measurements [22]. In this paper, we present research regarding the use of popular, low-cost Airly optic sensors (http://airly.eu, accessed on 29 July 2021). These sensors were used in air quality studies in Niedzica (Poland) [23] and for health risk research [24]. There was another evaluation of Airly sensors in 2018 as part of the LIFE Integrated Project *"Implementation of Air Quality Plan for Małopolska Region—Małopolska in a healthy atmosphere"*. Tests were performed according to the PN-EN 12341:2014-07 and PN-EN 16450:2017-05 standards based on an agreement concluded between the Lesser Poland Voivodeship Main Inspectorate of Environmental Protection Provincial Inspectorate for Environmental Protection in Krakow, AGH University of Science and Technology, and the community of Dobczyce, in cooperation with the Krakow Smog Alert Association. The results obtained for seven different LCS pairs of sensors (each pair produced by a different company) showed that only Airly devices provided adequate results. After calibration, the sensors provided satisfactory results in relation to the reference stations [25]. The measurement quality of the Airly sensors was also examined by AIRLAB during the Microsensors Challenge 2019, using the SET method [26]; their accuracy was 7.6 out of 10 [27]. Airly sensors were calibrated in the laboratory before the study. Measurements from a sensor in a particular location were assigned a dynamic calibration factor (created by machine learning (ML) and artificial intelligence (AI) algorithms) based on the characteristics of the location. Data are available to the public after ML/AI corrections.

Grassroots movements and local authority activities allowed for a significant improvement in air quality in Krakow [28]; however, air pollution problems still occur during spring and autumn. In this paper, we investigate whether the remaining problem seems to be mainly associated with the transport of pollutants from neighboring villages and towns. The study presented in this paper used popular, low-cost Airly sensors (for more details see http://airly.eu, accessed on 29 July 2021) to analyze whether there was a transmission of pollutants from neighboring areas to the city of Krakow in spring 2021. The impact of air pollution on human health (2005–2020) in the city of Krakow is well described by Bokwa [29] and by Traczyk and Gruszecka-Kosowska [12]. Our research focuses on the spatial-temporal analysis of the PM1, PM2.5, and PM10 contents obtained from approximately 100 Airly sensors (data available from: https://map.airly.org/ [30], accessed on 29 July 2021) located in Krakow and its surroundings. As a result of the COVID-19 pandemic, it was possible to study the effect of heating with the use of solid fuel more precisely, because of the limited traffic as compared to regular years. Research carried out by the Department of the City Traffic Engineer shows that the traffic volume during the coronavirus pandemic (at the 15 main city intersections) was as much as 40% lower as compared to pre-pandemic years [31]. We analyzed the changes in pollution levels in the city center over the last 10 years and during spring in 2021 in the city and its surroundings.

Our goals were to study:

1. The temporal and spatial distribution of air pollution in Krakow and nearby areas. We wanted to check whether pollution from heating households with fossil fuels in neighboring towns and villages migrates to Krakow and increases the level of pollution in the city. Our goal was to identify the main sources of pollution in the vicinity of Krakow and to assess the scale of the problem;
2. The effectiveness of the anti-smog policy that was gradually introduced in Krakow. Local authorities and organizations have been working since 2011 to change the air quality in Krakow. We wanted to study if these activities are related to changes in the PM concentration over the years.

The 2021 spring period was exceptionally favorable for this type of research (temporal and spatial distribution of PM). Overall, it was a period of abnormally low temperatures for this time of year, mainly due to the polar vortex disturbances. The simultaneous influx of cold Arctic air and the relatively large insolation caused successive periods of warmer temperatures and significant rapid cooling. In warmer periods, the air was cleaned of the so-called low-emission pollution, which is the main source of smog in and around Krakow [32]. With rapid cooling, there was an abrupt increase in low emissions from poor-quality heating systems often found outside the urban area, in a region not covered by the pro-clean-air legislation. This significantly helped the identification of local sources of pollution, which, in conditions of long-term, constant pollution of the environment in winter, could be hidden by the high background level.

2. Materials and Methods

Measurements of PM1, PM2.5, and PM10 from 90 optical Airly PM Sensors were used. Figure 1 shows the locations of the sensors used in this research. The receivers were selected to cover both the city of Krakow and its surroundings. Airly sensors are one of the most popular in the Krakow area and can provide a high density of spatial data [33]. As a result of their convenience and relatively low price, they are also gaining popularity throughout the country [23,24] and around the world [34]. Airly sensors for PM measurements are optical sensors that measure light scattering. The accuracy of the results from such devices is strongly related to the physical properties of the particles, which can vary depending on location and season [35]. The manufacturer of Airly sensor states that the PM1 measurement range is 0–500 µg/m^3 (5 µg/m^3 accuracy in the range 0–100 µg/m^3 and 10 µg/m^3 in the range 101–500 µg/m^3), and the PM2.5 and PM10 measurement range is 0–1000 µg/m^3 (10 µg/m^3 accuracy in the range 0–100 µg/m^3, 10% in the range 101–500 µg/m^3, and 20% in the range 501–1000 µg/m^3). Aside from PM concentrations, the basic Airly sensor measures pressure in hPa (in the range 700–1200 hPa, with an accuracy of 1 hPa), the temperature in °C (in the range −40–80 °C, with an accuracy of 0.5 °C), and the humidity (in the range 0–100%, with an accuracy of 3%). The smallest measurement interval is 5 minutes. The samples are sent to the database via GSM protocol and are available from the Airly website or API. Sensors can use a solar power supply [36].

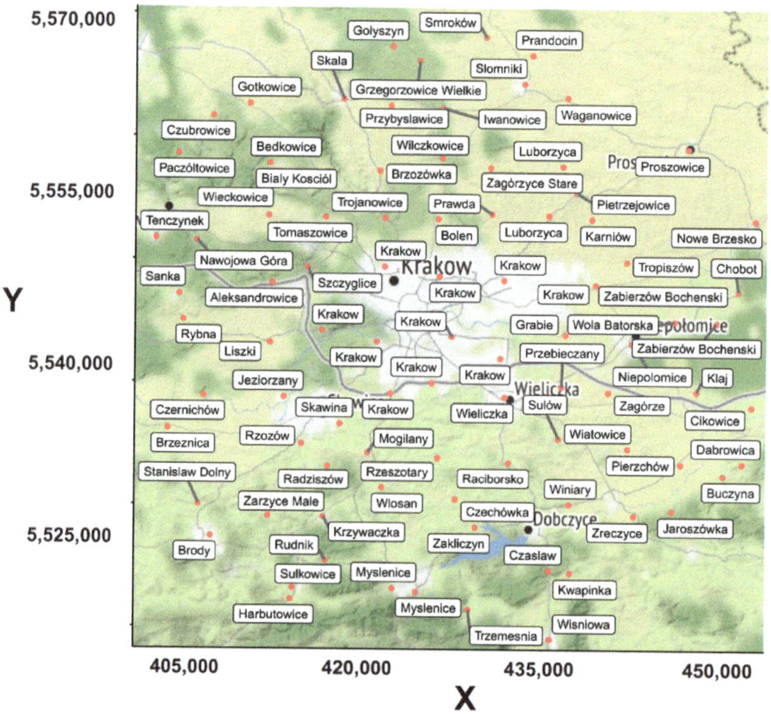

Figure 1. Locations of the Airly sensors used in the research (red dots) with the names of cities, towns, and villages in which they are located.

PM2.5 concentration data from the last 10 years (2010–2019) were used to determine trends using the Seasonal and Trend Decomposition with Loess method (STL), which was introduced by Cleveland et al. [37]. Raw data are difficult to interpret directly as, on certain days, there may be abnormally high PM concentration values. Unprocessed identification of such data does not necessarily translate into the general trend. Standard trend estimation methods (linear, polynomial, running average, etc.) can lead to incorrect conclusions as a result of not including seasonality or nonlinear relations. STL allows for data decomposition into trend, seasonal cycle, and even reminder component, which includes unusual data observations. The use of targeted data processing techniques can significantly improve data interpretation [38,39]. Data came from the Polish Chief

Inspectorate For Environmental Protection database [40]. These data came from one to three sensors located in the city center. In the years in which more than one sensor was available, the results were averaged; the data for days in which no reading was available were recovered by linear interpolation. The major law changes were highlighted, beginning in 2012, when Krakow Smog Alert initiated an informative campaign. In 2013, grassroots civil demonstrations and protests occurred. This was the beginning of various similar events in Krakow. Thanks to public involvement and the actions of the authorities, it was possible to enforce a law prohibiting the use of solid fuels for heating. Unfortunately, the law was successfully challenged. Thanks to Environmental Law changes in Poland in 2015, it was possible to pass a second law prohibiting the use of solid fuels in Krakow in 2016 [41], which, after the adaptation period, went fully into force in 2019. During this period, various other changes were conducted to Polish regulations, including changing the emission norms in 2018. Aside from changes to the law and the education process, other factors must be considered, such as the anomalously warm winter of 2014/2015.

Despite the ban on the use of solid fuels in Krakow, a periodic increase in air pollution could be observed in the city, especially during late autumn, winter, and early spring. This situation was even observed during the COVID-19 pandemic, during which traffic was almost 40% lower than before. Thanks to the dense distribution of LCS Airly sensors, it was possible to analyze data from Krakow and its surroundings. Airly sensors located up to 30 kilometers from the center of Krakow were used for these tests. There are about 364 such sensors in the analyzed area, but they are distributed very irregularly. In certain areas, sensors are a matter of meters apart, in others, there is not a single sensor for several kilometers. To enable the correct distribution, efforts were made to select their location so that they would comprise a quasi-regular measurement grid. It is difficult to obtain a regular sensor grid using Airly API. This is because anyone can buy an Airly sensor, making their distribution irregular. The area of study (Figure 1) was divided into a regular, 100-point grid, according to the X and Y axes. In this area, 364 LCSs were available. We designed the algorithm to search for the sensor located closest to each of the 100 regularly distributed points in this area. For further analysis, we used no more than one sensor per point. The algorithm was written in R as follows:

1. Define the area of investigation;
2. Use function *makegrid* and divide into 100 regularly distributed points;
3. Read all sensors' geographical positions;
4. Use *k-nearest-neighbours* to find the distance from sensors to a particular grid point;
5. Assign unique index number to sensors in relation to distance to grid points;
6. Choose the sensor closest to the grid point;
7. Exclude the sensors assigned to a particular grid point if the distance between them is less than $\frac{1}{4}$ of the distance between the neighboring grid points;
8. Save assigned sensor index numbers, and X and Y positions.

A satisfactory network of 90 sensors was, thus, obtained. We obtained an even distribution of measurement points, which we were able to fit into the open Airly license. To check if the Airly sensors measurements were sufficiently legible for use in the analysis, we compared them with the readings of the closest government sensors. We averaged the Airly measurements in 24-h periods and compared them with analogical scale measurements from government stations. Then, we calculated the differences between the government and Airly sensor measurements.

The general analysis of smog conditions was carried out using charts and maps created in R by fitting a thin plate spline surface to our data. This method is a special case of kriging; however, it guarantees a "smooth" surface [42], which is desirable when analyzing the dispersion propagation of pollutants. This step was performed for the identification of particularly interesting spatial-time snapshots (see the animation in the Supplementary Materials). Thereafter, we took a closer look at these data, comparing them with the temperature data.

To observe the inflow and outflow of air pollution in Krakow, the analysis of temperature changes in relation to the feeling of relative cold was performed. To observe the effect of household heating when the temperature drops below the comfort zone, KDE plots of PM were prepared. For two chosen dates, maps of PM1, PM2.5, and PM10 together with temperature were constructed. For grid preparation, the minimal curvature method was used, with grid spacing of approximately 500 meters in the X and Y directions.

3. Results

3.1. Historical Data Analysis Using Official Government Data from Reference Stations

The STL decomposition of PM2.5 obtained from official government sensors is shown in Figure 2. It is clearly visible that education, grassroots initiatives, and multilevel political involvement is crucial. Along with the increase in environmental awareness in society over the years, changes in the law and a stable decrease in emissions were visible. However, selective bans on the use of solid fuels in individual cities, without taking into account surrounding towns and cities, may cause a temporary excessive increase in the concentration of PM in the city, where the prohibition law is in force.

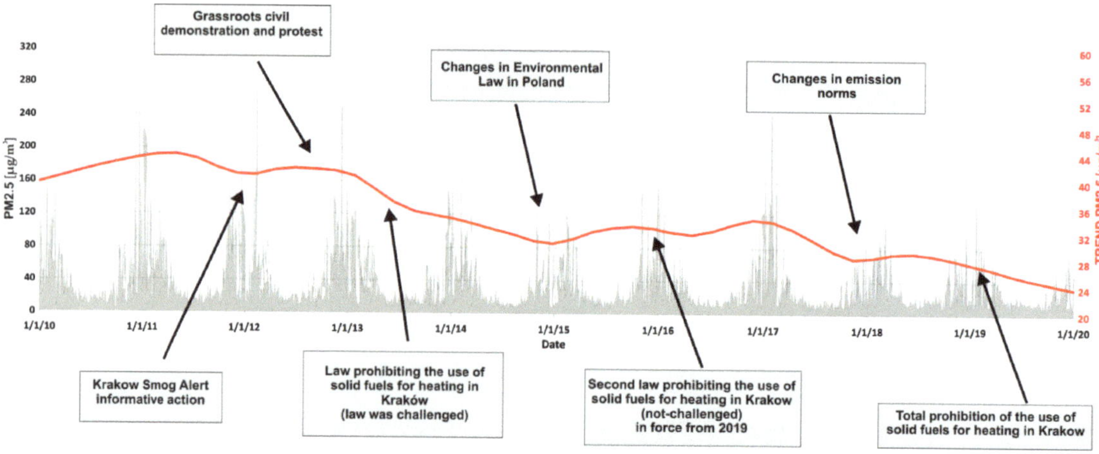

Figure 2. Krakow's PM2.5 concentration (grey) and its STL trend (red) in the years 2010–2019 together with major law changes and grassroots action.

3.2. Low-Cost Sensors for Inflow and Outflow Monitoring of PM1, PM2.5, and PM10

Table 1 shows a summary of the basic statistical analysis for all data. The data were characterized by a compact distribution, i.e., outliers were rare. The mean value of PM10 for this period (7 March to 16 April 2021) was lower than the EU standard (1-year averaged—40 μg/m^3). PM2.5 was slightly higher than that proposed by the EU standard (3-year averaged—25 μg/m^3); however, the median for PM2.5 was approximately the same as the standard.

Table 1. Summary of basic statistics for the measured parameters from Airly sensors in spring in Krakow, 2021, during the COVID-19 pandemic.

	Temp (°C)	Pressure (hPa)	Humidity (%)	PM1	PM2.5	PM10
					μg/m³	
Min	−7.37	994.40	17.65	0.02	0.21	0.28
1st Qu	1.02	1012.80	63.47	9.62	14.69	17.21
Median	3.72	1017.50	76.19	16.00	25.14	31.49
Mean	4.81	1016.80	74.39	18.35	29.72	37.12
3rd Qu	7.71	1021.10	87.25	23.36	37.68	49.56
Max	28.46	1032.70	100.00	148.54	305.25	376.18

For a better understanding of the air pollution problem in Krakow, an analysis of 960 h was conducted. Table 2 shows the correlation coefficients between all measured parameters for 90 detectors, for all time points. It was clear to see that PM1, PM2.5, and PM10 were correlated.

Table 2. Correlation coefficients for the measured parameters from Airly sensors for all data points.

		Temp (°C)	Pressure (hPa)	Humidity (%)	PM1	PM2.5	PM10
						μg/m³	
Temp (°C)		1.000	0.048	−0.583	−0.369	−0.357	−0.374
Pressure (hPa)		0.048	1.000	0.111	0.241	0.258	0.235
Humidity (%)		−0.583	0.111	1.000	0.342	0.341	0.354
PM2.5	μg/m³	−0.369	0.241	0.342	1.000	0.997	0.997
PM1		−0.357	0.258	0.341	0.997	1.000	0.995
PM10		−0.374	0.235	0.354	0.997	0.995	1.000

From all 90 sensors, sensor 36808 in Niepołomice was the closest to a government sensor. The government station is located on 3 May Street in Niepołomice (SE part of the investigated area). Table 3 shows statistics for the calculated time-series difference between 24-h averaged measurements of PM10 from government stations (GOV) and the Airly sensor in Niepołomice. The correlation coefficient for these measurements between 7 March and 16 April 2021 was 0.93. A comparison between the 24-h averaged measurements of Airly sensor 36808 and the government sensor located on 3 May street is shown in Figure 3.

Table 3. Statistics for the difference in PM10 measurements between government station and Airly sensor 36808 for 24-h averaged periods between 7 March and 16 April 2021.

	Min	1st Qu	Median	Mean	3rd Qu	Max
GOV and 36808 sensors difference (μg/m³)	−19.157	−7.425	−3.593	−3.001	1.069	17.407

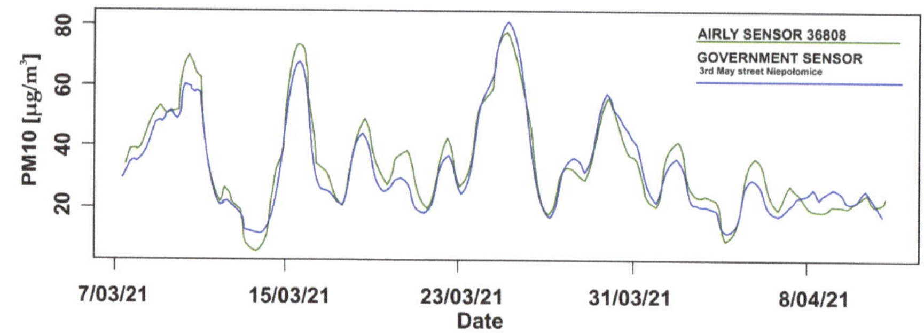

Figure 3. Twenty-four-hour averaged measurements of PM10 from government stations (blue) and the Airly sensor (green) in Niepołomice.

3.3. Analysis of Air Pollution Inflow and Outflow According to Temperature Changes

Figure 4 shows the temperature measurements from 7 March to 16 April 2021 together with the STL trend. Our aim was to show the inflow and outflow of air pollution in Krakow. To indicate the proper time and date, it was necessary to deeply study the temperature changes, as this is one of the main factors affecting whether people heat their households or not. It was possible to designate 10 temperature intervals: the first was the 7 to 10 March, which was stable at around 0 °C; the second was from the 11 to 14 March, during which the average temperature trend increase was visible; the next was a slowly decreasing period from 14 to 21 March, at the end of which the temperature reached its minimum value; the 4th period was up until the 2 April, when temperature stably increased (excepting a 1-day trend break); the next period was from the 2 to 5 April, wherein the temperature decreased to around 1–2 °C and stabilized on this value for 5 days; then, there were four three-day periods, in which the temperature first rose, then stabilized, before dropping again, and stabilizing once more in the last period. The 11th of March was chosen to study the effect of the air pollution outflow, and the 18th of March was chosen to study the effect of the air pollution inflow. Firstly, those days were characterized by an upward or downward temperature trend. Secondly, according to the relationship between the Predicted Mean Vote (PMV) and Perceived Temperature (PT) indicator in the study by Jendritzky et al. [43], it was shown that most people classified temperature between 0 and 20 °C as comfortable, and when the temperature decreases below 0 °C, most people classify it as cold. Figure 5 shows the increase in PM concentrations when the temperature dropped below the aforementioned comfortable temperature. In the first-row kernel density estimate plots for PM1, PM2.5, and PM10 versus temperature from the 18th of March, 18:00 was plotted. The second row contains the same plots but from 24:00. At 18:00, the temperature was above zero, and for all temperature points, the distribution of PM was similar. At 24:00, the distortion in the distribution symmetry is visible. White arrows indicate the increase in PM in the air. This is in clear accordance with the PMV and PT study.

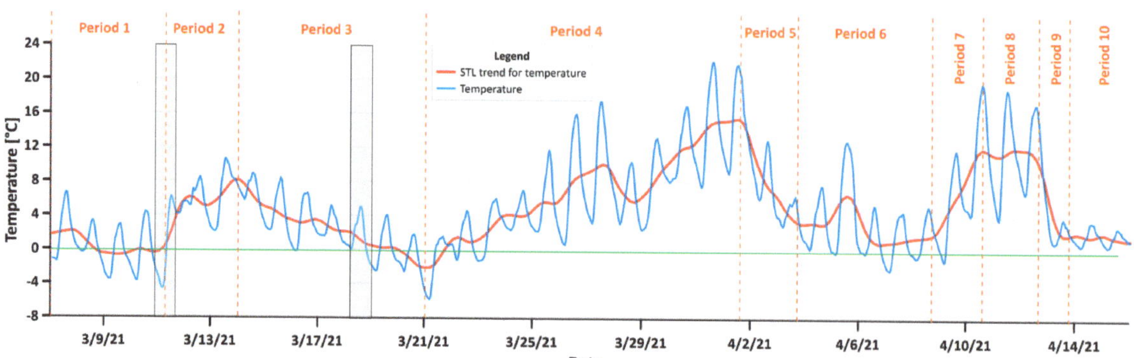

Figure 4. The temperature in Krakow between the 7th of March and 16th April (blue) and the temperature STL trend (red).

For the air pollution inflow, measurements from 12:00, 18:00, and 24:00 were used. For the air pollution outflow, measurements from 00:00, 04:00, and 08:00 were used. This is because people begin to heat their houses after sunset, which, in March, is around 17:00. To observe the effect of inflow, we analyzed midday, when the PM concentration was expected to be low at almost every sensor, then at 18:00, when the PM should have been visible in the surroundings of Krakow, and then at midnight, when the inflow effect was expected to be visible in the city. For outflow, we began at midnight, when the concentration of PM in the city due to inflow was expected to be high. We then proceeded in 4-h intervals to investigate the change.

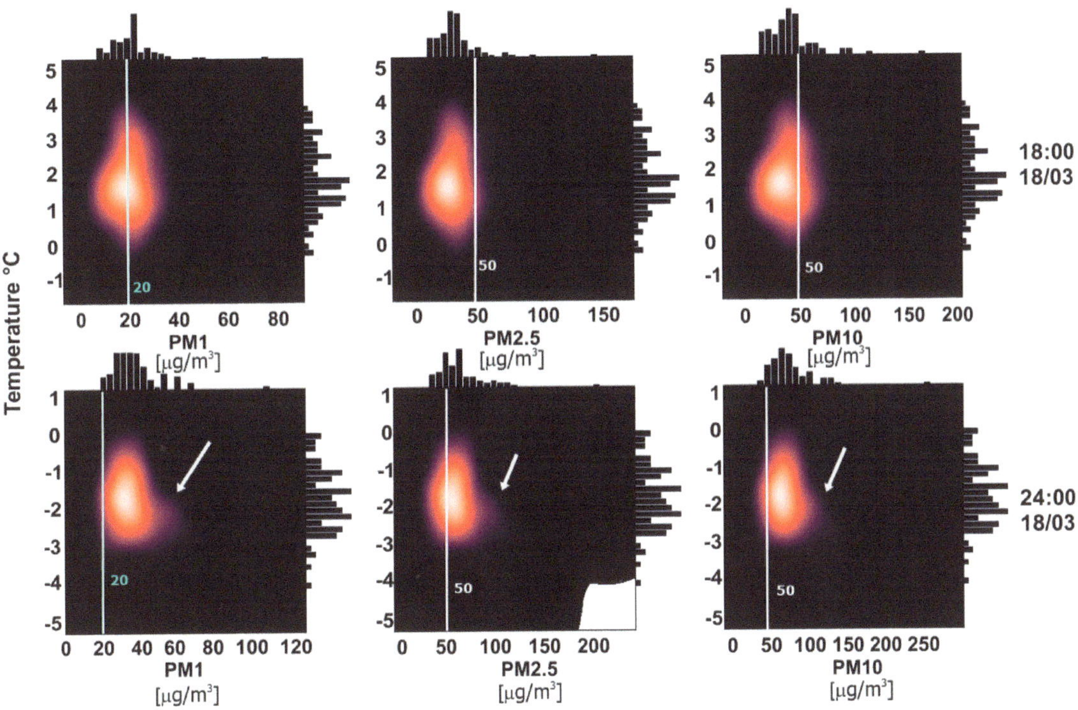

Figure 5. Kernel density estimate plots of PM1, PM2.5, PM10 and temperature at 18:00 and 24:00 on the 18th of March. Arrows indicate the increase in PM related to temperature and also to PT and PMV.

Figure 6 shows the air pollution inflow to Krakow, while Figure 7 shows the air pollution outflow. In the first row, the PM1 concentration was presented for three different hours, the second presents similar maps for PM2.5, the third shows PM10, and the fourth presents the temperature. Crosses represent LCS. Considering the inflow of air pollutants into Krakow from surrounding towns and cities, it is clear to see that, at noon, Krakow city was the warmest area on the map; however, the PM concentrations were low in and around the city. At 18:00, the temperature was stable; this was similar at every point with an average of approximately 1 °C. In certain places outside Krakow, air pollution was starting to increase. These were places where the temperature was lower than in the surrounding receivers, close to the thermal comfort limit (see white arrows in Figure 6). However, despite a significant increase in PM in neighboring towns and villages, the air quality in Krakow remained good. This situation changed with time. At midnight, it could be clearly seen that the pollutants diffused into Krakow; however, the level of PM contamination was still lower in the city of Krakow. The maps show that PM10 pollution increased the most, and PM1 pollution the least.

In the case of an outflow of pollutants, it is again visible in the concentration analysis that, at midnight, the accumulation of pollutants was the result of their diffusion from neighboring municipalities. The 11th of March was chosen because, on this day, the temperature in the morning began to rise above the comfort temperature. In this way, we were able to exclude the additional influx of pollutants resulting from heating houses in the morning.

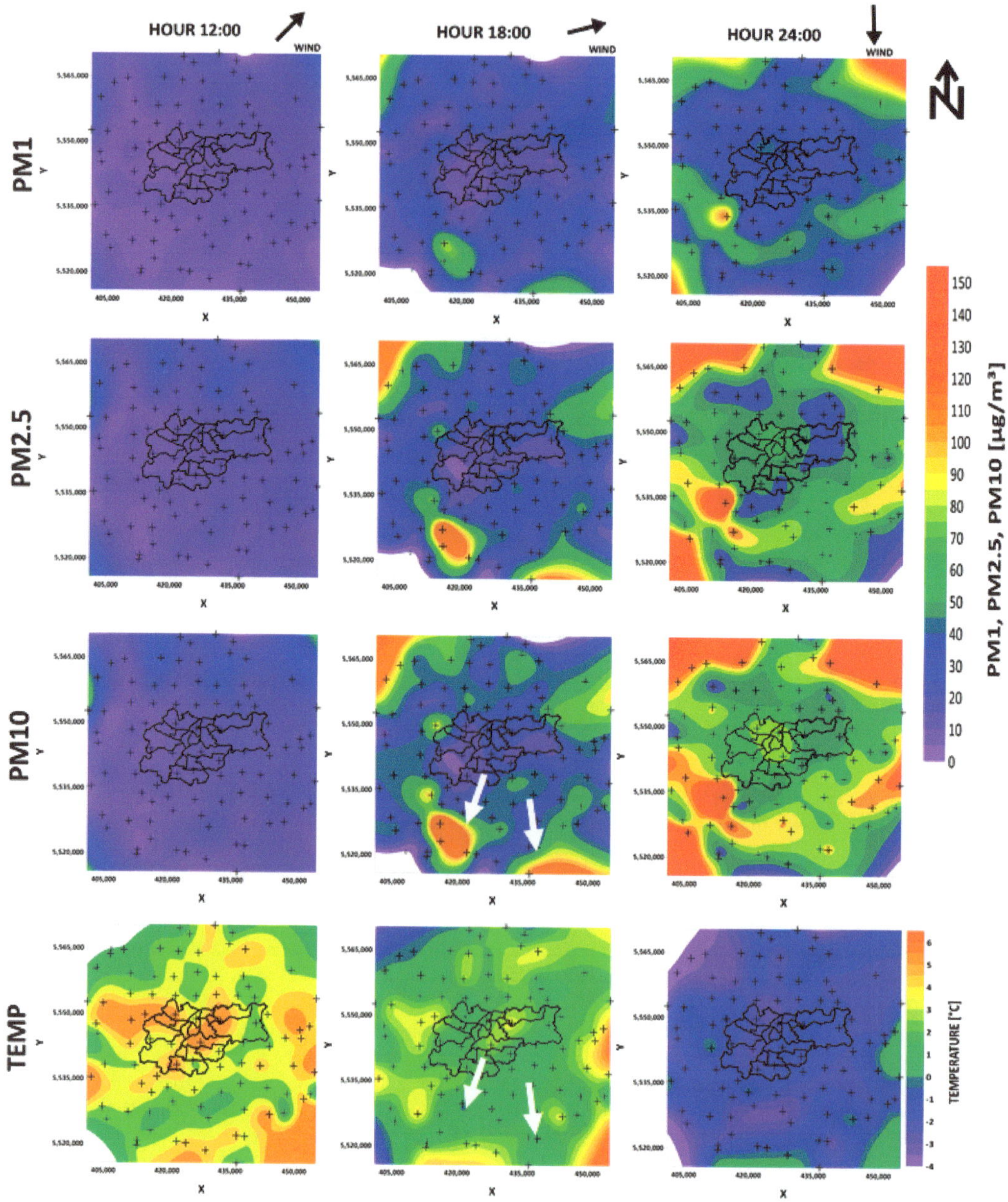

Figure 6. Temperature maps of air pollutions entering (PM1, PM2.5, PM10) Krakow on 18 March 2020. White arrows indicate areas in which the temperature was below the comfort zone.

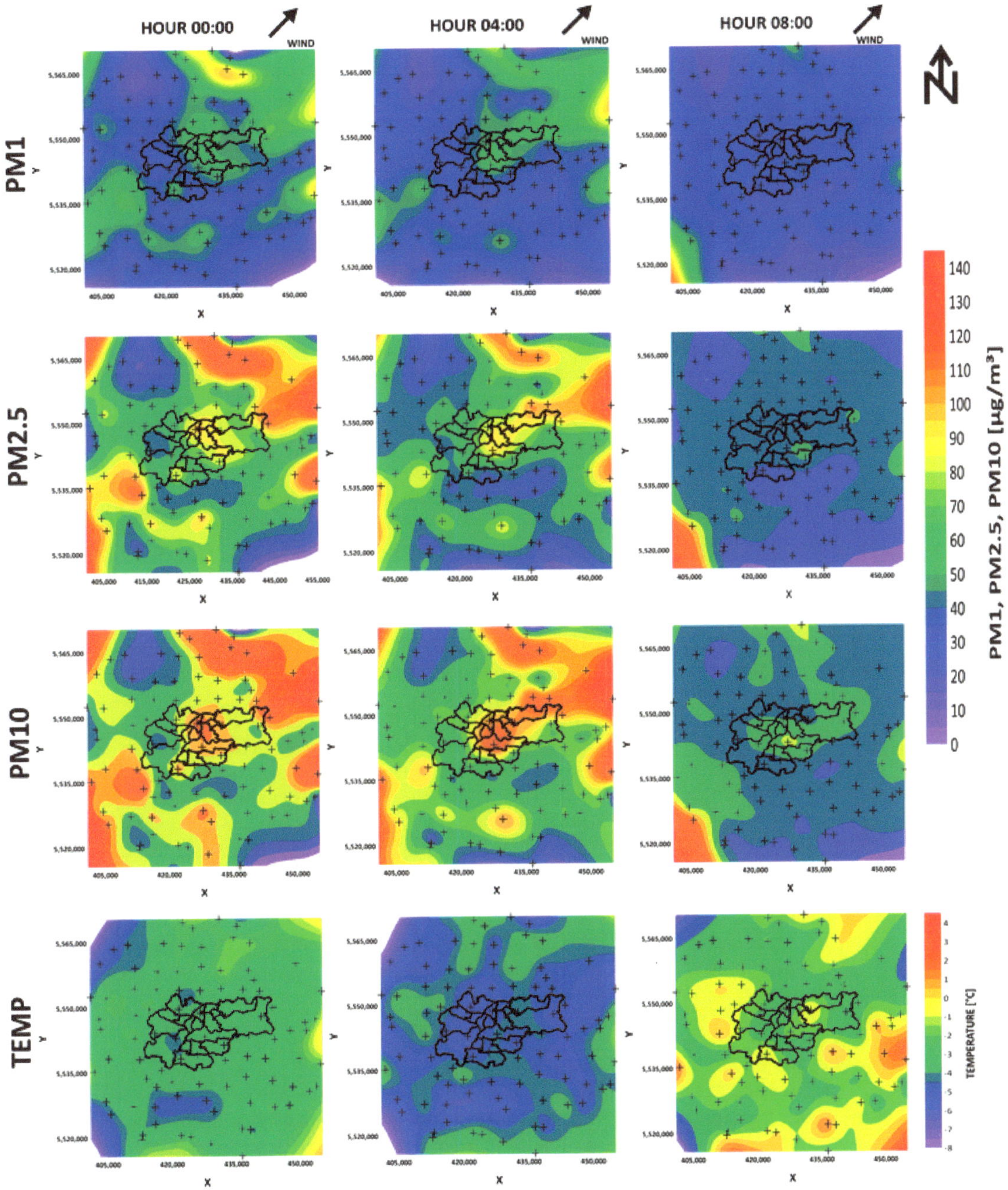

Figure 7. Temperature maps of air pollution leaving (PM1, PM2.5, PM10) Krakow on 11 March 2020.

4. Discussion

The statistics in Table 1 demonstrate that the distribution of Airly data from March to April 2021 was compact. There were no outliers and abnormally high or low indications. This was probably related to the Airly data preparation, as their measurements are filtered and corrected by their internal machine-learning and artificial intelligence algorithms. According to the correlation coefficients values in Table 2, it is clearly visible that PM1, PM2.5, and PM10 were correlated. This was expected as the individual particulate matter of a given fraction also contains information about particles of a smaller fraction. In this case, the PM2.5 data were used to prepare an animation (see the Supplementary Materials) to show the hourly changes in spring 2021. It was, therefore, also justified to use the PM2.5 historical data from the government stations to show the general trend of air pollution contamination in Krakow over the last decade.

Airly sensor 36808 in Niepołomice provided comparable results (averaged over a 24-h period) to the closest government station. They were highly correlated, i.e., the correlation coefficient was 0.93 for the investigated period and the average difference between their measurements was 3 $\mu g/m^3$. There were days (see Figure 3) in which the Airly sensors inflated or underestimated the results; however, the general measurement distribution was similar to that provided by the government sensors. This was sufficient to trace relative spatial-temporal changes with Airly LCS measurements. Aside from the comparative analysis of indications from government stations and Airly sensors, research conducted by the Marshal's Office in the Małopolska Region showed that Airly sensors give reliable measurements [25], better than other tested devices. Airly provides results that are already processed and corrected with the use of their machine-learning algorithms. However, LCS have their limitations and can easily be affected by various external factors. It can be difficult to correct raw data, while taking into account all factors. On the other hand, official government stations did not provide sufficient spatial coverage for our study. The presented accuracy of Airly sensors was sufficient to track the relative changes in air pollution in Krakow. The potential anomalous indication of a single station was easy to eliminate by analyzing the distribution. We did not notice such indications on our maps.

To analyze the inflow (Figure 6) and outflow (Figure 7) of air pollution in Krakow, we chose days in which the temperature changed from comfortable to cold (inflow), and from cold to comfortable. It was practical to find periods in which the temperature changed from cold to comfortable or vice versa. Whether people start to heat their homes basically depends on their perception of the temperature, so there was no reason to choose the 28th of March for example, which was characterized by a break in the upward trend, as even if the temperature dropped, it was still perceived as comfortable. This effect is clearly visible in Figure 5, which shows the kernel density estimate plots for different PM for different hours in relation to temperature. All PM concentrations rose in areas where the temperature dropped below 0 °C. This effect can also be observed in the maps. Air pollution started to increase in certain places outside Krakow at around 18:00. On the temperature maps, there were places where the temperature was lower than in the surrounding receivers. Aside from the relatively low value, the temperature in these places was very close to the thermal comfort limit (see Figure 6 white arrows). This observation could be very important in the future for forecasting air pollution.

Figures 6 and 7 contain wind arrow indicators; however, the wind speed was low on these days. The maps presented in Figures 6 and 7, and in the Supplementary animation, demonstrate that the process responsible for air pollution arriving to the city was related to both pollution blowing in on the wind and particular matter from surrounding towns and villages in the hills around diffusing and settling in Krakow, which is located in the valley. This valuable observation will allow for the better modeling of pollutant transport in the future.

PM contamination increased in Krakow depending on the fraction size. The highest increase could be seen for PM10, and the lowest for PM1. This may be because the lighter fractions could more easily remain at high altitudes and were not measured by devices located at a height of 1.5–8 m. The other reason may be related to the sources of air pollution. We considered fossil fuel use for household heating, so higher PM10 contamination was expected. In the case of pollution outflow, the city was most exposed to heavy PM10 dust, which, due to its size, is present at highest concentrations in the morning. The city of Krakow, as a result of its location in the valley, does not have favorable conditions for the outflow of pollutants, which was clearly shown by the distributions on the maps. We can see in Figure 7 that PM10 contamination remained in the city the longest and higher concentrations coincided with the course of rivers. Rivers tend to have an erosive base and usually flow at their lowest point. Perhaps it is because of their location that air pollution stays in these areas the longest. Another reason may be the increased presence of water vapor and the formation of mists in those areas. This may act to impede air movement and, consequently, the migration of heavier fractions of air pollution.

5. Conclusions

Our research clearly showed that a dense network of Airly low-cost sensors aids in the spatial and time analyses of air pollution inflow and outflow in Krakow. The ongoing COVID-19 pandemic allowed us to analyze the effect of pollutant diffusion from neighboring municipalities to Krakow, without introducing noise resulting from car traffic. It was shown that, apart from the daily temperature changes themselves, perceptible thermal comfort is an important factor. The subjective feeling of cold influences whether the inhabitants of neighboring municipalities heat their houses or not.

We utilized the STL decomposition method, the 10-year trend of PM2.5 concentration, the PM2.5 concentration, and data concerning the law and educational changes. Aside from physical factors such as the warm winter in 2014/2015, it was demonstrated that the Krakow and Małopolska Voivodeship authorities performed effective steps to improve air quality in the city. Moreover, it was demonstrated that the education and engagement of the local community were effective and important in this regard. A downward trend was visible from the time when the Krakow Smog Alert began its informative campaign in 2012.

The law prohibiting the use of solid fuels for heating in Krakow city brought about the intended reduction in PM. Unfortunately, as a result of its geographic location and the lack of similar bans in neighboring municipalities, Krakow is still exposed to pollution that exceeds air quality standards. Of course, long-term low emissions and its downwind transport completely fill urban areas over time; however, in the transition seasons, in which increased emissions last several days, the air in the city was of radically better quality than in the outskirts.

Our study clearly showed that the influx of PM1, PM2.5, and PM10 by diffusion was the greatest from the following towns: from the west—Rybna, Czernichów, and Brzeźnica; from the southwest—Brody, Skawina, Rzozów, Radziszów, Krzywaczka, Czechówka, and Zakliczyn; from the southeast—Dobczyce, Czasław, Kwapinka, Wieliczka, and Niepołomice; from the northeast—Proszowice, Waganowice, Słomniki, and Prandocin; from the northwest—Więckowice, Paczółtowice, Czubrowice, Gotkowice, Skała, and Gołyszyn. It also seems that the PM may be transported from sources a greater distance to the south, perhaps from Podhale, but this will be the focus of future research. The longest-lasting deposited dust was that of the PM10 fraction.

Supplementary Materials: Animation is available online at https://www.mdpi.com/article/10.3390/s21155208/s1, Animation S1: PM2.5 changes in Krakow and its neighborhood between 7 March and 16 April 2021.

Author Contributions: Conceptualization, T.D. and M.Z.; methodology, T.D. and M.Z.; validation, T.D. and M.Z.; formal analysis, T.D. and M.Z.; investigation, T.D. and M.Z.; resources, T.D. and M.Z.; data curation, T.D. and M.Z.; writing—original draft preparation, T.D. and M.Z.; writing—review and editing, T.D. and M.Z.; visualization, T.D. and M.Z.; supervision, T.D. and M.Z.; project administration, T.D. and M.Z. All authors have read and agreed to the published version of the manuscript.

Funding: This research was supported by AGH—University of Science and Technology, Faculty of Geology, Geophysics and Environmental Protection—as a part of the statutory project No. 11.11.140.613.

Institutional Review Board Statement: Not applicable.

Informed Consent Statement: Not applicable.

Data Availability Statement: 1. Publicly available datasets from Airly sensors were analyzed in this study. This data can be found here: (https://map.airly.org/, accessed on 29 July 2021). Airly API documentation is available here: (https://developer.airly.org/en/docs, accessed on 29 July 2021). 2. Publicly available datasets from the Chief Inspectorate For Environmental Protection database were analyzed in this study. This data can be found here: (http://powietrze.gios.gov.pl/pjp/home, accessed on 29 July 2021). API documentation is available here: (http://powietrze.gios.gov.pl/pjp/content/api, accessed on 29 July 2021).

Conflicts of Interest: The authors declare no conflict of interest.

References

1. Cohen, A.J.; Brauer, M.; Burnett, R.; Anderson, H.R.; Frostad, J.; Estep, K.; Balakrishnan, K.; Brunekreef, B.; Dandona, L.; Dandona, R.; et al. Estimates and 25-year trends of the global burden of disease attributable to ambient air pollution: An analysis of data from the Global Burden of Diseases Study. *Lancet* **2017**, *389*, 1907–1918. [CrossRef]
2. Hime, N.J.; Marks, G.B.; Cowie, C.T. A Comparison of the Health Effects of Ambient Particulate Matter Air Pollution from Five Emission Sources. *Int. J. Environ. Res. Public Health* **2018**, *15*, 1206. [CrossRef] [PubMed]
3. Weinmayr, G.; Romeo, E.; De Sario, M.; Weiland, S.K.; Forastiere, F. Short-term effects of PM10 and NO_2 on respiratory health among children with asthma or asthma-like symptoms: A systematic review and meta-analysis. *Environ. Health Perspect.* **2010**, *118*, 449–457. [CrossRef]
4. Dai, L.; Zanobetti, A.; Koutrakis, P.; Schwartz, J.D. Associations of fine particulate matter species with mortality in the United States: A multicity time-series analysis. *Environ. Health Perspect.* **2014**, *122*, 837–842. [CrossRef]
5. Samoli, E.; Stafoggia, M.; Rodopoulou, S.; Ostro, B.; Declercq, C.; Alessandrini, E.; Diaz, J.; Karanasiou, A.; Kelessis, A.G.; Le Tertre, A.; et al. Associations between fine and coarse particles and mortality in Mediterranean cities: Results from the MED-PARTICLES Project. *Environ. Health Perspect.* **2013**, *121*, 932–938. [CrossRef]
6. Raaschou-Nielsen, O.; Andersen, Z.J.; Beelen, R.; Samoli, E.; Stafoggia, M.; Weinmayr, G.; Hoffmann, B.; Fischer, P.; Nieuwenhuijsen, M.J.; Brunekreef, B.; et al. Air pollution and lung cancer incidence in 17 European cohorts: Prospective analyses from the European Study of Cohorts for Air Pollution Effects (ESCAPE). *Lancet Oncol.* **2013**, *14*, 813–822. [CrossRef]
7. MacIntyre, E.A.; Gehring, U.; Molter, A.; Fuertes, E.; Klumper, C.; Kramer, U.; Quass, U.; Hoffmann, B.; Gascon, M.; Brunekreef, B.; et al. Air pollution and respiratory infections during early childhood: An analysis of 10 European birth cohorts within the ESCAPE Project. *Environ. Health Perspect.* **2014**, *122*, 107–113. [CrossRef]
8. Cesaroni, G.; Forastiere, F.; Stafoggia, M.; Andersen, Z.J.; Badaloni, C.; Beelen, R.; Caracciolo, B.; de Faire, U.; Erbel, R.; Eriksen, K.T.; et al. Long term exposure to ambient air pollution and incidence of acute coronary events: Prospective cohort study and meta-analysis in 11European cohorts from the ESCAPE Project. *BMJ* **2014**, *348*, f7412. [CrossRef]
9. Schikowski, T.; Adam, M.; Marcon, A.; Cai, Y.; Vierkotter, A.; Carsin, A.E.; Jacquemin, B.; Al Kanani, Z.; Beelen, R.; Birk, M.; et al. Association of ambient air pollution with the prevalence and incidence of COPD. *Eur. Respir. J.* **2014**, *44*, 614–626. [CrossRef] [PubMed]
10. Pedersen, M.; Giorgis-Allemand, L.; Bernard, C.; Aguilera, I.; Andersen, A.M.; Ballester, F.; Beelen, R.M.; Chatzi, L.; Cirach, M.; Danileviciute, A.; et al. Ambient air pollution and low birthweight: A European cohort study (ESCAPE). *Lancet Respir. Med.* **2013**, *1*, 695–704. [CrossRef]
11. Thurston, G.D.; Kipen, H.; Annesi-Maesano, I.; Balmes, J.; Brook, R.D.; Cromar, K.; De Matteis, S.; Forastiere, F.; Forsberg, B.; Frampton, M.W.; et al. A joint ERA/ATS policy statement: What constitutes an adverse health effect of air pollution? An analytical framework. *Eur. Respir. J.* **2017**, *49*, 1600419. [CrossRef] [PubMed]
12. Traczyk, P.; Gruszecka-Kosowska, A. The Condition of Air Pollution in Kraków, Poland, in 2005–2020, with Health Risk Assessment. *Int. J. Environ. Res. Public Health* **2020**, *17*, 6063. [CrossRef] [PubMed]
13. World Health Organization. *WHO Ambient Air Pollution Database May 2016*; World Health Organization: Geneva, Switzerland, 2016.
14. Krakowski Alarm Smogowy. Krakowski Alarm Smogowy z Tytułem Człowieka Roku Polskiej Ekologi. Available online: https://www.gramwzielone.pl/walka-ze-smogiem/21385/krakowski-alarm-smogowy-z-tytulem-czlowieka-roku-polskiej-ekologi (accessed on 1 June 2021).
15. Krakowski Alarm Smogowy. Chcemy oddychać [PL]. Available online: https://www.pol-int.org/pl/salon/chcemy-oddychac-pl (accessed on 1 June 2021).

16. Directive 2008/50/EC of the European Parliament and of the Council of 21 May 2008 on Ambient Air Quality and Cleaner Air for Europe. Available online: http://eur-lex.europa.eu/legal-content/en/ALL/?uri=CELEX:32008L0050 (accessed on 23 June 2021).
17. European Commission. Fitness Check of the Ambient Air Quality Directives. Directive 2004/107/EC Relating to Arsenic, Cadmium, Mercury, Nickel and Polycyclic Aromatic Hydrocarbons in Ambient Air and Directive 2008/50/EC on Ambient Air Quality and Cleaner Air for Europe. SWD (2019) 427 Final. Brussels, Commission Staff Working Document. 28 November 2019. Available online: https://ec.europa.eu/environment/air/pdf/SWD_2019_427_F1_AAQ%20Fitness%20Check.pdf (accessed on 23 June 2021).
18. Chief Inspectorate for Environmental Protection. Characteristics of the Zone for Air Quality Assessment, Zone PL 1201. Available online: http://powietrze.gios.gov.pl/pjp/zone/characteristic/PL1201/2019/true?lang=en (accessed on 23 June 2021).
19. Chief Inspectorate for Environmental Protection. Characteristics of the Zone for Air Quality Assessment, Zone PL 1203. Available online: http://powietrze.gios.gov.pl/pjp/zone/characteristic/PL1203/2019/true?lang=en (accessed on 23 June 2021).
20. Chief Inspectorate for Environmental Protection. PMs Measuring in the Air. Available online: http://www.gios.gov.pl/pl/aktualnosci/391-pomiary-pylu-zawieszonego-w-powietrzu (accessed on 23 June 2021).
21. European Commission Directorate-General for Environment. Air Quality Standards. Available online: https://ec.europa.eu/environment/air/quality/standards.htm (accessed on 23 June 2021).
22. Gerboles, M.; Spinelle, L.; Borowiak, A. Measuring Air Pollution with Low-Cost Sensors. European Commission, JRC107461. 2017. Available online: https://ec.europa.eu/environment/air/pdf/Brochure%20lower-cost%20sensors.pdf (accessed on 23 June 2021).
23. Adamiec, E.; Dajda, J.; Gruszecka-Kosowska, A.; Helios-Rybicka, E.; Kisiel-Dorohinicki, M.; Klimek, R.; Pałka, D.; Wąs, J. Using Medium-Cost Sensors to Estimate Air Quality in Remote Locations. Case Study of Niedzica, Southern Poland. *Atmosphere* **2019**, *10*, 393. [CrossRef]
24. Kowalski, P.A.; Szwagrzyk, M.; Kielpinska, J.; Konior, A.; Kusy, M. Numerical analysis of factors, pace and intensity of the corona virus (COVID-19) epidemic in Poland. *Ecol. Inform.* **2021**, *61*, 101284. [CrossRef] [PubMed]
25. Bartyzel, J.; Frączkowski, T.; Pindel, A.; Łyczko, P.; Musielok, M.; Zięba, D.; Dworakowska, A. Report on the Second Series of Tests Comparative Dust Measuring Devices Suspended PM10 (Non-Reference Devices and Without Demonstrated Equivalence to Devices Reference). Marshal's Office of the Małopolska Region. Available online: https://powietrze.gios.gov.pl/pjp/documents/download/105407 (accessed on 23 June 2021). (In Polish)
26. AIRLAB Solution Pour Notre Air. Microsensors Challenge 2019. Available online: http://www.airlab.solutions/sites/default/files/presse/brochure_2019_gb-version%2010.02_0.pdf (accessed on 23 June 2021).
27. Fishbain, B.; Lerner, U.; Castell, N.; Cole-Hunter, T.; Popoola, O.; Broday, D.M.; Iñiguez, T.M.; Nieuwenhuijsen, M.; Jovasevic-Stojanovic, M.; Topalovic, D.; et al. An evaluation tool kit of air quality micro-sensing units. *Sci. Total Environ.* **2017**, *575*, 639–648. [CrossRef] [PubMed]
28. Jasiński, R.; Galant-Gołębiewska, M.; Nowak, M.; Kurtyka, K.; Kurzawska, P.; Maciejewska, M.; Ginter, M. Emissions and Concentrations of Particulate Matter in Poznan Compared with Other Polish and European Cities. *Atmosphere* **2021**, *12*, 533. [CrossRef]
29. Bokwa, A. Environmental impact of long-term air pollution changes in Krakow, Poland. *Polish J. Environ. Stud.* **2008**, *5*, 673–686.
30. Intelligent Air Quality Monitoring System. Map of Air Quality by Airly. Available online: https://map.airly.org (accessed on 1 June 2021).
31. City Traffic Engineer of Krakow. *Wpływ Stanu Zagrożenia Epidemicznego Na Natężenia Ruchu Drogowego w Krakowie*; Department of City Traffic Engineer of Krakow: Krakow, Poland, 2020; pp. 1–4.
32. Adamczyk, J.; Piwowar, A.; Dzikuć, M. Air protection programmes in Poland in the context of the low emission. *Environ. Sci. Pollut. Res.* **2017**, *24*, 16316–16327. [CrossRef] [PubMed]
33. Banach, M.; Długosz, R.; Pauk, J.; Talaśka, T. Hardware Efficient Solutions for Wireless Air Pollution Sensors Dedicated to Dense Urban Areas. *Remote Sens.* **2020**, *12*, 776. [CrossRef]
34. Tucker, C. Polish-US Startup Airly Raises €1.7 Million for Global Air Quality Platform. Available online: https://www.eu-startups.com/2020/10/polish-us-startup-airly-raises-e1-7-million-for-global-air-quality-platform/ (accessed on 1 June 2021).
35. Karagulian, F.; Barbiere, M.; Kotsev, A.; Spinelle, L.; Gerboles, M.; Lagler, F.; Redon, N.; Crunaire, S.; Borowiak, A. Review of the Performance of Low-Cost Sensors for Air Quality Monitoring. *Atmosphere* **2019**, *10*, 506. [CrossRef]
36. Airly. Airly Air Quality Sensors PRODUCT CARD. Available online: https://www.danintra.dk/pdf/Danintra_Airly.pdf (accessed on 1 June 2021).
37. Cleveland, R.B.; Cleveland, W.S.; McRae, J.E.; Terpenning, I.J. STL: A seasonal-trend decomposition procedure based on loess. *J. Off. Stat.* **1990**, *6*, 3–33.
38. Zaręba, M.; Danek, T.; Zając, J. On Including Near-surface Zone Anisotropy for Static Corrections Computation—Polish Carpathians 3D Seismic Processing Case Study. *Geosciences* **2020**, *10*, 66. [CrossRef]
39. Zaręba, M.; Laskownicka, A.; Zając, J. The use of S-guided CREP methodology for advanced seismic structure enhancing processing. *Acta Geophys.* **2019**, *67*, 1711–1719. [CrossRef]
40. Chief Inspectorate for Environmental Protection. Interfejs Programistyczny Aplikacji (API). Available online: https://powietrze.gios.gov.pl/pjp/content/api (accessed on 1 May 2021).

41. Marshal's Office of the Małopolska Region. Uchwała Nr XVIII/243/16 Sejmiku Województwa Małopolskiego z Dnia 15 Stycznia 2016 r. w Sprawie Wprowadzenia Na Obszarze Gminy Miejskiej Kraków Ograniczeń w Zakresie Eksploatacji Instalacji, w Których Następuje Spalanie Paliw. Available online: http://edziennik.malopolska.uw.gov.pl/ActDetails.aspx?year=2016&poz=812 (accessed on 1 June 2021).
42. Nychka, D.; Furrer, R.; Paige, J.; Sain, S. Fields: Tools for spatial data. In *R Package Version 12.3*; University Corporation for Atmospheric Research: Boulder, CO, USA, 2017. [CrossRef]
43. Jendritzky, G.; Maarouf, A.; Staiger, S. Looking for a Universal Thermal Climate Index UTCI for Outdoor Applications. In Proceedings of the Windsor-Conference on Thermal Standards, Windsor, UK, 5–8 April 2001.

sensors MDPI

Article

Evaluation of Solid Particle Number Sensors for Periodic Technical Inspection of Passenger Cars

Anastasios Melas, Tommaso Selleri, Ricardo Suarez-Bertoa and Barouch Giechaskiel *

European Commission, Joint Research Centre (JRC), 21027 Ispra, Italy; Anastasios.Melas@ec.europa.eu (A.M.); Tommaso.Selleri@ec.europa.eu (T.S.); Ricardo.Suarez-Bertoa@ec.europa.eu (R.S.-B.)
* Correspondence: Barouch.Giechaskiel@ec.europa.eu; Tel.: +39-0332-78-5312

Abstract: Following the increase in stringency of the European regulation limits for laboratory and real world automotive emissions, one of the main transport related aspects to improve the air quality is the mass scale in-use vehicle testing. Solid particle number (SPN) emissions have been drastically reduced with the use of diesel and gasoline particulate filters which, however, may get damaged or even been tampered. The feasibility of on-board monitoring and remote sensing as well as of the current periodical technical inspection (PTI) for detecting malfunctioning or tampered particulate filters is under discussion. A promising methodology for detecting high emitters is SPN testing at low idling during PTI. Several European countries plan to introduce this method for diesel vehicles and the European Commission (EC) will provide some guidelines. For this scope an experimental campaign was organized by the Joint Research Centre (JRC) of the EC with the participation of different instrument manufacturers. Idle SPN concentrations of vehicles without or with a malfunctioning particulate filter were measured. The presence of particles under the current cut-off size of 23 nm as well as of volatile particles during idling are presented. Moreover, the extreme case of a well performing vehicle tested after a filter regeneration is studied. In most of the cases the different sensors used were in good agreement, the high sub-23 nm particles existence being the most challenging case due to the differences in the sensors' efficiency below the cut-off size.

Keywords: periodical technical inspection; in-use vehicle emissions; particle number; diffusion charger; condensation particle counter; sub-23 nm particles

1. Introduction

Strong scientific evidence on adverse health effects of particulate matter (PM) [1] has driven regulators to implement stricter limits to vehicles equipped with combustion engines because they were considered an important contributor of PM. In the European Union (EU), additionally to PM mass, a solid particle number (SPN) limit for particles >23 nm (SPN_{23}) is also imposed to vehicles equipped with diesel and gasoline direct injection engines [2,3]. The SPN_{23} limit drove to the implementation of very efficient particulate filters. For example, diesel particulate filters have typically >99% solid particle number reduction efficiency [4], and gasoline particulate filters can also exceed >90% efficiency [5,6]. The stricter PM regulations in combination with the efficient PM exhaust after-treatment systems have resulted in reduced urban PM levels over the last years in European cities [7,8].

One issue that still remains open is the durability of exhaust after-treatment and tampering. Particulate filters become more efficient after their usage due to the formation of a soot cake and ash accumulation on their surface that traps soot particles [9]. Indeed, after the regeneration of a filter, the efficiency drops down significantly and increases de novo after the accumulation of soot [10]. However, defects of the particulate filters may reduce their trapping efficiency [11]. For example, laboratory and commercial fleet DPF (diesel particulate filter) failure studies have shown that uncontrolled filter regeneration with high temperature peaks in combination with the presence of ash may provoke thermal damages at the DPF's substrate and more specifically pinholes, melts, and cracks [12–14]. Additionally, DPF tampering by vehicle owners has been reported with the aim of reducing fuel consumption and the need to perform the periodic regeneration [15]. Although these cases are a small percentage of the fleet, they can contribute significantly to the total fleet emissions. For example, a study of 300 diesel Euro 5 and Euro 6 vehicles sampled from the Belgian commercial fleet showed that 15% of high SPN emitters may increase the fleet emissions by a factor of 30 [16]. Another study found that 10% of highest SPN emissions can be responsible for 85% of the fleet emissions [17]. The determination of high emitters effect on the national fleet emissions may depend on several factors that increase the uncertainty but considering the DPF and GPF (gasoline particulate filter) efficiency, it is undeniable that vehicles with malfunctioning or removed particulate filters dominate the SPN emissions.

At EU level, the conformity of the vehicles to the emission limits over their useful lifetime (currently 160,000 km) is checked via the in-service conformity (ISC) testing (up to 100,000 km or 5 years). ISC testing is done to well-maintained vehicles following the type approval procedures. While ISC is conducted by the vehicle manufacturers and the type approval authorities, market surveillance, which was recently introduced in the regulation, can be done by independent institutes at a wider range of test conditions. A few well maintained vehicles with up to 160,000 km on the odometer are also selected [18]. Market surveillance is a very useful tool in order to detect defeat devices as well as assess the durability of after-treatment exhaust. Due to the high cost of laboratory and on-road testing, these tests cannot be applied to a mass scale. Thus, tampered or badly maintained vehicles are not controlled.

For large scale fleet monitoring on-board monitoring (OBM), remote sensing and periodical technical inspection (PTI) are the most appropriate tools. The idea of OBM is similar to the on-board diagnostics (OBD) of the vehicle [19]. However, instead of only checking the malfunctions of the vehicle, sensors are used to monitor the actual emissions. This concept has been successfully applied to heavy-duty vehicles in China [20]. In Europe discussions are on-going for OBM introduction in the Euro 7 regulation, but at the moment there are no robust particle number sensors. Remote sensing is wide spread [21,22]. From the big amount of data, durability issues at vehicle model level can be identified. It is also possible to identify high emitters. The application to particle number though is very limited [23].

In the framework of PTI, an opacity measurement is implemented for controlling the particulate filter but as modern engine and filter technologies have become very efficient, there are concerns on the sensitivity of this method [16,24]. Over the last years an informal technical working group for new periodic technical inspection (NPTI) procedures has been formed, aiming to develop methodologies for detecting DPF and de-NO_x aftertreatment technologies malfunctions. A methodology that seems to be very efficient for detecting tampered of malfunctioning particulate filters is the SPN measurement at idling. Especially for diesel vehicles the idling SPN concentrations may correlate well with SPN emissions during regulatory tests. Two studies that correlated regulatory tests measured in the laboratory and during on-road tests with low idling emissions (mostly extracted by the same tests) found that diesel vehicles that complied with regulation limits (6×10^{11} #/km) had $<1 \times 10^5$ #/cm^3 low idling concentrations [25,26]. In another study, idling concentration in the order of 2.5×10^5 #/cm^3 corresponded to $>10^{12}$ #/km and was proposed as a possible limit for PTI [24]. Low idling diesel SPN concentrations $>1 \times 10^5$ #/cm^3 are correlated to $>6 \times 10^{11}$ #/km that is the current regulatory limit [17]. Instead, for gasoline vehicles the correlation is more difficult to be done [25] because they mainly produce particles during fuel enrichments [27] and not necessarily at idle. Nevertheless, the uncertainty of the methodology and the instrumentation has also to be taken into account in the determination of a limit.

Switzerland was the first country to introduce a high idling test for particles >23 nm (SPN_{23}) for non-road machineries. For light and heavy duty vehicles the Netherlands was the pioneer and a PTI procedure measuring solid particles >23 nm will be introduced on the 1 July 2022. Belgium will also introduce the SPN measurement in the PTI in July 2022 and Germany will follow in January 2023. Emission limits, applicable vehicles, and testing procedures have differences from one country to the other. For example, the Netherlands has a limit of 10^6 #/cm^3 for Euro 5 and 6 vehicles and a 15 s measurement, while Germany 2.5×10^5 #/cm^3 only for Euro 6 vehicles with three repeats of 30 s measurements. A recent proposal from the VERT (Verification of Emission Reduction Technologies) association suggested a lower limit of 5×10^4 #/cm^3 both for diesel and gasoline vehicles equipped with particulate filter [28]. VERT proposes the performance of three 15 s measurements. In parallel to national initiatives also the European Commission is preparing a harmonized procedure but each member state will have the possibility to introduce the procedure as an additional measure within their own national competence. An important aspect of introducing a PTI test is the characteristics of the sensors used to perform the measurement. The sensors must be robust enough for the garage environment, simple in operation for non-expert staff, and of low cost. This large-scale production of PTI sensors needs some compromise regarding technical specifications that fulfill the required preciseness for detecting faulty or removed particulate filters. The main specifications of PTI devices that count solid particles are their efficiency to remove volatile particles and their lower detection size (that should be around 23 nm). Their background level and their maximum concentration are important aspects as well.

The approaches to these requirements are numerous. The heart of the sensors is the particle detector, which is typically based on optical particle counting after condensation of an alcohol on the pre-existing particles or measurement of electrical current after diffusion charging of the pre-existing particles. Condensation particle counters (CPCs) have a heated section where aerosol particles are exposed to supersaturated vapors and a colder section where vapors condense on particles and grow them to sizes that are detectable with optical methods. Different working fluids can be used for this application, most typically butanol and isopropanol. CPCs usually operate near ambient temperatures but recently also high temperature CPCs have been developed [29], but without any commercial system available at the time of writing. Their counting efficiency is near to unity at large sizes, their cut-off size may be influenced by the nature of the particles [30]. Diffusion chargers (DCs) utilize electrical detection of particles. Particles are charged by a corona charger which is typically unipolar but also systems with bipolar charging have been developed. After the corona charger, an electro-precipitator removes all free ions and finally the particles' current is measured either with an electrometer or with a faraday cage [31,32]. Diffusion chargers do not use working fluids and can operate also at higher temperatures [33]. Compared to CPCs, DCs can measure higher particle concentrations but have higher background levels. Finally, their counting efficiency does not reach a plateau region as the CPC at large sizes due to the dependency of the charging efficiency on particles' size.

In order to remove volatile particles and measure only solids, three different technologies are typically used: (i) heated (or evaporation) tube [34], (ii) thermodesorption using a thermal denuder [35], (iii) catalytic stripper that oxidizes hydrocarbons and optionally also traps sulfur compounds [36–38]. A review on volatile removal technologies can be found in [39].

Some systems have a sampling line; some others dilute at the sampling point. While differences to these approaches have been discussed in the literature due to different agglomeration and thermophoretic losses [40], in the PTI testing, where the exhaust gas temperatures are low and the concentrations also low, the differences should be small. High dilution may also be important in avoiding volatile artifacts when measuring particles below 23 nm [41]. For PTI systems though the dilution depends more on the upper limit they are designed to reach. Diffusion chargers usually have a high upper concentration limit and can measure even without dilution.

In the context of introducing a harmonized PTI procedure in the EU, the Joint Research Centre (JRC) of the European Commission performed an experimental campaign using PTI sensors from different manufacturers. Testing aimed to study the specifications of the sensors used for PTI applications and the level of particle emissions at low idling of different vehicles. A reference instrument compliant to the technical requirements of the type approval regulation was used for comparison with the different sensors. Different cases were studied; vehicles without a filter or malfunctioning filter, low idling after a DPF regeneration. Finally, sub-23 nm and volatile particles were measured in order to study their effect on the SPN sensors performance. The paper is divided in (i) the experimental section where we present the specifications of the sensors, the procedures, and the vehicles; (ii) the results were we focus on the performance of the sensors; (iii) the discussion section where we identify all parameters that are important for future regulation; and (iv) the conclusions.

2. Materials and Methods

2.1. Experimental Setup and Procedure

Tests were performed in the vehicle emissions laboratory of the Joint Research Centre (JRC). The tested vehicles were placed in a laboratory with temperature varying from 20 to 27 °C. Figure 1 presents the experimental setup. All tests were performed at low idling with sampling directly from a depth of 30 cm in the tailpipe. Each of the six PTI sensors was measuring in parallel with reference systems.

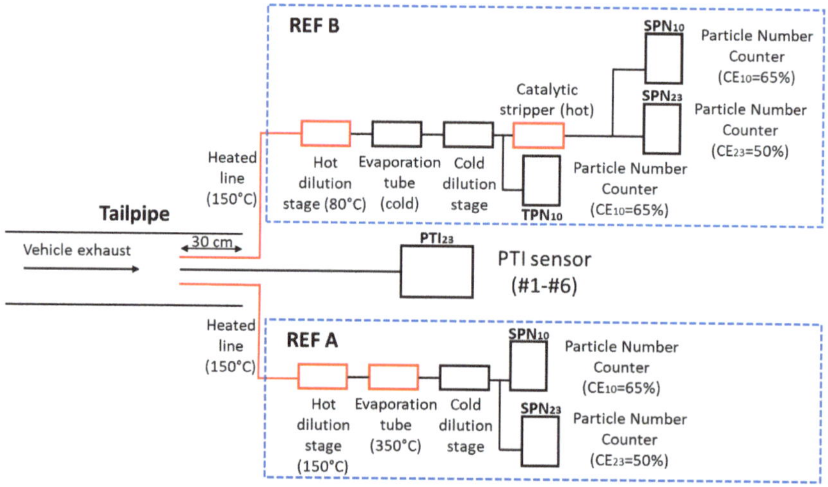

Figure 1. Schematic of the experimental setup. In red we show heated parts. Either 'REF A' or 'REF B' setups were employed while SPN-PTI sensors #1 to #6 were measuring sequentially.

The reference system (Nanomet 1, Testo, Titisee-Neustadt, Germany) was based on the technical requirements of the type approval EU regulation (2017/1151) and the Particle Measurement Programme (PMP) recommendations [42]. For this reason, quite often it is called the PMP system. It consisted of a 1 m hot sampling line at 150 °C, a hot dilution stage at 150 °C (dilution around 50:1), an evaporation tube operating at 350 °C, a secondary dilution stage (dilution around 5:1) at ambient temperature and a TSI (Shoreview, MN, USA) model 3790 Condensation Particle Counter (CPC) with 50% counting efficiency at 23 nm, CE_{23} = 50%. Additionally, a TSI model 3792 CPC with 65% efficiency at 10 nm, CE_{10} = 65%, was employed in parallel in order to measure solid particle number down to 10 nm. This reference setup will be called from now on 'REF A' and the SPN measurements with the 23 nm and 10 nm CPCs as SPN_{23} and SPN_{10}, respectively. The ratio $(SPN_{10} - SPN_{23})/SPN_{23}$ will be called sub-23 nm fraction.

The SPN concentrations were calculated with the Particle Concentration Reduction Factor (PCRF) that included the dilution factor and the average particle losses of particles with size 100 nm, 50 nm, 30 nm. The particle losses at 50 nm were 5% more than at 100 nm ($PCRF_{50}/PCRF_{100} = 1.05$) and at 30 nm 29% more ($PCRF_{30}/PCRF_{100} = 1.29$). The average PCRF was 300 in all tests. The sampling line did not add significant particle losses; for inlet flow 1.5 lpm and sampling line with length 1 m the diffusion losses of particles with size 23 nm is ~2%. The reference CPCs have been used in inter-laboratory exercises and during linearity checks for concentrations up to 10^4 #/cm^3 they had a slope between 0.9 and 1.1 and $R^2 > 0.99$ [43,44]. For CPC concentrations >10^4 #/cm^3, (i.e., measured concentrations >3×10^6 #/cm^3) the accuracy of the reference system was not in the ±10% and the measurement uncertainty increased. These cases are indicated in the "Results" section. For SPN_{10} no additional correction was performed for the diffusion losses of sub-23 nm particles, as prescribed in the Global Technical Regulation (GTR) 15. Particle losses at 15 nm were higher than the average PCRF by a factor of 2.2.

For some tests, the reference system setup was modified: while the hot sampling line was kept at 150 °C, the primary hot dilution was set to 80 °C, and the evaporation tube was switched off. Downstream of the system, a TSI model 3792E CPC with 65% efficiency at 10 nm was measuring total particle number (both solid and semi-volatile) emissions (TPN_{10}). With these sampling conditions nucleation of volatiles was probably suppressed, but nucleation of semi-volatiles was possible. A portion of the diluted aerosol flow was driven to a catalytic stripper model CS015 from Catalytic instruments (Rosenheim, Germany) with wall temperature 375 °C and then to the 23 nm CPC 3790 and 10 nm CPC 3792 in order to measure SPN_{23} and SPN_{10}, respectively. Henceforth, this setup is called 'REF B' while the ratio $(TPN_{10} - SPN_{10})/SPN_{10}$ is called volatile fraction. The PCRF of this setup was 225 (due to different temperatures used). The SPN_{23} and SPN_{10} measured downstream of the catalytic stripper and were additionally corrected by a factor of 1.4 to take into account the catalytic stripper's particle losses at sizes 30 nm, 50 nm, and 100 nm.

The systems in setup 'REF A' or 'REF B', whichever was applicable, were measuring and logging continuously the SPN_{23} and SPN_{10} (and TPN_{10} with 'REF B') concentrations during low idling. The sampling point of both 'REF A' and 'REF B' was, similarly to the PTI sensors, 30 cm inside the tailpipe. After the ignition of the vehicle's engine the PTI sensors were measuring sequentially for a predetermined time period that lasted from 15 (Sensors #1 to #4) to 45 (Sensor #5) seconds (3 repetitions of 15 s) according to the recommendations of the country of homologation that they followed. Sensors #1 to #4 had also a stabilization time of 15 s before measuring. Sensor #6 had only a continuous measurement option but its measurement time period was chosen to be similar to the rest of sensors and between 15 and 45 s. Before each measurement Sensors #1 to #4 performed an automatic or semi-automatic zero offset and leakage test. For Sensors #5 and #6 and the CPCs used at 'REF A' or 'REF B', whichever was applicable, the zero offset was checked before testing with a HEPA filter.

The measurement order of the PTI sensors changed from one test to another in order to have different concentration levels for all sensors. The duration of each idling test lasted from 10 to 30 min. Measurements were performed also during the cold start of the vehicles in order to have a wider range of concentrations. In some cases, the vehicle was switched off and on several times. The PTI sensors were compared against 'REF A' or 'REF B'. Due to the absence of any possibility for post-process alignment, the data alignment was done with a timer during the test. Experimental time started when 'REF A' or 'REF B' logging started and for each PTI sensor we recorded the time of measurement. No on-board diagnostics (OBD) measurements were available during the testing campaign.

2.2. PTI Sensors

Sensors #1 to #5 were provided from the manufacturers to JRC for the testing campaign. The companies in alphabetical order were: Capelec (Montpelier, France) and Pegasor (Tampere, Finland), DEKATI (Kangasala, Finland), Mahle (Stuttgart, Germany), TEN (Baambrugge, The Netherlands), and TSI (Aachen, Germany). Some of them were commercially available while others prototypes. Sensor #6 was owned by JRC and it was the NPET of TSI, homologated for PTI measurements of non-road mobile machinery in Switzerland.

2.2.1. Sampling and Measurement Technologies

Table 1 presents the PTI devices that were tested. For each device we report whether there was a heated sampling line or not, the dilution ratio (if applicable), the technology for removal of volatiles, the principle of particle detection, and finally the regulations each sensor complied with. Some of the sensors may comply also with other regulations (e.g., DE, BE) but we only report the country in which they applied for or obtained homologation at the time that this paper was written. Sensors #3 and #5 were prototypes and no specific country of homologation was defined.

Table 1. Sampling and measurement technologies used at the PTI sensors.

PTI	Sampling Line	Dilution (Temp.)	Volatile Particle Remover	Particle Detector	Certification
#1	Heated (75 °C)	No	Thermal denuder (150 °C)	DC	NL
#2	Heated (90 °C)	Venturi (150 °C)	Evaporation tube (200 °C)	DC	NL
#3	Heated (60 °C)	No	Evaporation tube (300 °C)	DC	N/A
#4	Heated (70 °C)	200:1 (ambient)	Evaporation tube (250 °C)	CPC	NL
#5	Not heated	20:1 (ambient)	Catalytic stripper (350 °C)	CPC	N/A
#6	Not heated	10:1 (ambient)	Catalytic stripper (350 °C)	CPC	CH

CH = Switzerland; CPC = Condensation Particle Counter; DC = Diffusion Charger; N/A = not available; NL = Netherlands.

Sensors #1 to #4 had a heated line at different temperatures in the range of 60 °C to 90 °C. Sensors #5 and #6 diluted the aerosol at the sampling point with a bifurcated flow diluter that filters part of the inlet flow and uses it as dilution air. Sensors #1 and #3 had no dilution while Sensor #2 had very low dilution. Sensor #4 diluted the aerosol flow 200:1 by using two ejector diluters with dilution ratio ~14. In order to remove volatile particles Sensor #1 included a thermal denuder, Sensors #2, #3, #4 included a heated or evaporation tube, and Sensors #5, #6 a catalytic stripper. Sensors #1 to #3 used a diffusion charger (DC) and #4 to #6 a condensation particle counter (CPC) as particle detectors.

All Sensors except from Sensor #3 were equipped with a water trap downstream the sampling line at the inlet of the device. All DC-based sensors used a unipolar charger to charge the particles. Sensor #2 sampled the aerosol using the Venturi effect while particles were charged after being mixed with a particle-free flow of positive ions generated by a corona charger. An ion trap collected ions that did not attach on particles while the particle number concentration was calculated by the escaping current which was continuously measured [45]. Sensor #3 used a diffusion charger, an ion trap, a diffusion particle collector, and an electrical detector while it operated at low pressure. Sensor #4 inlet flow was 1.2 lpm. It used a mixing type CPC [46] with cut-off size at 10 nm while particles in the range 10–23 nm were removed with a diffusion screen placed upstream the CPC. Sensors #5 and #6 had an inlet flow of 0.7 lpm but only 0.1 lpm was driven to the CPC. The rest of the inlet flow bypassed the CPC. The CPCs of the PTI sensors operated with isopropanol. For Sensors #4 and #6 the operator had to fill the working fluid while Sensor #5 incorporated a bag filled with the working fluid that had to be filled after certain number of measurements.

2.2.2. Calibration Values

Table 2 summarizes the Swiss and Dutch technical requirements regarding the counting efficiency of the sensor at different particle sizes, the linearity of the sensor at a specific particle size, and finally the efficiency of removing tetracontane (C40) particles which are considered to represent (semi)volatile particles. VERT proposes the same specifications as Dutch regulation but additionally requires an additional efficiency of <2 at 200 nm. The German regulation will set the same specifications as those required in the current European regulation for SPN-PEMS (Regulation 2017/1154).

Table 2. Requirements for PTI sensors at different regulations (CH, NL, VERT) and calibration values of the PTI sensors as provided by the manufacturers.

	Counting Efficiency				Linearity (80 nm)	VRE
	23 nm	50 nm	80 nm	200 nm	Polydisperse	30 nm Tetracontane
CH	<0.50	*	0.70–1.30	<1.30		>90% (<10^5 #/cm^3)
NL	0.20–0.60	0.60–1.30	0.70–1.30	-	0.75–1.25	>95% (<10^5 #/cm^3)
VERT	0.20–0.60	0.60–1.30	0.70–1.30	<2.00	0.75–1.25	>95% (<10^5 #/cm^3)
#1	0.34	0.75	1.00	-	1.03 (80 nm)	>95% (10^4 #/cm^3)
#2	0.47	0.86	1.12	-	0.99 (76 nm)	>95% (10^5 #/cm^3)
#3	0.43	0.76	1.00	1.67	0.99 (37–56 nm)	100% (>10^4 #/cm^3)
#4	0.40	0.90	1.00	1.15	0.998 (poly)	99.9% (3.5 × 10^4 #/cm^3)
#5	0.55	0.95	1.02 (70 nm)	1.04	N/A	N/A
#6	0.33	0.55 (41 nm)	-	-	1.04 (no size info)	>99%

* >0.4 at 41 nm. VRE = Volatile Removal Efficiency. In brackets the concentration of volatile particles; N/A = not available.

Table 2 presents also the counting efficiency, linearity, and volatile particle removal efficiency of the PTI sensors of the specific system that was used at the JRC campaign. The specification of the sensors changes according to the applicable regulation. The reported values were provided by the PTI sensors manufacturers and not tested by us.

2.3. Vehicles

In this study, six vehicles were tested at low idling. Table 3 presents for each vehicle, the model year and the Euro emissions standard it fulfilled, the existence of a particulate filter, the mileage, the engine displacement/power, and the fuel that was used. The notation we use is 'V' and the number of the vehicle. All vehicles were light duty homologated as M1 (=passenger cars) from Euro 3 to Euro 6d regulations. Five diesel and one gasoline with direct injection engine were tested. Note that for V4 an engine out flow was extracted and driven to the tailpipe where it was mixed with the DPF-out flow in order to simulate a malfunctioning DPF.

Table 3. Main characteristics of tested vehicles.

Code	Euro	Fuel	Year	Mileage (km)	Engine Displacement (cm^3)	Power (kW)	Particulate Filter
V1	6b	Diesel	2017	23,540	1.560	88	Yes
V2	6d	Diesel	2019	4.100	1.999	132	Yes
V3	4	Diesel	2009	209,000	1.997	100	Yes
V4	6d	Diesel	2020	4.200	1.968	110	Yes [1]
V5	5b	Gasoline DI	2012	151,831	1.197	77	No
V6	3	Diesel	2004	286,000	2.993	150	No

[1] An engine-out flow was available and mixed with the DPF-out flow. DI = Direct Injection. DPF = Diesel Particulate Filter.

3. Results

3.1. Vehicles without Particulate Filter

Figure 2 presents the low idling emissions of two vehicles without particulate filter; a diesel (V6) and a gasoline direct injection (V5). Figure 2a plots the SPN_{23} and SPN_{10} emissions of V6 measured with the setup 'REF A'. Both SPN_{23} and SPN_{10} of V6 low idling emissions were ~10^7 #/cm^3. This is in agreement with previous studies that reported >10^7 #/cm^3 [47,48] for diesel vehicles without DPF. Note that the reference system measured >3×10^6 #/cm^3 so the SPN_{23} was probably underestimated. Some PTI devices reported SPN emissions that were higher than the limit proposed by the (corresponding) regulation they follow. Specifically, Sensors #1 and #2 report up to 2×10^6 #/cm^3 twice as high the Dutch limit, and the Sensor #5 up to 5×10^5 #/cm^3. This is not the upper concentration limit of the sensors but the threshold value they use to report fail of the vehicle. In the cases that PTI sensors reached this upper limit we added a red circle (see Figure 1a). Even if PTI sensors underestimated in some cases the SPN_{23}, they all reported that the vehicle failed to comply with regulation due to SPN concentrations higher than 2×10^6 #/cm^3 (Sensor #5 has a limit at 5×10^5 #/cm^3). Thus, all sensors detected that this vehicle was a high emitter. Sensor #2 was not available during this test.

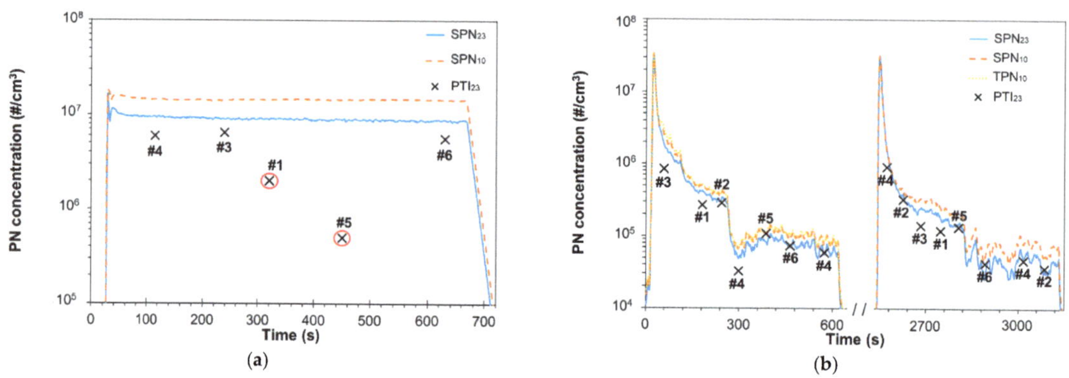

Figure 2. Particle number concentrations during low idling with cold start engine of direct injection vehicles without particulate filter: (**a**) diesel V6; (**b**) gasoline V5. Points in red circle show that the PTI sensors reported a threshold concentration that corresponds in automatic failure (2×10^6 #/cm^3 for NL or 5×10^5 #/cm^3 for DE).

Figure 2b reports the emissions of V5 after two engine ignitions; one with cold engine and one with hot that was performed approximately 30 min after the vehicle was switched off. During the first test the setup 'REF B' was used and SPN_{23}, SPN_{10} and TPN_{10} were measured, while during the second test only SPN_{23} and SPN_{10} were measured with 'REF A'. SPN_{23} emissions were initially >>10^6 #/cm^3 while after the first 300 s they were ~3×10^5 #/cm^3. The 10 nm to 23 nm concentration (SPN_{10-23}) was ~35% more than SPN_{23} throughout the test (see also Section 3.6). After 600 s of idling, the concentration decreased to 6.5×10^4 #/cm^3. A very similar concentrations profile was also observed during the second test, the lowest SPN_{23} being 3.5×10^4 #/cm^3. Interestingly, the concentration was not stable throughout the test but fluctuated significantly.

In general, all sensors were precise enough. The difference of Sensors #2, #5, and #6 to SPN_{23} were within ±19% and of Sensor #1 within ±34%. The highest differences compared to SPN_{23} were 54% and 38% for instruments #3 and #4, respectively. The variability of the emissions (defined as standard deviation of SPN_{23} divided by average SPN_{23} for the specific time period) was 15% and 6% for these two tests, respectively. The measurement with the device #4 was repeated at a more stable idling emissions point and the deviation with 'REF B' system decreased to 7%. During the second test, the highest PTI sensors deviation occurred for Sensor #4 but the SPN_{23} standard deviation was >40%.

Even after 400 s the SPN_{23} standard deviation was >10% that resulted in high uncertainty for the PTI sensors results that were measuring only for a frame of 15 s (except for Sensor #5 and Sensor #6).

3.2. Malfunctioning DPF

Figure 3 plots the emissions of two vehicles with reduced DPF efficiency. Figure 3a presents the SPN_{23} and SPN_{10} emissions of V3 (setup 'REF A') which was a Euro 4 with mileage >200,000 km. Initially, SPN_{23} emissions were higher than 10^6 #/cm^3 and gradually decreased to 8.5×10^5 #/cm^3. After 380 s, SPN_{23} decreased steeply to ~3.8×10^5 #/cm^3 and then stabilized. When SPN emissions stabilized, $SPN_{10\text{–}23}$ was ~350% more than SPN_{23} for V3 (see Section 3.6). Initially, SPN_{10} and TPN_{10} concentrations were $>3 \times 10^6$ #/cm^3 but after the first 380 s they decreased to their accurate measurement range. The PTI sensors had a good agreement with the SPN_{23} measurements when emissions stabilized except for Sensor #2 and to a lesser degree for Sensor #1 which overestimated the concentrations. In one case, a PTI sensor indicated that emissions were higher than 2×10^6 #/cm^3 that results an immediate failure in the NL regulation. For this case we added a red circle in Figure 3a.

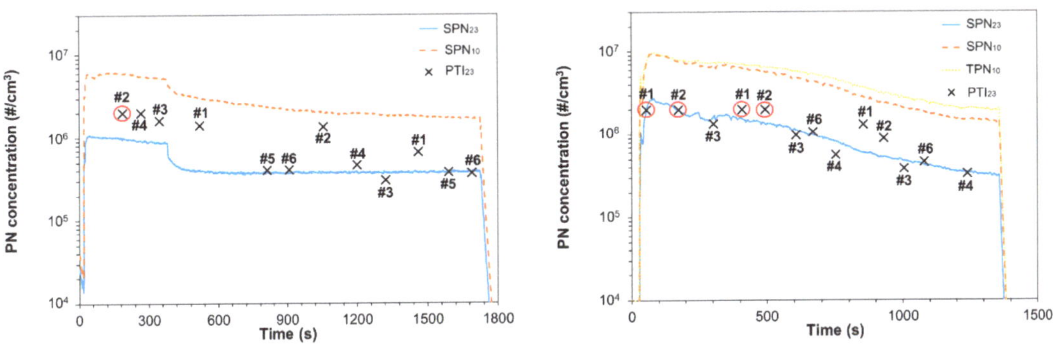

(a) (b)

Figure 3. Particle number concentration during low idling with cold start engine for: (**a**) Euro 4 diesel V3; (**b**) Euro 6d with DPF bypass V4. Points in red circle show that the PTI sensors reported a threshold concentration that corresponds in automatic failure (2×10^6 #/cm^3 for NL or 5×10^5 #/cm^3 for DE).

Figure 3b plots the SPN_{23}, SPN_{10} and TPN_{10} emissions of V4 (setup 'REF B'). For this vehicle an engine-out flow was bypassed and mixed with the flow downstream of the DPF. The concentration was initially 2.8×10^6 #/cm^3 and gradually decreased down to 2.7×10^5 #/cm^3. The SPN_{23} concentration range spanned over the limits imposed by both Netherlands (1.0×10^6 #/cm^3) and Germany (2.5×10^5 #/cm^3). Engine- and DPF-out SPN_{23} emissions of V4 were found to be ~5×10^6 #/cm^3 and ~5×10^3 #/cm^3, respectively. Thus, the filter bypass applied in our study reduced the DPF efficiency from ~99.9% to ~93%. Similar to Figure 3a, when PTI sensors reported the failure value of 2×10^6 #/cm^3 we added a red circle. When concentrations were near the Dutch limit, the PTI sensors were accurate detecting those cases where the emissions exceeded the limit. Specifically, Sensors #1, #2, #3, #4, and #6 detected all the cases that SPN_{23} was higher than the Dutch limit (1.0×10^6 #/cm^3). Sensor #5 had a limit of 5×10^5 #/cm^3. When SPN_{23} decreased to values lower than 1.0×10^6 #/cm^3, the sensors were still in good accuracy but in two cases Sensors #1 and #2 overestimated SPN_{23} by 119% and 68%, respectively. Sensor #5 did not measure in the presented test. Its efficiency when measuring the PN emissions of V4 will be discussed in Section 3.6.. Note that SPN_{10} and TPN_{10} measurements during the first ~600 s had high uncertainties due to elevated concentration values ($>3 \times 10^6$ #/cm^3).

3.3. After DPF Regeneration

Figure 4a plots the SPN_{23} and SPN_{10} low idle concentrations of diesel vehicle V2 ('REF A') and compares them to PTI sensors measurements. After the first cold start engine ignition, two more ignitions at hot engine conditions followed (around 1300 s and 2400 s, respectively). The engine remained switched off only for few minutes before the two hot engine ignitions. The concentration was initially very high, $>10^6$ #/cm^3, and gradually decreased two orders of magnitude. The profile indicates that the DPF efficiency was increasing during the test and thus, our measurements were performed right after a regeneration. No OBD was available to confirm our assumption.

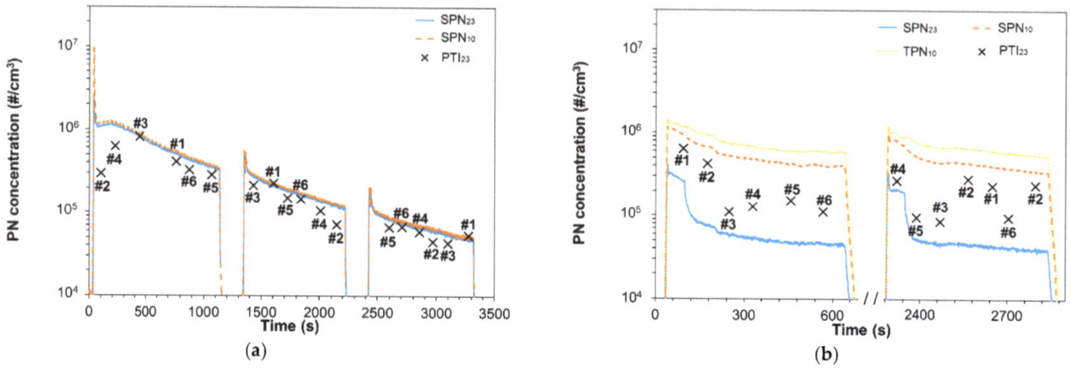

Figure 4. Particle number concentration during low idling: (**a**) After regeneration: V2 cold start and two hot starts. Hot engine ignitions were done few minutes after switching off the engine; (**b**) High sub-23 nm fraction: V1 cold start and one hot start.

Sensors #1 and #3 were very precise while the rest of the sensors underestimated the SPN_{23} emissions. During the first seconds of the first ignition, Sensors #2 and #4 underestimated SPN_{23} significantly but their deviation decreased at the second and third tests at levels near the accuracy requirement of NL regulation (~25%).

3.4. High Sub-23 nm Fraction

Figure 4b plots SPN_{23}, SPN_{10}, and TPN_{10} emissions of V1 (setup 'REF A') and compares them to the PTI sensors measurements. Initially the SPN_{23} concentration was $>2 \times 10^5$ #/cm^3. After ~80 s the SPN_{23} decreased steeply to ~6×10^4 #/cm^3 and then gradually down to 4×10^4 #/cm^3. The concentration of particles below 23 nm were almost seven times the concentration of particles >23 nm. This means that the mean particles size was below 23 nm. All PTI sensors overestimated significantly the SPN_{23} emissions. A second measurement was performed with hot engine ~25 min after switching off the engine. The results were similar to the first measurement. The sensor with the smallest deviation was Sensor #3 (85–91% difference), while the rest deviated 222–433% (Sensor #1), 464–515% (Sensor #2), 28–141% (Sensor #4), 83–226% (Sensor #5), and 128–144% (Sensor #6).

3.5. Total Particles

TPN_{10} concentrations (solid and volatile particles) were measured with setup "REF B" for V1, V4, and V5. Table 4 summarizes the volatiles fraction calculated as the absolute value of the ratio of SPN_{10}-TPN_{10} to SPN_{10}. Volatiles fraction is very low for the gasoline V5 (6%), while for the diesel vehicles V1 and V4 are 46% and 56%, respectively. The volatile fraction at low idling was low for these three vehicles. The highest fraction was detected for V4 that had also an engine-out flow that did not pass through the diesel oxidation catalyst (DOC) that oxidizes hydrocarbons.

Table 4. Mean concentrations and sub-23 nm and volatile fractions.

Vehicle	Comment	SPN_{23} (#/cm³)	SPN_{10} (#/cm³)	Sub-23 nm Fraction	TPN_{10} (#/cm³)	Volatiles Fraction
V1	DPF (high sub-23)	4.0×10^4	3.5×10^5	775%	4.5×10^5	47%
V2	DPF (after regen.)	4.8×10^4	5.0×10^4	5%	-	
V3	DPF (old)	3.8×10^5	1.7×10^6	346%	-	
V4	DPF (bypass)	2.7×10^5	7.0×10^5	158%	1.1×10^6	57%
V5	G-DI (no filter)	7.6×10^4	1.0×10^5	35%	1.1×10^5	6%
V6	No DPF	8.9×10^6	1.4×10^7	63%	-	

Sub-23 nm fraction = $(SPN_{10} - SPN_{23})/SPN_{23}$. Volatile fraction = $(TPN_{10} - SPN_{10})/SPN_{10}$.

3.6. Summary of Results

Table 4 summarizes the SPN_{23}, SPN_{10}, TPN_{10}, the sub-23 nm fraction of solid particles emitted by the tested vehicles defined as SPN_{10-23}/SPN_{23}, and the volatile fraction. The stabilized parts of the concentrations were used to calculate the fractions. The standard deviation of the calculated fractions was <3% except for V5 where standard deviation was 7%.

The SPN_{23} levels were from 4.0×10^4 #/cm³ (V1) up to 8.9×10^6 #/cm³ for the vehicle without DPF (V6). The vehicle with a fraction of the exhaust bypassing the DPF had a concentration of 2.7×10^5 #/cm³ (V4) and the high mileage DPF vehicle a slightly higher (V3). Both of them were higher than Germany's limit (2.5×10^5 #/cm³), but lower than the Dutch limit (1.0×10^6 #/cm³).

The sub-23 nm fraction was very high for V1, V3, and V4 (>150%). In these cases, the inclusion of sub-23 nm particles in the regulation may change the status of a vehicle from 'pass' to 'fail'. For example, V1 had SPN_{23} 4.0×10^4 #/cm³, but SPN_{10} 3.5×10^5 #/cm³, which is higher than the Germany's limit of 2.5×10^5 #/cm³. V4's SPN_{23} was at the limit, but SPN_{10} exceeded the limit by far. Such high sub-23 nm fractions indicate the importance of the cut-off size of the PTI sensors. This was clear with the tests of V1 that had the higher sub-23 nm fraction (Figure 4b).

Figure 5 correlates the PTI sensors to SPN_{23} measurements. Additionally, we plot the SPN_{10} and, when available, the TPN_{10} in order to study the effect of sub-23 nm and volatile particles on the performance of PTI sensors. The vertical solid lines divide the number concentrations for different vehicles. V6 is not plotted because most sensors saturated.

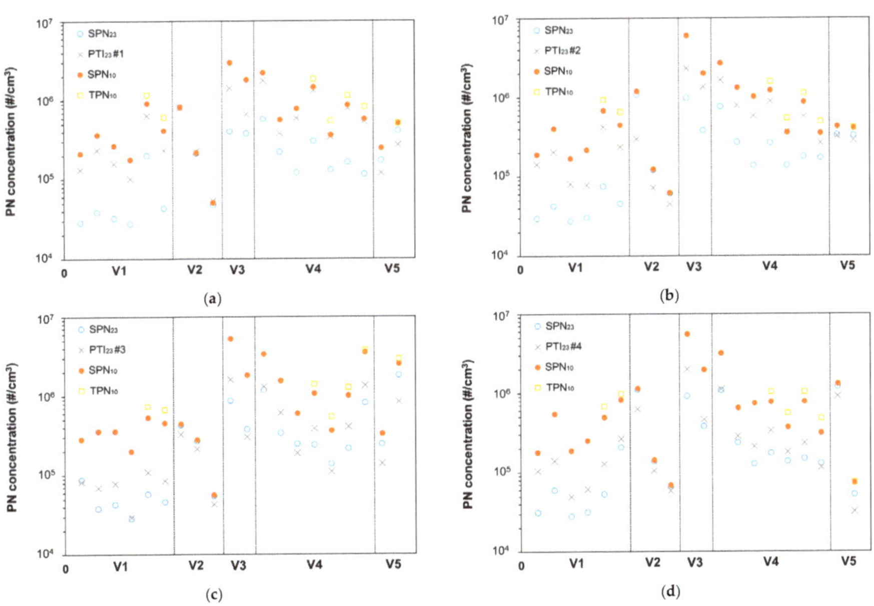

Figure 5. *Cont.*

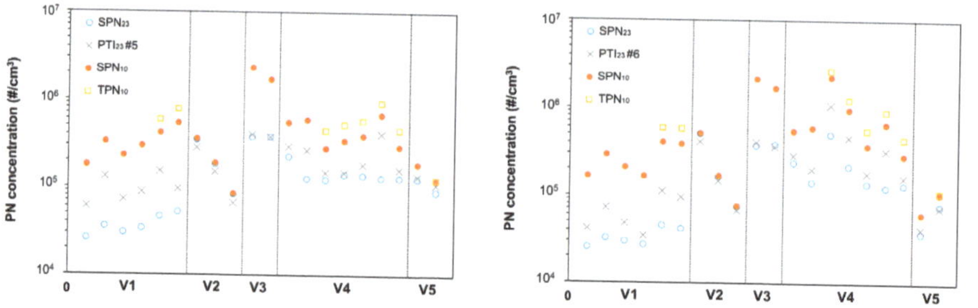

(e) (f)

Figure 5. Summary of PTI sensors measurements and comparison with SPN_{23} measured with a PMP compliant system. SPN_{10} and TPN_{10} (when available) are also provided. (**a**) PTI #1; (**b**) PTI #2; (**c**) PTI #3; (**d**) PTI #4; (**e**) PTI #5; (**f**) PTI #6.

Figure 6a plots the deviation of the PTI sensors compared to the SPN_{23} concentrations in function of the sub-23 nm fraction. When the sub-23 nm fraction is <100% the accuracy of the PTI sensors is very good. When the sub-23 nm fraction is 200% or more the deviations become bigger, especially for Sensors #1 and #2. CPC based sensors and DC-based Sensor #3 were less influenced by the presence of small particles. Figure 6b plots the deviation of the PTI sensors against the SPN_{23} measured with the reference system. Measurement uncertainty is in general higher at lower concentrations but there is no clear trend between the measured concentration and the sensors' deviation. The scatter is due to the sensitivity of the sensors to the sub-23 nm fraction as described in Figure 6a.

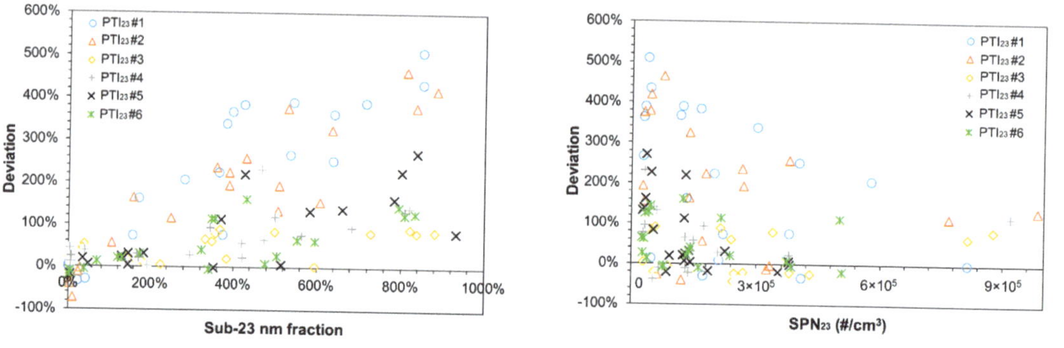

(a) (b)

Figure 6. Summary of the deviation of the PTI sensors plotted against: (**a**) the sub-23 nm fraction $(SPN_{10} - SPN_{23})/SPN_{23}$); (**b**) the solid particle number of particles down to 23 nm (SPN_{23}).

4. Discussion

This is one of the first studies that assessed sensors for the PTI of vehicles. Previous studies used prototypes [48–50], or the Swiss approved sensor [24]. Here from the six sensors tested, one had the approval (certificate) for the Swiss PTI (Sensor #6), two were prototypes without a specified country for homologation (Sensors #3 and #5), and three sensors (Sensors #1, #2 and #4) had either approval (certificate) for PTI testing or were ready for approval in Netherlands. All of the sensors fulfilled the technical requirements of the countries they had the approval from, or the VERT recommendations. In principle, depending on the regulation, an uncertainty of ±30% is expected from the technical specifications (see Table 2). Even though such small differences were indeed seen, there were some cases of much higher differences (up to 5 times higher). The key message of this study was that the reason of these high differences was the high (or low) sensitivity of the sensors to particles smaller than 23 nm (Figure 6a), which is the current lower size in the regulations. Most importantly, these sub-23 nm particles were "solid" and not volatiles. The implications of this finding will be discussed in more details below.

4.1. The Role of Sub-23 nm Particles in the PTI Sensors Deviation

Up to 160% sub-23 nm fraction could be handled acceptably by the PTI sensors (Figure 6a). At 200% sub-23 nm value, two DC−based sensors started deviating by >150%. The CPC-based systems had high deviations (>100%) at >400% sub-23 nm values. The declared counting efficiencies of the PTI sensors at 23 nm presented in Table 2 do not justify these high differences. Thus, the rationale for the differences observed is possibly due to the efficiency of the sensors at sizes below 23 nm. The counting efficiency of CPCs at sizes below the cut-off size decreases steeper than diffusion chargers. A recent study on the uncertainty of regulatory particle number measurements [40] found that at 50 nm a PMP system (CPC-based) and a portable emissions measurement system, SPN-PEMS (both CPC- and DC-based), have very similar efficiencies (~90%). Instead, at 15 nm the PMP system would typically measure in the range 16–23% while CPC-based PEMS ~24% and DC-based ~33%. Thus, a DC-based PEMS may measure even double sub-23 nm particles concentration compared to a high losses PMP system with 16% efficiency at 15 nm. Even if PTI sensors are not necessarily equal to PEMS systems the aforementioned differences give important input on the differences observed in this study. A previous study [47] has calculated the possible under- and over-estimation of SPN_{23} of PTI instruments as a factor of the geometric mean diameter. The upper maximal SPN_{23} measurement deviation for geometric mean diameters in the range 35–77 nm was estimated to be 18% to 84%; higher geometric mean diameters and lower geometric standard deviations resulted higher deviations. In addition to the sub-23 nm effect on the sensors' accuracy we also studied possible linearity issues (Figure 6b) but no clear trend was observed between deviation and concentration. Moreover, it was not clear whether the sensors that overestimated the emissions were affected by volatile particles. The total particles were measured for the two diesel vehicles with high sub-23 nm fraction. They were approximately 50% higher than the solid particles. As the concentration of solid particles below 23 nm was very high, we believe that any volatile particles would be mainly condensed on the existing solid nanoparticles, rather than forming a separate volatile nucleation mode. During idling of diesel engines, the air to fuel ratio is very high and small volatile particles fraction is emitted compared to other engine operation conditions [51]. More studies are needed to assess the volatile removal efficiency of the sensors under realistic and extreme conditions (i.e., with existence of nucleation mode particles).

4.2. Sub-23 nm Particles at Idling

The second point that needs to be discussed is what are the particles below 23 nm and whether their concentration is high. Diesel engines typically produce size distributions with geometric mean diameters in the range of 50–70 nm [42]. Thus, in general they have a low sub-23 nm particles fraction. Formation of sub-23 nm particles have been recently reported during urea or ammonia injection [52]. These particles were also found to carry high charge at high exhaust gas temperatures [53]. In our study, half of the diesel vehicles were not equipped with SCR (selective catalytic reduction for NO_x). For the tests with the vehicles with SCR (V2, V4), due to the low exhaust gas temperatures at idling, we believe that no urea injection took place and due to the low exhaust gas temperature no such particles were formed even if any ammonia desorbed from the catalyst. Indeed, V2 had very low sub-23 nm values while particle emissions of V4 mainly originated by the engine-out flow and, thus, they were not influenced by possible urea injection. Furthermore, we did not observe any significant variation of the sub-23 nm value that would indicate release of ammonia (Figure 3b). Another case with high concentration of solid particles below 23 nm is idling. This has been shown and confirmed repeatedly in the literature and it is assumed that they are heavy polyaromatic hydrocarbons (PAHs) that cannot evaporate at 350 °C [54]. The concentration of these particles was extremely high for vehicle V1 (7.5 times of SPN_{23}), but still high for V3 and V4 (1.5 to 3.5 times of SPN_{23}). If we also consider the particle losses of sub-23 nm particles due to diffusion (not corrected in this study), their fraction would be even higher (1.5–2 times). For the remaining diesel vehicles, the fraction of particles below 23 nm was <65%. What is important to note is that the high concentration of particles below 23 nm at idling does not extrapolate to other engine operation modes or the type approval cycle. Dedicated tests with V1 showed that the SPN_{23} type approval cycle emissions were 1.48×10^{11} #/km, but for the same cycle the SPN_{10} emissions were 1.85×10^{11} #/km. Thus, the approximately 700% higher SPN_{10} idle concentration corresponded to only 27% higher SPN_{10} cycle emissions. There was also no correlation between SPN_{23} idle concentrations and sub-23 nm fraction (see Table 4). Combining this lack of correlation with the lack of correlation of idle sub-23 nm fraction and type approval cycle sub-23 nm fraction, it can be concluded that the PTI sensors need to avoid counting this fraction.

4.3. The Importance of PTI Sensors Efficiency in the Sub-23 nm Size Region

The third question that needs to be answered is whether this sensors' concentration uncertainty at low particle sizes is important. For vehicles having low idle emissions (i.e., $<5 \times 10^4$ #/cm^3) an error on the order of 5 times (e.g., V1), will bring the result close to the German limit. For a vehicle close to the German limit an error on the order of 3 is still below the Dutch limit (e.g., V3 or V4). When discussing limits, the uncertainty of the whole procedure should also be taken into account. The idle concentration can give an estimation of the type approval cycle emissions (factor 10^7 cm^3/km), but this factor has an uncertainty margin of at least 2 (for diesel vehicles); for gasoline vehicles the factor is much higher [25]. Thus, a vehicle that is close to the type approval limit (6×10^{11} #/km), taking into account the factor 2 would have idle concentration of up to 1.2×10^5 #/cm^3. By setting a limit of 2.5×10^5 #/cm^3 (German regulation) an additional factor of 2 is permitted for the PTI sensors uncertainty. In our study two vehicles were close to the German limit: V3 (old DPF) and V4 (bypassed DPF). All PTI sensors correctly identified that idling emissions of these two vehicles were $>2.5 \times 10^5$ #/cm^3, but in many cases the SPN_{23} emissions were significantly overestimated. V3 with idle concentration of 3.8×10^5 #/cm^3 was precisely assessed by Sensors #5 and #6 (within 2%). The average deviation of Sensors #3 and #4 was 32% and 70%, respectively. Sensors #1 and #2 overestimated >150% due to the high fraction of particles below 23 nm (346%). SPN_{23} of V4, when reducing the DPF efficiency from 99.9% to 93%, was 2.7×10^5 #/cm^3 and the sub-23 nm fraction was 158%. Similar to V1 all PTI sensors overestimated the SPN_{23}; on average (for each sensor's measurements) Sensors #3–#6 were within 52% while Sensors #1 and #2 overestimated by 273% and 187%, respectively.

One vehicle had higher emissions than the Dutch limit: V6 (no DPF). This vehicle had emissions close to 1×10^7 #/cm^3. Except from Sensor #5 that reported the failure threshold value in the German regulation (5×10^5 #/cm^3), all sensors detected that this vehicle had $>2 \times 10^6$ #/cm^3 that results in an immediate fail in the Dutch regulation. The idle concentrations of $<5 \times 10^4$ #/cm^3 of V2 were measured accurately by all sensors within 20% (sub-23 nm fraction 5%). On the other hand, the idle concentrations of V1, which were at the same levels (4×10^4 #/cm^3) were not determined accurately by all sensors due to the high sub-23 nm fraction (770%). Sensors #3 and #6 measured $<1 \times 10^5$ #/cm^3, but Sensors #1, #2, #4 and #5 above; Sensors #1 and #2 even above the German limit in some cases. V1 type approval cycle emissions are well below the limit. Thus, with the current Dutch regulation technical requirements, the German limit might result in some false "fails". Any limit at this or lower level needs more rigorous characterization of the cut-off curve of the sensors.

4.4. Gasoline Vehicles

The discussion focused on diesel vehicles, because the upcoming Dutch, German, and Belgian PTI regulations will apply only to diesel vehicles. The reason is that tampering or malfunction of DPFs will have a significant impact on the emissions, because the engine out emissions are very high (around 10^{14} #/km) [24]. On the other hand, the SPN_{23} emissions of gasoline vehicles even without any particulate filter are near the regulation limit (around 10^{12} #/km), while modern gasoline vehicles may emit one order of magnitude lower SPN_{23} [3,55]. Thus, the detection of existence or malfunctioning of the filter is very difficult. A previous study also showed that is difficult to find a good correlation between idle concentration and type approval emissions [25]. More studies in this direction are necessary.

5. Conclusions

In this study, the measurement of low idling emissions of different vehicles was performed with six SPN sensors designed for periodical technical inspection (PTI) applications and a reference system that measured >23 nm (SPN_{23}) and >10 nm (SPN_{10}), and in some cases also the total particle number >10 nm (TPN_{10}). Our scope was twofold; to evaluate the efficiency of the PTI_{23} sensors in the context of the limits set by different current or future PTI regulations and to provide input on the procedures. The cases we studied were: high sub-23 nm particles and volatiles fraction, emissions after a DPF regeneration, and vehicles without particulate filter or with a malfunctioning filter.

SPN_{23} low idling emissions of a diesel vehicle without a DPF were around 1×10^7 #/cm^3, one order of magnitude higher than the Dutch limit, and easily detectable by all sensors. For malfunctioning DPFs we found emissions SPN_{23} slightly higher than the German limit of 2.5×10^5 #/cm^3. In one case (V4), the DPF efficiency of a well performing vehicle was controllable reduced from ~99.9% to ~93% and SPN_{23} emissions were 2.7×10^5 #/cm^3. SPN_{23} emissions were very high after a DPF regeneration (even > 3.8×10^5 #/cm^3) and gradually decreased to <1×10^5 #/cm^3 showing the necessity of a short conditioning (e.g., some minutes of driving) of the vehicle in these cases. Finally, the SPN_{23} low idling emissions of a GDI vehicle without a filter were much lower (<1×10^5 #/cm^3) than the currently proposed limits pointing the necessity of performing more studies on both the procedures and the PTI limit for gasoline vehicles.

Our results suggest that PTI requirements for PN measurements may be met by both CPC- and DC-based sensors. All sensors detected high emitters (>1×10^6 #/cm^3) and for low sub-23 nm fractions their accuracy was within 50% in most of the cases. The highest deviations of the PTI_{23} sensors were observed when the sub-23 nm fraction was high. The SPN_{10-23} was even 775% higher than SPN_{23} in one case (V1), much higher than typical values for diesel vehicles, showing that diesel engines may emit high concentrations of nonvolatile nucleation particles during idling. Two out of the three DC-based sensors (Sensors #1 and #2) were mostly affected by the presence of sub-23 nm particles and overestimated significantly SPN_{23} resulting in false 'fails' in case that a limit in the order of 2.5×10^5 #/cm^3 will be imposed. The volatile particles down to 10 nm were ~50% for two diesel vehicles and 6% for a gasoline (G-DI) vehicle more than SPN_{10}. In these three cases no correlation was found between volatiles and sensors deviation.

Author Contributions: Conceptualization, A.M. and B.G.; methodology, A.M. and B.G.; formal analysis, A.M.; writing—original draft preparation, A.M.; writing—review and editing, B.G., T.S., R.S.-B. All authors have read and agreed to the published version of the manuscript.

Funding: This research received no external funding.

Institutional Review Board Statement: Not applicable.

Informed Consent Statement: Not applicable.

Data Availability Statement: Data available upon request from the corresponding author.

Acknowledgments: The authors would like to acknowledge the following companies for providing the sensors of this study (in alphabetical order): Capelec (Montpelier, France) and Pegasor (Tampere, Finland), DEKATI (Kangasala, Finland), Mahle (Stuttgart, Germany), TEN (Baambrugge, The Netherlands), and TSI (Aachen, Germany).

Conflicts of Interest: The authors declare no conflict of interest. The opinions expressed in the manuscript are those of the authors and should in no way be considered to represent an official opinion of the European Commission. Mention of trade or commercial products does not constitute endorsement or recommendation by the authors or the European Commission.

References

1. Pope, C.A., III; Burnett, R.T.; Thun, J.M.; Calle, E.E.; Krewski, D.; Ito, K.; Thurston, G.D. Lung Cancer, Cardiopulmonary Mortality, and Long-Term Exposure to Fine Particulate Air Pollution. *J. Am. Med. Assoc.* **2002**, *287*, 1132–1141. [CrossRef]
2. Giechaskiel, B.; Maricq, M.; Ntziachristos, L.; Dardiotis, C.; Wang, X.; Axmann, H.; Bergmann, A.; Schindler, W. Review of Motor Vehicle Particulate Emissions Sampling and Measurement: From Smoke and Filter Mass to Particle Number. *J. Aerosol Sci.* **2014**, *67*, 719–749. [CrossRef]
3. Giechaskiel, B.; Joshi, A.; Ntziachristos, L.; Dilara, P. European Regulatory Framework and Particulate Matter Emissions of Gasoline Light-Duty Vehicles: A Review. *Catalysts* **2019**, *9*, 586. [CrossRef]
4. Mayer, A.; Matter, U.; Scheidegger, G.; Czerwinski, J.; Wyser, M.; Kieser, D.; Weidhofer, J. Particulate Traps for Retro-Fitting Construction Site Engines VERT: Final Measurement and Implementation. *SAE Tech. Pap.* **1999**, 90092. [CrossRef]
5. Adam, F.; Olfert, J.; Wong, K.; Kunert, S.; Richter, J.M. Effect of Engine-out Soot Emissions and the Frequency of Regeneration on Gasoline Particulate Filter Efficiency. *SAE Tech. Pap.* **2020**. [CrossRef]
6. Boger, T.; Glasson, T.; Rose, D.; Ingram-Ogunwumi, R.; Wu, H. Next Generation Gasoline Particulate Filters for Uncatalyzed Applications and Lowest Particulate Emissions. *SAE Tech. Pap.* **2021**. [CrossRef]
7. Font, A.; Guiseppin, L.; Blangiardo, M.; Ghersi, V.; Fuller, G.W. A Tale of Two Cities: Is Air Pollution Improving in Paris and London? *Environ. Pollut.* **2019**, *249*, 1–12. [CrossRef] [PubMed]
8. Murzyn, F.; Sioutas, C.; Cavellin, L.D.; Joly, F.; Baudic, A.; Mehel, A.; Cuvelier, P.; Varea, E.; Rouland, B.P. Assessment of Air Quality in Car Cabin in and around Paris from On-Board Measurements and Comparison with 2007 Data. *J. Aerosol Sci.* **2021**, *158*, 105822. [CrossRef]
9. Konstandopoulos, A.G.; Kostoglou, M.; Skaperdas, E.; Papaioannou, E.; Zarvalis, D.; Kladopoulou, E. Fundamental Studies of Diesel Particulate Filters: Transient Loading, Regeneration and Aging. *SAE Tech. Pap.* **2000**. [CrossRef]
10. Giechaskiel, B. Particle Number Emissions of a Diesel Vehicle during and between Regeneration Events. *Catalysts* **2020**, *10*, 587. [CrossRef]
11. Yamada, H. Improving Methodology of Particulate Measurement in Periodic Technical Inspection with High-Sensitivity Techniques: Laser Light Scattering Photometry and Particle Number Method. *Emiss. Control Sci. Technol.* **2019**, *5*, 37–44. [CrossRef]
12. Merkel, G.A.; Cutler, W.A.; Warren, C.J. Thermal Durability of Wall-Flow Ceramic Diesel Particulate Filters. *SAE Tech. Pap.* **2001**. [CrossRef]
13. Maier, N.; Nickel, K.G.; Engel, C.; Mattern, A. Mechanisms and Orientation Dependence of the Corrosion of Single Crystal Cordierite by Model Diesel Particulate Ashes. *J. Eur. Ceram. Soc.* **2010**, *30*, 1629–1640. [CrossRef]
14. Yang, K.; Fox, J.T.; Hunsicker, R. Characterizing Diesel Particulate Filter Failure during Commercial Fleet Use Due to Pinholes, Melting, Cracking, and Fouling. *Emiss. Control Sci. Technol.* **2016**, *2*, 145–155. [CrossRef]
15. UN ECE. Tampering of Emissions Control Systems. Available online: https://unece.org/DAM/trans/doc/2019/wp29grpe/GRPE-78-04e.pdf (accessed on 10 December 2021).
16. Boveroux, F.; Cassiers, S.; Buekenhoudt, P.; Chavatte, L.; De Meyer, P.; Jeanmart, H.; Verhelst, S.; Contino, F. Feasibility Study of a New Test Procedure to Identify High Emitters of Particulate Matter during Periodical Technical Inspection. *SAE Tech. Pap.* **2019**. [CrossRef]
17. Burtscher, H.; Lutz, T.; Mayer, A. A New Periodic Technical Inspection for Particle Emissions of Vehicles. *Emiss. Control Sci. Technol.* **2019**, *5*, 279–287. [CrossRef]
18. Michael, C.; Victor, V.M.; Pierre, B.; Carsten, G.; Jelica, P.; Dario, M.; Robert, L.; Barouch, G.; Massimo, C.; Marcos, O.G.; et al. *Joint Research Centre 2019 Light-Duty Vehicles Emissions Testing, JRC122035*; Publications Office of the European Union: Luxembourg, 2020.
19. Posada, F.; Bandivadekar, A. Global Overview of On-Board Diagnostics (OBD) Systems for Heavy-Duty Vehicles. Available online: https://theicct.org/sites/default/files/publications/ICCT_Overview_OBD-HDVs_20150209.pdf (accessed on 10 December 2021).
20. Zhang, S.; Zhao, P.; He, L.; Yang, Y.; Liu, B.; He, W.; Cheng, Y.; Liu, Y.; Liu, S.; Hu, Q.; et al. On-Board Monitoring (OBM) for Heavy-Duty Vehicle Emissions in China: Regulations, Early-Stage Evaluation and Policy Recommendations. *Sci. Total Environ.* **2020**, *731*, 139045. [CrossRef]
21. Carslaw, D.; Rhys-Tyler, G. New Insights from Comprehensive On-Road Measurements of NOx, NO2 and NH3 from Vehicle Emission Remote Sensing in London, UK. *Atmos. Environ.* **2013**, *81*, 339–347. [CrossRef]

22. Gruening, C.; Bonnel, P.; Clairotte, M.; Giechaskiel, B.; Valverdre, V.; Zardini, A.; Carriero, M. *Potential of Remote Sensing Devices (RSDs) to Screen Vehicle Emissions*; Publications Office of the European Union: Luxembourg, 2019. Available online: https://op.europa.eu/en/publication-detail/-/publication/9f6af994-fa26-11e9-8c1f-01aa75ed71a1/language-en (accessed on 10 December 2021).
23. Ropkins, K.; DeFries, T.H.; Pope, F.; Green, D.C.; Kemper, J.; Kishan, S.; Fuller, G.W.; Li, H.; Sidebottom, J.; Crilley, L.R.; et al. Evaluation of EDAR Vehicle Emissions Remote Sensing Technology. *Sci. Total Environ.* **2017**, *609*, 1464–1474. [CrossRef]
24. Kadijk, G.; Elstgeest, M.; Ligterink, N.E.; Van der Mark, P.J. *Investigation into a Periodic Technical Inspection (PTI) Test Method to Check for Presence and Proper Functioning of Diesel Particulate Filters in Light-Duty Diesel Vehicles–Part 2*; TNO: Hague, The Netherlands, 2017.
25. Giechaskiel, B.; Lähde, T.; Suarez-Bertoa, R.; Valverdre, V.; Clairotte, M. Comparisons of Laboratory and On-Road Type-Approval Cycles with Idling Emissions. Implications for Periodical Technical Inspection (PTI) Sensors. *Sensors* **2020**, *20*, 5790. [CrossRef]
26. Giechaskiel, B.; Lähde, T.; Suarez-Bertoa, R.; Clairotte, M.; Grigoratos, T.; Zardini, A.; Perujo, A.; Martini, G. Particle Number Measurements in the European Legislation and Future JRC Activities. *Combust. Engines* **2018**, *174*, 3–16. [CrossRef]
27. Giechaskiel, B.; Valverdre, V.; Kontses, A.; Melas, A.; Martini, G.; Balazs, A.; Andersson, J.; Samaras, Z.; Dilara, P. Particle Number Emissions of a Euro 6d-Temp Gasoline Vehicle under Extreme Temperatures and Driving Conditions. *Catalysts* **2021**, *11*, 607. [CrossRef]
28. Mayer, A.; Lutz, T.; Spielvogel, J.; Franken, O.; Heuser, M. *PTI by Particle Count PN at Low Idle*; TA-024/21; VERT Association: Niederweningen, Switzerland, 2021.
29. Kupper, M.; Kraft, M.; Boies, A.; Bergmann, A. High-Temperature Condensation Particle Counter Using a Systematically Selected Dedicated Working Fluid for Automotive Applications. *Aerosol Sci. Technol.* **2020**, *54*, 381–395. [CrossRef]
30. Giechaskiel, B.; Wang, X.; Horn, H.G.; Spielvogel, J.; Gerhart, C.; Southgate, J.; Jing, L.; Kasper, M.; Drossinos, Y.; Krasebrink, A. Calibration of Condensation Particle Counters for Legislated Vehicle Number Emission Measurements. *Aerosol Sci. Technol.* **2009**, *43*, 1164–1173. [CrossRef]
31. Fierz, M.; Meier, D.; Steigmeier, P.; Burtscher, H. Aerosol Measurement by Induced Currents. *Aerosol Sci. Technol.* **2014**, *48*, 350–357. [CrossRef]
32. Schriefl, M.A.; Bergmann, A. Design Principles for Diffusion Chargers Sensing Particle Number Concentration. In Proceedings of the 2016 IEEE Sensors, Orlando, FL, USA, 30 October–3 November 2016.
33. Rüggeberg, T.; Fierz, M.; Burtscher, H.; Melas, A.D.; Deloglou, D.; Papaioannou, E.; Konstandopoulos, A.G. Application of DC-Sensors to Measure Ultrafine Particles from Combustion Engines. In Proceedings of the 23rd Transport and Air Pollution Conference, Thessaloniki, Greece, 15–17 May 2019.
34. Giechaskiel, B.; Drossinos, Y. Theoretical Investigation of Volatile Removal Efficiency of Particle Number Measurement Systems. *SAE Int. J. Engines* **2010**, *3*, 1140–1151. [CrossRef]
35. Burtscher, H.; Baltensperger, U.; Bukowiecki, N.; Cohn, P.; Hüglin, C.; Mohr, M.; Matter, U.; Nyeki, S.; Schmatloch, V.; Streit, N.; et al. Separation of Volatile and Non-Volatile Aerosol Fractions by Thermodesorption: Instrumental Development and Applications. *J. Aerosol Sci.* **2001**, *32*, 427–442. [CrossRef]
36. Khalek, I.S.; Kittelson, D.B. Real Time Measurement of Volatile and Solid Exhaust Particles Using a Catalytic Stripper. In Proceedings of the International Congress & Exposition, Detroit, MI, USA, 27 February–2 March 1995.
37. Amanatidis, S.; Ntziachristos, L.; Giechaskiel, B.; Katsaounis, D.; Samaras, Z.; Bergmann, A. Evaluation of an Oxidation Catalyst ('Catalytic Stripper') in Eliminating Volatile Material from Combustion Aerosol. *J. Aerosol Sci.* **2013**, *57*, 144–155. [CrossRef]
38. Melas, A.D.; Koidi, V.; Deloglou, D.; Daskalos, E.; Zarvalis, D.; Papaioannou, E.; Konstandopoulos, A.G. Development and Evaluation of a Catalytic Stripper for the Measurement of Solid Ultrafine Particle Emissions from Internal Combustion Engines. *Aerosol Sci. Technol.* **2020**, *54*, 704–717. [CrossRef]
39. Giechaskiel, B.; Melas, A.D.; Lähde, T.; Martini, G. Non-Volatile Particle Number Emission Measurements with Catalytic Strippers: A Review. *Vehicles* **2020**, *2*, 342–364. [CrossRef]
40. Giechaskiel, B.; Lähde, T.; Melas, A.D.; Valverdre, V.; Clairotte, M. Uncertainty of Laboratory and Portable Solid Particle Number Systems for Regulatory Measurements of Vehicle Emissions. *Environ. Res.* **2021**, *197*, 111068. [CrossRef] [PubMed]
41. Yamada, H.; Funato, K.; Sakurai, H. Application of the PMP Methodology to the Measurement of Sub-23 Nm Solid Particles: Calibration Procedures, Experimental Uncertainties, and DATA Correction Methods. *J. Aerosol Sci.* **2015**, *88*, 58–71. [CrossRef]
42. Giechaskiel, B.; Mamakos, A.; Andersson, J.; Dilara, P.; Martini, G.; Schindler, W.; Bergmann, A. Measurement of Automotive Nonvolatile Particle Number Emissions within the European Legislative Framework: A Review. *Aerosol Sci. Technol.* **2012**, *46*, 719–749. [CrossRef]
43. Giechaskiel, B.; Lähde, T.; Schwelberger, M.; Kleinbach, T.; Roske, H.; Teti, E.; van den Bos, T.; Neils, P.; Delacroix, C.; Jakobsson, T.; et al. Particle Number Measurements Directly from the Tailpipe for Type Approval of Heavy-Duty Engines. *Appl. Sci.* **2019**, *9*, 4418. [CrossRef]
44. Terres, A.; Giechaskiel, B.; Nowak, A.; Ebert, V. Calibration Uncertainty of 23nm Engine Exhaust Condensation Particle Counters with Soot Generators: A European Automotive Laboratory Comparison. *Emiss. Control Sci. Technol.* **2021**, *7*, 124–136. [CrossRef]
45. Amanatidis, S.; Maricq, M.M.; Ntziachristos, L.; Samaras, Z. Application of the Dual Pegasor Particle Sensor to Real-Time Measurement of Motor Vehicle Exhaust PM. *J. Aerosol Sci.* **2017**, *103*, 93–104. [CrossRef]

46. Wang, J.; McNeill, V.F.; Collins, D.R.; Flagan, R.C. Fast Mixing Condensation Nucleus Counter: Application to Rapid Scanning Differential Mobility Analyzer Measurements. *Aerosol Sci. Technol.* **2002**, *36*, 678–689. [CrossRef]
47. Bainschab, M.; Schriefl, M.A.; Bergmann, A. Particle Number Measurements within Periodic Technical Inspections: A First Quantitative Assessment of the Influence of Size Distributions and the Fleet Emission Reduction. *Atmos. Environ.* **2020**, *8*, 100095. [CrossRef]
48. Jarosinski, W.; Wisniowski, P. Verifying the Efficiency of a Diesel Particulate Filter Using Particle Counters with Two Different Measurements in Periodic Technical Inspection of Vehicles. *Energies* **2021**, *14*, 5128. [CrossRef]
49. Suarez-Bertoa, R.; Lähde, T.; Giechaskiel, B. Verification of NPTI-Instruments for Diesel and Petrol Vehicles—First Results. Presented at the ETH Conference on Nanoparticles, Zurich, Switzerland, 21 June 2018. Available online: https://www.nanoparticles.ch/archive/2018_Suarez_PR.pdf (accessed on 10 December 2021).
50. Schriefl, M.A.; Bergmann, A.; Fierz, M. Design Principles for Sensing Particle Number Concentration and Mean Particle Size with Unipolar Diffusion Charging. *IEEE Sens. J.* **2019**, *19*, 1392–1399. [CrossRef]
51. Burtscher, H. Physical Characterization of Particulate Emissions from Diesel Engines: A Review. *J. Aerosol Sci.* **2005**, *36*, 896–932. [CrossRef]
52. Amanatidis, S.; Ntziachristos, L.; Giechaskiel, B.; Bergmann, A.; Samaras, Z. Impact of Selective Catalytic Reduction on Exhaust Particle Formation over Excess Ammonia Events. *Environ. Sci. Technol.* **2014**, *48*, 11527–11534. [CrossRef] [PubMed]
53. Mamakos, A.; Schwelberger, M.; Fierz, M.; Giechaskiel, B. Effect of Selective Catalytic Reduction on Exhaust Nonvolatile Particle Emissions of Euro VI Heavy-Duty Compression Ignition Vehicles. *Aerosol Sci. Technol.* **2019**, *53*, 898–910. [CrossRef]
54. De Filippo, A.; Maricq, M.M. Diesel Nucleation Mode Particles: Semivolatile or Solid? *Environ. Sci. Technol.* **2008**, *42*, 7957–7962. [CrossRef]
55. Giechaskiel, B.; Lähde, T.; Drossinos, Y. Regulating Particle Number Measurements from the Tailpipe of Light-Duty Vehicles: The next Step? *Environ. Res.* **2019**, *172*, 1–9. [CrossRef] [PubMed]

sensors

MDPI

Article

Development of Air Quality Boxes Based on Low-Cost Sensor Technology for Ambient Air Quality Monitoring

Paul Gäbel [1], Christian Koller [2] and Elke Hertig [1,*]

1 Regional Climate Change and Health, Faculty of Medicine, University of Augsburg, Universitätsstraße 2, 86159 Augsburg, Germany; paul.gaebel@med.uni-augsburg.de
2 Faculty of Design, Hochschule München, Lothstraße 34, 80335 Munich, Germany; ckoller@hm.edu
* Correspondence: elke.hertig@med.uni-augsburg.de

Abstract: Analyses of the relationships between climate, air substances and health usually concentrate on urban environments because of increased urban temperatures, high levels of air pollution and the exposure of a large number of people compared to rural environments. Ongoing urbanization, demographic ageing and climate change lead to an increased vulnerability with respect to climate-related extremes and air pollution. However, systematic analyses of the specific local-scale characteristics of health-relevant atmospheric conditions and compositions in urban environments are still scarce because of the lack of high-resolution monitoring networks. In recent years, low-cost sensors (LCS) became available, which potentially provide the opportunity to monitor atmospheric conditions with a high spatial resolution and which allow monitoring directly at vulnerable people. In this study, we present the atmospheric exposure low-cost monitoring (AELCM) system for several air substances like ozone, nitrogen dioxide, carbon monoxide and particulate matter, as well as meteorological variables developed by our research group. The measurement equipment is calibrated using multiple linear regression and extensively tested based on a field evaluation approach at an urban background site using the high-quality measurement unit, the atmospheric exposure monitoring station (AEMS) for meteorology and air substances, of our research group. The field evaluation took place over a time span of 4 to 8 months. The electrochemical ozone sensors (SPEC DGS-O3: R^2: 0.71–0.95, RMSE: 3.31–7.79 ppb) and particulate matter sensors (SPS30 PM1/PM2.5: R^2: 0.96–0.97/0.90–0.94, RMSE: 0.77–1.07 $\mu g/m^3$/1.27–1.96 $\mu g/m^3$) showed the best performances at the urban background site, while the other sensors underperformed tremendously (SPEC DGS-NO2, SPEC DGS-CO, MQ131, MiCS-2714 and MiCS-4514). The results of our study show that meaningful local-scale measurements are possible with the former sensors deployed in an AELCM unit.

Keywords: electrochemical sensors; metal oxide semiconductor sensors; particulate matter sensors; urban air quality; smart environment monitoring (SEM)

1. Introduction

Worldwide, one of the greatest and most challenging problems is the degradation of air quality, especially when caused by human activities. In the year 2019, according to the World Health Organization (WHO), 99% of the Earth's population was living in regions where the air quality guideline levels given by the WHO were not achieved [1].

Recently, the WHO global air quality guidelines were updated based on the latest systematic reviews of exposure-response studies [2]. For the classical air pollutants, which include particulate matter (PM), ozone (O_3), nitrogen dioxide (NO_2), carbon monoxide (CO) and sulfur dioxide (SO_2), an improved assessment of their adverse effects on health in low-, mid- and high-income countries took place in the last 15 years, achieved, for instance, through the progress in measurements, modelling, data availability and exposure analysis [2]. Olschewski et al. have shown that high-mortality events of patients with cancer are strongly linked with above-average levels of NO_2 and PM2.5 during unfavorable weather conditions in late winter until spring, which leads to the accumulation of polluted air [3]. Generally, weather conditions and the atmospheric state influence the transport, mixing ratio, transformation and deposition of air substances; hence they are important factors defining the air quality level [4–6]. Many studies show that short- as well as long-term exposures to O_3 are associated with a higher morbidity and mortality [7–9]. Moreover, major evidence exists that cardiovascular mortality and morbidity is related to PM exposure [10–12]. CO, a highly poisonous gas [13], is another air pollutant detrimental to human health, being associated with acute respiratory and cardiovascular diseases, for example [14,15]. The combination of the high impact of ambient air pollution on human health with a still not satisfactory spatial resolution of air quality monitoring worldwide points at the need for an increase in measurement resolution of health-deteriorating air pollutants. The lack of high-resolution air pollutant monitoring is due to the high initial and maintenance costs of regular devices for monitoring air pollution [16–18].

Therefore, it is reasonable to aim at developing custom multipollutant air-quality monitoring systems to characterize the variation of those classic pollutants in a high spatial resolution and which are also far less cost-intensive compared to official measurement devices. In the last decade, an increasing number of companies started to produce low-cost sensors (LCS) or even whole sensor systems (SSys), which include supporting components (e.g., enclosure, power supply, hardware and software for data treatment). This also sparked the endeavors of researchers to build low-cost air quality monitoring networks as a complementary source of information [16]. While these manufactured sensor systems were evaluated, for instance, through collocation experiments [18], researchers also evaluated the sole sensors or sensor systems embedded in their own custom-built measurement systems [19–22]. In this work, we do the latter and introduce the atmospheric exposure low-cost monitoring (AELCM) device, which can deliver high spatial and temporal resolution data of air pollutant concentrations and climate parameters for Internet of Things (IoT) applications using an Arduino and Pycom microcontroller board.

Generally, lower-cost monitoring devices for gaseous pollutants in ambient air use electrochemical sensors (EC) or metal oxide sensors (MOS), while optical particle counters (OPC) are used the most for the detection of PM [23]. All these sensor types are incorporated in the first version of the AELCM, which also includes a low-cost sensor for the measurement of temperature and humidity inside the box. The sensor output can be highly dependent on the variation of these meteorological variables [24–26]. The incorporated sensors in the AELCM box are supposed to measure the classic pollutants PM2.5, NO_2, O_3 and CO. While low-cost sensors offer many potential advantages, there are also major disadvantages. Their overall short lifespan, lack of long-term stability, signs of inter-sensor unit variability, cross-sensitivity and the need to calibrate them using reference instruments because of the lack of sufficient conversion formulas for individually purchased sensors are detrimental to their overall use in a systematic and beneficial way [16].

In addition to a technical presentation of the AELCM, we also want to address the performance of the used sensors (some of them are rarely evaluated, such as the digital gas sensors manufactured by SPEC Sensors) in the context of some of the issues these sensors entail. For this purpose, we are using the reference measurements of the atmospheric exposure monitoring station (AEMS) for air substances and meteorological variables, while our AELCM boxes are mounted next to the station.

2. Materials and Methods

2.1. AELCM Design and Collocation Experiment

The first version of the atmospheric exposure low-cost monitoring box uses an Arduino Mega 2560 Rev3 board for the collection of sensor measurements. Moreover, the Arduino device is used for the control of the other embedded hardware in the AELCM box. The board uses an ATmega 2560, which is a power-efficient and performant 8-bit microcontroller [27].

The Arduino Mega 2560 provides a wide variety of features, while the most important features for our measurement system were the 16 MHz crystal oscillator, an available USB connection, the high number of digital input/output pins and analogue input pins, as well as the amount of memory. For communication with external devices, such as a secure digital (SD) card module (which enables storage of measurement data on a memory card) or a real-time clock (RTC), the board provides SPI, four UARTs (hardware serial ports) and I2C. The synchronous serial data protocol called the serial peripheral interface (SPI) is used to communicate with the SD card module and to control it. The universal asynchronous receiver-transmitter (UART) protocol and the serial communication bus I2C (inter-integrated circuit) are two additional communication protocols being used. The connection between the board and its external devices was mostly realized through different self-designed printed circuit boards (PCBs). These custom-built PCBs included a so-called bus board, a sensor board and a communication board. The PCBs have individual slots for different hardware (for instance, sensors, SD card module, data modem and optional GPS module).

Data transmissions and regular syncs of the external Adafruit DS3231 RTC are realized through a GPy board by Pycom based on the Espressif ESP32 SoC (System on Chip) [28]. The external real-time clock is used to guarantee that the system time of an AELCM unit is still precise, even if there is poor network connectivity. Next to location tracking, time synchronizations in countries with a lack of IoT infrastructure could be realized through the optional GPS module (Adafruit Mini GPS PA1010D). The GPy is a triple-bearer, offering connectivity in form of WiFi, Bluetooth and cellular LTE-CAT-M1/NB1. We are using long-term evolution for machines (LTE-M) to enable the AELCM box to gain a connection to the internet, which requires an IoT Nano–SIM. We are employing an IoT SIM card by the carrier 1NCE. Likewise, for transmitting data successfully we needed a communication device that supports modern security standards, such as TLS version 1.2 (Transport Layer Security). In addition, a minimal peak current consumption during transmissions was needed. Both requirements are features of the GPy. Two uninterruptable power supplies (Adafruit PowerBoost 1000C) provide current of up to 1 ampere at a voltage between 5 and 5.2 volts (V) for the GPy and the rest of the measurement system via the mains or batteries separately [29]. The separation exists to avoid a possible instability of an AELCM unit that is due to voltage drops caused by potential high current consumption greater than 1 ampere during data transmissions. Moreover, it provides more flexibility with respect to the operation of metal oxide sensors, which consume vastly more current by design because they use heaters in contrast to electrochemical gas sensors. The integrated LCD is used for more comfortable deployment in the field, confirming if the boot up of an AELCM unit is successful or if hardware components are failing.

The housing, shown in Figure 1b, is a modified ABS plastic box, which is actively ventilated through a fan to probe the ambient air. The fan functions as an exhaust fan, removing air from inside the box. The resulting air flow direction is toward the bottom left of the box. The fan guarantees a constant air flow over the gas sensors and the meteorological sensor, which are described in Table 1.

The revolutions per minute (RPM) of the fan blades are 2000 RPM, causing a flow velocity between 0.8 and 0.9 m/s at the end of the bottom exhaust pipe of an AELCM box [39]. 3D printing was an essential tool for customizing the box and securing the used hardware.

The gas sensors in Table 1 can be separated into two categories: electrochemical gas sensors (SPEC DGS) and metal oxide gas sensors (MiCS-2714, MiCS-4514, MQ131). Electrochemical gas sensors generate a measurable current while in contact with their target gas [40]. The amount of measured current depends on the concentration of the target gas. This relationship is utilized to estimate gas concentrations. Metal oxide sensors on the other hand feature a semiconductor layer. The conductivity of this layer changes with the fractions of the gases in the air to which a metal oxide type gas sensor is sensitive [16]. We can measure this change and translate it into estimations of gas concentrations of a specific gas. For the measurement of PM, an optical particle counter (OPC) was used, called SPS30. Through the interaction of laser beams and particles, resulting in scattered light inside a measurement cell collected by a photodetector, the mass concentration of suspended matter with diameters smaller than or equal to specific aerodynamic diameters (PM2.5 and PM1) can be estimated [38].

(**a**) (**b**)

Figure 1. Photographs of the AEMS and an AELCM unit, which is mounted on the fence next to the AEMS: (**a**) the stationary air and climate measurement station of the Chair for Regional Climate Change and Health, Faculty of Medicine, University of Augsburg; and (**b**) the housing of an AELCM unit consists of a weather-proof NEMA (National Electrical Manufacturers Association) enclosure (ABS Plastic Case, PN: NBB-22251; L × W × H: 37 cm × 27 cm × 15 cm). The unit includes a sensor board (equipped with gas sensors and a meteorological sensor) connected to a bus board, which in turn is connected to an Arduino Mega 2560 Rev3 and an external RTC (A). The communication board is equipped with an antenna-equipped Pycom GPy and an SD card module (B). An uninterruptable power supply for the communication device (GPy) and a second one for the measurement system and its external devices (LCD, fan and SD card module) are each connected to a LiPo battery (C). A liquid crystal display (LCD) shows the status and functionality of the AELCM unit (D). The DC PWM fan inside the enclosure (ODROID-H2 DC Fan; L × H × D: 92 mm × 92 mm × 25 mm) makes use of the custom air inlet (top right) and custom air outlet (bottom left) for the exchange of air (E). The SPS30 located at the bottom of the box directly coupled to the ambient air through a short plastic pipe (F). Both uninterruptable power supplies are connected to a two-port USB charger plugged into a regular power socket (G).

Table 1. Overview of the specifications of the sensors that can be used in the AELCM unit.

Measured Variable	Sensor	Manufacturer	Measuring Range	Accuracy * (Repeatability) * [Precision] *	Approx. Price (Euro) 2020
Temperature Humidity	BME280 [30]	Bosch	−40–65 °C 0–100%	±0.5–±1.5 °C ±3% [1]	5
O_3	MQ131 [31]	Winsen	0.01–1 ppm	/	20
NO_2	MiCS-2714 [32]	SGX Sensortech	0.05–10 ppm	/	10
CO	MiCS-4514 [33]	SGX Sensortech	1–1000 ppm	/	14
O_3	DGS-O3 968-042 [34]	SPEC Sensors	0–5 ppm	±15% (<±3%)	80
NO_2	DGS-NO2 968-043 [35]	SPEC Sensors	0–5 ppm	±15% (<±3%)	80
CO	DGS-CO 968-034 [36]	SPEC Sensors	0–1000 ppm	±15% (<±3%)	80
$PM_{1/2.5}$	SPS30 [37,38]	Sensirion	0–1000 μg/m^3	[±10 μg/m^3 at 0 to 100 μg/m^3] [±10% at 100 to 1000 μg/m^3]	32

* This information is given if the manufacturer's sensor data sheet provided it. [1] 20–80%RH, 25 °C, including hysteresis.

The measurement of the meteorological parameters of relative humidity and temperature are relevant for analyzing LCS output. The neglect of meteorological conditions can lead to a sensor output and in consequence to model-estimated concentration levels, which can be questionable as mentioned in the introduction.

In our configuration, the measurements of these two variables taken by a BME280 are an integral part of the model-based output adjustments of the sensors embedded in an AELCM unit. The possible influence of these parameters on the sensor outputs given by electrochemical- and metal oxide-type sensors are also evident in the official documents of manufacturers who produce these types of sensors [31,40].

The Arduino board, a device with a 5 V logic level, was interfaced with the digital gas sensors (DGS) by SPEC sensors using UART hardware serial ports and logic-level converters, given that these sensors are devices operating on 3.3 V. These units give gas concentrations in parts per billion (ppb) and were rarely analyzed in regards of their performance in reports or other scientific literature in the past compared to the electrochemical sensors of the manufacturer Alphasense (e.g., B4-series and A4-series), which are up-to-date, common sights in experiments or low-cost sensor systems for gaseous air pollutants [22,23,41]. Therefore, we have decided to evaluate the performance of these sensors as well. The communication between the SPS30, BME280, Adafruit ADS1115, Adafruit DS3231 and Arduino board was realized by I2C. The analogue sensors MQ131, MiCS-2714 and MiCS-4514 were connected to the Adafruit ADS1115 analog-to-digital converter (ADC) to read their outputs.

AELCM units were collocated with the air and climate measurement station of the Chair of Regional Climate Change and Health next to the University Hospital Augsburg in an urban background setting shown in Figure 1a named AEMS (location: 48° 23.04′ N, 10° 50.53′ E; Germany). The measurement station and its instruments are provided by the company Horiba (APOA-370, principle: non-dispersive ultra-violet-absorption method (NDUV), APNA-370, principle: chemiluminescence method (CLD), APMA-370, principle: non-dispersive infrared (NDIR) absorption method, APDA-372, principle: optical light-scattering). The AEMS provides measurements of the air constituents O_3, nitrogen oxides (NO, NO_2 and NO_x), CO and fine dust (PM1, PM2.5, PM4 and PM10). Additionally, the AEMS is equipped with a Lufft WS600-UMB, which measures the meteorological variables air temperature, relative humidity, precipitation intensity, precipitation type, precipitation amount, air pressure, wind direction and wind speed.

An internal datalogger called the μIO-Expander averages the gas concentration measurements to 3 min means, while the concentrations of PM fractions are rolling means, averaged over 15 min time intervals by the datalogger. Measurements of an AELCM unit take place every 10 s and get saved on an SD card inside an SD card module.

The Arduino board averages the collected data to 3 min means and sends them to the GPy via a software serial port established on the Arduino board. This step was necessary because all hardware serial ports (UART) are occupied by digital gas sensors. The GPy sends the averaged data in real time to our database, being embedded in the server infrastructure of the University of Augsburg.

2.2. Data and Data Treatment

We deployed three AELCM units with slightly different sensor configurations and different deployment dates, which are summarized in Table 2. The general analysis period ended on 24 October 2021. The longest time series is available for the unit AELCM003, which was deployed in late February 2021. The units AELCM004 and AELCM005 were deployed in June 2021. The different deployment dates are a result of the development of a custom surface-mounted device (SMD) socket for the MiCS-2714 and MiCS-4514, which was not completed by the end of February 2021 (Figures S1.1 and S1.2). A custom SMD socket was developed to avoid manually soldering the MiCS devices to the sensor board, possibly damaging them in the process.

Table 2. General information about the deployed AELCM units' individual LCS configurations including the amount of available data for the first atmospheric exposure monitoring network (AEMN) experiment for every deployed AELCM unit since their individual deployment dates.

No. AELCM Unit	Deployment Date	Missing Sensors *	Available Data Logger/Database
003	26 February 2021	MiCS-2714 MiCS-4514	100%/95.56%
004	4 June 2021 21 June 2021	DGS-CO	100%/94.95%
005	4 June 2021	DGS-CO	n. A./96.91%

* The standard sensor equipment of the first version of an AELCM unit is shown in Table 1. Therefore, missing sensors relate to the introduced sensors in Table 1.

Eventually, we decided to deploy the first AELCM box called AELCM003 without MiCS sensors. Originally, the box AELCM004 was deployed at the same date as the unit AELCM005, but we had to readjust the SMD sockets for the MiCS sensors multiple times for this box, so that the official measurement start was 21st of June for the AELCM004 unit.

The issues with the current MiCS socket made us redesign it. The latest MiCS socket board could not be implemented in the first version of the AELCM units anymore because it was not compatible with the current sensor boards in the already deployed units (Figures S1.3 and S1.4). We also decided not to employ any SPEC DGS-CO devices in later units anymore after looking into its initial performance in the field experiment. The reasoning behind this decision is given in Section 3.

Overall, saving the measurement data with the microSD breakout by the manufacturer SparkFun on a SD memory card worked well. We could retrieve all the measurement data for the AELCM units 003 and 004. Unfortunately, the SD memory card of AELCM005 stopped functioning, so the raw data for this measurement system were not available (n. A.). Nevertheless, we could still use the 3 min averages saved to our database.

During the field experiment, the GPy boards were able to send most of the averaged measurement data to our database. Approximately 95% or more of the averaged data were successfully sent. The missing averages in the database can be explained with ongoing network issues on the carrier side or with a rare problem, where the GPy is losing its internet connection and cannot recover solely with a board reset. Only a manual power cycle resolved this issue. We recommend integrating a metal–oxide–semiconductor field-effect transistor (MOSFET) controlled by a microcontroller or microcontroller board, such as in our case an Arduino board, to power-cycle the GPy automatically in those cases.

While called SPEC DGS-O3, this EC sensor has a 1:1 sensitivity to NO_2; hence, it must be deployed with a NO_2 sensor to estimate the concentration of O_3 correctly [34]. Thus, the AEMS measurements for NO_2 were subtracted from the measurements of the DGS-O3 units. Most of the data sheets of the different employed gas sensors recommend a warm-up time before actually using the sensor output for estimates. The digital gas sensors by SPEC sensors show a startup output profile after getting powered. The length of the output stabilization process may depend on the length of time the sensor was unpowered or the sensor type [34–36]. The observed stabilization periods of the digital outputs were different between the DGS used in the field experiment, but in general were shorter than a day. On that account, the first 24 h of measurements of the SPEC DGS devices were not included in this analysis. Another important detail is that the digital gas sensors and even the AEMS gas measurement devices show a slight degree of noise in their measurements. This is a normal feature inherent to their technical design and becomes most evident in negative gas-concentration measurements (DGS, AEMS). Consequently, negative hourly means of gas concentrations were flagged and set to zero ppb.

The noise levels for the AEMS gas measurement devices have been kept between ± 0.5 ppb (APOA, APNA) and ± 10 ppb (APMA) through recalibrations. Subtracting the AEMS NO_2 levels from the output of the DGS-O3 could also result in negative ppb concentrations, showing that there are limitations in sensing lower levels of O_3 and/or NO_2

with the DGS-O3. A warm-up phase also exists for the MQ131 according to its data sheet. Because we do not know how long a sensor was not operated after being manufactured and stored at a sensor distributor's warehouse, we decided to exclude a week of sensor data from the MQ131 before evaluating its actual output. This amount of removed data is related to a storage time of more than six months for the MQ131 [31]. The data sheets for the MiCS sensors do not provide any information about a needed warm-up duration. Consequently, we also removed the first week of measurements of the MiCS sensors. Our decision is based on the data sheet of the MQ131 (another metal oxide sensor), which was described earlier. All low-cost gas sensor data were averaged to hourly means, while the sensor data of the low-cost PM sensor SPS30 were averaged to 15 min means. Since the SD memory card data were not available for the unit AELCM005, we used the transmitted 3 min averages for this unit. For this measurement system, hourly means and 15 min means were only calculated, when all data were available for the averaging process, otherwise an hourly mean or 15 min mean was flagged as a missing value. Furthermore, we limited the evaluation of the SPS30 to its output for the fractions PM1 and PM2.5 because of its technical limitation to provide proper estimates for the fractions PM4 and PM10 [38]. We also removed the first 24 h of measurements for the SPS30 after observing on multiple occasions that there is an initial stabilization period (the PM2.5, PM4 and PM10 measurements become identical after some time) after powering this device.

The 3 min averages for the gaseous air pollutants given by the reference measurement station AEMS were averaged to hourly means. Because the concentrations of the PM fractions are provided as rolling averages by the AEMS over a time span of 15 min, we extracted the data at the minutes 0, 15, 30 and 45 for every hour (e.g., 14:00, 14:15, 14:30 and 14:45 CET) and used these as a reference basis for the evaluation of the SPS30. The system time of the AEMS is CET, while the system time of the AELCM units is UTC. Subsequently, we adjusted the time stamps of the AELCM measurement data to CET. Portions of the reference data had to be excluded or were not available. The reason is, on the one hand, regular maintenance work involving the power grid of the University Hospital, where no power is available to operate the measurement station. On the other hand, the AEMS must be regularly maintained and checked (e.g., feeding of reference gas or reference dust, recalibration of measurement devices and filter exchange) to guarantee a reliable operation and high-quality data output.

2.3. *Methods*

2.3.1. Calibration

We built multiple linear regression models (MLR) for the LCS to calibrate the hourly means of the gaseous sensor data and 15 min means of PM sensor data. We did not expect that adding every possible predictor in our MLR models would result in a significant improvement of the model, so we selected a reasonable set of predictors using the following steps: (1) using a Spearman rank correlation to obtain a first general impression about the strength of the relationship between low-cost sensor output and reference measurements; (2) building MLR models (calibration functions) for each sensor based on their data output considering environmental influences on sensor output and reference measurements (AEMS); and (3) predictor selection for the final regression models for every individual LCS deployed based on the found models in step (2). The last step is realized through a stepwise regression involving a sequential replacement algorithm and an out-of-sample (OOS) approach using the RMSE as an evaluation parameter. The sequential replacement algorithm was provided by the package leaps in statistics software R [42]. The final regression models are introduced in Section 2.3.4 for every LCS. For the development of the MLR models in step (2), we start with a common set of predictors for each model, which includes the sensor data output, atmospheric humidity and air temperature. This can be regarded as the simplest model realization (baseline model, see Equation (3)). We adjust this baseline model where necessary, considering the usual MLR model assumptions, including the inspection of the residuals.

It must be mentioned that a common problem of a linear regression on atmospheric time-series data with a high temporal resolution is also apparent in this analysis. A standard assumption for building an MLR model is independent residuals, which is violated in this work. Additionally, multicollinearity (e.g., between air temperature and relative humidity) is another common issue within the subject of modelling air pollution based on LCS data. Air temperature also has a positive effect on the buildup of O_3 near the surface. Nevertheless, we are using MLR models because MLR is still the most common basic approach for building calibration functions to process LCS data given by new measurement systems such as our AELCM unit before developing or approaching more sophisticated calibration models in future works [23,41,43]. We thoroughly evaluate every final regression model and estimate the RMSE and coefficient of determination as recommended in other literature [23].

2.3.2. Evaluation Statistics

To obtain a first guess about the (monotonic) relationship between the raw *LCS* output and the reference data given by the *AEMS* and the meteorological parameters, we use the Spearman rank correlation [44]. The Spearman rank correlation is defined as:

$$r_{rank} = \frac{Cov(R(LCS_{Raw}), R(AEMS_{ref}))}{\sigma_{R(LCS_{Raw})}\sigma_{R(AEMS_{ref})}}, \tag{1}$$

given by the standardized covariance of the ranked data values of the raw *LCS* measurements $(R(LCS_{Raw}))$ and ranked observations of the *AEMS* $(R(AEMS_{ref}))$. The standard deviations that belong to these ranked data values are $\sigma_{R(LCS_{Raw})}$ and $\sigma_{R(AEMS_{ref})}$, respectively.

For evaluating the calibration functions, we use two measures of the model fit: coefficient of determination (R^2) and the root-mean-square error (*RMSE*) [44]. R^2 can be understood as a measure of the proportion of variation of the predictand accounted for or described by a regression model. The mean squared error (*MSE*) is the arithmetic average of the squared difference between predictions and observations, reflecting the average forecast accuracy. The *MSE* is defined as:

$$MSE = \frac{1}{N}\sum_{i=1}^{N}\left(LCS_{adj_i} - AEMS_{ref_i}\right)^2, \tag{2}$$

where LCS_{adj_i} and $AEMS_{ref_i}$ describe the ith pair of N pairs of model-adjusted *LCS* measurements and observations of the *AEMS*. The *MSE* is particularly sensitive to outliers caused by the squaring of errors. We use the *RMSE*, which can be expressed as the square root of the *MSE*, $RMSE = \sqrt{MSE}$, to describe the error between calibration (adjusted) data and reference data.

2.3.3. Stepwise Regression

For a given air pollutant LCS, we are building linear regression models (MLR) as calibration functions for the LCS output. Hence, we are using the raw LCS output and the meteorological measurements of air temperature and relative humidity as predictors and the AEMS output as the predictand. We are screening the group of predictor candidates using a stepwise regression method to decide upon which and how many predictors to use. A stepwise regression is an automated filtering procedure that follows systematic rules in adding or removing variables with predictive power from a regression model according to a selection criterion [44,45]. After every regression step (final found p predictor model), an out-of-sample (OOS) procedure is performed.

The autocorrelation functions for the hourly and 15 min means of the AEMS data show an autocorrelation pattern for every pollutant (Figure S2.1). This indicates that the observational data at the analyzed time scales were dependent. Cerqueira et al. show that an OOS approach is an appropriate choice for model validation under this circumstance [46].

For this reason, we rely on the OOS approach to evaluate the models, considering the apparent dependence structure of the data.

Summarizing this method, which is described in detail in [46], a random point *t* in time (e.g., 10 September 2021 12:00:00 CET) of the time series *ts* is chosen to separate the training and evaluation data. The previous window with reference to *t* comprising 60% of *ts* is used for training and the following window of 10% of *ts* is used for testing. For 10 repetitions, we receive 10 randomly chosen dates *t*, which separate the training and evaluation sets. The sizes of the training and evaluation sets depend on the length of the available LCS time series and reference data (see Table 2 and Section 2.2). Finally, considering the average, minimum and maximum RMSE for the training and evaluation period for every *p* predictor model, we have chosen the final regression equation for a LCS (S3). In the case different predictive variables for a *p* predictor model were chosen between repetitions, the most chosen predictor combination defined the final *p* predictor regression model. In addition, results of a fivefold cross validation (CV) are included in the supplement of this study because of its common use in model evaluation (S4). The conclusions drawn from the CV approach are similar to the findings for the OOS approach in this study.

2.3.4. Regression Models

First, we introduce the baseline linear regression model:

$$LCS_{adj} = b_0 + b_1 LCS_{Raw} + b_2 T + b_3 RH, \tag{3}$$

where LCS_{Raw} is the raw output of the low-cost pollutant sensor, while T and RH are temperature and relative humidity, respectively, given by the meteorological low-cost sensor inside the AELCM unit. LCS_{adj} is the model-predicted low-cost sensor output. Transformations of predictors and/or predictands are the result of a lack of homoscedasticity and/or Gaussian distributed residuals. Based on the calibration strategy introduced in Section 2.3.1, we build the following models, where:

$$[DGS\text{-}O3]_{adj} = b_0 + b_1[DGS\text{-}O3]_{Raw} + b_2\frac{[DGS\text{-}O3]^2_{Raw} - 1}{2} + b_3 T + b_4 RH, \tag{4}$$

is the final calibration function for all deployed *DGS-O3* units. $DGS\text{-}O3_{Raw}$ is corrected by $AEMS_{ref}$ for NO_2 considering that the *DGS-O3* has a 1:1 sensitivity to O_3 and NO_2 ($DGS\text{-}O3_{Ox}$), so that:

$$[DGS\text{-}O3]_{Raw} = [DGS\text{-}O3]_{Ox} - [AEMS\text{-}NO2]_{ref}, \tag{5}$$

while for the *DGS-NO2*, we used the following regression models:

$$\log\left([DGS\text{-}NO2]_{adj}\right) = b_0 + b_1[DGS\text{-}NO2]_{Raw} + b_2 T + b_3 RH, \tag{6}$$

$$[DGS\text{-}NO2]_{adj} = b_0 + b_1[DGS\text{-}NO2]_{Raw} + b_2 T + b_3 RH, \tag{7}$$

where the former calibration function was used for unit AELCM003 and the latter one for the other deployed AELCM units. For the deployed *DGS-CO* device the regression model

$$[DGS\text{-}CO]_{adj} = b_0 + b_1[DGS\text{-}CO]_{Raw} + b_2 T + b_3 RH, \tag{8}$$

was chosen. For all *MQ131*, *MiCS-2714* and *MiCS-4514* sensors we used the following calibration functions:

$$[MQ131]_{adj} = b_0 + b_1[MQ131]_{Raw} + b_2 T + b_3 RH, \tag{9}$$

$$[MiCS\text{-}2714]_{adj} = b_0 + b_1[MiCS\text{-}2714]_{Raw} + b_2 T + b_3 RH, \tag{10}$$

$$\log\left([MiCS\text{-}4514]_{adj}\right) = b_0 + b_1[MiCS\text{-}4514]_{Raw} + b_2T + b_3RH. \quad (11)$$

For adjusting the measured concentrations of a *SPS30* for the fractions PM1 and PM2.5, we used the regression model:

$$\log\left([SPS30]_{adj}\right) = b_0 + b_1\log([SPS30]_{Raw}) + b_2T + b_3RH, \quad (12)$$

for the unit AELCM003, while using the calibration function:

$$\log\left([SPS30]_{adj}\right) = b_0 + b_1\log([SPS30]_{Raw}) + b_2RH, \quad (13)$$

for the other units.

3. Results and Discussion

The environmental conditions and pollution concentrations based on hourly (15 min) means are provided in Table 3. Erroneous measurements of the ambient air temperature and relative humidity were removed. At least 85% of the data of both variables had to be available for the calculation of the mean values.

Table 3. Statistics based on the hourly (15 min) means of the different atmospheric variables measured by the AEMS between the 27 February 2021 and 24 October 2021. For O_3, NO_2 and CO hourly gas [ppb] and for PM 15 min concentration means [μg/m^3] were used. The temperature [°C] and relative humidity [%] statistics are based on hourly means.

Atmospheric Variable	Min	25th Percentile	Mean	75th Percentile	Max
O_3	0.00	16.79	27.79	38.01	85.65
NO_2	0.38	2.59	6.24	8.39	36.99
CO	93.44	141.55	176.89	196.69	1366.96
PM1	0.20	3.21	7.46	10.36	44.23
PM2.5	0.32	4.14	8.72	11.91	136.51
Temperature	−4.27	8.21	12.94	17.46	31.68
Relative Humidity	17.94	58.54	71.15	86.22	95.46

In our work, not every LCS shows the premise of being a good-quality source of information according to the employed evaluation parameters. It was evident that some LCS could not reflect the patterns in the reference data well or at all, making them factually useless. Table 4 gives a general overview about the suitability of the deployed LCS in building a low-cost monitoring network for air constituents.

Generally, the results of our field experiment indicate that only the electrochemical sensor SPEC DGS-O3 and the PM sensor SPS30 are reasonable choices for LCS used in a low-cost atmospheric exposure monitoring network.

The other electrochemical sensors, DGS-NO2 and DGS-CO, showed no capability to measure the given concentrations at the measurement station according to the coefficient of determination and the Spearman rank correlation. The correlations were close to zero for two DGS-NO2 devices. For the longest running DGS-NO2, the Spearman rank correlation only amounted to 0.18. The low predictive power of these NO_2 sensors is also reflected in the coefficients of determination (R^2: 0.28–0.59). Since they are not able to sense their target gas at all or only in a very weak way, they cannot describe the variability of ambient NO_2 to a satisfying degree. The DGS-CO even showed a negative correlation (Rs: −0.25), implying that it only produced noise. Thus, we have decided not to employ any SPEC DGS-CO devices in later AELCM units anymore after looking into their initial performance in the field experiment and testing different units, which all showed the same behavior as the SPEC DGS-CO operated in the field experiment. During a field experiment done by the Air Quality Sensor Performance Evaluation Center (AQ-SPEC), out of the DGS-CO,

DGS-NO2 and DGS-O3 measurement units, only for the DGS-CO was there evidence that the LCS is able to sense ambient CO [47]. The concentration levels of ambient CO at the University Hospital Augsburg were, overall, lower in our field experiment compared to the measured concentrations in the field experiment of AQ-SPEC. This implies that the DGS-CO (at least the ones purchased from sensor distributors) is not useable for sensing lower concentrations of CO (Figure S6.2).

Table 4. Evaluation statistics for every LCS used in the AELCM units: Spearman rank correlation (Rs), root-mean-square error (RMSE) and the coefficient of determination (R^2). For the calculation of R^2 and RMSE, all model-adjusted LCS data (hourly gas [ppb] and 15 min PM concentration means [μg/m^3]), while for Rs all raw LCS data for every individual sensor are used (excluding the data taken during the warm-up duration). The AEMS data are used as reference. For all LCS data, besides the SPS30, the RMSE is in ppb. For SPS30 data, the RMSE is in μg/m^3.

No. AELCM Unit	DGS-O3	DGS-NO2	DGS-CO	MQ131 (O_3)	MiCS-2714 (NO_2)	MiCS-4514 (CO)	SPS30 (PM1/PM2.5)
003	Rs: 0.90	Rs: 0.18	Rs: −0.25	Rs: −0.55	/	/	Rs: 0.97/0.94
	RMSE: 6.74	RMSE: 3.47	RMSE: 53.67	RMSE: 6.53	/	/	RMSE: 1.07/1.96
	R^2: 0.80	R^2: 0.59	R^2: 0.20	R^2: 0.81	/	/	R^2: 0.96/0.90
004	Rs: 0.50	Rs: −0.02	/	Rs: −0.26	Rs: 0.28	Rs: 0.39	Rs: 0.97/0.95
	RMSE: 7.79	RMSE: 3.61	/	RMSE: 7.81	RMSE: 4.16	RMSE: 30.53	RMSE: 0.77/1.27
	R^2: 0.71	R^2: 0.35	/	R^2: 0.71	R^2: 0.15	R^2: 0.66	R^2: 0.97/0.94
005	Rs: 0.97	Rs: −0.01	/	Rs: −0.28	Rs: 0.53	Rs: −0.49	Rs: 0.96/0.94
	RMSE: 3.31	RMSE: 3.80	/	RMSE: 6.31	RMSE: 3.51	RMSE: 44.95	RMSE: 0.87/1.51
	R^2: 0.95	R^2: 0.28	/	R^2: 0.83	R^2: 0.40	R^2: 0.27	R^2: 0.96/0.91

In contrast to the results of the AQ-SPEC field evaluation experiment, we found that most of the deployed DGS-O3 were able to detect changes in O_3 concentration levels, albeit with differences, as can be seen in the evaluation statistics in Table 4. The raw output of the DGS-O3 deployed in AELCM005 shows the strongest relationship with the reference measurements (Rs: 0.97) and the smallest error (RMSE: 3.31 ppb), and provides the highest coefficient of determination (R^2: 0.95). As a result, it is evident that the DGS-O3 deployed with AELCM005 shows the best level of agreement with the reference measurements, while the other DGS-O3 units show a lesser performance (RMSE: 6.74–7.79 ppb; R^2: 0.71–0.80). Therefore, we conclude that every DGS-O3 (or any other promising LCS) must be individually screened (evaluated) and calibrated before considering them for use in any monitoring application, since even identical sensors show different characteristics in response to the same environmental conditions, such as temperature, humidity and ambient O_3 concentrations (Figure S7.1; Table 3).

Figures 2 and 3 show the model-adjusted and the raw and model-adjusted hourly means of measured O_3 concentrations in ppb given by SPEC DGS-O3 units deployed with the low-cost monitoring systems AELCM003, AELCM004 and AELCM005 plotted against the hourly means of the reference measurements of ambient O_3 given by the AEMS. Additionally, the time series data in Figure 2 were smoothed using rolling 24 h averages to make deviations from the reference data more detectable. The figures for the other sensors in Supplement S6 are presented in the same manner. The AELCM boxes were deployed at different points in time because of the reasons mentioned in Section 2.2. Generally, box AELCM003 offers the longest time-series data for most of the LCS. The top panel in Figure 2 belongs to AELCM003 and indicates that the found model for the DGS-O3 describes the variability in the reference measurements reasonably well. On the other hand, quite large discrepancies between the model-adjusted LCS output and reference measurements can also be found (top panels in Figures 2 and 3), resulting in a rather high error expressed through an RMSE of 6.74 ppb. The calculated average of the RMSE for the repeated holdout procedure (10 evaluation periods) is also high and amounts to 8.21 ppb (Table S3.1). Therefore, we doubt that quantitative predictions for this DGS-O3 are possible with the current model approach. Qualitative predictions in the form of an estimation of trends in the O_3 concentration levels could still be possible with the current model approach but must be investigated in future works. The DGS-O3 deployed with AELCM004 was the worst performing sensor out of the three DGS-O3 considering the reference measurements, which is evident from the deviations between the rolling 24 h averages in the middle panel of Figure 2. The higher deviations compared to the DGS-O3 deployed in AELCM003 and AELCM005 are also shown through the RMSE statistic, which is 7.79 ppb. Considering the strength of the deviations and the only moderate relationship with the reference measurements (Rs: 0.50), a qualitative and quantitative use of the DGS-O3 deployed with AELCM004 with the current model is questionable. This is also evident from the scatterplot in the middle panel of Figure 3. Looking at the bottom panels of Figures 2 and 3, a far more positive outlook for this LCS is provided through the results for the DGS-O3 of AELCM005. The DGS-O3 deployed in AELCM005 exhibits the strongest relationship (Rs: 0.97; R^2: 0.95) and the smallest deviation from the reference measurements (RMSE: 3.31 ppb). In addition, the calculated root-mean-square errors for the predictions indicate the robustness of the final regression model (Table S3.1). The model robustness is evident by comparing the determined average, minimum and maximum RMSE for the training periods with their corresponding counterparts of the evaluation periods, which produce quite similar errors. For the selected training periods, the average RMSE is 3.09 ppb, while the individual root-mean-square errors lie between 2.49 and 3.65 ppb for the 10 training periods. For the selected evaluation periods, the average RMSE is 2.67 ppb, while the individual root-mean-square errors lie between 2.42 and 3.09 ppb for the 10 evaluation periods (Table S3.1). We would consider the DGS-O3 deployed with AELCM005 as the most promising LCS for realizing reasonable predictions of ambient O_3 concentration using an MLR model.

Figure 3 shows clearly how differently identical LCS units can behave under the same environmental conditions and pollution concentrations. While the bottom panel shows a strong linear relationship between the raw LCS data and the reference measurements, the top panel suggests a strong nonlinear relationship with the reference. The middle panel indicates highly noisy raw LCS data of little quality. The problem of inter-sensor unit variability is thus apparent.

After discussing the electrochemical sensors, we are looking with more detail at the deployed metal oxide sensors.

While for the model-adjusted MQ131 sensor output moderate to high coefficients of determination were determined (R^2: 0.71–0.83), the raw sensor output only shows a weak-to-moderate relationship with the reference measurements (Rs: -0.55–-0.28).

A negative relationship exists because the sensor resistance increases with higher concentrations of O_3 [31]. The only moderate relationship for the MQ131 with its target gas reference measurements is also indicated in the stepwise selection process of predictor variables during the initial model-building process. During the model-training process, none of the MQ131 sensors could be selected as best 1-predictor model in our stepwise regression. Thus, a meteorological variable or two meteorological variables, in this case relative humidity or/and temperature, yielded a better performance than the sensor output itself. As a result, for most training periods, relative humidity was chosen as the best variable with the most predictive power in a one-predictor model. The stronger relationship between meteorological variables and ambient O_3 is also reflected in the Spearman rank correlation between LCS measurements and reference measurements (relative humidity Rs: -0.76–-0.80; temperature Rs: 0.47–0.74; Table S5.1). Based on our analyses, we question the usefulness of the MQ131 in a low-cost monitoring network.

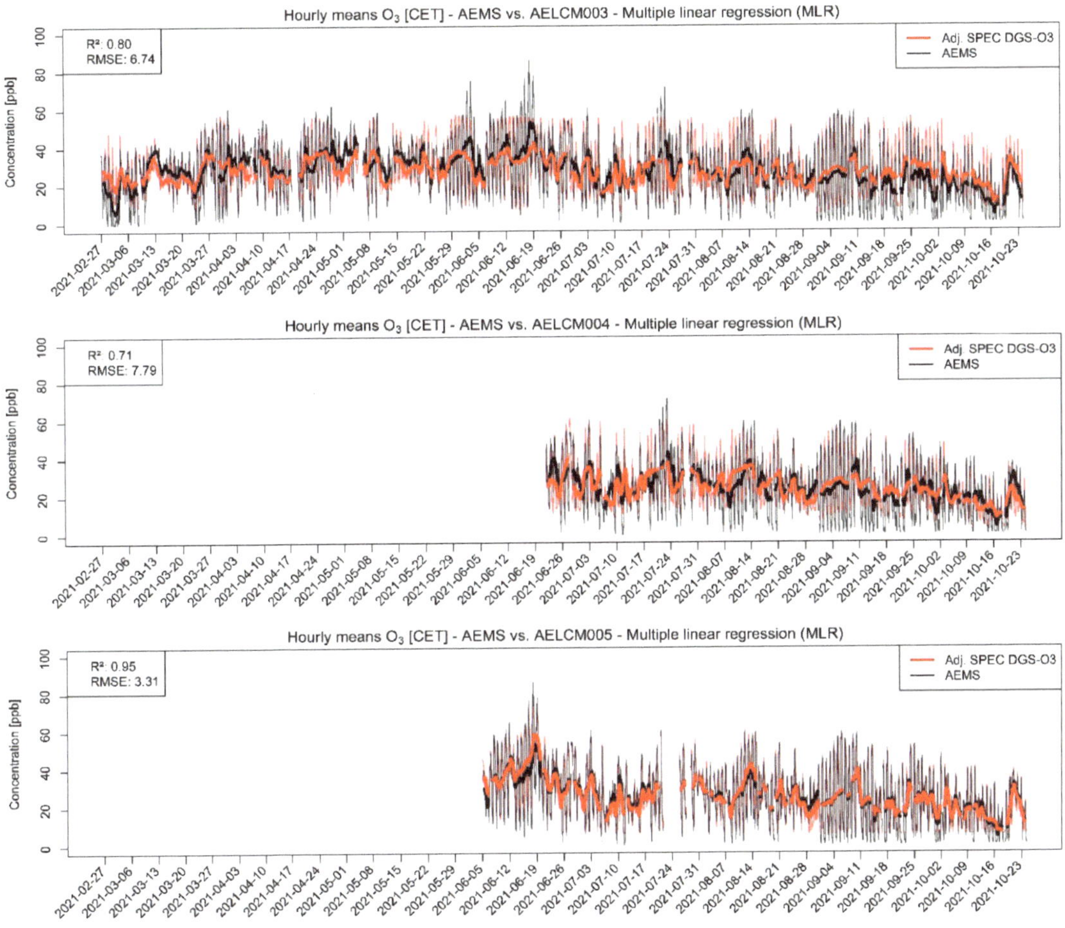

Figure 2. Model-adjusted hourly concentration means of different deployed SPEC DGS-O3 using multiple linear regression vs. hourly concentration means of reference measurements given by the AEMS for ambient O_3 in ppb. The RMSE is in ppb. For smoothing the hourly concentration means, a rolling 24 h average is used: (**Top**) SPEC DGS-O3 deployed in AELCM003; (**Middle**) SPEC DGS-O3 deployed in AELCM004; and (**Bottom**) SPEC DGS-O3 deployed in AELCM005.

We draw the same conclusion for the MiCS LCS. Neither the MiCS-2714 nor the MiCS-4514 showed promising results in our field experiment considering the evaluation parameters shown in Table 4. After gaining a first impression about the usefulness of these sensors (MiCS-2714 Rs: 0.28–0.53; MiCS-4514 Rs: −0.49–0.39), their sensor output as a predictor for ambient NO_2 and ambient CO is rather questionable. This is also reflected in the quality of the model-adjusted output of these sensors. The coefficients of determination are quite low (MiCS-2714 R^2: 0.15–0.40; MiCS-4514 R^2: 0.27–0.66). On the one hand, the reason could be that the overall concentration levels during the measurement campaign were too low for the sensors to sense their target gases. According to their data sheets, the MiCS-2714 has a detection range between 0.05 and 10 ppm, while the MiCS-4514 has a detection range between 1 and 1000 ppm. An ambient hourly mean of NO_2 of 50 ppb was never reached during the measurement period of these sensors.

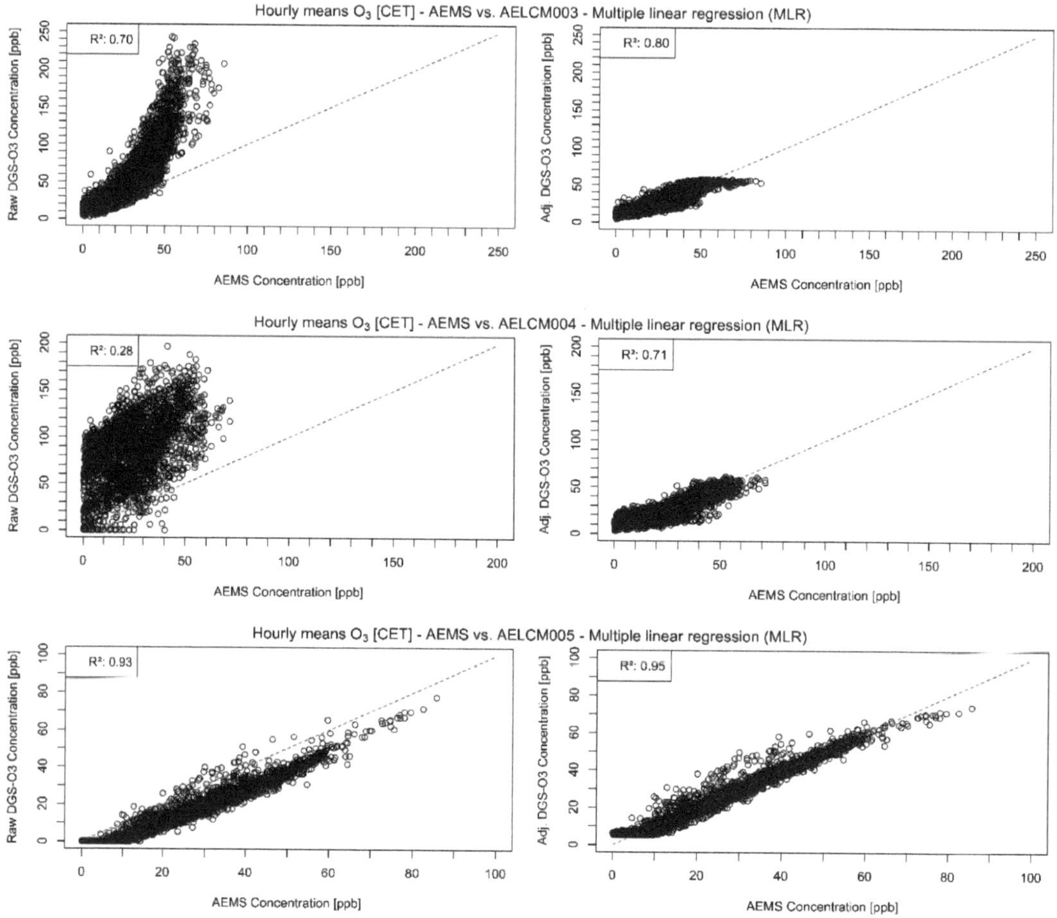

Figure 3. Raw (**left**) and model-adjusted (**right**) hourly concentration means of different deployed SPEC DGS-O3 vs. hourly concentration means of reference measurements given by the AEMS for ambient O_3 in ppb: (**Top**) SPEC DGS-O3 deployed in AELCM003; (**Middle**) SPEC DGS-O3 deployed in AELCM004; and (**Bottom**) SPEC DGS-O3 deployed in AELCM005. Multiple linear regression is used for adjusting the raw LCS data.

For the analysis periods of the deployed MiCS-4514, no hourly mean of ambient CO higher than 1 ppm was reached. Another reason could be that our former custom-built, initial prototype SMD socket for the MiCS sensors worked insufficiently, so the measurements were taken under unfavorable circumstances. The issues could be potentially resolved by our latest version, which needs to be investigated by another field experiment (Figures S1.3 and S1.4).

Considering the evaluation statistics given above for the deployed LCS, the metal oxide gas sensors performed the worst in our field experiment; therefore, we do not consider using them for future AELCM units.

Because of the poor performance of the used electrochemical and metal oxide NO_2 LCS, we used the NO_2 reference measurements of the AEMS to correct the raw measurements of the SPEC DGS-O3 units (see Section 2.2) before applying any calibration function.

In our analyses, the PM sensor SPS30 was the best performing LCS. Each of the SPS30 devices showed for PM with aerodynamic diameters smaller than or equal to 2.5 and 1 μm strong agreements with the reference measurements, indicated through the evaluation statistics in Table 4 and in the stepwise regression results (Tables S3.7 and S3.8). The raw LCS data show a strong relationship with the reference measurements of the AEMS for both PM classes, but also an increasing spread with increasing reference measurements (Figures 4 and 5). For PM2.5, the Spearman rank correlation lies between 0.94 and 0.95 and for PM1 between 0.96 and 0.97. The high level of agreement is also visible in the model-adjusted LCS output shown in Figures 6 and 7, in the RMSE values and in the R^2 values. Furthermore, the scatterplots show that the model-adjusted data align much better with the reference data. For PM2.5, the coefficients of determination are between 0.90 and 0.94 and for PM1 they are between 0.96 and 0.97. The model-adjusted LCS PM1 values describe the variability of the reference measurements slightly better than the model-adjusted LCS PM2.5 measurements.

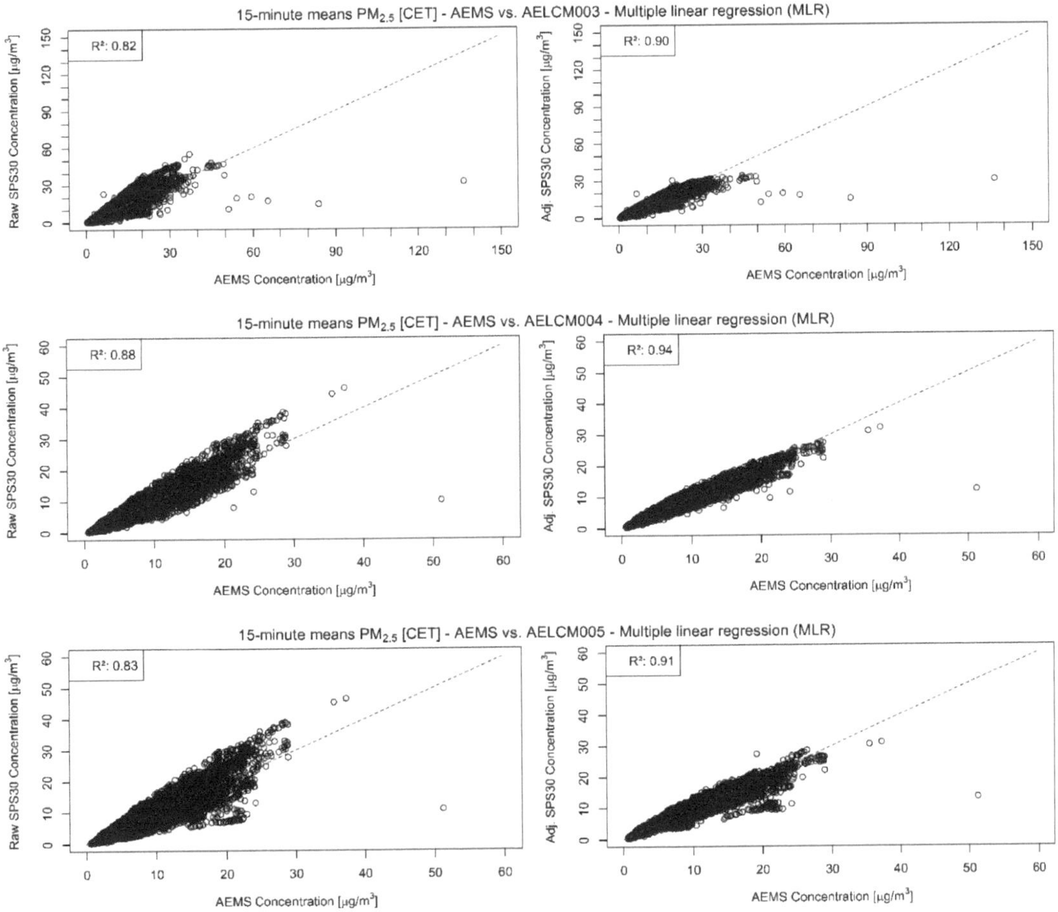

Figure 4. Raw (**left**) and model-adjusted (**right**) 15 min concentration means of different deployed SPS30 vs. 15 min concentration means of reference measurements given by the AEMS for PM2.5 in μg/m^3: (**Top**) SPS30 deployed in AELCM003; (**Middle**) SPS30 deployed in AELCM004; and (**Bottom**) SPS30 deployed in AELCM005. Multiple linear regression is used for adjusting the raw LCS data.

The slightly worse agreement of model-adjusted PM2.5 values with the reference measurements is also reflected in the higher errors. These discrepancies expressed through the RMSE are between 1.27 and 1.96 μg/m^3 but are between 0.77 and 1.07 μg/m^3 for the model-adjusted PM1 values. The calibration functions found for the deployed SPS30 devices create good estimates overall. However, the figures and scatterplots also show that for higher PM2.5 and PM1 concentrations the error between estimate and reference can increase, resulting in underestimations of particulate matter concentrations. This behavior is less apparent for the model-adjusted PM1 concentrations.

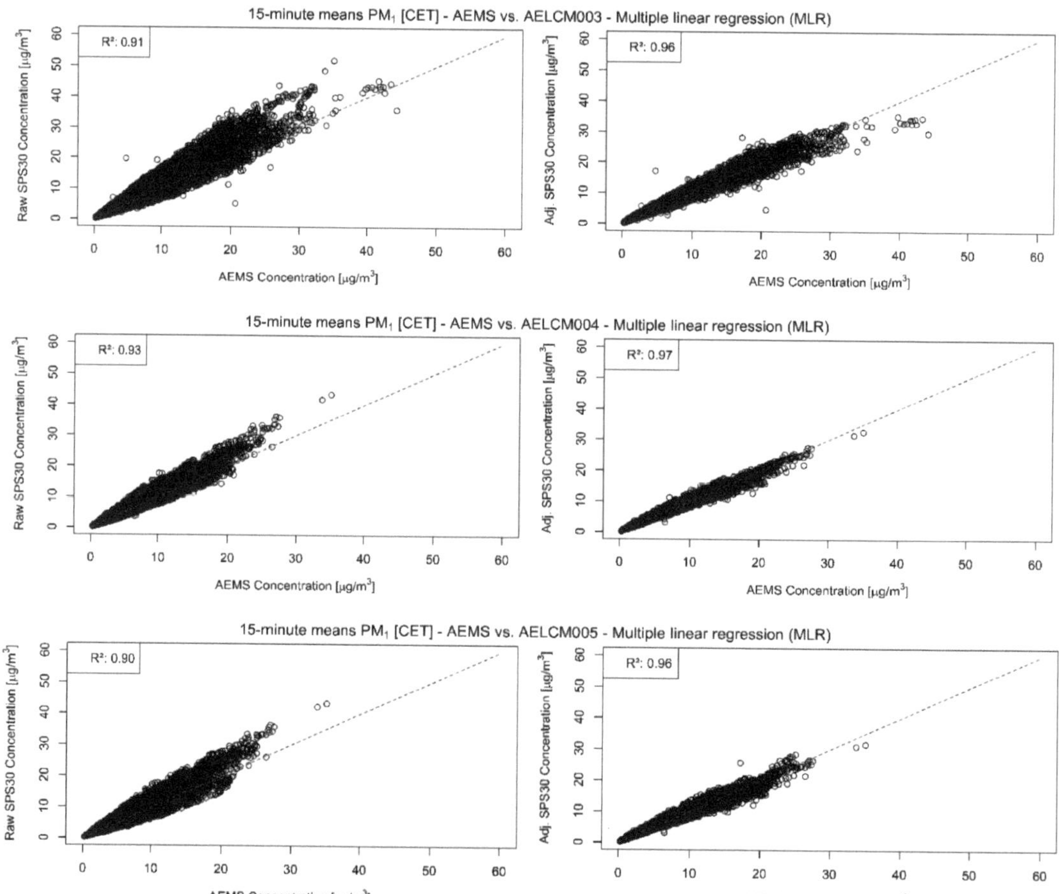

Figure 5. Raw (**left**) and model-adjusted (**right**) 15 min concentration means of different deployed SPS30 vs. 15 min concentration means of reference measurements given by the AEMS for PM1 in μg/m^3: (**Top**) SPS30 deployed in AELCM003; (**Middle**) SPS30 deployed in AELCM004; and (**Bottom**) SPS30 deployed in AELCM005. Multiple linear regression is used for adjusting the raw LCS data.

It should be noted that we cannot explain the high peaks in some of the PM reference measurements so far. After explicitly checking the provided logs of the datalogger for these measurements producing the peaks in Figure 6, we could not find any error messages related to them, so therefore they appear to be regular measurements. Because the deployed SPS30 consistently show great agreement with the reference measurements and are deployed with the AELCM boxes just next to the AEMS, we assume that the probed air of the AEMS during these moments was highly different to the probed air by every single deployed SPS30. One of the reasons could be that suspended matter in the form of pollen could have entered the AEMS locally, which affected the measurements in these points in time, while the measurements of the deployed SPS30 did not get affected.

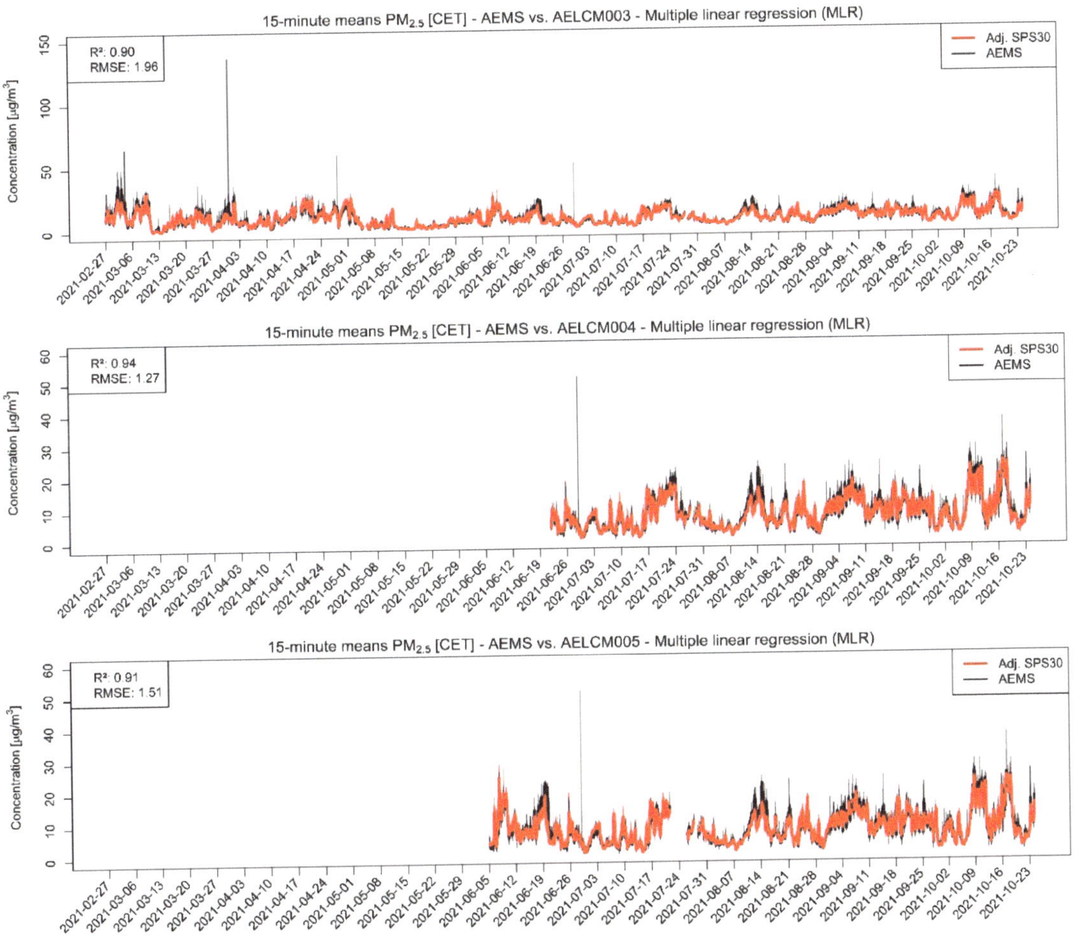

Figure 6. Model-adjusted 15 min concentration means of different deployed SPS30 using multiple linear regression vs. 15 min concentration means of reference measurements given by the AEMS for PM2.5 in µg/m^3. The RMSE is in µg/m^3. For smoothing the 15 min concentration means a rolling 24 h average is used: (**Top**) SPS30 deployed in AELCM003; (**Middle**) SPS30 deployed in AELCM004; and (**Bottom**) SPS30 deployed in AELCM005.

It should also be mentioned that for the introduced calibration functions for the raw PM1 measurements the difference between training errors and evaluation errors are generally quite low for all deployed SPS30. This indicates also that the current models seem to have a good skill for predicting PM1 concentrations, or rather for adjusting raw PM1 output of an SPS30 accordingly (Table S3.8). While the calculated RMSE values for the training and evaluation periods are higher for the model-adjusted PM2.5 measurements, we can also see similar magnitudes of errors between the training and evaluation periods (Table S3.7). Again, this indicates that the current models for adjusting raw PM2.5 concentrations given by an SPS30 are showing good skill.

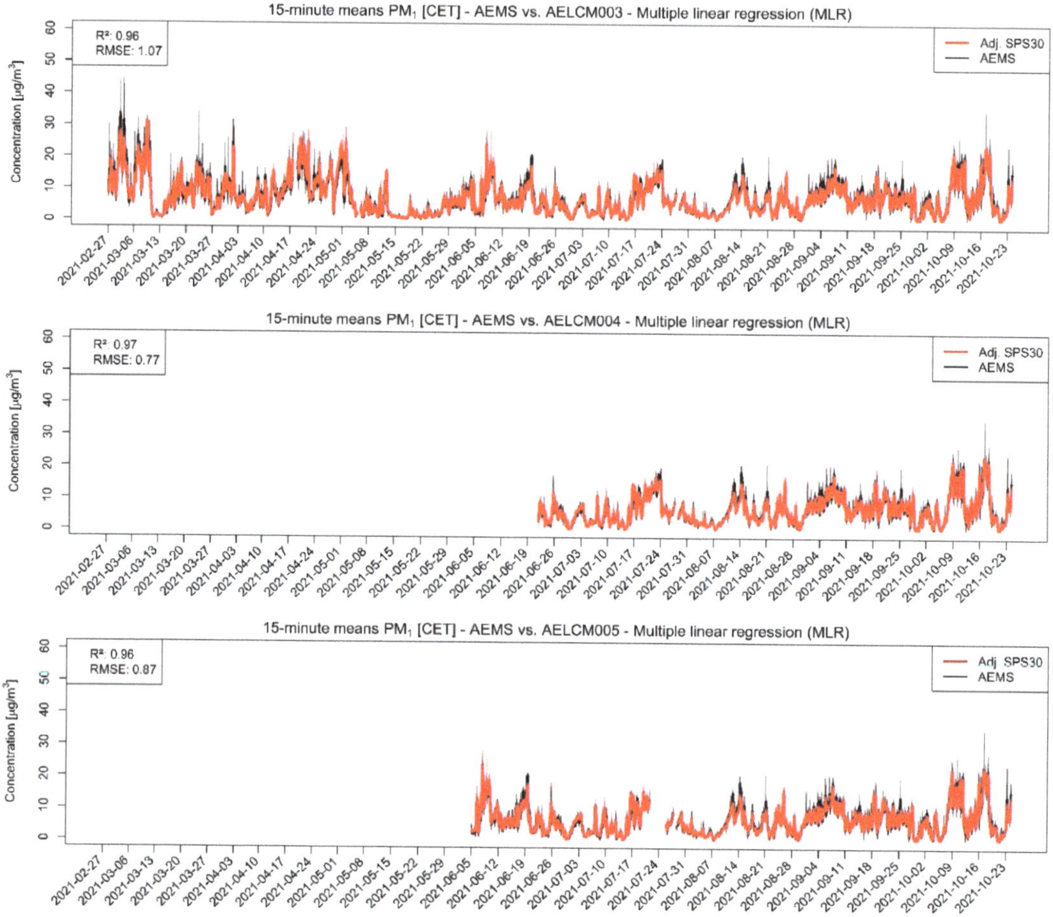

Figure 7. Model-adjusted 15 min concentration means of different deployed SPS30 using multiple linear regression vs. 15 min concentration means of reference measurements given by the AEMS for PM1 in µg/m^3. The RMSE is in µg/m^3. For smoothing the 15 min concentration means a rolling 24 h average is used: (**Top**) SPS30 deployed in AELCM003; (**Middle**) SPS30 deployed in AELCM004; and (**Bottom**) SPS30 deployed in AELCM005.

The United States Environmental Protection Agency (U.S. EPA) recognizes LCS as tools that are not useable for regulatory purposes in their current state. According to the U.S. EPA, LCS technology could be potentially used as a complementary source of information, including, for example, hot-spot localization, identification of potential sites for regulatory monitoring and better understanding of local air quality [48]. For that reason, the U.S. EPA started developing performance target reports for air-quality sensors used in nonregulatory applications. The first performance-target reports for O_3 and PM2.5 sensors were finished in 2021 [49,50]. These are based on the findings of scientific literature reviews, among other things, conducted by the U.S. EPA. We use both reports to further assess if the DGS-O3 and SPS30 could be used for nonregulatory purposes based on the calculated metrics (RMSE, R^2) in this study for hourly (15 min) model-adjusted (corrected) data. The base testing protocol for a field test recommends an RMSE smaller than or equal to 5 ppb and an R^2 value of at least 0.80 to consider an O3 LCS for nonregulatory supplemental and informational monitoring (NSIM) applications in ambient, outdoor and fixed-site environments [49]. For PM2.5, an RMSE smaller than or equal to 7 μg/m^3 and an R^2 value of at least 0.70 is needed to regard a PM LCS for NSIM purposes [50]. The DGS-O3 employed in AELCM003 and AELCM005 both fulfill the latter requirement (R^2: 0.80, 0.95), but only for the DGS-O3 of AELCM005 is the error in an acceptable range (RMSE: 3.31 ppb) for NSIM applications. Instead of 15 min means, the recommended performance metrics and target values for PM2.5 are based on 24 h averages. Considering the individual results for the model-adjusted 15 min means of all employed SPS30 (RMSE: 1.27–1.96 μg/m^3; R^2: 0.90–0.94), the calculated daily averages and their corresponding performance metrics and target values fulfill the PM2.5 NSIM requirements.

For both LCS, it must be noted that not all recommended requirements for base testing were fulfilled. For instance, three O_3 and PM LCS of the same model were not deployed at the same date (Table 2). Again, it must be said that the metrics were calculated using corrected LCS data. The U.S. EPA emphasizes in both target reports that test protocols were provided for estimating the out-of-the-box LCS performance and potential variation among identical sensors. Nevertheless, considering the recommended target values for base testing by the U.S. EPA and the calculated metrics in this study, only the SPS30 appears to be useable for NSIM applications. The DGS-O3 needs improvement to be regarded for NSIM purposes, considering its obvious strong inter-sensor unit variability.

4. Conclusions

In this work, besides introducing the technical aspects of our research group's own monitoring device, the atmospheric exposure low-cost monitoring (AELCM) for meteorological variables and air constituents, we also evaluated the performance of low-cost sensors for the air pollutants O_3, NO_2, CO and PM with aerodynamic diameters smaller than or equal to 2.5 and 1 μm during a field experiment in an urban background area.

The overall quality and quantity of data for different air substances (PM and O_3) and the stability of our measurement devices in the field over their individual measurement periods (4 to 8 months) show that our prototype AELCM devices are built on a solid foundation for actual individual exposure monitoring. Based on our gained experiences during the field experiment, we already included new features in our latest version of the AELCM, which will be used in future works. The new features are as follows: we eliminated the need of manual power cycling because of data modem instabilities by using a MOSFET. Additionally, it is possible to retrieve all measurement data using the file transfer protocol (FTP). This feature is especially useful for countries where the network infrastructure is not as developed for low-power data transmissions (in our instance LTE-M) as in Germany, but a degree of connectivity is needed without interrupting the measurement process of an AELCM box. Furthermore, the status of an SD memory card belonging to an AELCM unit can be checked in the database. Through this feature, we can act accordingly and replace a damaged card immediately. The already existing and latest feature set of our AELCM units and their flexibility given through a modular PCB design (easily switchable and stackable custom boards with different functions) make them promising devices for further research related to exposure monitoring.

To evaluate the sensors, we used the Spearman rank correlation for the raw measurements and a multiple linear regression approach for training and testing the sensor data. The basis of this evaluation process were the reference measurements given through high-quality measurement devices of the atmospheric exposure monitoring station (AEMS) belonging to the Chair for Regional Climate Change and Health of the University of Augsburg. Ultimately, we found that the deployed metal oxide gas sensors seem not to be useful for exposure monitoring given the circumstances in our field experiment. Either these sensors could not offer meaningful sensor output for the given pollution levels or our current sockets were too flawed in their design (not including the MQ131) to use these sensors to their full potential. The deployed electrochemical sensor for O_3 called SPEC DGS-O3 was the only electrochemical gas sensor that showed any degree of promise, but also strong inter-sensor unit variability. It must be noted, though, that this sensor must be combined with another NO_2 LCS to use its full potential, since its measurements get affected strongly by NO_2. Only two of these O_3 sensors showed promise under the aspect of at least qualitative predictions. We ordered this LCS (and every other LCS used in this study) from a distributor and not from the manufacturer directly.

The manufacturer SPEC Sensors offers individual calibrations of sensors from the same production batch. Considering the inter-sensor unit variability for the SPEC DGS-O3, it would be interesting to see if calibrated sensors from the same batch show a reduced inter-sensor variability and measurement error. The inter-sensor variability increases the challenge of calibrating the sensors. It hints at the possible difficulty of finding a generalized calibration model for the SPEC DGS-O3. The PM LCS called SPS30 shows very good calibration performance, given the reasonably small errors during the training and testing periods. As a result, we consider this sensor to be a good choice for any future AELCM unit.

Supplementary Materials: The following supporting information can be downloaded at https://www.mdpi.com/article/10.3390/s22103830/s1, Figure S1.1: Top view of the AELCM sensor board with MiCS sensor sockets for the MiCS-4514 on the left and MiCS-2714 on the right side at the bottom of the printed circuit board used in this study. Figure S1.2: Side view of the AELCM sensor board with the MiCS sensor sockets for the MiCS-4514 on the left and MiCS-2714 on the right side used in this study. Figure S1.3: Top view of the latest MiCS sensor socket board for the MiCS-2714 on the left and MiCS-4514 on the right side at the top of the printed circuit board. Figure S1.4: Side view of the latest MiCS sensor socket board for the MiCS-2714 and MiCS-4514. Figure S2.1: Autocorrelation coefficients for different lags of different air pollutants measured by the Atmospheric Exposure Monitoring Station (AEMS). The observation period is between the 27 February 2021 and 24 October 2021. The concentrations of the air pollutants ozone, nitrogen dioxide and carbon monoxide are measured in the measurement unit ppb (hourly means), while the particulate matter measurements are measured in the measurement unit $\mu g/m^3$ (15-min means). From top to bottom: O_3, NO_2, CO, PM2.5 and PM1. Table S3.1: Average, minimum and maximum RMSE for ambient ozone in parts per billion (ppb) and for each p predictor model regarding LCS SPEC DGS-O3 deployed with specific AELCM units. The performance estimation method is the repeated holdout procedure using reference data of the AEMS and model-calibrated (adjusted) LCS data based on hourly averages. Table S3.2: Average, minimum and maximum RMSE for ambient nitrogen dioxide in parts per billion (ppb) and for each p predictor model regarding LCS SPEC DGS-NO2 deployed with specific AELCM units. The performance estimation method is the repeated holdout procedure using reference data of the AEMS and model-calibrated (adjusted) LCS data based on hourly averages. Table S3.3: Average, minimum and maximum RMSE for ambient carbon monoxide in parts per billion (ppb) and for each p predictor model regarding LCS SPEC DGS-CO deployed with specific AELCM units. The performance estimation method is the repeated holdout procedure using reference data of the AEMS and model-calibrated (adjusted) LCS data based on hourly averages. Table S3.4: Average, minimum and maximum RMSE for ambient ozone in parts per billion (ppb) and for each p predictor model regarding LCS MQ131 deployed with specific AELCM units. The performance estimation method is the repeated holdout procedure using reference data of the AEMS and model-calibrated (adjusted) LCS data based on hourly averages. Table S3.5: Average, minimum and maximum RMSE for ambient nitrogen dioxide in parts per billion (ppb) and for each p predictor model regarding LCS SGX MiCS-2714 deployed with specific AELCM units. The performance estimation method is the repeated holdout procedure using reference data of the AEMS and model-calibrated (adjusted) LCS data based on hourly averages. Table S3.6: Average, minimum and maximum RMSE for ambient carbon monoxide in parts per billion (ppb) and for each p predictor model regarding LCS SGX MiCS-4514 deployed with specific AELCM units. The performance estimation method is the repeated holdout procedure using reference data of the AEMS and model-calibrated (adjusted) LCS data based on hourly averages. Table S3.7: Average, minimum and maximum RMSE for ambient particulate matter PM2.5 in microgram per cubic meter ($\mu g/m^3$) and for each p predictor model regarding SPS30 deployed with specific AELCM units. The performance estimation method is the repeated holdout procedure using reference data of the AEMS and model-calibrated (adjusted) LCS data based on 15-min averages. Table S3.8: Average, minimum and maximum RMSE for ambient particulate matter PM1 in microgram per cubic meter ($\mu g/m^3$) and for each p predictor model regarding SPS30 deployed with specific AELCM units. The performance estimation method is the repeated holdout procedure using reference data of the AEMS and model-calibrated (adjusted) LCS data based on 15-min averages. Table S4.1: Average, minimum and maximum RMSE for ambient ozone in parts per billion (ppb) and for each p predictor model regarding LCS SPEC DGS-O3 deployed with specific AELCM units. The performance estimation method is the 5-fold cross validation procedure using reference data of the AEMS and model-calibrated (adjusted) LCS data based on hourly averages. Table S4.2: Average, minimum and maximum RMSE for ambient nitrogen dioxide in parts per billion (ppb) and for each p predictor model regarding LCS SPEC DGS-NO2 deployed with specific AELCM units. The performance estimation method is the 5-fold cross validation procedure using reference data of the AEMS and model-calibrated (adjusted) LCS data based on hourly averages. Table S4.3: Average, minimum and maximum RMSE for ambient carbon monoxide in parts per billion (ppb) and for each p predictor model regarding LCS SPEC DGS-CO deployed with specific AELCM units. The performance estimation method is the 5-fold cross validation procedure using reference data of the AEMS and model-calibrated (adjusted) LCS data based on hourly averages. Table S4.4: Average, minimum and maximum RMSE for ambient ozone in parts per billion (ppb) and for each p predictor model regarding LCS MQ131 deployed with specific AELCM units. The performance estimation method is the 5-fold cross validation procedure using reference data of the AEMS and model-calibrated (adjusted) LCS data based on hourly averages. Table S4.5: Average, minimum and maximum RMSE for ambient nitrogen dioxide in parts per billion (ppb) and for each p predictor model regarding LCS SGX MiCS-2714 deployed with specific AELCM units. The performance estimation method is the 5-fold cross validation procedure using reference data of the AEMS and model-calibrated (adjusted) LCS data based on hourly averages. Table S4.6: Average, minimum and maximum RMSE for ambient carbon monoxide in parts per billion (ppb) and for each p predictor model regarding LCS SGX MiCS-4514 deployed with specific AELCM units. The performance estimation method is the 5-fold cross validation procedure using reference data of the AEMS and model-calibrated (adjusted) LCS data based on hourly averages. Table S4.7: Average, minimum and maximum RMSE for ambient particulate matter PM2.5 in microgram per cubic meter ($\mu g/m^3$) and for each p predictor model regarding SPS30 deployed with specific AELCM units. The performance estimation method is the 5-fold cross validation procedure using reference data of the AEMS and model-calibrated (adjusted) LCS data based on 15-min averages. Table S4.8: Average, minimum and maximum RMSE for ambient particulate matter PM1 in microgram per cubic meter ($\mu g/m^3$) and for each p predictor model regarding SPS30 deployed with specific AELCM units. The performance estimation method is the 5-fold cross validation procedure using reference data of the AEMS and model-calibrated (adjusted) LCS data based on 15-min averages. Table S5.1: Spearman rank correlations between different LCS output (MQ131 output, temperature and relative humidity) and reference measurements of ambient ozone for all AELCM units. As starting points for AELCM003, AELCM004 and AELCM005 we use the measured data at 5 March 2021, 28 June 2021 and 11 June 2021 respectively. The ending date is the 24 October 2021. Figure S6.1: Model-adjusted hourly concentration means of different deployed SPEC DGS-NO2 using multiple linear regression vs. hourly concentration means of reference measurements given by the AEMS for ambient nitrogen dioxide in ppb. For smoothing the hourly concentration means a rolling 24-h average is used: (Top) SPEC DGS-NO2 deployed in AELCM003; (Middle) SPEC DGS-NO2 deployed in AELCM004; (Bottom) SPEC DGS-NO2 deployed in AELCM005. Figure S6.2: Model-adjusted hourly concentration means of a SPEC DGS-CO deployed in AELCM003 using multiple linear regression vs. hourly concentration means of reference measurements given by the AEMS for ambient carbon monoxide in ppb. For smoothing the hourly concentration means a rolling 24-h average is used. Figure S6.3: Model-adjusted hourly concentration means of different deployed MQ131 using multiple linear regression vs. hourly concentration means of reference measurements given by the AEMS for ambient ozone in ppb. For smoothing the hourly concentration means a rolling 24-h average is used: (Top) MQ131 deployed in AELCM003; (Middle) MQ131 deployed in AELCM004; (Bottom) MQ131 deployed in AELCM005. Figure S6.4: Model-adjusted hourly concentration means of different deployed MiCS-2714 using multiple linear regression vs. hourly concentration means of reference measurements given by the AEMS for ambient nitrogen dioxide in ppb. For smoothing the hourly concentration means a rolling 24-h average is used: (Top) MiCS-2714 deployed in AELCM004; (Bottom) MiCS-2714 deployed in AELCM005. Figure S6.5: Model-adjusted hourly concentration means of different deployed MiCS-4514 using multiple linear regression vs. hourly concentration means of reference measurements given by the AEMS for ambient carbon monoxide in ppb. For smoothing the hourly concentration means a rolling 24-h average is used: (Top) MiCS-4514 deployed in AELCM004; (Bottom) MiCS-4514 deployed in AELCM005. Figure S7.1: Raw hourly means of ambient ozone measured by SPEC DGS-O3 units deployed in AELCM003, AELCM004 and AELCM005 compared with hourly means of ambient ozone measurements given by the reference station AEMS. For smoothing the hourly concentration means a rolling 24-h average is used.

Author Contributions: Conceptualization, P.G., C.K. and E.H.; methodology, P.G. and E.H.; coding, P.G.; data collection and analysis, P.G.; enclosure and PCB design, C.K.; assembly of AELCM units, C.K.; writing—original draft preparation, P.G.; writing—review and editing, E.H.; supervision, E.H.; project administration, P.G.; funding acquisition, E.H. All authors have read and agreed to the published version of the manuscript.

Funding: This work was supported by the Deutsche Forschungsgemeinschaft (DFG, German Research Foundation) under project number 408057478.

Informed Consent Statement: Not applicable.

Data Availability Statement: The study data are available upon request from the corresponding authors.

Conflicts of Interest: The authors declare no conflict of interest.

References

1. World Health Organization. WHO: Ambient (Outdoor) Air Pollution. 2021. Available online: https://www.who.int/news-room/fact-sheets/detail/ambient-(outdoor)-air-quality-and-health (accessed on 20 December 2021).
2. World Health Organization. *WHO Global Air Quality Guidelines: Particulate Matter (PM2.5 and PM10), Ozone, Nitrogen Dioxide, Sulfur Dioxide and Carbon Monoxide (No. 9789240034228)*; World Health Organization: Geneva, Switzerland, 2021.
3. Olschewski, P.; Kaspar-Ott, I.; Koller, S.; Schenkirsch, G.; Trepel, M.; Hertig, E. Associations between Weather, Air Quality and Moderate Extreme Cancer-Related Mortality Events in Augsburg, Southern Germany. *Int. J. Environ. Res. Public Health* **2021**, *18*, 11737. [CrossRef] [PubMed]
4. Russo, A.; Trigo, R.M.; Martins, H.; Mendes, M.T. NO_2, PM10 and O_3 urban concentrations and its association with circulation weather types in Portugal. *Atmos. Environ.* **2014**, *89*, 768–785. [CrossRef]
5. Russo, A.; Gouveia, C.; Levy, I.; Dayan, U.; Jerez, S.; Mendes, M.; Trigo, R. Coastal recirculation potential affecting air pollutants in Portugal: The role of circulation weather types. *Atmos. Environ.* **2016**, *135*, 9–19. [CrossRef]
6. Dayan, U.; Levy, I. Relationship between synoptic-scale atmospheric circulation and ozone concentrations over Israel. *J. Geophys. Res. Atmos.* **2002**, *107*, ACL 31-1–ACL 31-12. [CrossRef]
7. Strickland, M.J.; Darrow, L.A.; Klein, M.; Flanders, W.D.; Sarnat, J.A.; Waller, L.A.; Sarnat, S.E.; Mulholland, J.A.; Tolbert, P.E. Short-term associations between ambient air pollutants and pediatric asthma emergency department visits. *Am. J. Respir. Crit. Care Med.* **2010**, *182*, 307–316. [CrossRef]
8. Nuvolone, D.; Petri, D.; Voller, F. The effects of ozone on human health. *Environ. Sci. Pollut. Res.* **2018**, *25*, 8074–8088. [CrossRef]
9. Hertig, E.; Schneider, A.; Peters, A.; von Scheidt, W.; Kuch, B.; Meisinger, C. Association of ground-level ozone, meteorological factors and weather types with daily myocardial infarction frequencies in Augsburg, Southern Germany. *Atmos. Environ.* **2019**, *217*, 116975. [CrossRef]
10. Brook, R.D.; Rajagopalan, S.; Pope, C.A.; Brook, J.R.; Bhatnagar, A.; Diez-Roux, A.V.; Holguin, F.; Hong, Y.; Luepker, R.V.; Mittleman, M.A.; et al. Particulate Matter Air Pollution and Cardiovascular Disease. *Circulation* **2010**, *121*, 2331–2378. [CrossRef]
11. Li, J.; Liu, F.; Liang, F.; Huang, K.; Yang, X.; Xiao, Q.; Chen, J.; Liu, X.; Cao, J.; Chen, S.; et al. Long-Term Effects of High Exposure to Ambient Fine Particulate Matter on Coronary Heart Disease Incidence: A Population-Based Chinese Cohort Study. *Environ. Sci. Technol.* **2020**, *54*, 6812–6821. [CrossRef]
12. Hamanaka, R.B.; Mutlu, G.M. Particulate Matter Air Pollution: Effects on the Cardiovascular System. *Front. Endocrinol.* **2018**, *9*, 1–15. [CrossRef]
13. Varon, J.; Marik, P.E.; Fromm Jr, R.E.; Gueler, A. Carbon monoxide poisoning: A review for clinicians. *J. Emerg. Med.* **1999**, *17*, 87–93. [CrossRef]
14. Orazzo, F.; Nespoli, L.; Ito, K.; Tassinari, D.; Giardina, D.; Funis, M.; Cecchi, A.; Trapani, C.; Forgeschi, G.; Vignini, M.; et al. Air Pollution, Aeroallergens, and Emergency Room Visits for Acute Respiratory Diseases and Gastroenteric Disorders among Young Children in Six Italian Cities. *Environ. Health Perspect.* **2009**, *117*, 1780–1785. [CrossRef] [PubMed]
15. Bell, M.L.; Peng, R.D.; Dominici, F.; Samet, J.M. Emergency Hospital Admissions for Cardiovascular Diseases and Ambient Levels of Carbon Monoxide. *Circulation* **2009**, *120*, 949–955. [CrossRef] [PubMed]
16. Lewis, A.; Peltier, W.R.; von Schneidemesser, E. Low-Cost Sensors for the Measurement of Atmospheric Composition: Overview of Topic and Future Applications (World Meteorological Organization). Available online: https://www.ccacoalition.org/en/resources/low-cost-sensors-measurement-atmospheric-composition-overview-topic-and-future (accessed on 28 December 2021).
17. Li, J.; Mattewal, S.K.; Patel, S.; Biswas, P. Evaluation of Nine Low-cost-sensor-based Particulate Matter Monitors. *Aerosol Air Qual. Res.* **2020**, *20*, 254–270. [CrossRef]

18. Jiao, W.; Hagler, G.; Williams, R.; Sharpe, R.; Brown, R.; Garver, D.; Judge, R.; Caudill, M.; Rickard, J.; Davis, M.; et al. Community Air Sensor Network (CAIRSENSE) project: Evaluation of low-cost sensor performance in a suburban environment in the southeastern United States. *Atmos. Meas. Tech.* **2016**, *9*, 5281–5292. [CrossRef]
19. Mead, M.; Popoola, O.; Stewart, G.; Landshoff, P.; Calleja, M.; Hayes, M.; Baldovi, J.; McLeod, M.; Hodgson, T.; Dicks, J. The use of electrochemical sensors for monitoring urban air quality in low-cost, high-density networks. *Atmos. Environ.* **2013**, *70*, 186–203. [CrossRef]
20. Mueller, M.; Meyer, J.; Hueglin, C. Design of an ozone and nitrogen dioxide sensor unit and its long-term operation within a sensor network in the city of Zurich. *Atmos. Meas. Tech.* **2017**, *10*, 3783–3799. [CrossRef]
21. Piedrahita, R.; Xiang, Y.; Masson, N.; Ortega, J.; Collier, A.; Jiang, Y.; Li, K.; Dick, R.P.; Lv, Q.; Hannigan, M.; et al. The next generation of low-cost personal air quality sensors for quantitative exposure monitoring. *Atmos. Meas. Tech.* **2014**, *7*, 3325–3336. [CrossRef]
22. Cross, E.S.; Williams, L.R.; Lewis, D.K.; Magoon, G.R.; Onasch, T.B.; Kaminsky, M.L.; Worsnop, D.R.; Jayne, J.T. Use of electrochemical sensors for measurement of air pollution: Correcting interference response and validating measurements. *Atmos. Meas. Tech.* **2017**, *10*, 3575–3588. [CrossRef]
23. Karagulian, F.; Barbiere, M.; Kotsev, A.; Spinelle, L.; Gerboles, M.; Lagler, F.; Redon, N.; Crunaire, S.; Borowiak, A. Review of the Performance of Low-Cost Sensors for Air Quality Monitoring. *Atmosphere* **2019**, *10*, 506. [CrossRef]
24. Helm, I.; Jalukse, L.; Leito, I. Measurement Uncertainty Estimation in Amperometric Sensors: A Tutorial Review. *Sensors* **2010**, *10*, 4430–4455. [CrossRef] [PubMed]
25. Wang, C.; Yin, L.; Zhang, L.; Xiang, D.; Gao, R. Metal Oxide Gas Sensors: Sensitivity and Influencing Factors. *Sensors* **2010**, *10*, 2088–2106. [CrossRef] [PubMed]
26. Crilley, L.R.; Shaw, M.; Pound, R.; Kramer, L.J.; Price, R.; Young, S.; Lewis, A.C.; Pope, F.D. Evaluation of a low-cost optical particle counter (Alphasense OPC-N2) for ambient air monitoring. *Atmos. Meas. Tech.* **2018**, *11*, 709–720. [CrossRef]
27. Arduino LLC. Mega 2560 Rev3. Available online: https://docs.arduino.cc/hardware/mega-2560 (accessed on 5 January 2022).
28. Pycom Limited. GPy. Available online: https://docs.pycom.io/gitbook/assets/specsheets/Pycom_002_Specsheets_GPy_v2.pdf (accessed on 5 January 2022).
29. Adafruit Industries LLC. Adafruit PowerBoost 1000C. Available online: https://cdn-learn.adafruit.com/downloads/pdf/adafruit-powerboost-1000c-load-share-usb-charge-boost.pdf (accessed on 27 February 2022).
30. Sensortec, B. BME280 Integrated Environmental Unit. Available online: https://www.bosch-sensortec.com/media/boschsensortec/downloads/product_flyer/bst-bme280-fl000.pdf (accessed on 13 January 2022).
31. Winsen. MQ131 Ozone Gas Sensor (Model: MQ131 Low Concentration). Available online: https://www.winsen-sensor.com/d/files/PDF/Semiconductor%20Gas%20Sensor/mq131-(low-concentration)-ver1_3-manual.pdf (accessed on 13 January 2022).
32. SGX Sensortech. Data Sheet MiCS-2714. Available online: https://www.sgxsensortech.com/content/uploads/2014/08/1107_Datasheet-MiCS-2714.pdf (accessed on 13 January 2022).
33. SGX Sensortech. Data Sheet MiCS-4514. Available online: https://www.sgxsensortech.com/content/uploads/2014/08/0278_Datasheet-MiCS-4514.pdf (accessed on 28 December 2021).
34. SPEC Sensors. Digital Gas Sensor—Ozone. Available online: https://www.spec-sensors.com/wp-content/uploads/2017/01/DGS-O3-968-042_9-6-17.pdf (accessed on 13 January 2022).
35. SPEC Sensors. Digital Gas Sensor—Nitrogen Dioxide. Available online: https://www.spec-sensors.com/wp-content/uploads/2017/01/DGS-NO2-968-043_9-6-17.pdf (accessed on 13 January 2022).
36. SPEC Sensors. Digital Gas Sensor—Carbon Monoxide. Available online: https://www.spec-sensors.com/wp-content/uploads/2017/01/DGS-CO-968-034.pdf (accessed on 13 January 2022).
37. Sensirion. Datasheet SPS30. Available online: https://www.sensirion.com/fileadmin/user_upload/customers/sensirion/Dokumente/9.6_Particulate_Matter/Datasheets/Sensirion_PM_Sensors_Datasheet_SPS30.pdf (accessed on 13 January 2022).
38. Sensirion. Sensor Specification Statement. Available online: https://sensirion.com/media/documents/B7AAA101/61653FB8/Sensirion_Particulate_Matter_AppNotes_Specification_Statement.pdf (accessed on 13 January 2022).
39. Hardkernel, Co., Ltd. 92 × 92 × 25 mm DC Cooling Fan w/PWM, Speed Sensor (Tacho). Available online: https://www.hardkernel.com/shop/92x92x25mm-dc-cooling-fan-w-pwm-speed-sensor-tacho/ (accessed on 21 April 2022).
40. SPEC Sensors. SPEC Sensor Operation Overview. Available online: http://www.spec-sensors.com/wp-content/uploads/2016/05/SPEC-Sensor-Operation-Overview.pdf (accessed on 14 January 2022).
41. Han, P.; Mei, H.; Liu, D.; Zeng, N.; Tang, X.; Wang, Y.; Pan, Y. Calibrations of Low-Cost Air Pollution Monitoring Sensors for CO, NO_2, O_3, and SO_2. *Sensors* **2021**, *21*, 256. [CrossRef]
42. Miller, A. *Subset Selection in Regression*, 2nd ed.; Chapman and Hall/CRC: Boca Raton, FL, USA, 2002.
43. Bigi, A.; Mueller, M.; Grange, S.K.; Ghermandi, G.; Hueglin, C. Performance of NO, NO_2 low cost sensors and three calibration approaches within a real world application. *Atmos. Meas. Tech.* **2018**, *11*, 3717–3735. [CrossRef]
44. Wilks, D.S. *Statistical Methods in the Atmospheric Sciences*; Academic Press: Cambridge, MA, USA, 2011; Volume 100.
45. Venables, W.N.; Ripley, B.D. *Modern Applied Statistics with S*; Springer: New York, NY, USA, 2002.
46. Cerqueira, V.; Torgo, L.; Mozetič, I. Evaluating time series forecasting models: An empirical study on performance estimation methods. *Mach. Learn.* **2020**, *109*, 1997–2028. [CrossRef]

47. AQ-SPEC. South Coast Air Quality Management District; South Coast Air Quality Management District Air Quality Sensor Performance Evaluation Reports. Available online: http://www.aqmd.gov/aq-spec/evaluations#&MainContent_C001_Col00=2 (accessed on 22 February 2022).
48. US Environmental Protection Agency (EPA). Air Sensor Performance Targets and Testing Protocols. Available online: https://www.epa.gov/air-sensor-toolbox/air-sensor-performance-targets-and-testing-protocols (accessed on 30 April 2022).
49. Duvall, R.; Clements, A.; Hagler, G.; Kamal, A.; Kilaru, V.; Goodman, L.; Frederick, S.; Johnson Barkjohn, K.; VonWald, I.; Greene, D. *Performance Testing Protocols, Metrics, and Target Values for Ozone Air Sensors: Use in Ambient, Outdoor, Fixed Site, Non-Regulatory and Informational Monitoring Applications*; U.S. Environmental Protection Agency, Office of Research and Development: Washington, DC, USA, 2021.
50. Duvall, R.; Clements, A.; Hagler, G.; Kamal, A.; Kilaru, V.; Goodman, L.; Frederick, S.; Barkjohn, K.; VonWald, I.; Greene, D. *Performance Testing Protocols, Metrics, and Target Values for Fine Particulate Matter Air Sensors: Use in Ambient, Outdoor, Fixed Sites, Non-Regulatory Supplemental and Informational Monitoring Applications*; U.S. Environmental Protection Agency, Office of Research and Development: Washington, DC, USA, 2021.

sensors

MDPI

Article

Assessment of Individual-Level Exposure to Airborne Particulate Matter during Periods of Atmospheric Thermal Inversion

Rok Novak [1,2,*], Johanna Amalia Robinson [1,2,3], Tjaša Kanduč [1], Dimosthenis Sarigiannis [4,5,6] and David Kocman [1]

1. Department of Environmental Sciences, Jožef Stefan Institute, 1000 Ljubljana, Slovenia
2. Jožef Stefan International Postgraduate School, 1000 Ljubljana, Slovenia
3. Center for Research and Development, Slovenian Institute for Adult Education, Ulica Ambrožiča Novljana 5, 1000 Ljubljana, Slovenia
4. Environmental Engineering Laboratory, Department of Chemical Engineering, Aristotle University of Thessaloniki, 54124 Thessaloniki, Greece
5. HERACLES Research Centre on the Exposome and Health, Center for Interdisciplinary Research and Innovation, 54124 Thessaloniki, Greece
6. Department of Science, Technology and Society, University School of Advanced Study IUSS, 27100 Pavia, Italy

* Correspondence: rok.novak@ijs.si

Abstract: Air pollution exposure is harmful to human health and reducing it at the level of an individual requires measurements and assessments that capture the spatiotemporal variability of different microenvironments and the influence of specific activities. In this paper, activity-specific and general indoor and outdoor exposure during and after a period of high concentrations of particulate matter (PM), e.g., an atmospheric thermal inversion (ATI) in the Ljubljana subalpine basin, Slovenia, was assessed. To this end, personal particulate matter monitors (PPM) were used, worn by participants of the H2020 ICARUS sampling campaigns in spring 2019 who also recorded their hourly activities. ATI period(s) were determined based on data collected from two meteorological stations managed by the Slovenian Environmental Agency (SEA). Results showed that indoor and outdoor exposure to PM was significantly higher during the ATI period, and that the difference between mean indoor and outdoor exposure to PM was much higher during the ATI period (23.0 $\mu g/m^3$) than after (6.5 $\mu g/m^3$). Indoor activities generally were associated with smaller differences, with cooking and cleaning even having higher values in the post-ATI period. On the other hand, all outdoor activities had higher PM values during the ATI than after, with larger differences, mostly >30.0 $\mu g/m^3$. Overall, this work demonstrated that an individual-level approach can provide better spatiotemporal resolution and evaluate the relative importance of specific high-exposure events, and in this way provide an ancillary tool for exposure assessments.

Keywords: personal exposure; particulate matter; atmospheric thermal inversion; personal monitoring; exposure assessment

1. Introduction

Airborne particulate matter (PM) negatively impacts human health, reduces life expectancy, and increases mortality, and is a particularly important health risk in urban environments as traffic and other factors additionally contribute to higher concentrations of PM and other pollutants [1–5].

A common approach to assessing exposure is using monitoring stations that measure outdoor concentration levels of various pollutants and require compliance with regulatory protocols, which makes them the reference standard in an urban environment for evaluating long-term trends, outdoor concentrations, and city-wide exposure assessments [6]. On the other hand, they are expensive to operate, physically large, and consequently limited in number and coverage. While there are several options to use data from monitoring stations to estimate indoor exposure [7], static outdoor stations are not able to capture the variability of exposure based on an individual's activities and daily movement trajectory [8]. Multiple studies have shown that collecting data on air quality and exposure on an individual level, in contrast to city-wide monitoring, provides higher spatiotemporal granularity to observe individual-level exposure and daily fluctuations in diverse indoor and outdoor urban settings, including the impact of atmospheric thermal inversions (ATIs) [9–11]. Exposure assessments based on individual-level measurements usually show higher recorded values than estimates based on data from monitoring stations [12,13], and therefore assessments that use community average concentrations of PM can underestimate the health burden of air pollution [14]. Personal monitoring devices can be used to estimate negative health outcomes from exposure to higher PM concentrations [15] as well as the importance of socioeconomic variables, e.g., sociodemographic status, urban mobility, and living conditions, when assessing exposure to PM [16]. To further explore the applicability of PM monitors in individual-level exposure research, their performance should be assessed within a period that would show distinct differences in exposure during activities and in microlocations that the individual records. A period with persistent ATIs in a suitable wintertime environment (e.g., an alpine basin) could provide the necessary conditions, as it is characterized by two clearly delimited periods of high and low concentrations of PM.

ATIs that occur in urban environments, as a consequence of atypical temperature gradients, produce a "cap" which reduces the diffusion of dust, smoke, and other air pollutants [17] and can cause concentrations of air pollutants to increase, with a high level of spatiotemporal variability throughout the urban environment [18,19]. An elevated level of exposure during ATIs can lead to detrimental health effects, mostly as an increase in the incidence of acute respiratory diseases, asthma, and cardiovascular diseases [20–22]. However, increased exposure on an individual level during ATIs is still poorly understood.

A unique set of conditions present in a subalpine basin (Ljubljana, Slovenia), e.g., concave shape, extended periods of anticyclonic conditions, and drag associated with the complex topography, resulting in frequent foggy days and ATIs, which in turn cause a buildup of PM [23]. The Ljubljana basin experiences frequent short-lived inversions in all seasons, though persistent inversions occur mostly in the colder part of the year [24]. Meteorological conditions are a driving factor in determining air quality in Ljubljana, and can surpass the importance of emission ceilings [25]. Although air quality has been improving in most European cities over the past decade, Ljubljana, as of 27 July 2022, ranks 279 out of 344 cities from the European Environment Agency (EEA) member countries in terms of air quality, with an average PM concentration of 15.7 $\mu g/m^3$, measured in 2020 and 2021, labeled as "poor air quality" by the EEA [26]. ATIs, compounded by the poor air quality in Ljubljana during winter, temporarily increase exposure to PM and offer a distinctive perspective on high-exposure events in urban environments, which individual-level monitoring could help to assess in more detail.

This study used next-generation sensing and monitoring (NGSM) technology—a wearable PM monitor—to determine how these types of devices could provide fine-grained spatiotemporal resolution of personal exposure to PM_{10} in a period of persistent ATIs and immediately after, based on individual activities and microlocations. Individual-level exposure assessments were based on data obtained from personal PM monitors (PPM) used as part of the ICARUS H2020 project [27], where participants carried the devices for one week in a heating and non-heating season, and additionally provided hourly data on their activities, transport mode, and microlocations [28].

2. Materials and Methods

2.1. Collecting Particulate Matter Data

For assessing exposure on an individual level a wearable device called the PPM (shown in Figure 1) was used, which provided data on PM_1, $PM_{2.5}$, and PM_{10} concentrations, ambient temperature, relative humidity, and location/GPS data with minute resolution. The devices were designed and constructed for the ICARUS project [29] by IoTech Telecommunications, Thessaloniki, Greece [30] and are based on the Arduino platform and the Plantower, Beijing, China, pms5003 sensor [31,32]. To determine whether the device provided data that was fit for purpose and accurate, a validation was conducted by collocating the PPM with a GRIMM (Durag Group, Hamburg, Germany) model 11-A (1.109) aerosol spectrometer, which showed that the PPM had relatively high accuracy and was fit for purpose, further described in Novak et al. [33].

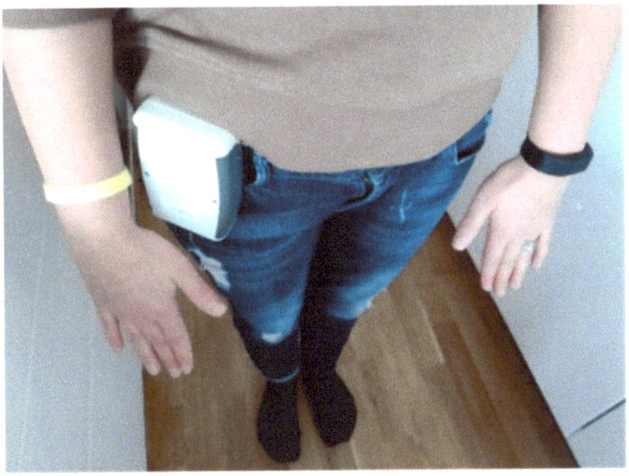

Figure 1. PPM device (white box attached to clothes) worn by a participant.

Participants were instructed to wear the PPM for the entire duration of the study or have it placed near them if they performed sedentary or stationary activities, e.g., office work or sleeping. The data were collected on an internal SD card and exported via a web app/portal and stored on a local drive. Each participant had to fill out a time activity diary (TAD) and indicate what the characteristic activity was that they were performing each hour of the day for seven days. Data were used from the ICARUS heating season sampling campaign, which took place from 16 February 2019 to 12 March 2019.

To observe the trend of PM_{10} concentrations in Ljubljana during and after the period with persistent ATIs, city-wide data on PM_{10} concentrations (30 min values) and meteorological conditions (temperature and wind speed) were provided by the Slovenian Environmental Agency (SEA) [34] from the urban background reference station in the Bežigrad district of Ljubljana. Preliminary observations showed that a period with persistent ATIs could have occurred in Ljubljana in the same period as the heating season sampling campaign. Data from the monitoring stations were collected for the period from 10 February 2019 to 15 March 2019 to provide additional context.

2.2. Determining ATIs

ATIs were determined by analyzing temperature gradients between stations at different elevations, per the Largeron and Staquet [35] pseudo-vertical temperature gradient method (TGM), which presupposes two assumptions: (1) horizontal homogeneity of the temperature field and (2) the quasi-linearity of the temperature profile. When these considerations are met, the ratio of the temperature and height difference between the stations ($\Delta T/\Delta z$) can be used to determine the stability of the boundary layer when the inversion occurs [35]. Kikaj et al. [23] determined that these assumptions were met for low- and medium-lying stations in and around the Ljubljana basin in the colder months of the year. High-elevation stations showed moderate correlation coefficients when calculating horizontal air temperature homogeneity and were, in the scope of this paper, only used to estimate the height of the inversion layer by determining if the inversion persisted up to the height of the station.

Measurements were collected from three automatic weather stations (AWS), one low-lying station located in the center of Ljubljana (station Bežigrad—AWS-B), one medium-lying station situated on a hill at the border of the basin (Topol—AWS-T) and one high-lying station at the northern border of the basin (Krvavec—AWS-K), shown in Table 1 with their respective elevations, coordinates, and collected parameters. The stations cover the central, western, and northwestern parts of the Ljubljana basin, as shown in Figure 2. All stations measure and report air temperature at 2 m above ground, at 7:00, 14:00, and 21:00, each day.

Table 1. Automatic weather stations (AWS) used to determine ATIs, their locations, elevations, and parameters collected.

Station	Meters above Sea Level	Coordinates	Parameters
AWS-B	299 m	46.0654 N, 14.5123 E	Temperature, PM_{10}
AWS-T	692 m	46.0940 N, 14.3713 E	Temperature
AWS-K	1742 m	46.2978 N, 14.5335 E	Temperature

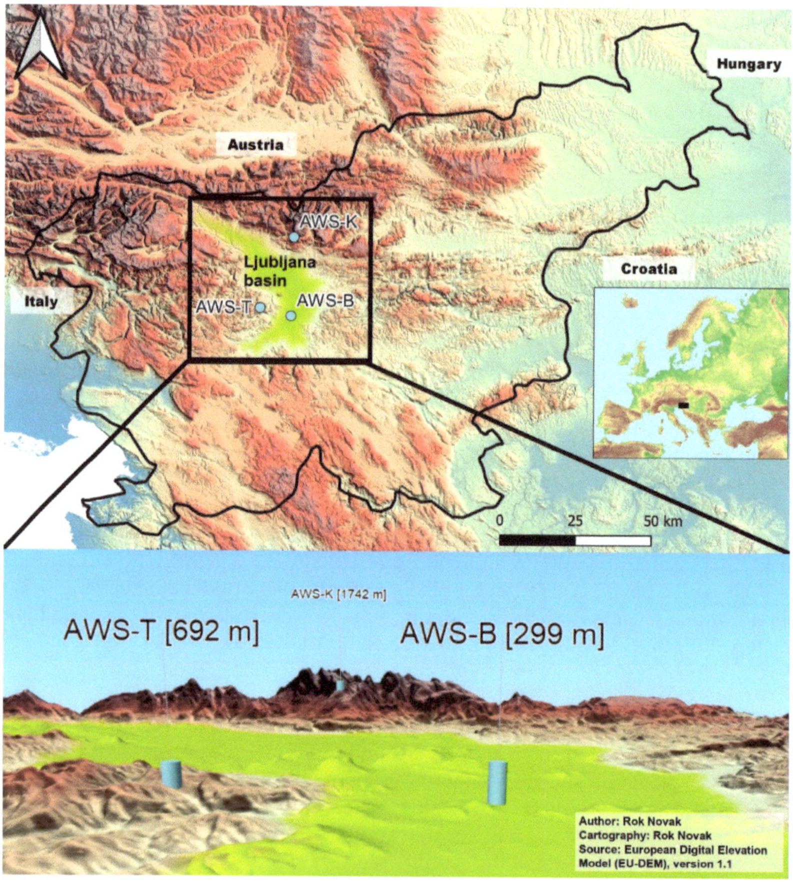

Figure 2. Geographical locations of automatic weather stations Bežigrad, Topol, and Krvavec. Top: Locations on a topographical map of Slovenia and neighboring countries with the location of the Ljubljana basin. Bottom: 3D visualization of locations and their respective elevations with vertical exaggeration (2), designed with the Qgis2threejs plugin [36] in QGIS 3.20.1-Odense [37].

As per the TGM, the ratio of the temperature difference between stations AWS-B and AWS-T to their difference in elevation ($\Delta T/\Delta z$) was used to indicate the stability of the boundary layer and consequently when an inversion occurred, as shown in Equation (1):

$$\frac{\Delta T}{\Delta z} = \frac{\Delta T_{T,B}}{\Delta z_{T,B}} \times 10^3 \tag{1}$$

where $\Delta T_{T,B}$ is the temperature difference between the AWS-T and AWS-B station and $\Delta z_{T,B}$ is the height difference between AWS-T and AWS-B station.

Positive values indicate an ATI, and periods when values consistently show $\Delta T/\Delta z > 0$ for at least 72 consecutive hours indicate persistent inversions, as shown in Equation (2):

$$\frac{\Delta T}{\Delta z} > 0 \text{ for } \geq 72 \text{ h} \tag{2}$$

The definition of 72 h for a persistent ATI is based on the criteria set by Largeron and Staquet [35].

2.3. Treatment of Data from the PPMs and TADs

After harmonizing the data sets of the PPM and TAD (described in detail in Novak et al. [38]), data were selected based on a specific set of criteria: (a) they were part of the heating period data set, (b) data were available for the period when ATIs occurred, (c) the PPM consistently provided data, and (d) the TADs were filled out.

A key procedure was to assign an indoor/outdoor label to each minute value. As the GPS data provided by the PPM did not provide accurate enough spatial resolution to determine if the person was indoors or outdoors, data on activities in the TAD and temperature measured by the PPM were used as follows:

- Outdoor activities in the TAD were: using a bicycle, walking, running outdoors, participating in outdoor sports activities, and three generic labels: "Home.OUT", "Office.OUT", and "Other.OUT". For indoor activities, there were similar generic labels included, "Home.IN", "Office.IN", and "Other.IN", as well as resting and sleeping indoors, playing, indoor sporting activities, cooking, cleaning, and smoking indoors. More specific activities were included in the generic labels;
- Primarily, activity and microlocation labels from the TADs were used to determine if the person was indoors or outdoors. To further refine the accuracy of the indoor/outdoor variable, ambient temperature data recorded by the PPM were used;
- During the observed period from late February to early March in 2019, the outdoor temperatures as measured by AWS-B never exceeded 19 °C in Ljubljana. Using this value as the highest base value gave an approximate highest possible temperature for outdoor activities, though it did have some drawbacks as the device could be exposed to direct sunlight and show higher values than those recorded at automatic weather stations;
- The PPM was collocated with a reference instrument (Testo SE & Co. KGaA, Lenzkirch, Germany, Testo 435-2 sensor with an external IAQ probe [39]) to assess the accuracy of the temperature measurements. Results showed that the PPM had a very high correlation (0.98) with the values recorded by the reference instrument, though the values consistently showed 4.5 °C higher values than the reference instrument. Though the PPM had precise values, they were not accurate. There are several possible causes, most probably due to the positioning of the sensor enclosed in the device close to a warm rechargeable battery. Temperature difference was even higher during the first half hour of charging the battery. This does not affect the outdoor measurements as it is reasonable to assume that the device was not charged during outdoor activities;
- Considering the above (the temperature in Ljubljana never exceeding 19 °C and the 4.5 °C (offset) higher values of the PPM), activities were removed from the outdoor

category if they had a temperature above 23.5 °C, and similarly removed from the indoor category if the temperatures were below 23.5 °C.

In some cases, certain inputs in the TADs could overlap between indoor and outdoor microlocations, where a person selected an indoor activity because it represented a majority of the hour, though they spent an amount of time in that same hour outside, e.g., preparing a meal for 40 min then going for a walk would be indicated as an indoor activity for this hour even though the person spent a third of the time outdoors. Using the temperature correction improved the accuracy of the activity dataset.

2.4. Data Selection and Evaluation

Exposure was calculated for the period between 16–22 February 2019 and 23 February to 12 March 2019, the first being the period with a persistent ATI event and two days of latency for the PM concentrations as observed at the AWS-B, and the second being the period after the inversion dispersed. Exposure in each respective domain and time period was calculated based on Equation (3):

$$E_{d,p} = \frac{\sum_{i=1}^{n} m_i}{n} \tag{3}$$

where E is the cumulative exposure, d indicates the domain (indoor, outdoor, or activity) and p the period (during ATI or post-ATI) of exposure, m_i represents each respective minute measurement in the spatiotemporal period, and n the number of measurements made in that period. A cumulative exposure approach was used to determine the baseline differences between the ATI and post-ATI periods, inter-activity differences, and how cumulative exposure assessments based on personal monitors fared in contrast to an assessment based on data collected from monitoring stations. Minute values of PM_{10} collected from the PPMs carried by the participants were aggregated and averaged based on each respective domain (temporal, spatial, activity).

The periods chosen were determined based on the results of the ATI calculations, wind speed data, and data on the height of the inversion layer. Only certain activities from the collected dataset were considered in the scope of this paper; some were removed due to the unavailability of the data in both periods, e.g., smoking and burning of incense/candles. After eliminating participants who didn't have any PM_{10} data or empty TADs, the period with persistent ATIs had fewer individuals available (3) than the post-ATI period (24). Some of the data were removed from certain participant datasets if they did not meet the required temperature criteria. Mean values with standard deviation were calculated for all indoor and outdoor activities and plotted in boxplots. A one-way ANOVA was performed on the final dataset, in combination with a Tukey's HSD (honestly significant difference) post-hoc test for pairwise comparisons.

3. Results and Discussion

3.1. ATIs

Figure 3 shows the $\Delta T/\Delta z$ values calculated for the period from 10 February to 15 March 2019 to provide some context for the observed ATIs and patterns in the fluctuations of $\Delta T/\Delta z$ [°C/km]. An ATI is defined by $\Delta T/\Delta z > 0$ °C/km, which is present several times in Figure 3, most frequently in the mornings (colored black) when the temperatures in the valley were still lower than in higher elevations. This pattern indicates diurnal inversions, formed every night due to radiative cooling of the soil, producing a very stable surface layer particularly associated with alpine valleys [35].

A persistent ATI, defined as $\Delta T/\Delta z > 0$ °C/km for at least 72 consecutive hours, as indicated in Equation (2), is present in only one period, from 16 February to 19 February, shaded gray in Figure 3, with an exception on 18 February at 14:00 with a $\Delta T/\Delta z$ of -0.25 °C/km. As this individual observation indicated only a small negative number, the decision was made to include this and the next three measurements in the period of the persistent ATI, as the daily average still showed a high $\Delta T/\Delta z$, as well as the three days following the end of this period. Importantly, the presence of the inversion is key, and the strength of the inversion does not play a vital role in the scope of this research. The days between 20 February and 22 February still experienced frequent inversions and primarily showed high levels of PM_{10} concentrations, and so were consequently included in the post-ATI period to better capture the true exposure associated with ATIs. Persistent ATIs occur in Ljubljana multiple times per year; based on Equation (2) and data collected from SEA [34] there were three periods in 2019 with persistent ATIs: 16 to 19 February, 22 to 26 October, and 23 to 27 December.

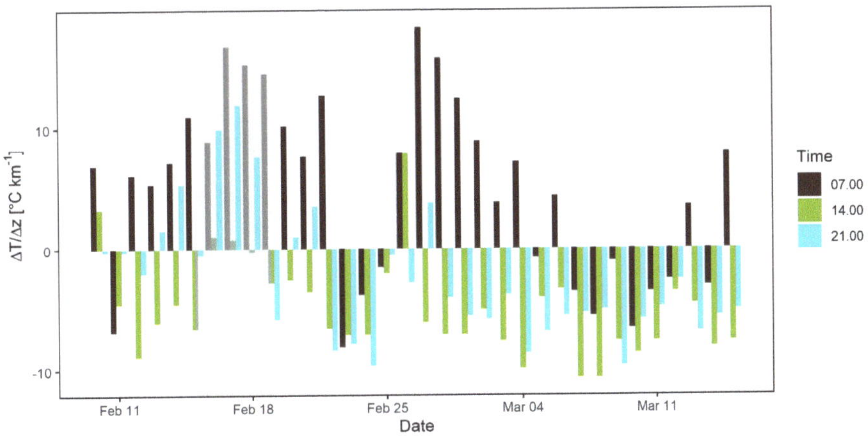

Figure 3. $\Delta T/\Delta z$ between Bežigrad and Topol stations from 10 February to 15 March 2019. Period with persistent ATI shaded in grey.

The period following the persistent ATI shows sporadic occurrences of inversions during morning and evening measurements, which completely stop on 22 February. After 25 February there is a period of diurnal ATIs in the morning hours with the highest $\Delta T/\Delta z$ value on 27 February at 18.3 °C/km. Data from AWS-B also showed that on 23 February average wind speeds increased from 0.2 to 5.0 m/s and average temperatures decreased from 9.1 °C to 1.1 °C.

Estimate of Boundary Layer Height

Data collected from the AWS-K station showed that the inversion layer did not surpass the height of the station itself (1742 m) in any of the measurements made at 12:00 in the period between 10 February and 15 March 2019. On the other hand, there were several instances in the 7:00 measuring interval, most common during the period with persistent ATIs. A case of inversion that stood out happened on 17 February 2019 at the 7:00 interval, when the AWS-B station measured a temperature of −2.1 °C, and AWS-K 6.7 °C (a difference of 8.8 °C). Moreover, observing data from the highest-lying station in Slovenia (Kredarica, at elevation 2513 m.a.s.l., some 60 km distance from AWS-B), revealed a temperature of 0.8 °C, indicating that the boundary layer was above this height on 17 February 2019 at 7:00.

3.2. PM Measurements at the Monitoring Station

As evident in Figure 4, the concentrations of PM_{10} as measured at the AWS-B started increasing as the ATIs became more frequent, peaking on 20 February, a day after the period with persistent ATIs ended. This shows a latency effect of rising PM concentrations in relation to ATIs, as the inversions continued during morning and evening measuring intervals and affected the concentrations of PM_{10}. The highest value of PM_{10} was recorded on 20 February, with 75 $\mu g/m^3$, which decreased rapidly and reached its lowest point three days later on 23 February, with 11 $\mu g/m^3$. Mean values for the high-PM period

(16–22 February, shaded green in Figure 4) and low-PM period (23 February–15 March, shaded blue in Figure 4) were 47.7 μg/m^3 and 23.2 μg/m^3, respectively, shown with dashed lines in Figure 4 for both periods.

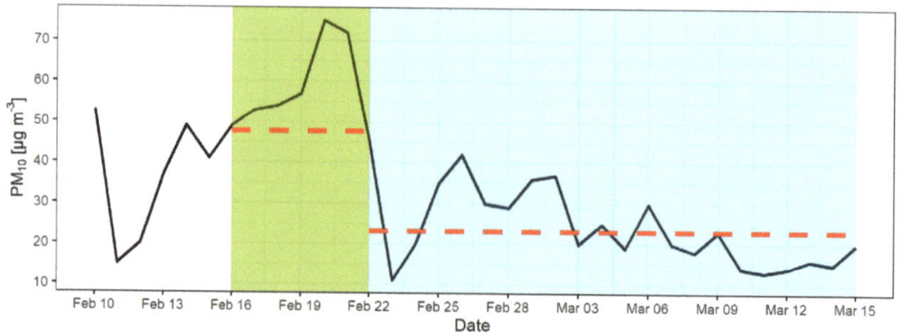

Figure 4. Measured daily PM$_{10}$ concentrations in the observed period from 10 February to 15 March 2019, collected from AWS-B. Persistent ATI period with a latent increase of PM concentrations shaded green, the post-ATI period included in the analysis shaded blue. Dashed red lines show mean PM$_{10}$ values for each period.

3.3. Data Collected from PPMs and TADs

The entire ICARUS dataset (in Ljubljana) consisted of 1,439,231 observations of 107 variables, which was refined to 136,115 observations of 32 variables for the purposes of this paper, including timestamps, indoor and outdoor activities, PM$_{10}$ concentrations, and temperature. Next, the data were separated into four groups—indoor and outdoor during the ATI (with 10,622 and 1931 observations, respectively), and indoor and outdoor post-ATI period (with 59,719 and 6664 observations, respectively).

As evident in Tables 2 and 3, there are certain activities that have a large number of recorded instances, e.g., resting, sleeping, cycling, and generic "home", "office", and "other", and some activities that have few instances, e.g., sports and running. The lowest number of instances (61) are recorded for running during the period with persistent ATIs, which is due to the fact that the recorded temperatures for most of the running period were higher than the elimination criteria for outdoor activities (23.5 °C). Almost all the values for the number of instances per activity in the post-ATI period were higher, as this period was longer and included more participants. Note that the "office" activity in Table 3 shows the number of instances for outdoor activities during the individual's work hours.

Table 2. Number of instances for each indoor activity during the ATI period and in the post-ATI period.

Period	Cleaning	Cooking	Home	Office	Other	Resting	Sleeping	Sports
ATI	395	472	8535	1319	632	1414	3638	142
post-ATI	2060	3686	41,320	11,346	4921	15,525	19,287	373

Table 3. Number of instances for each outdoor activity during the ATI period and in the post-ATI period.

Period	Bicycle	Foot	Home	Office	Other	Running	Sports
ATI	1057	538	184	1170	366	61	295
post-ATI	1312	1822	1583	967	3479	294	1825

3.4. Exposure Assessment

Figure 5 shows that exposure to PM_{10} calculated based on Equation (3), as measured by the PPM, was higher indoors and outdoors during the persistent ATI event, compared to the post-ATI period. During the ATI period participants were exposed to a mean PM_{10} concentration of 43.5 μg/m^3 ($\sigma \pm$ 26.8 μg/m^3) indoors, and 66.5 μg/m^3 ($\sigma \pm$ 23.5 μg/m^3) outdoors, which is in stark contrast with the post-ATI period where indoor and outdoor exposures were 31.2 μg/m^3 ($\sigma \pm$ 56.8 μg/m^3) and 37.7 μg/m^3 ($\sigma \pm$ 96.1 μg/m^3), respectively. As determined by the ANOVA test (and subsequently the Tukey's HSD test), the differences between the means were statistically significant, and the results show that there was a real difference between all four microlocation combinations (indoors during and after ATI, and outdoors during and after ATI). This result shows that elevated levels of PM_{10} outdoors impacts the cumulative exposure to PM_{10} indoors and outdoors. Moreover, these results show that the difference in indoor and outdoor exposure was much higher during the period of ATIs (23.0 μg/m^3) than after (6.5 μg/m^3), which indicates that exposure to PM can be influenced by high-exposure events during specific activities and in specific microlocations during a period with persistent ATIs.

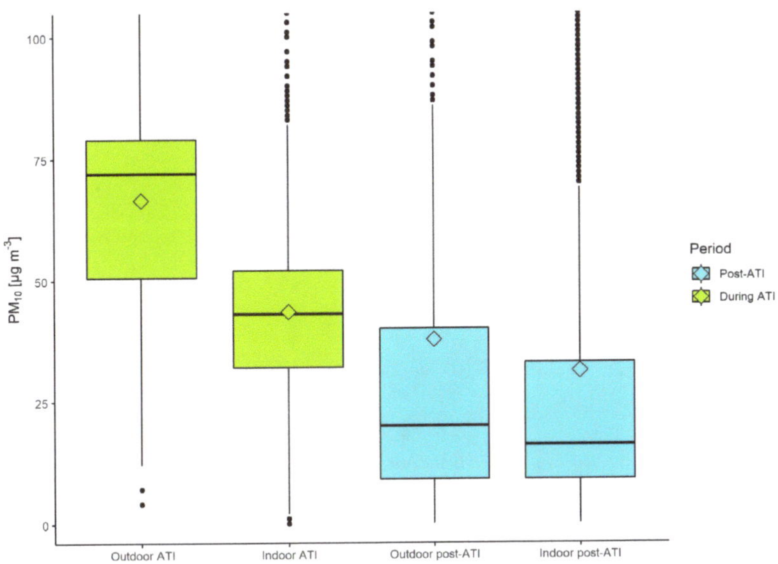

Figure 5. Calculated exposure to PM_{10} indoors and outdoors in the period with a persistent ATI event (in green) and the post-ATI period (blue), collected from the PPM devices. Values above 100 μg/m^3 were removed from the plot to better visualize the differences between mean (diamond) and median values (line).

Comparing the results obtained from the AWS-B monitoring station with the PPM data shows that the cumulative outdoor exposure assessment during the period with persistent ATIs yields a similar result regardless of which method was used (57.9 μg/m^3 for the monitoring station and 66.5 μg/m^3 for the PPM). Outdoor exposure in the post-ATI period shows a moderately different result, where the monitoring station showed a mean value of 23.2 μg/m^3 and the PPM 37.7 μg/m^3. This discrepancy could be a consequence of the PPM better capturing the actual individual exposure when the participant moved throughout the city, e.g., elevated levels of PM in some areas due to the street canyon effect [40], urban green spaces and foliage [41,42], construction sites [43], or a specific action that the person was performing. Outdoor data from the PPM showed 181 instances (3% of all recordings) of PM_{10} concentrations $\geq$200 μg/m^3 in the post-ATI period, while there were only 2 in the period with ATIs. The ability of the PPM to capture actual individual exposure is further illustrated by comparing indoor data from the PPM with monitoring stations, which showed lower concentrations for the PPM compared to AWS-B during the ATI period (43.5 μg/m^3 and 57.9 μg/m^3, respectively), and higher in the post-ATI period (31.2 μg/m^3 for the PPM data, 23.2 μg/m^3 for monitoring station).

Exposure to PM_{10} (calculated using Equation (3)) while performing different indoor activities varied between 15.0 μg/m^3 for indoor sports during the post-ATI period and 69.2 μg/m^3 for non-determined other indoor activities during the ATI period, as shown in Figure 6. Indoor activities during the ATI period had mostly the highest mean values of PM_{10}, with the exception of cleaning and cooking, which were almost the same as in the post-ATI period. The largest difference was for indoor sporting activities, which had a difference of 32.9 μg/m^3 between the two periods. Engaging in sporting activities indoors can prompt the person in the enclosed space to open windows or doors to cool down and ventilate the room, consequently causing an influx of air with a higher concentration of PM_{10} during an ATI event [44]. The next-highest difference was for other indoor activities (69.2 μg/m^3 during ATI and 38.7 μg/m^3 during post-ATI period), which often included a combination of different already-listed activities and various others. This difference is more difficult to explain due to the variability of different activities, which might include dust resuspension, use of incense, having an open window, etc.

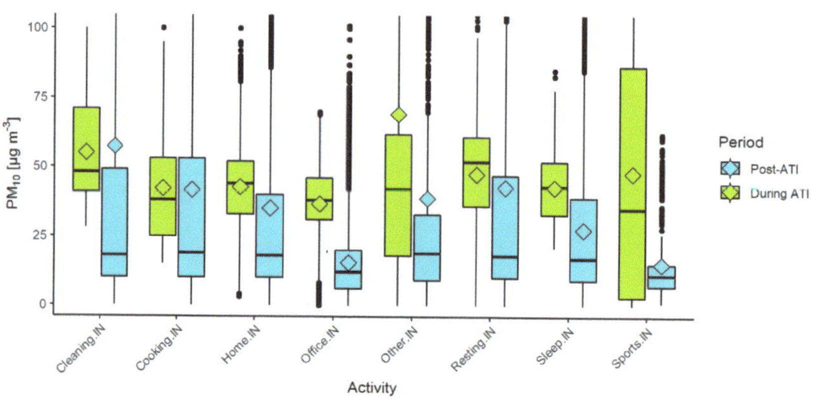

Figure 6. Exposure to PM_{10} for different indoor activities, during the ATI event (green) and after (blue), collected with the PPMs. Plot limited to 100 μg/m^3 on the y axis to better illustrate the differences.

On the other hand, mean concentrations of PM_{10} for the "cleaning" activity were higher in the post-ATI period by 2.3 μg/m^3, and only slightly lower for cooking (difference of 0.5 μg/m^3), though the median shows much lower values. Cooking and cleaning are important sources of indoor PM, and sporadic high emission events such as frying of food or dust resuspension during cleaning can increase exposure to a higher concentration than exposure during ATI events [45–47]. These events are not captured by monitoring stations and can even be missed by stationary indoor sensors if they are not present in all rooms. Individual-level monitoring shows data on a very granular level and includes specific high-emission events.

Figure 7 illustrates the differences between the distribution, mean, and median values of PM_{10} for all outdoor activities during the ATI period and post-ATI. Almost all outdoor activities show higher recorded values during the ATI period than the post-ATI period. The differences range from 15.7 μg/m^3 for running to 45.9 μg/m^3 for outdoor sports that don't include running, with the exception of the "home" activity, which was higher during the post-ATI period by 1.7 μg/m^3. A possible explanation for a higher concentration would be that outdoor activities around the home can include gardening, burning of gardening and agricultural residues [48], and home-improvement activities that often include some kind of construction (sanding wood, mixing cement, demolishing objects, etc.) which present a source of particulate matter [49,50]. Such activities can be expected to occur more frequently during the post-ATI period with clearer weather and could elevate the concentrations of PM_{10}. Participants could also differently interpret specific outdoor "home" activities, e.g., resting on a semi-enclosed balcony. Although the mean values do not differ much for the "Home.OUT" activity, the median values show that there is still a difference between these two time periods and indicate that the higher mean value of outdoor activities in the post-ATI period could be influenced by a few high-exposure events. For illustration, if the high-exposure events (>100 $\mu g/m^3$) are removed, the mean value decreases from 61.9 $\mu g/m^3$ to 30.1 $\mu g/m^3$, while the median value drops from 33 $\mu g/m^3$ to 27 $\mu g/m^3$.

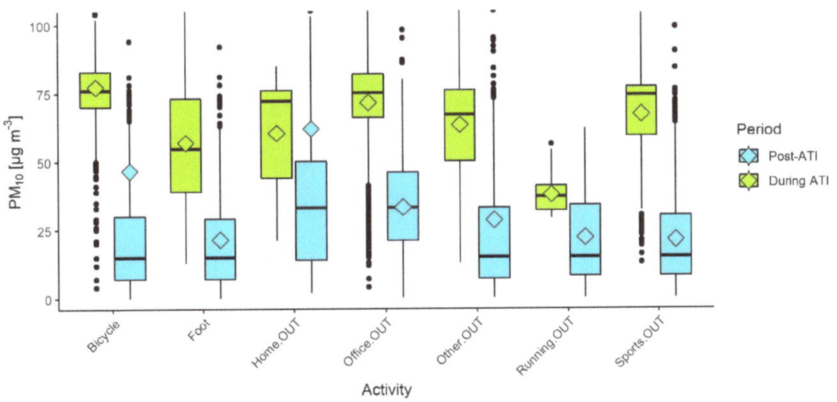

Figure 7. Exposure to PM_{10} for different outdoor activities, during (green) and after the period with persistent ATI (blue), collected with the PPMs. Plot limited to 100 $\mu g/m^3$ on the y axis to better illustrate the differences between mean and median values.

Riding a bicycle shows the highest recorded mean value of PM_{10} among all pre-classified activities, with 77.2 $\mu g/m^3$. As cyclists cover large distances and areas in an urban environment compared to pedestrians, they could potentially cross through more areas with higher concentrations of PM_{10}, e.g., construction areas, heavy traffic, and intersections [51], and increase their exposure based on extreme values, high-exposure events, and higher spatiotemporal variations [52]. Moreover, cyclists are regularly forced to share traffic lanes with motor vehicles, which increases their exposure to PM [53]. A risk-benefit balance assessment between active travel-related physical activity and exposure to air pollution shows that in areas with $PM_{2.5}$ concentrations of >100 $\mu g/m^3$, harms would exceed benefits after 90 min of bicycling per day or more than 10 h of walking per day [54]. On the other hand, our research revealed that there is a fairly large discrepancy between the mean and median value for cycling in the post-ATI period, which is a consequence of several brief high-exposure events (>100 $\mu g/m^3$) that represent 4.3% of the recorded values for the cycling activity. If these specific events are omitted, the mean value decreases from 46.5 $\mu g/m^3$ to 19.9 $\mu g/m^3$, which is close to the median value of 15 $\mu g/m^3$ (14 $\mu g/m^3$, after high-exposure values are removed).

As evident for the "foot" (walking) activities in the post-ATI period in Figure 7, pedestrians had a smaller difference between their mean and median value of exposure of 1.4 $\mu g/m^3$. Moreover, the walking activity had a smaller number of high-exposure events of >100 $\mu g/m^3$ (0.9% of all recorded values), which had a lower mean value of 181.3 $\mu g/m^3$, compared to values for cycling >100 $\mu g/m^3$ with a mean value of 633.4 $\mu g/m^3$. A similar trend is present for the period with ATIs. Pedestrians are exposed to varying concentrations of PM throughout the urban environment based on different types of road, traffic volume, time of day, and season [55], or specific high-exposure events, e.g., queuing by or walking across a crosswalk [56], which influence their cumulative exposure. As they move slower and cover less distance than cyclists, the variability can be lower, moreover, they are less frequently exposed to direct traffic exhaust than cyclists. Exposure of pedestrians is influenced by background concentrations and on smaller local roads by the pedestrians themselves who resuspend dust and might increase the concentrations of coarser fractions of PM [57].

These results illustrate how the difference in exposure between the ATI and the post-ATI period for the outdoor activities is larger than for the indoor activities, indicating that specific activities and the associated sources of PM increase exposure indoors.

4. Limitations

Certain limitations were observed in this study. Data in the ICARUS sampling campaign in Ljubljana were collected in only one non-heating season at the end of February and the beginning of March, which resulted in only a single period with persistent ATIs. This research could be further improved by analyzing multiple periods in different years and in different locations/cities, though this would present the logistical challenge of organizing yearly sampling campaigns with hundreds of participants. The sampling campaign began just as the period of persistent ATIs started, which prevented any comparisons with data prior to the ATI period.

Personal monitors based on low-cost sensors often have issues regarding their usability, data accuracy, and technical malfunctions. The PPMs used in this study frequently stopped working, did not record data, had poor accuracy of GPS data, ran out of battery, and showed data that were clearly erroneous, which resulted in data loss and also increased the workload of researchers and field workers. PM values recorded for certain activities, e.g., running, could be erroneous due to the aforementioned issues.

An additional limitation of the study was the manual logging of activity data by the participants, who frequently logged data for several days at once and sometimes mistakenly chose the wrong activity, which increased the possibility of errors. The activities did not have the same frequency in the two periods (during and post-ATI), and some were recorded only in one period and consequently eliminated from this analysis.

5. Conclusions

Within the scope of this research, an analysis of the applicability of personal PM monitor-based individual-level exposure assessments for capturing the spatiotemporal variability of individual exposure profiles was made. Two contrasting periods in terms of meteorological conditions and air quality—a period with persistent atmospheric thermal inversions (ATIs) and a post-ATI period—were used to determine how the aforementioned approach can assess exposure during specific activities and in specific microlocations. Data were collected on indoor and outdoor activities performed by participants in Ljubljana. Exposure was compared by observing the statistical values of the recorded data in the two distinct periods and comparing it with data collected from monitoring stations.

Results showed that the difference in indoor and outdoor exposure was much higher during the period of ATIs (23.0 $\mu g/m^3$) than after (6.5 $\mu g/m^3$). Indoor activities generally showed less difference in mean and median values, with cooking and cleaning having higher values in the post-ATI period than during the ATI. On the other hand, almost all outdoor activities had higher PM values during the ATI than after. Several conclusions can be drawn from these results:

1. Periods with persistent ATIs present a fitting opportunity to assess the applicability of personal monitors to capture the spatiotemporal variability of indoor and outdoor exposure. A clear distinction in terms of PM concentrations between the two periods provides an opportunity to observe how high-exposure events can influence cumulative exposure;
2. Exposure to PM_{10} is higher during periods with persistent ATIs, when ambient concentrations increase due to specific meteorological conditions. This is evident indoors and outdoors and for almost all activities, except for a few that are mainly influenced by the PM_{10} associated with the respective activity. Indoor concentrations

are lower than the outdoor concentrations during the period with ATIs, though they are still higher than indoor and outdoor concentrations in the post-ATI period;

3. Using activity data enables an individual-level scale analysis of exposure and illustrates that the influence of activities on exposure indoors should not be disregarded when assessing cumulative exposure. Activities can directly, e.g., cooking and cleaning, or indirectly reduce air quality, e.g., opening a window during a period with poor outdoor air quality;
4. Measuring exposure on an individual level is necessary to capture high-exposure events in microlocations. These results showed that several high-exposure events can greatly raise exposure levels. Additionally, personal monitors can detect trends and show how specific routines influence exposure;
5. These measurements confirm that there are high levels of exposure indoors even in high-income countries that mostly don't use solid fuels for cooking and heating. A better understanding of activity-specific exposure could provide a basis for policies that can more accurately address exposure to poor air quality.

Overall, this study demonstrated that utilizing personal monitors in exposure assessments can provide better spatiotemporal resolution and capture specific high-exposure events. These devices provide an ancillary tool that can indicate trends and guide further research.

Future work should include more detailed activities and a better spatiotemporal resolution. Personal monitors could be further improved to better record, store, harmonize and transfer data, detect outliers, have on-the-fly calibration options, and integrate multiple devices. Reducing the proportion of data that are recorded by human input via an approach with automated activity recognition could improve exposure assessments [58]. Exposure models that rely solely on outdoor measuring stations or indoor stationary devices fail to capture high-exposure events and could be improved by integrating data from personal monitors. Moreover, data from personal monitors could be integrated into agent-based models to supplement other data sources [59], e.g., monitoring stations, statistical and demographic data, etc.

This research addressed exposure to particulate matter, though there are numerous other air pollutants that could be further investigated by employing personal monitors. Moreover, current AQ guidelines often do not include indoor environments or individual-level exposure. Results obtained in the scope of this research should be further developed and transferred into policy, to include approaches that utilize data on a personal scale and the specifics related to human behavior in urban environments.

Author Contributions: Conceptualization, R.N. and D.K.; methodology, R.N.; software, R.N.; validation, R.N. and D.K.; formal analysis, R.N.; investigation, R.N., D.K., J.A.R. and T.K.; resources, R.N., D.K., J.A.R. and T.K.; data curation, R.N.; writing—original draft preparation, R.N.; writing—review and editing, R.N., D.K., J.A.R. and T.K.; visualization, R.N., J.A.R. and T.K.; supervision, D.K.; project administration, D.K. and D.S.; funding acquisition, D.S. All authors approved the content of the manuscript. All authors have read and agreed to the published version of the manuscript.

Funding: This work has received funding from the European Union's Horizon 2020 Programme for Research, Technological Development, and Demonstration, under grant agreement No. 690105 (Integrated Climate forcing and Air pollution Reduction in Urban Systems (ICARUS)). This work reflects only the authors' views, and the European Commission is not responsible for any use that may be made of the information it contains. Funding was received from the Young Researchers Program and the P1-0143 program "Cycling of substances in the environment, mass balances, modelling of environmental processes and risk assessment", both funded by the Slovenian Research Agency.

Institutional Review Board Statement: Ethical approval for the ICARUS project in Slovenia was obtained from the National Medical Ethics Committee of the Republic of Slovenia (approval nr. 0120-388/2018/6 on 22 August 2018). The data in this paper were selected only from participants in Slovenia.

Informed Consent Statement: Informed consent was obtained from all subjects involved in the study.

Data Availability Statement: Data available on request due to privacy restrictions.

Acknowledgments: The authors would like to thank the participants from all the cities that generously provided their time and effort to this research.

Conflicts of Interest: The authors declare no conflict of interest.

References

1. Anenberg, S.C.; Achakulwisut, P.; Brauer, M.; Moran, D.; Apte, J.S.; Henze, D.K. Particulate Matter-Attributable Mortality and Relationships with Carbon Dioxide in 250 Urban Areas Worldwide. *Sci. Rep.* **2019**, *9*, 11552. [CrossRef] [PubMed]
2. Pope, C.A.; Ezzati, M.; Dockery, D.W. Fine-Particulate Air Pollution and Life Expectancy in the United States. *N. Engl. J. Med.* **2009**, *360*, 376–386. [CrossRef] [PubMed]
3. Kim, S.-Y.; Pope, A.C.; Marshall, J.D.; Fann, N.; Sheppard, L. Reanalysis of the Association between Reduction in Long-Term PM2.5 Concentrations and Improved Life Expectancy. *Environ. Health* **2021**, *20*, 102. [CrossRef] [PubMed]
4. Gutman, L.; Pauly, V.; Orleans, V.; Piga, D.; Channac, Y.; Armengaud, A.; Boyer, L.; Papazian, L. Long-Term Exposure to Ambient Air Pollution Is Associated with an Increased Incidence and Mortality of Acute Respiratory Distress Syndrome in a Large French Region. *Environ. Res.* **2022**, *212*, 113383. [CrossRef]
5. Wu, Y.; Wang, W.; Liu, C.; Chen, R.; Kan, H. The Association between Long-Term Fine Particulate Air Pollution and Life Expectancy in China, 2013 to 2017. *Sci. Total Environ.* **2020**, *712*, 136507. [CrossRef] [PubMed]
6. *Directive 2008/50/EC of the European Parliament and of the Council of 21 May 2008 on Ambient Air Quality and Cleaner Air for Europe*; EEA: Copenhagen, Denmark, 2008; Volume 152.
7. de Ferreyro Monticelli, D.; Santos, J.M.; Goulart, E.V.; Mill, J.G.; da Silva Corrêa, J.; dos Santos, V.D.; Reis, N.C. Comparison of Methods for Assessment of Children Exposure to Air Pollution: Dispersion Model, Ambient Monitoring, and Personal Samplers. *Air Qual. Atmos. Health* **2022**, *15*, 645–655. [CrossRef]
8. Park, Y.M.; Kwan, M.-P. Individual Exposure Estimates May Be Erroneous When Spatiotemporal Variability of Air Pollution and Human Mobility Are Ignored. *Health Place* **2017**, *43*, 85–94. [CrossRef]
9. Menon, J.S.; Nagendra, S.M.S. Personal Exposure to Fine Particulate Matter Concentrations in Central Business District of a Tropical Coastal City. *J. Air Waste Manag. Assoc.* **2018**, *68*, 415–429. [CrossRef]
10. Georgiadis, P.; Stoikidou, M.; Topinka, J.; Kaila, S.; Gioka, M.; Katsouyanni, K.; Sram, R.; Kyrtopoulos, S.A. Personal Exposures to PM(2.5) and Polycyclic Aromatic Hydrocarbons and Their Relationship to Environmental Tobacco Smoke at Two Locations in Greece. *J. Expo. Anal. Environ. Epidemiol.* **2001**, *11*, 169–183. [CrossRef]
11. Choi, H.; Perera, F.; Pac, A.; Wang, L.; Flak, E.; Mroz, E.; Jacek, R.; Chai-Onn, T.; Jedrychowski, W.; Masters, E.; et al. Estimating Individual-Level Exposure to Airborne Polycyclic Aromatic Hydrocarbons throughout the Gestational Period Based on Personal, Indoor, and Outdoor Monitoring. *Environ. Health Perspect.* **2008**, *116*, 1509–1518. [CrossRef]
12. Mohammed, M.O.A.; Song, W.-W.; Ma, W.-L.; Li, W.-L.; Ambuchi, J.J.; Thabit, M.; Li, Y.-F. Trends in Indoor–Outdoor PM2.5 Research: A Systematic Review of Studies Conducted during the Last Decade (2003–2013). *Atmos. Pollut. Res.* **2015**, *6*, 893–903. [CrossRef]
13. Dessimond, B.; Annesi-Maesano, I.; Pepin, J.-L.; Srairi, S.; Pau, G. Academically Produced Air Pollution Sensors for Personal Exposure Assessment: The Canarin Project. *Sensors* **2021**, *21*, 1876. [CrossRef] [PubMed]
14. Jerrett, M.; Burnet, R.T.; Ma, R.; Pope, C.A.I.; Krewski, D.; Newbold, B.; Thurston, G.; Shi, Y.; Finkelstein, N.; Calle, E.E.; et al. Spatial Analysis of Air Pollution and Mortality in Los Angeles. *Epidemiology* **2006**, *17*, S69. [CrossRef]
15. Tsou, M.-C.M.; Lung, S.-C.C.; Shen, Y.-S.; Liu, C.-H.; Hsieh, Y.-H.; Chen, N.; Hwang, J.-S. A Community-Based Study on Associations between PM2.5 and PM1 Exposure and Heart Rate Variability Using Wearable Low-Cost Sensing Devices. *Environ. Pollut.* **2021**, *277*, 116761. [CrossRef] [PubMed]
16. Liang, L.; Gong, P.; Cong, N.; Li, Z.; Zhao, Y.; Chen, Y. Assessment of Personal Exposure to Particulate Air Pollution: The First Result of City Health Outlook (CHO) Project. *BMC Public Health* **2019**, *19*, 711. [CrossRef] [PubMed]
17. Niedźwiedź, T.; Łupikasza, E.B.; Małarzewski, Ł.; Budzik, T. Surface-Based Nocturnal Air Temperature Inversions in Southern Poland and Their Influence on PM10 and PM2.5 Concentrations in Upper Silesia. *Appl. Clim.* **2021**, *146*, 897–919. [CrossRef]
18. Tunno, B.J.; Michanowicz, D.R.; Shmool, J.L.C.; Kinnee, E.; Cambal, L.; Tripathy, S.; Gillooly, S.; Roper, C.; Chubb, L.; Clougherty, J.E. Spatial Variation in Inversion-Focused vs. 24-h Integrated Samples of PM2.5 and Black Carbon across Pittsburgh, PA. *J. Expo. Sci. Environ. Epidemiol.* **2016**, *26*, 365–376. [CrossRef]
19. Wallace, J.; Corr, D.; Kanaroglou, P. Topographic and Spatial Impacts of Temperature Inversions on Air Quality Using Mobile Air Pollution Surveys. *Sci. Total Environ.* **2010**, *408*, 5086–5096. [CrossRef]
20. Trinh, T.T.; Trinh, T.T.; Le, T.T.; Nguyen, T.D.H.; Tu, B.M. Temperature Inversion and Air Pollution Relationship, and Its Effects on Human Health in Hanoi City, Vietnam. *Env. Geochem. Health* **2019**, *41*, 929–937. [CrossRef]
21. Beard, J.D.; Beck, C.; Graham, R.; Packham, S.C.; Traphagan, M.; Giles, R.T.; Morgan, J.G. Winter Temperature Inversions and Emergency Department Visits for Asthma in Salt Lake County, Utah, 2003–2008. *Environ. Health Perspect.* **2012**, *120*, 1385–1390. [CrossRef]
22. Abdul-Wahab, S.A.; Bakheit, C.S.; Siddiqui, R.A. Study the Relationship between the Health Effects and Characterization of Thermal Inversions in the Sultanate of Oman. *Atmos. Environ.* **2005**, *39*, 5466–5471. [CrossRef]

23. Kikaj, D.; Vaupotič, J.; Chambers, S.D. Identifying Persistent Temperature Inversion Events in a Subalpine Basin Using Radon-222. *Atmos. Meas. Tech.* **2019**, *12*, 4455–4477. [CrossRef]
24. Ogrin, M. Značilnosti temperaturnih inverzij. *Geogr. Obz.* **2005**, *2*, 24–27.
25. Faganeli Pucer, J.; Štrumbelj, E. Impact of Changes in Climate on Air Pollution in Slovenia between 2002 and 2017. *Environ. Pollut.* **2018**, *242*, 398–406. [CrossRef]
26. European City Air Quality Viewer—European Environment Agency. Available online: https://www.eea.europa.eu/themes/air/urban-air-quality/european-city-air-quality-viewer (accessed on 22 February 2022).
27. ICARUS2020 ICARUS2020. Available online: https://icarus2020.eu/ (accessed on 28 January 2022).
28. Robinson, J.A.; Novak, R.; Kanduč, T.; Maggos, T.; Pardali, D.; Stamatelopoulou, A.; Saraga, D.; Vienneau, D.; Flückiger, B.; Mikeš, O.; et al. User-Centred Design of a Final Results Report for Participants in Multi-Sensor Personal Air Pollution Exposure Monitoring Campaigns. *Int. J. Environ. Res. Public Health* **2021**, *18*, 12544. [CrossRef] [PubMed]
29. Kocman, D.; Kanduč, T.; Novak, R.; Robinson, J.A.; Mikeš, O.; Degrendele, C.; Sáňka, O.; Vinkler, J.; Prokeš, R.; Vienneau, D.; et al. Multi-Sensor Data Collection for Personal Exposure Monitoring: ICARUS Experience. *Fresenius Environ. Bull.* **2022**, *31*, 8349–8354.
30. IoTECH Telecommunications | LoRaWANTM Network Provider | LoRaWAN LPWAN Connectivity in Greece | Internet of Things | Smart Cities | IoT Connectivity. Available online: https://iotech.gr/ (accessed on 27 July 2022).
31. PMS5003—Laser PM2.5 Sensor-Plantower Technology. Available online: https://www.plantower.com/en/products_33/74.html (accessed on 27 July 2022).
32. Plantower-Pms5003-Manual Annotated. Available online: https://usermanual.wiki/Pdf/plantowerpms5003manualannotated.626592918/view (accessed on 3 September 2019).
33. Novak, R.; Kocman, D.; Robinson, J.A.; Kanduč, T.; Sarigiannis, D.; Horvat, M. Comparing Airborne Particulate Matter Intake Dose Assessment Models Using Low-Cost Portable Sensor Data. *Sensors* **2020**, *20*, 1406. [CrossRef]
34. Slovenian Environmental Agency—Air Quality Data. Available online: https://www.arso.gov.si/en/air/data/ (accessed on 6 October 2021).
35. Largeron, Y.; Staquet, C. Persistent Inversion Dynamics and Wintertime PM10 Air Pollution in Alpine Valleys. *Atmos. Environ.* **2016**, *135*, 92–108. [CrossRef]
36. Qgis2threejs—QGIS Python Plugins Repository. Available online: https://plugins.qgis.org/plugins/Qgis2threejs/ (accessed on 1 August 2022).
37. QGIS Association QGIS.Org. Available online: https://qgis.org/ (accessed on 1 August 2022).
38. Novak, R.; Petridis, I.; Kocman, D.; Robinson, J.A.; Kanduč, T.; Chapizanis, D.; Karakitsios, S.; Flückiger, B.; Vienneau, D.; Mikeš, O.; et al. Harmonization and Visualization of Data from a Transnational Multi-Sensor Personal Exposure Campaign. *Int. J. Environ. Res. Public Health* **2021**, *18*, 11614. [CrossRef]
39. Testo 435-2—Indoor Air Quality Meter | Testo International. Available online: https://www.testo.com/en/testo-435-2/p/0563-4352 (accessed on 2 August 2022).
40. Miao, C.; Yu, S.; Hu, Y.; Bu, R.; Qi, L.; He, X.; Chen, W. How the Morphology of Urban Street Canyons Affects Suspended Particulate Matter Concentration at the Pedestrian Level: An in-Situ Investigation. *Sustain. Cities Soc.* **2020**, *55*, 102042. [CrossRef]
41. Miao, C.; Yu, S.; Hu, Y.; Liu, M.; Yao, J.; Zhang, Y.; He, X.; Chen, W. Seasonal Effects of Street Trees on Particulate Matter Concentration in an Urban Street Canyon. *Sustain. Cities Soc.* **2021**, *73*, 103095. [CrossRef]
42. Bi, S.; Dai, F.; Chen, M.; Xu, S. A New Framework for Analysis of the Morphological Spatial Patterns of Urban Green Space to Reduce PM2.5 Pollution: A Case Study in Wuhan, China. *Sustain. Cities Soc.* **2022**, *82*, 103900. [CrossRef]
43. Yan, H.; Ding, G.; Feng, K.; Zhang, L.; Li, H.; Wang, Y.; Wu, T. Systematic Evaluation Framework and Empirical Study of the Impacts of Building Construction Dust on the Surrounding Environment. *J. Clean. Prod.* **2020**, *275*, 122767. [CrossRef]
44. Slezakova, K.; Peixoto, C.; Pereira, M.d.C.; Morais, S. Indoor Air Quality in Health Clubs: Impact of Occupancy and Type of Performed Activities on Exposure Levels. *J. Hazard. Mater.* **2018**, *359*, 56–66. [CrossRef] [PubMed]
45. Zhang, L.; Ou, C.; Magana-Arachchi, D.; Vithanage, M.; Vanka, K.S.; Palanisami, T.; Masakorala, K.; Wijesekara, H.; Yan, Y.; Bolan, N.; et al. Indoor Particulate Matter in Urban Households: Sources, Pathways, Characteristics, Health Effects, and Exposure Mitigation. *Int. J. Environ. Res. Public Health* **2021**, *18*, 11055. [CrossRef]
46. Abdel-Salam, M.M.M. Indoor Particulate Matter in Urban Residences of Alexandria, Egypt. *J. Air Waste Manag. Assoc.* **2013**, *63*, 956–962. [CrossRef] [PubMed]
47. Park, S.; Song, D.; Park, S.; Choi, Y. Particulate Matter Generation in Daily Activities and Removal Effect by Ventilation Methods in Residential Building. *Air Qual. Atmos. Health* **2021**, *14*, 1665–1680. [CrossRef]
48. Noblet, C.; Besombes, J.-L.; Lemire, M.; Pin, M.; Jaffrezo, J.-L.; Favez, O.; Aujay-Plouzeau, R.; Dermigny, A.; Karoski, N.; Van Elsuve, D.; et al. Emission Factors and Chemical Characterization of Particulate Emissions from Garden Green Waste Burning. *Sci. Total Environ.* **2021**, *798*, 149367. [CrossRef]
49. Latif, M.T.; Baharudin, N.H.; Velayutham, P.; Awang, N.; Hamdan, H.; Mohamad, R.; Mokhtar, M.B. Composition of Heavy Metals and Airborne Fibers in the Indoor Environment of a Building during Renovation. *Environ. Monit. Assess.* **2011**, *181*, 479–489. [CrossRef] [PubMed]
50. Abdel Hameed, A.A.; Yasser, I.H.; Khoder, I.M. Indoor Air Quality during Renovation Actions: A Case Study. *J. Environ. Monit.* **2004**, *6*, 740–744. [CrossRef]

51. Bergmann, M.L.; Andersen, Z.J.; Amini, H.; Khan, J.; Lim, Y.H.; Loft, S.; Mehta, A.; Westendorp, R.G.; Cole-Hunter, T. Ultrafine Particle Exposure for Bicycle Commutes in Rush and Non-Rush Hour Traffic: A Repeated Measures Study in Copenhagen, Denmark. *Environ. Pollut.* **2022**, *294*, 118631. [CrossRef] [PubMed]
52. Ueberham, M.; Schlink, U.; Dijst, M.; Weiland, U. Cyclists' Multiple Environmental Urban Exposures—Comparing Subjective and Objective Measurements. *Sustainability* **2019**, *11*, 1412. [CrossRef]
53. Qiu, Z.; Wang, W.; Zheng, J.; Lv, H. Exposure Assessment of Cyclists to UFP and PM on Urban Routes in Xi'an, China. *Environ. Pollut.* **2019**, *250*, 241–250. [CrossRef]
54. Tainio, M.; de Nazelle, A.J.; Götschi, T.; Kahlmeier, S.; Rojas-Rueda, D.; Nieuwenhuijsen, M.J.; de Sá, T.H.; Kelly, P.; Woodcock, J. Can Air Pollution Negate the Health Benefits of Cycling and Walking? *Prev. Med.* **2016**, *87*, 233–236. [CrossRef]
55. Alas, H.D.; Stöcker, A.; Umlauf, N.; Senaweera, O.; Pfeifer, S.; Greven, S.; Wiedensohler, A. Pedestrian Exposure to Black Carbon and PM2.5 Emissions in Urban Hot Spots: New Findings Using Mobile Measurement Techniques and Flexible Bayesian Regression Models. *J. Expo. Sci. Environ. Epidemiol.* **2021**, *32*, 1–11. [CrossRef] [PubMed]
56. Wang, Z.; He, H.; Zhao, H.; Peng, Z. Spatiotemporal Analysis of Pedestrian Exposure to Submicron and Coarse Particulate Matter on Crosswalk at Urban Intersection. *Build. Environ.* **2021**, *204*, 108149. [CrossRef]
57. Qiu, Z.; Xu, X.; Song, J.; Luo, Y.; Zhao, R.; Zhou, B.X.W.; Li, X.; Hao, Y. Pedestrian Exposure to Traffic PM on Different Types of Urban Roads: A Case Study of Xi'an, China. *Sustain. Cities Soc.* **2017**, *32*, 475–485. [CrossRef]
58. Novak, R.; Kocman, D.; Robinson, J.A.; Kanduč, T.; Sarigiannis, D.; Džeroski, S.; Horvat, M. Low-Cost Environmental and Motion Sensor Data for Complex Activity Recognition: Proof of Concept. *Eng. Proc.* **2020**, *2*, 54. [CrossRef]
59. Chapizanis, D.; Karakitsios, S.; Gotti, A.; Sarigiannis, D.A. Assessing Personal Exposure Using Agent Based Modelling Informed by Sensors Technology. *Environ. Res.* **2021**, *192*, 110141. [CrossRef]

sensors

MDPI

Article

Enhanced Ambient Sensing Environment—A New Method for Calibrating Low-Cost Gas Sensors

Hugo Savill Russell [1,2,3,*], Louise Bøge Frederickson [1,2,3], Szymon Kwiatkowski [3], Ana Paula Mendes Emygdio [4], Prashant Kumar [4], Johan Albrecht Schmidt [3], Ole Hertel [2,5] and Matthew Stanley Johnson [3,6]

1 Department of Environmental Science, Aarhus University, DK-4000 Roskilde, Denmark
2 Danish Big Data Centre for Environment and Health (BERTHA), Aarhus University, DK-4000 Roskilde, Denmark
3 AirLabs, Nannasgade 28, DK-2200 Copenhagen N, Denmark
4 Global Center for Clean Air Research (GCARE), School of Sustainability, Civil and Environmental Engineering, Faculty of Engineering and Physical Sciences, University of Surrey, Surrey GU2 7XH, UK
5 Department of Ecoscience, Aarhus University, DK-4000 Roskilde, Denmark
6 Department of Chemistry, University of Copenhagen, DK-2100 Copenhagen Ø, Denmark
* Correspondence: hugo.russell@envs.au.dk

Abstract: Accurate calibration of low-cost gas sensors is, at present, a time consuming and difficult process. Laboratory calibration and field calibration methods are currently used, but laboratory calibration is generally discounted due to poor transferability, and field methods requiring several weeks are standard. The Enhanced Ambient Sensing Environment (EASE) method described in this article, is a hybrid of the two, combining the advantages of a laboratory calibration with the increased accuracy of a field calibration. It involves calibrating sensors inside a duct, drawing in ambient air with similar properties to the site where the sensors will operate, but with the added feature of being able to artificially increases or decrease pollutant levels, thus condensing the calibration period required. Calibration of both metal-oxide (MOx) and electrochemical (EC) gas sensors for the measurement of NO_2 and O_3 (0–120 ppb) were conducted in EASE, laboratory and field environments, and validated in field environments. The EC sensors performed marginally better than MOx sensors for NO_2 measurement and sensor performance was similar for O_3 measurement, but the EC sensor nodes had less node inter-node variability and were more robust. For both gasses and sensor types the EASE calibration outperformed the laboratory calibration, and performed similarly to or better than the field calibration, whilst requiring a fraction of the time.

Keywords: low-cost sensors; metal oxide sensor; electrochemical sensor; calibration protocol; calibration

1. Introduction

Poor air quality (AQ) constitutes a global public health emergency. Estimates of the global death toll caused by air pollution reach up to 9 million per year, or ∼1 in 6 deaths, and this number is increasing [1–4]. The impact of poor AQ on the cardiopulmonary system has been known for many years, but emerging studies link pollution exposure to a massive range of adverse health impacts, extending from dementia, Parkinson's disease, and cognitive impairment, to diabetes, obesity and issues with the reproductive system [2,5–7]. It may, in fact, be damaging every organ in the human body [8]. Nitrogen dioxide (NO_2) and Ozone (O_3) are gas phase pollutants, with high spatiotemporal variability in urban environments and known health impacts [9,10]. The World Health Organisation Global Air Quality Guideline for NO_2 is $10\,\mu g\,m^{-3}$ (∼5 ppb) as an annual mean, and for O_3 is $60\,\mu g\,m^{-3}$ (∼30 ppb) as an 8-h mean for peak season [9].

Further efforts to quantify the links between AQ and health, and to produce targeted solutions, particularly for individual pollutants, are hampered however by the scarcity of AQ monitoring [11,12]. Reference standard monitoring utilises large and expensive instruments based on chemiluminescence for NO_2 and UV photometry for O_3 measurements [13,14], while pollutant concentrations vary rapidly in time and space, their measurement has low spatiotemporal resolution in developed countries due to the size and cost of traditional reference standard monitoring stations [12], and can be almost non-existent in low and middle-income countries, where the greater burden of poor AQ is felt [2,4,10]. Modelling of air pollution levels, through parameterised semi-empirical models, is an alternative, but often challenged by lack of high quality input and calibrated data, as modelling can not stand alone, but should be used in combination with measurements (see Hertel et al., 2007 [15]).

Low-cost sensors (LCS) have the potential to revolutionise air quality monitoring. There is no agreed definition of a low-cost sensor, however, they are far cheaper, smaller and record with greater time-resolution than traditional methods [12,16–18]. This means that they can be deployed in greater numbers, used in mobile applications, and in areas where monitoring is not currently possible, all with high time-resolution. Hence hot-spots, point-sources, indoor concentrations and even personal exposure levels can be identified and measured [12]. Monitoring programmes and epidemiological studies could also be implemented in lower income countries, which until now have had to reply on findings extrapolated from studies in other areas, that may underestimate the effects of poor AQ [19].

Many companies are already producing LCS commercially, and this has led to their use in scientific studies and citizen science projects, as well as privately [20,21]. However, together with the many benefits LCS have over traditional methods, they also have outstanding issues with data quality. In particular, LCS suffer from issues with selectivity, sensitivity, and stability, all of which detract from their overall accuracy [12,17,22,23]. These issues are typical to all LCS but also depend on the sensor's principle of operation, and the context they are used in. In this study, we focus on gas sensors for NO_2 and O_3, based on Metal-Oxides (MOx) and Electrochemical (EC) cells. These are the most widely used LCS for NO_2 and O_3 measurement, with EC being more common [22,24]. They are both chemo-resistive sensors; MOx operate by measuring resistance change across a metal-oxide surface resulting from gas adsorption [25], and EC cells by using amperometry to measure the current of a redox reaction which is proportional to the gas concentration present in the air above the cell [26]. Further details of the individual sensors are given in Section 2. Both of these sensor types suffer particular issues with drift/ageing, cross-sensitivity between pollutants, and effects from temperature (T) [27–29].

Resultant data quality can be drastically improved with effective sensor calibration (and hardware approaches described in Section 2). Pre-deployment calibration is used to identify sources of error and develop calibration models that bring the sensors into best possible agreement with reference instruments [18]. Post-deployment calibration is necessary to maintain this during longer deployments, however, there is not currently a recognised standard procedure for calibration, and many commercial sensors are sold uncalibrated [22,30]. Calibration is individual and must be repeated for each LCS unit. Ideally each unit should also have a different calibration model for individual environments, as different co-pollutant levels and environmental factors will alter their response [24]. Once deployed, the drift in response will also be individual, dependent on both the LCS's original characteristics and it's environment. Therefore, individual re-calibration should also be performed when necessary [24,27].

The prevailing calibration methods are either Field (also called *co-location*) calibration or laboratory (Lab) calibration, both of which have known drawbacks [18,22]. Laboratory calibration can be performed relatively quickly, in any laboratory with appropriate equipment, and the range of pollutant concentrations, relative humidity (RH), and T levels, can be chosen. However, laboratory-based calibrations rarely perform well when validated in real field conditions [16,31] and, in a recent review of LCS for AQ monitoring, 90% of LCS studies surveyed (for NO_2 and O_3 measurement) utilised Field calibration [22]. The exact reason for this difference in efficacy has not been determined, however, it is thought to be due to meteorological and co-pollutant fluctuations in the field environment, which are not accurately represented in the laboratory, as well as the high cost associated with an effective laboratory calibration setup. There may also be contaminants in the laboratory environment for example volatile organic compounds (VOCs) and their oxidation products, not found in the field.

Field calibration is considered superior, particularly if calibration occurs in a similar geographical area to the actual measurement and in the same season of the year. However, it is time intensive, typically requiring several weeks to observe a comprehensive range of concentrations, particularly for more complex models, requires access to an official reference station or similar staging area, and leaves to chance whether or not the full range of pollutant concentrations is encountered [30,32–34]. Field calibration models may also not be transferable if moved between sites with different concentration profiles, co-pollutant matrices or prevailing meteorological conditions [22,35,36]. It has been claimed that the Laboratory and Field calibration methods are complementary and a combination of data from both methods is required for a full assessment of sensor performance and the production of a robust calibration model [16,26].

Within the different methods used to obtain calibration data (e.g., Field calibration and Laboratory calibration) there are many calibration models available, chiefly; Linear Regression (LR), Multivariate Linear Regression (MLR), or a range of Machine Learning (ML) algorithms, such as artificial neural networks, random forest and support vector regression, amongst others [22,30,36]. The most suitable calibration approach for all situations has not been determined. Complex ML algorithms often perform better than MLR when observing training data and if sensors are not moved after calibration, but can perform poorly after sensors are transferred to a different site [30,36]. This may be due to the complex models over-fitting specific aspects of the training site that are not directly related to pollutant concentrations, whereas, a simpler LR or MLR model may appear worse during training but not have a significant increase in error after transfer [33,37].

In this study, a 'hybrid' calibration method is described. This calibration method is a combination of both the Laboratory and the Field calibration methods, as the Hybrid calibration method draws ambient air into an insulated, well-mixed duct with a steady flow. This ensures that the air maintains similar properties to those observed outdoors. Inlets for adding NO_2 and O_3, as well as an activated carbon filter were also added to the duct. Thereby, sensor nodes placed inside the duct are exposed to ambient pollution levels, as well as artificially increased and lowered levels. The responses from the LCS are compared to reference monitors sampling the duct from the outside. This system is called the Enhanced Ambient Sensing Environment or EASE.

Previous works have been performed on the development of optimal laboratory calibration setups, but without the addition of ambient pollutants/conditions, this work was drawn on in the initial stages of the EASE design [38,39]. However, these typically focus on the manipulation and control of experimental conditions, whereas the main focus of the EASE setup is on preserving ambient conditions, whilst having the added advantage of manipulating the pollutant concentrations. This means that the LCS can be calibrated according to ambient concentrations, under realistic conditions in terms of RH, T, and the presence of a co-pollutant matrix, but with additional spiking of pollutants to ensure that the full span of pollutant concentrations is included in the calibration, within a short time-period. At points, filtration of the incoming air is also used, for identifying the zero/baseline response of the sensors under ambient conditions.

This study paves the way for future calibration methods that are faster and more accurate for real-world use. In this work, the EASE method is compared with Field and Laboratory calibration methods for 12 MOx and 15 EC sensors. Calibration was performed separately for the two sensor types and with differing setups. The main differences in the overall protocol is that the MOx sensors were EASE calibrated in Copenhagen, Denmark, and also Field calibrated and validated in Copenhagen, 3 km away, whereas, the EC sensors were calibrated in Copenhagen but were Field calibrated and validated in Surrey,

UK, and with a longer separation in time. Furthermore, the MOx nodes were laboratory calibrated at Copenhagen University whereas the EC sensors were laboratory calibrated by their manufacturer prior to node assembly.

The main aim of the work is to compare the three different calibration methods (Field, Lab and EASE) for both sensor types (MOx and EC) as well as comparison between the sensors. The EASE method gave better results than Laboratory calibration and had similar results to Field calibration, whilst requiring a fraction of the time. The EC sensors performed better than the MOx nodes for NO_2 measurement and similarly for O_3 measurement, but with less inter-variability between nodes.

2. Materials and Methods

In this section, the sensor node hardware is described, followed by the calibration setups and procedures for each of the three calibration methods, separately, for both node types. The data analysis was conducted in R [40], using the ggplot2 [41] and Openair [42] packages for data visualisation.

2.1. Sensor Nodes

In this article, 'sensor' refers to individual sensing devices, e.g., a metal-oxide chip, whereas 'sensor node' or 'node' refers to a complete package including sensors, housing, sampling system, and the ability to internally log or broadcast data.

Two types of sensor nodes were used in this study, both are prototypes developed by AirLabs ApS for monitoring urban air quality and called 'AirNodes'. Both nodes measure NO_2 and O_3, Generation 2 (Gen 2) AirNodes do so with MOx sensors, whereas Generation 5 (Gen 5) AirNodes utilise electrochemical cells. The relevant sensors to this study are detailled for both nodes in Table 1, schematics and images of the nodes are shown in Figures 1 and 2.

Table 1. Relevant sensors within the Gen 2 and Gen 5 nodes. The output column describes sensor output after processing and calibration.

Node	Sensor	Producer	Type	Output
Gen 2	MiCS-6814	SGX Sensortech/AirLabs	MOx	NO_2/ppb
Gen 2	MiCS-6814	SGX Sensortech/AirLabs	MOx	O_3/ppb
Gen 5	NO2-B43F	Alphasense/AirLabs	EC	NO_2/ppb
Gen 5	OX-B431	Alphasense/AirLabs	EC	O_3/ppb

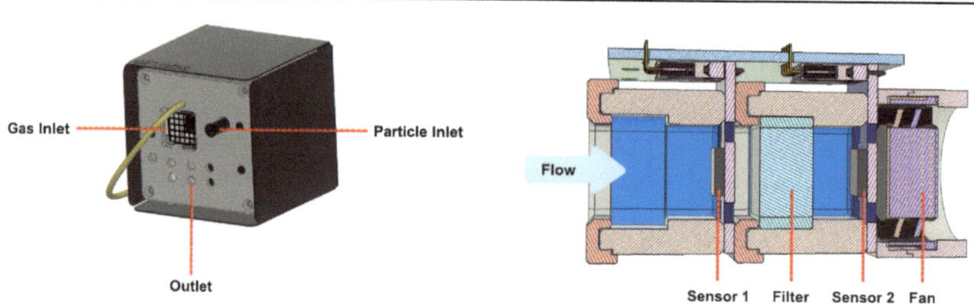

Figure 1. AirNode Gen 2 (**left**) and cross-section of the AirNode Gen 2 gas sampling system, showing both of the MOx cells in series, the filter between them, and the fan behind them (**right**).

Figure 2. Gen 5 node views, node with heat-shield installed (**left**), and cross-section of the node with sensor locations (**right**).

2.1.1. Gen 2 MOx Nodes

In the MOx node, the gas sensors used are MiCS-6814 metal-oxide sensors from SGX Sensortech [43]. Each sensor chip contains three sensing elements (s_1, s_2, s_3), each of these has a heating element, a magnified image of the sensor is shown in Figure A5. Sensing element three (pure WO_3) is optimised for measuring oxidising gasses and therefore its output is primarily used in this context. Element 2 is designed for the measurement of reducing gases and element 1 for ammonia. The sensing mechanism of a MOx is reliant on surface reactions, therefore the grain size, thickness and porosity of the MOx surface layer will alter the sensitivity and response rate of the MOx sensor to a pollutant. These effects are explained further in other studies [44–46].

The method used to create the MOx film will affect the microstructure of it's surface and therefore the performance, and even when the same method is used for sensor chips of the same model. the microstructure of each chip can vary. In use, the performance of individual sensor chips can be highly variable, meriting individual calibration. An image of the chips, showing surface structure, difference in drop size and damage to a sensing element is shown in Figure A5.

The chips are operated with temperature cycling operation (TCO), which is a technique developed by Schütze and coworkers [47]. Cycling the operating temperature means that at a certain point in the cycle the optimal temperature for binding of a specific gas will be reached, thus providing the highest sensitivity for the gas of interest. Species on the surface can also be burnt off at high temperatures, cleaning the surface. In the Gen 2 MOx sensor, the cycle is optimised for NO_2 and O_3 and the sensor output is recorded at the high and low points of each cycle. A schematic of the nodes is shown in Figure 1, two sensors (MOX1 and MOX2) measure the sampled air in series, with an integrated O_3 filter between them, this removes O_3 and therefore sensor 1 is exposed to both gasses, whereas sensor 2 (after the filter) is exposed to NO_2 without the presence of O_3. The output of both sensors is then used in determining the concentrations of the gasses, and the cross-sensitivity can be mitigated [48].

Each complete Gen 2 node is installed in a weatherproof enclosure ($88 \times 88 \times 90$ mm) with inlets, exhaust holes, and a fan for active sampling, illustrated in Figure 1. Active sampling was found to be integral to sensor performance. The sensing elements consume the gas of interest when measuring; if the air around the sensor were stagnant, a lower and less linear response would be observed than with active flow.

2.1.2. Gen 5 EC Nodes

The Gen 5 nodes (shown in in Figure 2) contain EC cells from Alphasense, a NO2-B43F cell [49], which has a MnO_2 filter that reduces O_3, and a OX-B431 cell [50] which does not have the filter and is sensitive to both NO_2 and O_3. The O_3 concentration can be found from the difference between cell responses (analogously to the MOx nodes). The EC cells do not rapidly consume the target gas and therefore active sampling is not included in the node design (schematics of the node are shown in Figure 2). The cells are sensitive to temperatures above 20 °C, however, and a sun shield is added to the node body to mitigate heat build-up. A temperature sensor is also included within the node and during this study the nodes internal temperature did not exceed the threshold of 20 °C at which a temperature correction is necessary. The node body dimensions are 190 × 105 × 70 mm.

Use of these sensors has been extensively reported in a number of studies [22,26,28], but in simple terms, each cell contains a working electrode (WE) and an auxiliary electrode (AE). The WE is exposed to the environment and is where the redox reactions occur, resulting in a change in current. The AE has the same structure as the WE but is not exposed to ambient air, and so is not affected by gas concentrations, only other environmental parameters such as temperature. The difference in output between the two electrodes, therefore, corresponds to changes in concentration at the EC cell surface. AirLabs have developed a printed circuit board (PCB) for converting the cell output from nA into mV, the board has a conversion rate of 0.735 mV/nA.

2.2. Calibration Method Overview

The MOx and EC nodes underwent separate calibration procedures, the timings of these are detailed in Table A4. In order to validate the methods, the nodes were co-located for several weeks at reference stations, this co-location was split into training for the Field calibration and a validation period. The same nodes were also calibrated over ∼3 days each in the EASE and Lab setups. The performance of the Field, EASE and Laboratory models over the validation period were then compared, a schematic of this is shown in Figure 3. The evaluation statistics for validation of the calibration models (for all methods) are calculated based on a comparison between the nodes and reference instruments during the Field validation period, with the different calibration models applied.

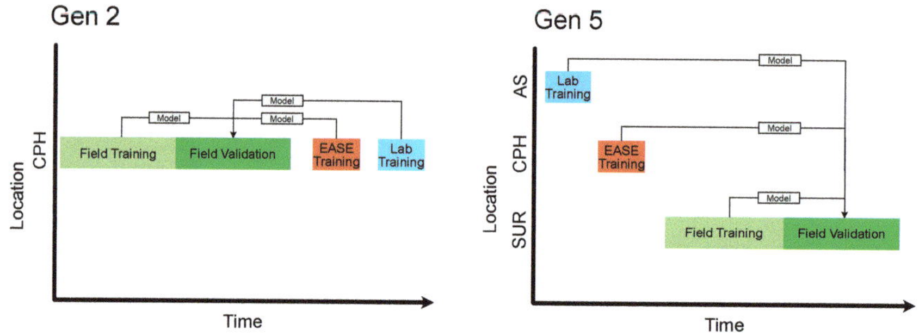

Figure 3. Schematic of calibration periods. The blocks sizes are not to scale for time, Field training should be 7 times the size of EASE and Laboratory training. Arrows indicate calibration models. The CPH location stands for Copenhagen, AS for Alphasense, and SUR for Surrey. Lab is short for laboratory.

A general overview of the methods is presented in Table 2, and the protocols for each node and calibration method are described below.

Table 2. Calibration method overview. A schematic showing the time-series for the different methods is included in Figure 3.

	Field	Laboratory	EASE
RH/%	Ambient	25, 50, 75	Ambient
T/°C	Ambient	10, 20	Ambient
[C]/ppb	Ambient	0-80-0	Ambient + 0, 40, 80
Time taken	∼3 weeks (preferably)	∼3 days	∼3 days
Resource intensity	Low (but requires station access)	High	Medium

2.2.1. MOx Laboratory Calibration Protocol

Laboratory calibration of the MOx nodes took place inside a 1 m^3 chamber, made from aluminium, stainless steel and Perspex, situated inside a larger climate-controlled chamber (Viessmann A/S), this setup is already partially described in Bulot et al. [51] (2020). A schematic of the Laboratory calibration setup is shown in Figure 4, it includes an ozone generator and O_2 flask (pure O_2), as well as an NO_2 flask (1–2.3% NO_2 in N_2) and mass flow controllers (MFCs, 0–100 ml min^{-1}). A Model 42i chemiluminescence NOx Analyser (Thermo-Fisher Scientific, Waltham, MA, USA) was used for NO_2 measurement as well as a direct absorbance analyser, a Model 405 nm (2B Technologies, Boulder, CO, USA). Ozone was measured with a BMT 930 UV photometer, and RH and T were monitored with an HTU21D digital sensor. In order to control RH levels, filtered, dry air was supplied to the chamber, via a MFC directly (lowering RH) or diverted through a Nafion membrane submerged in water (increasing RH). The chamber air was mixed with three fans in X,Y and Z directions, test mixing with CO_2 is shown in Figure A4.

The Laboratory testing protocol consisted of arranging the nodes inside the chamber and sealing it before steadily increasing pollutant concentrations (NO_2 or O_3 separately) from zero to ∼80 ppb, before allowing the concentration to steadily decay, at RH of 25, 50 and 75% (±15%), and T of 12 and 20 °C (±2 °C), respectively. Resulting in 6 concentration spikes for each pollutant in total, over the course of 3 days. An example time-series of of an O_3 concentration spike is shown in Figure 5.

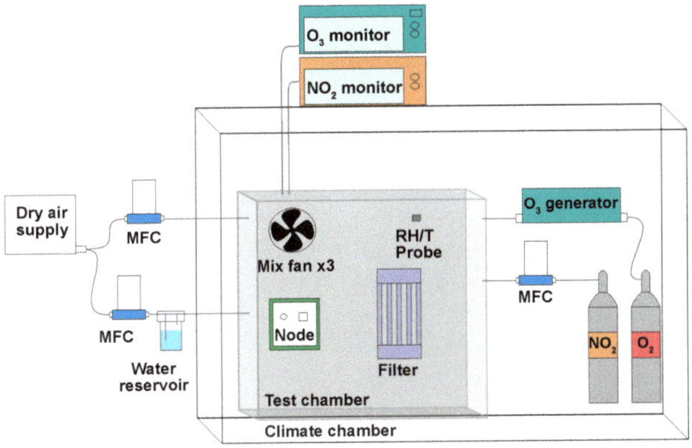

Figure 4. Laboratory calibration setup schematic for the Gen 2 MOx nodes.

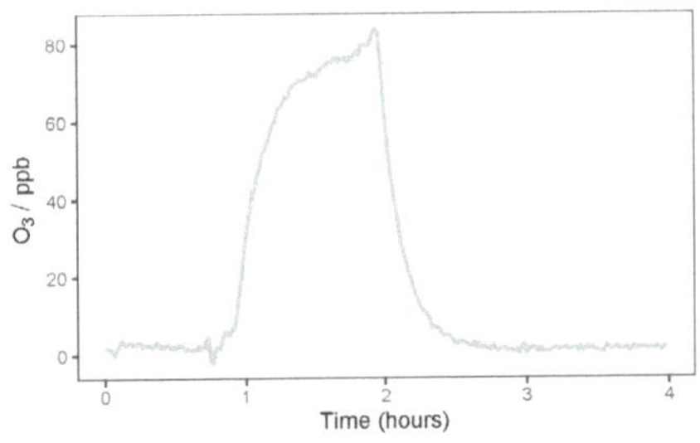

Figure 5. Example O_3 spike from the MOx Laboratory calibration.

2.2.2. MOx EASE Calibration Protocol

A schematic of the EASE calibration setup is shown in Figure 6. It consists of an insulated galvanised steel duct (32 × 32 cm cross-section), which acts as the mixing chamber, and is connected to a blower fan with 12 cm diameter ducting. The inlet to the system was extended 2 m away from the building through a window on the 5th floor. Inlets for the addition of NO_2 (50 ppm NO_2 in N_2 flask) and O_3 (pure O_2 passed through an O_3 generator) were also added to the main inlet and their flow controlled by MFCs (0–100 mL min^{-1}). The duct is equipped with flow, RH and T probes (LS control ES991 and ES989), mixing fans, and outlets for NO_2 and O_3 monitors (Model 405 nm NOx monitor and BMT 930 UV photometer). As the setup is insulated, and a relatively large throughput of air is used, the interior of the chamber is similar to the ambient air outdoors, however the setup has the advantage of being able to artificially increase (with gas flasks or generators) or decrease (with an activated carbon filter) pollutant concentrations. An example section of a concentration profile, with spikes and a 'rush-hour' period, is shown in Figure 7.

The chosen EASE calibration protocol required 48 h per pollutant (or 72 h total for both pollutants as the ambient measurements can be used for NO_2 and O_3 calibration), during which time the sensors were exposed to ambient pollutant concentrations throughout, except for zeroing periods and pollutant spikes, at low (~40 ppb), medium (~80 pbb) and high (~100 pbb) concentrations for NO_2 and O_3 independently. With each lasting ~45 min and occurring during non-rush hour periods. During the rush hour periods NO_2 and O_3 concentrations are not altered. The flow was held constant throughout and RH and T dictated by the ambient environment. A 24 h example section is shown in Figure 7.

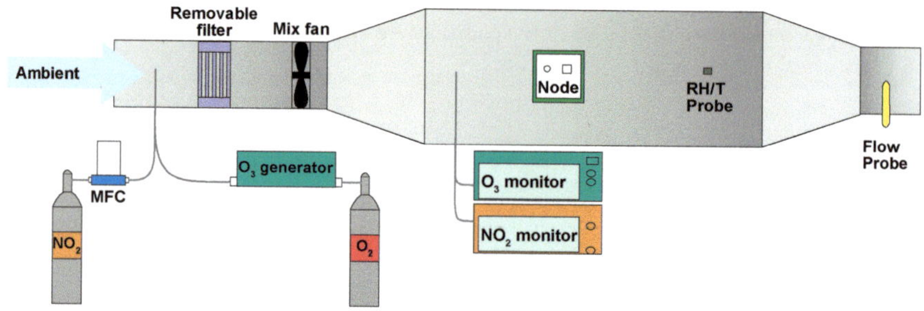

Figure 6. EASE calibration setup. This schematic is representative of the setup used for both MOx and EC calibration.

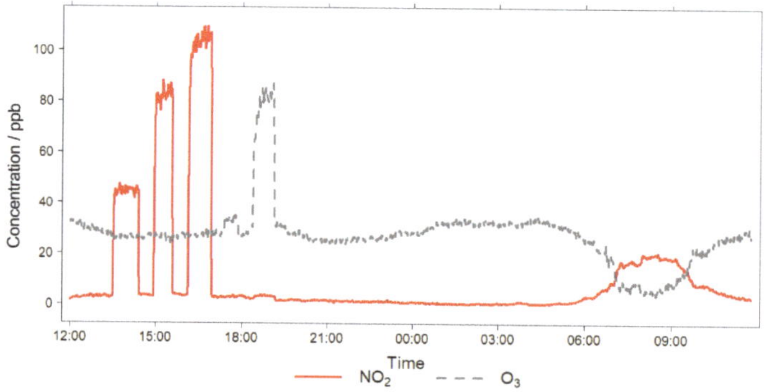

Figure 7. EASE calibration setup example concentrations, note the artificial spikes from 12:00 to 21:00, and morning rush hour period from 06:00 to 10:00.

2.2.3. MOx Field Calibration Protocol

The MOx Field calibration took place at the H. C. Andersens Boulevard (HCAB) roadside monitoring station in Copenhagen, Denmark, operated under the NOVANA program (the Danish National Monitoring Program for Water and Nature [52]). HCAB is a highly trafficked street with relatively high air pollution levels. The nodes were co-located at the station from 22nd December 2020 until 3rd February 2021 (43 days). The period before 12th January 2021 was designated as the training period (21 days) and post as the validation period (22 days). During the co-location mean RH was 80% (range: 58–96%), and mean T was 2.1 °C (range: −6.2–9.1 °C). A time-series of the reference NO_2 and O_3 concentrations during the co-location is shown in Figure 8. The nodes were installed at the same height and approximately the same distance from the traffic at HCAB as the inlet for the reference instruments. A schematic of the setup is also shown in Figure 8.

The reference instrument for NO_2 is a chemiluminescence NOx monitor (Teledyne API model T200). Reference O_3 measurements were recorded via UV-absorption at 254 nm with a Teledyne API model T400. The detection limit for both monitors is below 1 ppb, with precision of ±5%. Reference station data is recorded at 30 min time-resolution and therefore the LCS data was averaged to the same level for analysis.

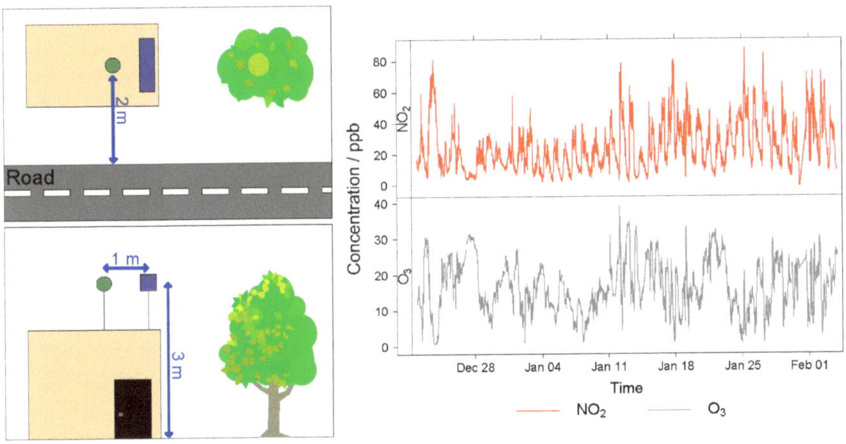

Figure 8. MOx Field calibration and validation setup (**left**) and pollutant time-series (**right**). In the setup schematic the node location is the blue rectangle and the reference inlet is the green circle.

2.2.4. MOx Calibration Models

As the Laboratory and EASE models in this study must inherently be transferable between sites, a simple MLR was chosen to produce the calibration models, to avoid overfitting to a specific environment [30,36]. An individual stepwise-selected model was used for each of the MOx nodes for all of the MOx calibration methods, where the input variables were the following, for predicting both NO_2 and O_3:

$$\begin{aligned} &P_{NO_2 \text{ or } O_3} = \\ &a_0 + a_1 * MOX1\_s_2\_high + a_2 * MOX1\_s_2\_low + a_3 * MOX1\_s_2\_offset + a_4 * MOX1\_s_2\_scale + \\ &a_5 * MOX1\_s_3\_high + a_6 * MOX1\_s_3\_low + a_7 * MOX1\_s_3\_offset + a_8 * MOX1\_s_3\_scale + \\ &a_9 * MOX2\_s_2\_high + a_{10} * MOX2\_s_2\_low + a_{11} * MOX2\_s_2\_offset + a_{12} * MOX2\_s_2\_scale + \\ &a_{13} * MOX2\_s_3\_high + a_{14} * MOX2\_s_3\_low + a_{15} * MOX2\_s_3\_offset + a_{16} * MOX2\_s_3\_scale + \\ &a_{17} * RH + a_{18} * T \end{aligned} \quad (1)$$

where a_0 is the offset and a_1–a_{16} are the calibration coefficients, and a_{17} and a_{18} are the temperature and relative humidity correction coefficients, calculated using the method of multiple least squares, separately for each MOx node. Due to the temperature cycle (described in Section 2.1.1), the MOx sensor provides several parameters obtained as raw data, as seen in Equation (1). For each sensor chip (s_2 and s_3), a scale and offset are provided for both the high- and low-temperature period in the temperature cycle, and the average conductance during the two temperature periods is also provided. Since two sensors are included in the MOx node, and measure the air in series, with an O_3 filter between them, the readings from both sensors (*MOX*1 and *MOX*2) are included. Parameters from sensing element two (s_2) and three (s_3) are both included as input variables in the model, even though only s_3 is optimised for oxidising gases. Parameter s_2 was included to check for interference, but if none were found (p-value > 0.05), the input variables were removed before doing the stepwise-selected model. The stepwise-selected model is determined based on the `step` function in R with the mode of stepwise search done with a 'forward' and 'backward' direction until the lowest Akaike Information Criterion (AIC) was found. Temperature and RH were included as input parameters in the models, but the stepwise-selection model disregarded them as inputs as they did not improve the final MOx node calibration models.

2.2.5. EC Laboratory Calibration Protocol

The EC cells (NO_2-B43F [49] and OX-B431 [50]) are produced by Alphasense and each cell is tested in a single pass setup, yielding zero current (nA) and sensitivity (nA ppm^{-1}) values, prior to dispatch [53]. Tests are conducted with a flow of 5 L min^{-1}, T of 22 ± 2 °C and RH of 45 ± 15%. Alphasense is confident in the linearity of cell response within the specified measurement range (0–20 ppm) and therefore only test at pollutant concentrations of zero ppb, and one known concentration (not disclosed). Tables of sensitivities and zero values for the cells are included in the Appendix A (Tables A1 and A2).

2.2.6. EC EASE Calibration Protocol

For the EC nodes, a scaled-up duct system was constructed (38 × 50 cm cross-section) to accommodate the larger node bodies, and different gas monitors were used, a Thermo Electron Model 42C chemiluminescence NOx Analyzer and a Thermo Scientific Model 49i Ozone Analyzer. Otherwise, the calibration followed a similar procedure to the MOx EASE calibration, except that ~120 ppb was used for the 'high' exposure level, and the inlet concentration was filtered during the artificial pollutant spikes for better consistency. An example plot of the full procedure is displayed in Figure 9. The EASE setup in Figure 6 is also representative of the EC calibration setup.

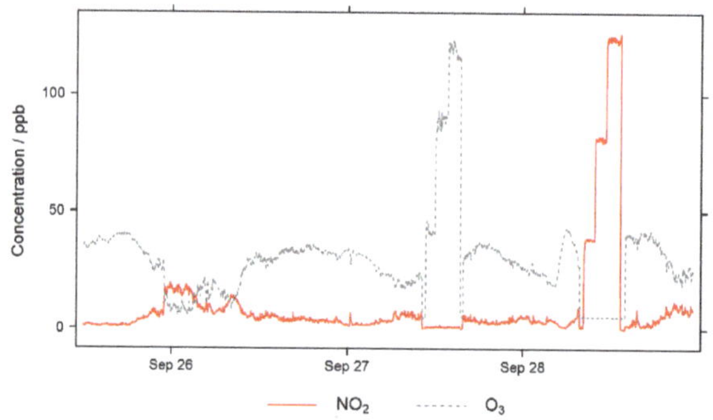

Figure 9. Example concentration profile from EC calibration in the EASE setup.

2.2.7. EC Field Calibration Protocol

The EC co-location took place at Surrey University, Guildford, United Kingdom, where the EC nodes were mounted in front of the Thomas Telford Building Air Quality Lab, on the university campus, at ~1 m from the ground. This is the same level as the intake of the reference monitors (Figure 10), which are operated by Surrey University. The co-location took place between 9th December 2021 and 23rd December 2021 (15 days), and was split into training and validation periods which consist of measurements before and after 15th December 2021, respectively. During the co-location internal node mean RH was 44% (range: 32–58%) and internal node mean T was 11.2 °C (range: 1.10–18.1 °C). Since the co-location took place on the university campus, low NO_2 levels were encountered, relative to the HCAB co-location. The O_3 levels were similar, a time-series of reference pollutant concentrations is shown in Figure 10. The reference instrument for NO_2 was a Serinus Ecotech 40 NOx monitor, whereas reference O_3 measurements were recorded via a Thermo Fisher Scientific 49i Ozone monitor. All reference data was obtained with 1-mintime resolution, and therefore the node data did not need to be aggregated.

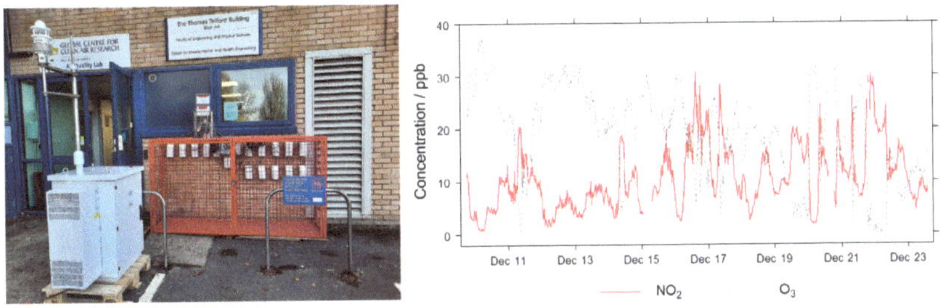

Figure 10. EC Field calibration and validation setup (**left**), and pollutant time-series (**right**).

2.2.8. EC Calibration Models

For the Laboratory calibration, the supplied coefficients (see Tables A1 and A2 in the Appendix A) and the recommended equations are utilised. Concentrations of NO_2 are predicted from the NO2-B43F cell (cell 2), using Equation (2) to find corrected WE output ($WE2_C$), from raw WE output ($WE2_v$), WE sensor zero ($WE2_0$), temperature dependent correction factor (n_{T_1}), raw AE output ($AE2_v$) and AE sensor zero ($AE2_0$). Which is then used in Equation (3) with the supplied sensitivity (S_2) and an offset (C_2) to convert the output into ppb. The sensitivities, sensor zeroes and temperature dependent correction factor are supplied by the manufacturer, the offset (if included) is determined by Field co-location. The sensitivities are multiplied by -10 due to the change in cell output when using the AirLabs PCB and not an Alphasense PCB, this is omitted from the equations for clarity.

$$WE2_C = (WE2_v - WE2_0) - n_{T_1}(AE2_v - AE2_0) \quad (2)$$

$$P_{NO_2} = WE2_C * S_2 + C_2 \quad (3)$$

Predicted O_3 concentration is calculated from the difference in response between the two cells. Corrected output for cell 1 WE ($WE1_C$) is determined in the same way as $WE2_C$, except that the variables relate to cell 1, as shown in Equation (4), and afterwards, the O_3 concentration can be determined based on Equation (5):

$$WE1_C = ((WE1_v - WE1_0) - n_{T_2}(AE1_v - AE1_0)) \quad (4)$$

$$P_{O_3} = (WE1_C * S_1 - WE2_C * S_2) + C_1 \quad (5)$$

Alterations were made to these stock equations after testing. The supplied temperature dependent correction factor (n_{T_1}) for NO_2 prediction was 1 if T is 0–10 °C and 0.6 for temperatures 10–20 °C, whereas n_{T_2} for O_3 prediction was 1.5 if T is below 10 °C and 1.7 for temperatures 10–20 °C, although when this was applied, the R^2 value of the model dropped drastically. When the difference between modelled and reference NO_2 or O_3 concentration was plotted against temperature for the modelled concentration without a temperature correction, there was no relation between the error and temperature in the range experienced at the test site, and therefore the n_{T_1} and n_{T_2} values were set to 1. This is in agreement with other deployments of the sensors, where a temperature dependence is not noticeable below 20 °C.

For the EASE and Field calibration models, similar equations are used, however the supplied sensitivities and zero values for the cells were not included. For prediction of NO_2 concentrations the voltage change in cell 2 is found from the difference between the WE and AE outputs, as shown in Equation (6), and the predicted O_3 concentration is found by subtracting cell 2 response from cell 1 response, as shown in Equation (7).

$$P_{NO_2} = (WE2_v - AE2_v) * S_N + C_N \quad (6)$$

$$P_{O_3} = ((WE1_V - AE1_V) - (WE2_V - AE2_V)) * S_O + C_O \quad (7)$$

The sensitivities, S_N and S_O, as well as offsets, C_N and C_O, are determined using the method of multiple least squares, separately for each EC node.

3. Results and Discussion

In the following section, the different calibration methods are compared, firstly for the MOx nodes, followed by the EC nodes, and finally the different sensor types are compared with each other. Results from the validation of the EASE setup against a reference station are included in the Appendix A, Appendix A.1.

In the review by Karagulian et al. (2019) [22], a good level of agreement for a sensor with a reference instrument is denoted by a R^2 value of >0.75 and a slope 'close' to 1.0, which we take to mean 1 ± 0.3, this definition will be used in the following analysis. The R^2 value gives a measure of the goodness of fit between variables but does not account for bias. Relative bias is denoted by a slope that diverges from 1. A non-zero intercept denotes absolute bias and impacts the limit-of-detection, this is also recorded and discussed for each method [22]. The root mean square error (RMSE), mean bias error (MBE), and mean absolute error (MAE) are also included as statistical indicators for the models.

3.1. Mox Node Results

Results for the validation of MOx sensor calibration with each method are displayed in Table 3 and selected statistical indicators are shown in Figure 11. Example time-series of NO_2 and O_3 for one of the nodes during the validation period are shown in Figures 12 and 13, respectively.

Table 3. Evaluation statistics for validation of the MOx nodes with all of the calibration methods, all values are means over all MOx nodes with their corresponding standard deviation of the results shown in brackets.

Pollutant	Method	R^2	Slope	Intercept/ppb	RMSE/ppb	MBE/ppb	MAE/ppb
NO_2	Laboratory	0.67 (0.22)	1.6 (0.39)	11 (21)	8.5 (3.0)	−16 (17)	21 (11)
	EASE	0.80 (0.065)	1.6 (0.63)	−11 (19)	6.8 (1.2)	−3.5 (13)	13 (2.7)
	Field	0.83 (0.12)	1.6 (0.24)	−6.5 (4.9)	6.2 (2.2)	−7.7 (3.2)	8.8 (2.9)
O_3	Laboratory	0.82 (0.11)	1.4 (0.54)	−6.4 (5.3)	3.2 (0.98)	5.2 (16)	11 (12)
	EASE	0.93 (0.062)	1.2 (0.16)	−2.2 (6.9)	1.9 (0.90)	−1.3 (4.5)	4.3 (1.4)
	Field	0.96 (0.037)	0.88 (0.12)	1.4 (0.9)	1.4 (0.67)	0.87 (2.8)	2.2 (2.3)

It was found that individual models of this prototype node were highly variable and some were unstable and gave poor results. Nodes that had training R^2 values < 0.75 and or validation R^2 < 0.1 were discounted from the study, consequently, 5 of the 12 nodes had to be discounted. The poor reliability of the nodes is presumably due to issues with the complex sampling/filtration system, or the fragile MOx chips, as some of the broken nodes had low sampling flows or damaged chips, an example chip is shown in Figure A5). After calibration, it was found that the response of the MOx cells changed dramatically at temperatures <0 °C, presumably due to perturbation of the TCO. Negative temperatures were present only in the Field validation period and not the Field training (or EASE or Laboratory training), and all models performed poorly when temperatures were <0 °C in the validation period, therefore, these data were also removed from the validation to better compare the models.

Overall, MOx sensor measurement of O_3 was superior to NO_2, with R^2 values of 0.96 vs. 0.83 in the Field validation, and better performance for all calibration methods. In terms of the different calibration methods, the Laboratory method was least successful, followed by the EASE method, with the Field being most effective. All six statistical indicators follow this pattern (Laboratory < EASE < Field) in terms of R^2 and slope difference from 1, intercept difference from zero, and size of RMSE, MBE, and MAE, except for Field NO_2 MBE, which is greater than EASE MBE, and slope which is the same for all methods. In the case of slope and intercept, a mean value can be misleading as these parameters can be above or below the optimum value (1 or 0, respectively), which is why the standard deviations and boxplots are included. One of the nodes calibrated for NO_2 with the Laboratory method had a negative slope (−0.9) for the validation, this pulled the mean slope value closer to 1 despite being a poor and anomalous result and therefore was removed from the analysis. Despite the EASE and Field methods performing better than the Laboratory method, they do not meet the requirements for 'good' sensor performance for NO_2 measurements, due to slope values that are too high (1.6). However, the EASE and Field calibrated O_3 measurements are well within the requirements, with particularly high R^2 values (0.93 and 0.96) and acceptable slope values (1.2 and 0.88).

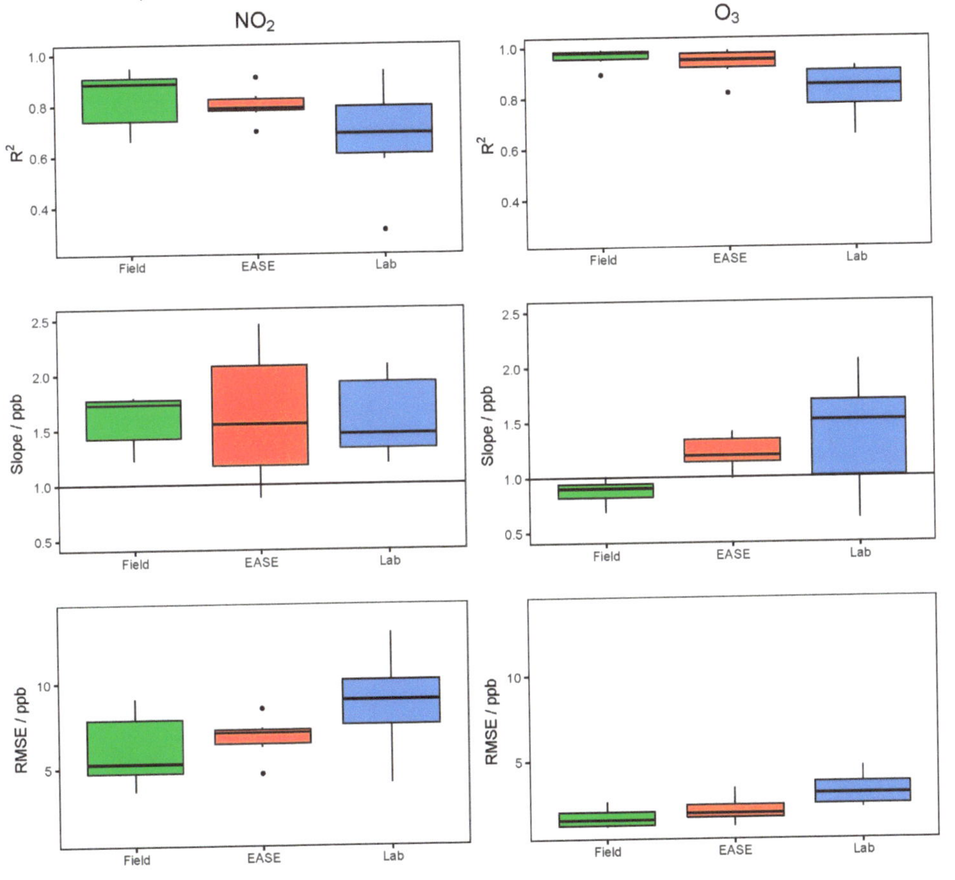

Figure 11. Comparison of R^2, slope, and RMSE for calibration model validation of MOx sensors with the different methods for both NO_2 (**left**) and O_3 (**right**). Lab is short for laboratory.

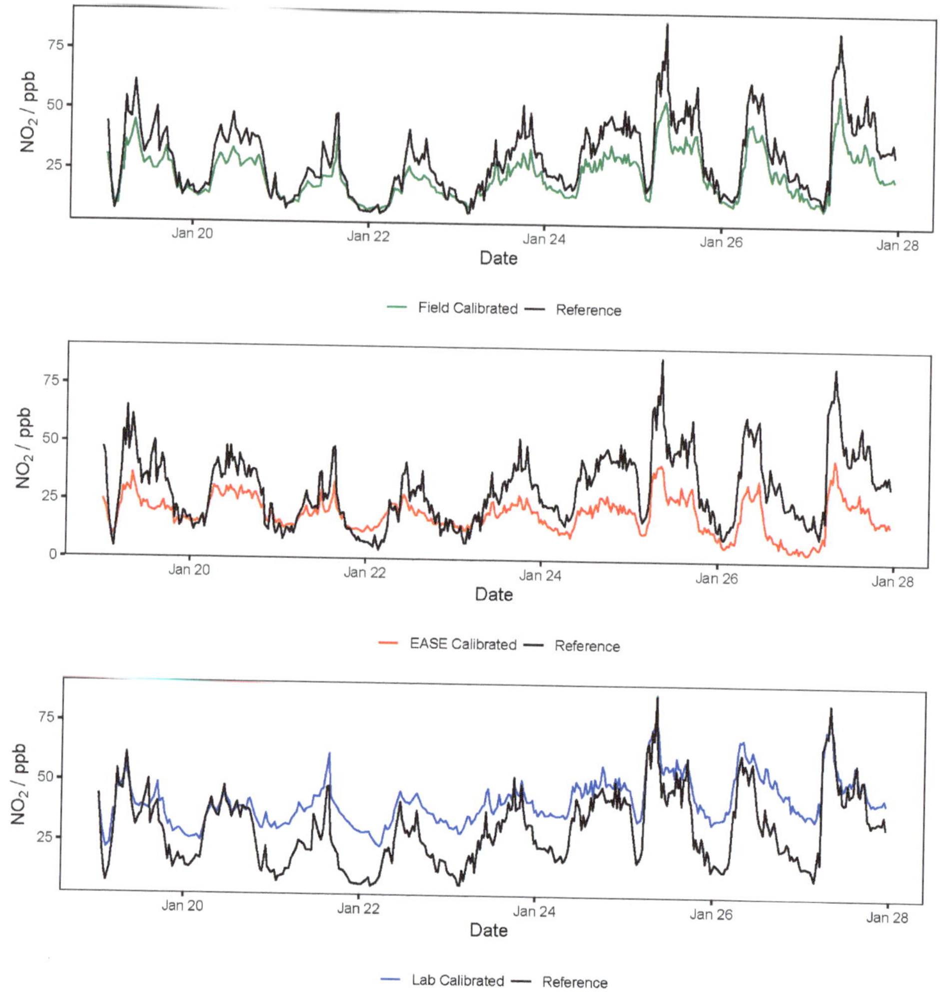

Figure 12. Example NO_2 time-series from a single MOx node during the validation of the different calibration methods, compared with data from the reference instrument. Lab is short for laboratory.

In conclusion, the EASE method outperforms the Laboratory method with similar overall results to the Field method, but the sensor hardware only performs well enough for O_3 measurement, as even the Field calibrated NO_2 measurements are not within the requirements for a good sensor, under these conditions. The MOx nodes also had poor reliability, with a large fraction being discounted, and did not perform well at negative temperatures (although they were not trained under negative temperatures).

3.2. EC Node Results

Results for validation of the EC sensors with all three methods are displayed in Table 4, and selected statistics are displayed in Figure 14. Example time-series of NO_2 and O_3 during the validation period are shown in Figures 15 and 16, respectively. Results from the calibration training periods are found in Table A6 in the Appendix A. The mean NO_2 concentration was low during the entire period (training and validation) at 10.4 ppb, and particularly low during the training period (7.7 ppb, vs. 13 ppb in the validation period). This means that at times the levels are near the limit of detection for the cells (reported previously as ∼4 ppb [54]). Therefore a lower R^2 value is recorded for the sensor output compared with the reference instruments in the training period, relative to the validation period (0.49 vs. 0.83), whereas the mean O_3 concentrations were greater throughout, and the R^2 for the O_3 measurement is similar for both periods (∼0.83) [55,56]. The Field calibration performs very well for O_3 (R^2 = 0.83, slope = 0.97) but over-predicts NO_2 concentrations in the validation period (slope = 1.4), which is most likely due to the short training period, with lower NO_2 levels than the validation period. As stated previously, ∼3 weeks or longer is recommended for Field calibration.

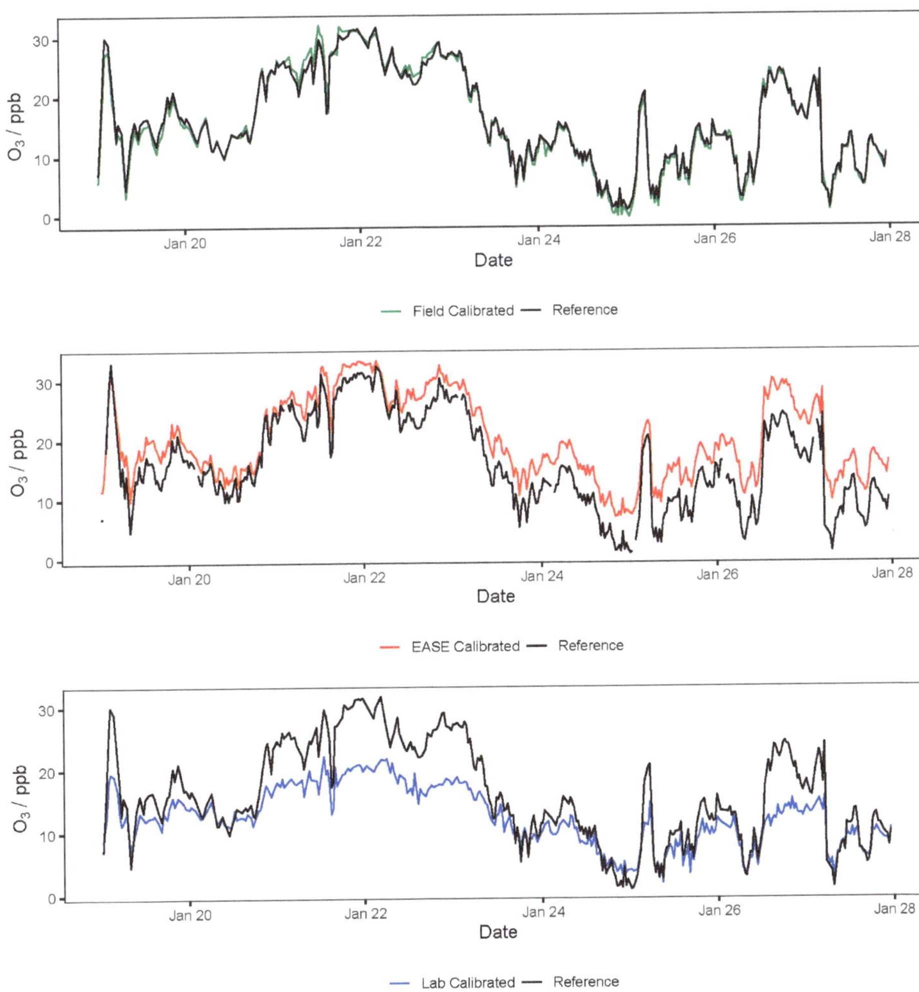

Figure 13. Example O_3 time-series from a single MOx node during the validation of the different calibration methods, compared with data from the reference instrument. Lab is short for laboratory.

Table 4. Evaluation statistics for validation of the EC nodes with all calibration methods. Note that the statistics for the Laboratory calibration are calculated after the Field offsets, shown in Table A3, have been applied. All values are means over all EC nodes with their corresponding standard deviation shown in brackets.

Pollutant	Method	R^2	Slope	Intercept/ppb	RMSE/ppb	MBE/ppb	MAE/ppb
NO_2	Laboratory	0.83 (0.025)	1.3 (0.11)	−2.7 (1.1)	2.6 (0.20)	−1.1 (0.42)	2.4 (0.19)
	EASE	0.83 (0.027)	1.2 (0.13)	−2.6 (5.2)	2.6 (0.22)	0.32 (4.2)	3.9 (2.2)
	Field	0.83 (0.027)	1.4 (0.17)	−2.6 (1.4)	2.6 (0.22)	−1.6 (0.49)	2.6 (0.39)
O_3	Laboratory	0.83 (0.027)	0.73 (0.10)	6.1 (2.3)	3.4 (0.27)	−2.8 (1.7)	4.5 (1.1)
	EASE	0.83 (0.027)	1.3 (0.17)	−8.3 (7.6)	3.4 (0.27)	2.7 (4.9)	5.3 (2.5)
	Field	0.83 (0.027)	0.97 (0.073)	−0.092 (1.7)	3.4 (0.27)	0.47 (0.95)	2.8 (0.29)

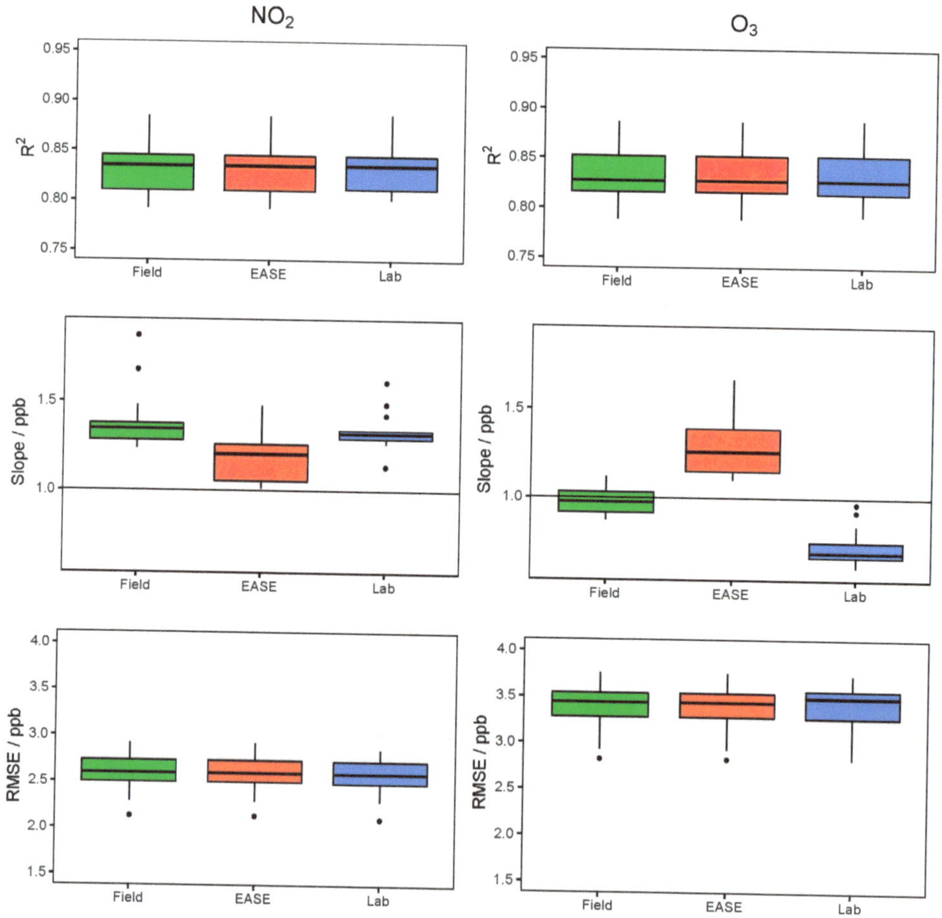

Figure 14. Comparison of R^2, slope, and RMSE for calibration model validation of EC sensors with the different methods for both NO_2 (**left**) and O_3 (**right**). Lab is short for laboratory.

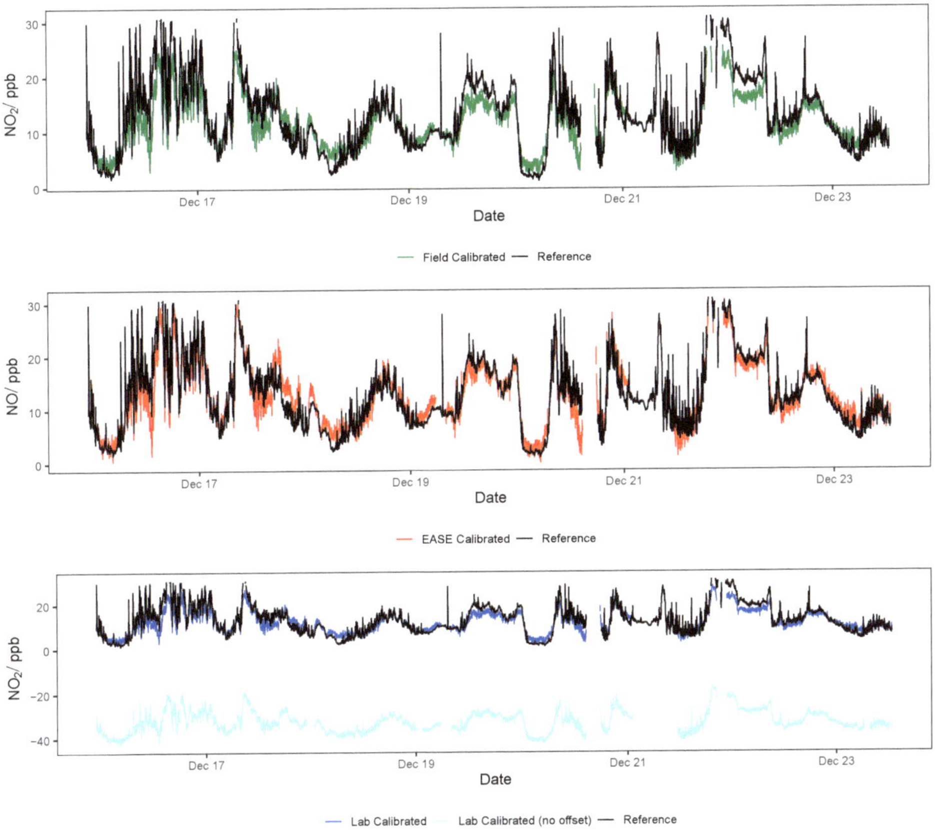

Figure 15. Example NO_2 time-series from a single EC node during the validation of the different calibration methods, compared with data from the reference instrument. Lab is short for laboratory.

The R^2 and RMSE values for validation of the EC nodes with different methods are essentially the same, unlike the MOx nodes for which R^2 differs between methods. This is because the EC cells have a more linear response to pollutant concentrations, and less output variables, and therefore a change in the slope between methods does not alter the coefficient of determination. This is also reflected by a generally better performance and lower inter-unit variability for the EC nodes.

It was found that the the Laboratory coefficients provided by the EC manufacturer (Alphasense) yield results that scale well with concentration increase in the Field, with a slope similar to the Field calibrations (∼1.3), however the intercept of the Laboratory predicted concentrations were either greatly above or below zero (between −175 and +126 ppb) when applied to the Field data. Consequently, a Field offset correction was also identified, based on the the Field calibration training period. The offsets are shown in the Appendix A in Table A3 and as an example by the light blue lines in Figures 15 and 16. This means that the pure Laboratory calibration would only be usable for measuring relative changes, and not absolute values, unless a short Field co-location is performed to determine their offset (as is the case for the statistics in Table 4 and the dark blue lines in Figures 15 and 16).

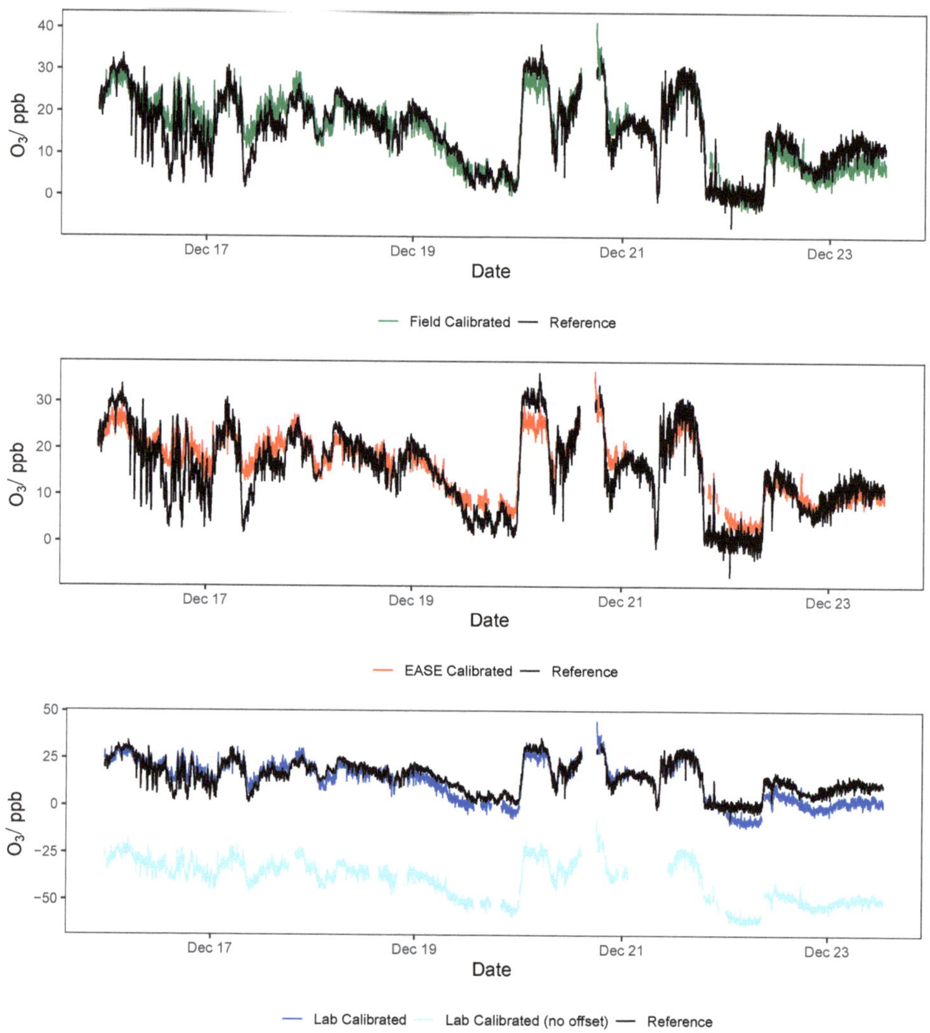

Figure 16. Example O_3 time-series from a single EC node during the validation of the different calibration methods, compared with data from the reference instrument. Lab is short for laboratory.

The zero coefficients supplied by Alphasense were included in the Laboratory calibration models, as described in Section 2.2.8, but this only partially reduced the range of intercepts encountered for the validation period (e.g., from 184 ppb down to 166 ppb for NO_2), compared with not using the zero coefficients, and just subtracting the AE from the WE response. The offset issue may be partly due to the use of a PCB designed by AirLabs in the Gen 5 EC node, which complicates the use of the 'zero' coefficients supplied by Alphasense, as the electronic offset may be altered. However, the large variability in offset between units suggests that individual offset calibration is necessary for each cell in the Field, regardless of the PCB used. In the application note from Alphasense, it is stated that large over/under-estimation may occur if using just Laboratory coefficients and that a secondary correction method is normally required [53], which was the case here. However, the scale for the models produced by the Laboratory calibration was impressive. Potentially a simple on-site zero air calibration, or calibration at a single fixed concentration, could be used together with the Laboratory calibration model coefficients to improve their results. This will be tested as part of a future study.

Temperature correction using the n_T coefficients supplied was also tested but did not improve the results. A correlation linking the difference between model predictions and reference measurements was not observed for NO_2 measurements during the Field validation. For O_3, a decrease in absolute difference was observed for increasing temperatures, meaning that the sensors appear to perform better for O_3 measurements at greater temperatures, however O_3 concentration also had a strong correlation with temperature, meaning that the increased absolute accuracy is due to the sensors having lower absolute error at greater concentrations. When temperature was included in the training of the MLR model for the Field method it did not improve results in the Field validation. This suggests that below 20 °C using the difference between the WE and AE is sufficient to account for any sensitivity that the EC sensors have towards temperature increase. This is also evidenced by the strong performance from the Field trained model (R^2 = 0.83, slope = 0.97), which does not include temperature.

In conclusion, the Field and EASE methods perform similarly for both NO_2 and O_3 measurements, all being in good agreement with the reference, except for Field NO_2 which has a slope slightly above the acceptable range, as described above. It is also clear to see from the time-series in Figures 15 and 16 that these methods result in models that track the observed data well. In this case, EASE performs essentially as well as the Field calibration, with a better slope for NO_2, worse for O_3, but meeting the criteria for both. This is despite the fact that EASE calibration was performed in CPH and the nodes validated in Surrey, UK, although it should also be noted that the Field calibration was shorter than is optimal. Overall we believe this study demonstrates EASE as a viable alternative to Field calibration for the EC nodes.

3.3. Node Comparison

As can be seen in Figures 8 and 10, the concentration profiles in the Field co-locations periods are very different. The HCAB co-location (MOx nodes) is longer and importantly for the NO_2 calibration has a large NO_2 concentration range during the Field training period (0–81 ppb). Meanwhile the Surrey Field training period (EC nodes) is short and has a lower mean NO_2 level (7.7 ppb), with a lower NO_2 concentration range (0.50–29 ppb), followed by a validation period with a slightly greater mean NO_2 concentration (13 ppb). A lower mean concentration can result in a lower R^2 value regardless of the sensor being tested, and validating sensors using a concentration range greater than the training range is not optimal [34,55,56]. Half-hourly data is the maximum resolution available from HCAB, whereas the Surrey co-location data had a one-minute time resolution, using greater time-resolution generally results in lower performance statistics. Therefore, it is difficult to compare the Field calibration of the sensors under these differing conditions. However, the EC nodes appear to have been trained and validated under more challenging conditions and yet still perform as well as or better than the MOx nodes.

Temperature is also a factor in the training/validation of the models, but again is difficult to compare. It was warmer during the Surrey co-location than the HCAB co-location (11 vs. 2.8 °C), but the sensors respond differently to temperature changes. The MOx sensors operate poorly below 0 °C, whereas the EC cells operate well at negative temperatures, but lose sensitivity at greater temperatures (particularly above 20 °C), which were not present during the co-location.

The laboratory MOx training consisted of a number of concentration, T and RH combinations (although temperatures were greater than those encountered in the field). In contrast, the EC Laboratory calibration consisted of a zero calibration and one other concentration/T/RH combination. This may partly explain the large offsets produced by the EC Laboratory method. In terms of the EASE training, the MOx nodes were trained and tested in the same geographical area, which is the preferred method for EASE calibration, whereas the EC nodes were trained in Copenhagen, Denmark and tested in Surrey, UK. Despite this, the EASE method performs well for the EC nodes.

It is clear that the EC Gen 5 nodes are more robust and less variable than the MOx nodes as none of them had to be discounted from the analysis, compared with 41% of the Gen 2 MOx nodes. It seems that the fan/filter system in the MOx nodes and sensors chips themselves are vulnerable to damage.

Overall, If using the best calibration method in each case (Field or EASE), the statistics for the different sensors appear similar, for NO_2 measurement, EC: $R^2 = 0.83$, slope = 1.2, MOx: $R^2 = 0.83$, slope = 1.6 and for O_3 measurement, EC: $R^2 = 0.83$, slope = 0.97, MOx: $R^2 = 0.96$, slope = 0.88. However, when all factors are taken into account, including the less optimal EC Field co-location and the unreliability of the MOx nodes, the EC nodes are judged as superior.

4. Conclusions

This study demonstrates how the EASE calibration method performs better than pure Laboratory calibration and similarly to a Field calibration (and in some cases better, e.g., EC NO_2), whilst requiring a fraction of the time, being completed in days instead of weeks. The EASE method even performed well when nodes were calibrated in Copenhagen, Denmark and validated in Surrey, UK, up to 3 months later, suggesting that using a site with similar characteristics (e.g., Urban, European) and at least within the same season, yields acceptable results. Although, we expect even better results from calibration at or nearer to the intended measurement site and directly before deployment.

Although the EASE method performed well under the circumstances in this study and can expose sensors to the full expected range of pollutant concentrations in a condensed period, it does not 'condense' the RH and T exposure. These meteorological parameters have less impact on the node output than concentration range, at least within certain ranges (e.g., T < 20 °C for EC and >0 °C for MOx), therefore in most cases, we expect that if the calibration is performed in the same season as the deployment an EASE calibration will be sufficient. However, the temperature during the co-location should be monitored and if it is outside of the optimal operating range for the sensors, then either a separate temperature correction should be applied, or data may need to be discounted. The EASE method will be best applied to shorter deployments, e.g., one season, but is still expected to perform better than Laboratory calibration over longer deployments. The issue with not capturing an appropriate range for meteorological factors is also present for Field calibration methods, as can be seen from the MOx data in this study. Theoretically, longer Field calibrations could solve this issue, for instance a year long co-location would ensure the sensors encounter a large meteorological range, but this is not viable for most LCS, and in particular for EC sensors, due to their overall lifetime and the drift they exhibit during deployment. We propose that additional EASE calibration periods during a deployment would be a better solution. The speed of an EASE calibration also means it is a viable option for conducting pre- and post-deployment calibration, and using the the combined calibration models to account for drift.

Laboratory calibration did not produce a calibration model meeting the requirements for a 'good' sensor performance in any of the cases in this study. A potential improvement would be to model the expected concentration, RH and T of the site during a deployment and then to use a range around them for the defined Laboratory parameters. Although this would not account for potential co-pollutant species present at the deployment site.

The secondary purpose of this study was to compare the Gen 2 MOx nodes and the Gen 5 EC nodes. It is clear from the results that despite the Field co-location being less optimal for the EC nodes (shorter, lower concentration range, lower mean NO_2 level, warmer), they out-perform the MOx nodes for NO_2 measurements, for O_3 the results from both sensor types were similar and rate as 'good' for Field and EASE calibration. However, the MOx nodes were less reliable.

The method is in its infancy but we expect that further testing and iteration of the procedure will improve the results, particularly when dealing with correlated or anti-correlated pollutants that the sensors are cross-sensitive to. The natural progression for this work would be to either install an EASE system directly at a reference station and use that for optimal EASE calibration for the surrounding area, or to build a mobile EASE system inside a vehicle that could be used for condensed calibration at the exact site at which nodes will be measuring, and to validate the sensors in the same setup. This could provide rapid and accurate calibration on-demand. Based on the results from this study, the method appears to perform well enough to invest in this.

Author Contributions: Conceptualisation, H.S.R. and J.A.S.; methodology, H.S.R., J.A.S., L.B.F., S.K. and M.S.J.; validation, H.S.R. and L.B.F.; software, H.S.R., L.B.F. and S.K.; formal analysis, H.S.R., L.B.F. and S.K.; investigation, H.S.R., L.B.F., S.K. and A.P.M.E.; resources, M.S.J., O.H., J.A.S. and P.K.; data curation, H.S.R., L.B.F., S.K. and A.P.M.E.; writing—original draft preparation, H.S.R., S.K. and L.B.F.; writing—review and editing, H.S.R., L.B.F., S.K., J.A.S., O.H., M.S.J. and P.K.; visualization, H.S.R. and L.B.F.; supervision, J.A.S., O.H., M.S.J. and P.K.; project administration, H.S.R.; funding acquisition, O.H., M.S.J. and P.K. All authors have read and agreed to the published version of the manuscript.

Funding: H.S.R., L.B.F. and O.H. were supported by BERTHA—The Danish Big Data Centre for Environment and Health, funded by the Novo Nordisk Foundation Challenge Programme (grant NNF17OC0027864).

Data Availability Statement: All raw data is available upon request.

Acknowledgments: The authors acknowledge AirLabs for the loan of the Gen 5 AirNodes and ACTRIS-DK for some of the reference instruments. The authors thank Archie Waller for their help. Prashant Kumar acknowledges the support from Innovate UK funded project 'MyGlobalHome' the pilot demonstrator under the Technology Strategy Board File Reference of 106168.

Conflicts of Interest: M.S.J., J.A.S. and S.K. are employees of AirLabs, who manufacture the sensor nodes. H.S.R. and L.B.F. are partially funded by AirLabs.

Abbreviations

The following abbreviations are used in this manuscript:

AE	Auxiliary Electrode
AQ	Air Quality
EASE	Enhanced ambient Sensing Environment
EC	ElectroChemical
HCAB	H. C. Andersens Boulevard
LCS	Low-Cost Sensor
LR	Linear Regression
MAE	Mean Absolute Error
MBE	Mean Bias Error
MFC	Mass Flow Controller
ML	Machine Learning
MLR	Multivariate Linear Regression
MOx	Metal-Oxide
PCB	Printed Circuit Board
RH	Relative Humidity
RMSE	Root Mean Square Error
T	Temperature
TCO	Temperature Cycling Operation
WE	Working Electrode

Appendix A. Equipment Tests and Training Parameters

Appendix A.1. EASE

Prior to use, the EASE setup was compared against data from the urban-background monitoring station operated under the NOVANA program (the Danish National Monitoring Program for Water and Nature [52]) on the roof of the neighbouring building, (H.C. Ørsted Institute), to confirm that it would correctly track fluctuations in ambient pollution levels. The system was in good agreement for trends in NO_2 (R^2 = 0.77) and O_3 (R^2 = 0.84), shown in Figure A1, although with lower values for both pollutants. This may be due to reaction on the walls of the system or the difference in position compared with the reference monitors. T and RH were higher and lower, respectively, in the EASE setup than the reference station but followed the same trends (R^2 of 0.79 and 0.84, respectively, shown in Figure A2) and were in good agreement with a RH and T probe mounted 2 m outside the building. The mixing and homogeneity of the setup was also probed and found to be satisfactory (results displayed in Figure A3).

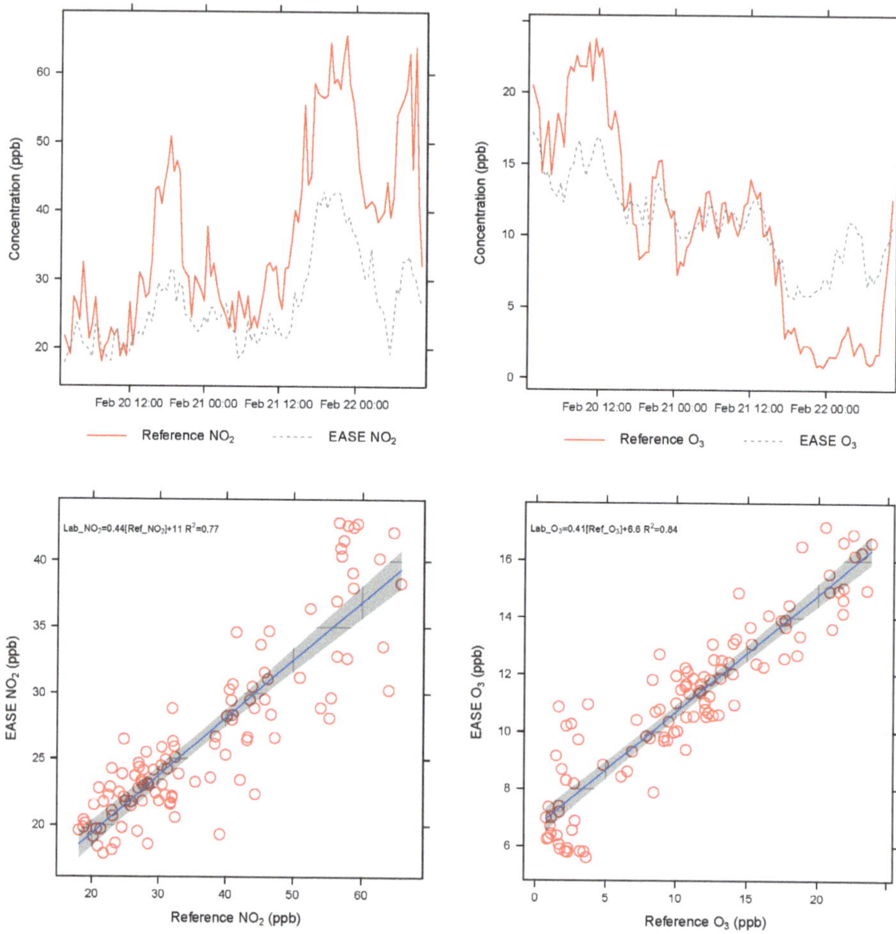

Figure A1. Time-series (**above**) and scatter-plots (**below**) showing the correspondence between the EASE duct and nearby reference station for NO_2 (**left**) and O_3 (**right**) measurements.

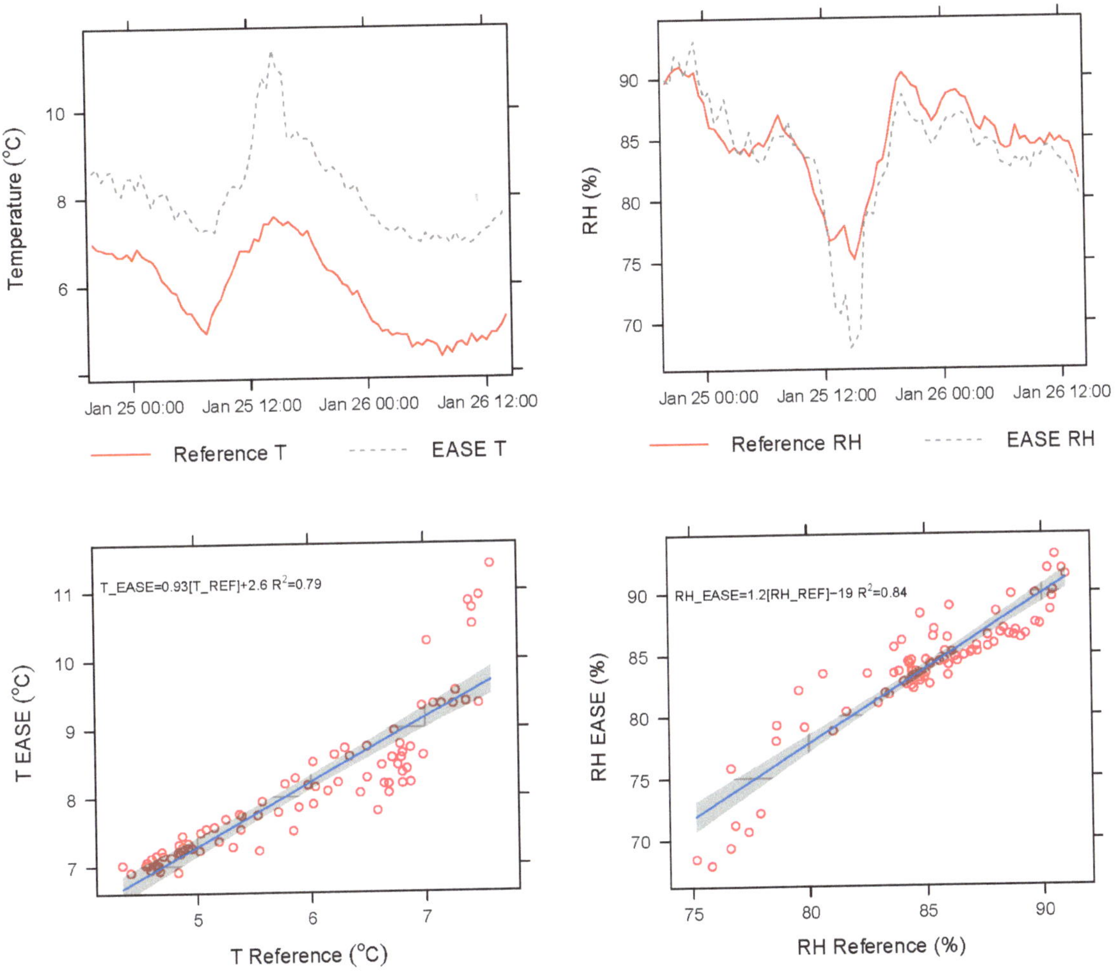

Figure A2. Time-series (**above**) and scatter-plots (**below**) showing the correspondence between the EASE duct and nearby reference station for T (**left**) and RH (**right**) measurements.

Table A1. Sensitivity and zero currents for NO2-B43F cells, provided by Alphasense.

Node	NO2-B43F Cell Serial	WE Zero Current/nA	AE Zero Current/nA	Sensitivity/ (nA ppm^{-1})
ANG500012	202750622	32.16	18.29	−386.68
ANG500149	202750349	19.86	2.84	−338.6
ANG500151	202750350	25.54	18.60	−348.85
ANG500174	202750153	30.58	11.35	−400.24
ANG500194	202750122	30.90	8.51	−380.38
ANG500208	202750105	31.21	12.30	−412.38
ANG500218	202750639	33.73	18.92	−379.75
ANG500219	202750638	33.42	21.44	−384.32
ANG500224	202055609	9.46	9.14	−341.13

Table A1. *Cont.*

Node	NO2-B43F Cell Serial	WE Zero Current/nA	AE Zero Current/nA	Sensitivity/ (nA ppm^{-1})
ANG500225	202055604	25.22	17.66	−319.37
ANG500245	202240647	52.97	20.49	−393.62
ANG500252	202750601	24.91	17.34	−355.63
ANG500254	202750119	31.84	14.19	−374.39
ANG500255	202750110	35.63	15.76	−368.87
ANG500259	202750112	35.94	10.72	−382.58

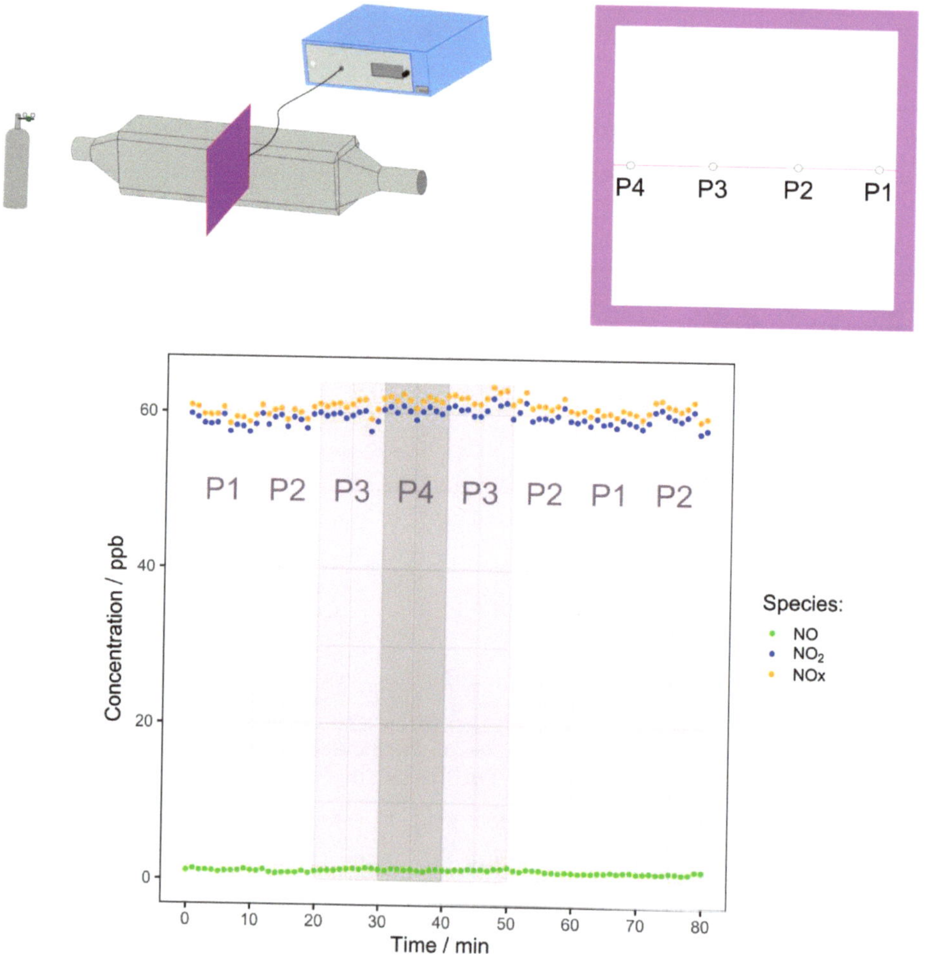

Figure A3. Results of mixing test in EASE duct, the sampling line was moved between points P1 and P4 shown in the schematic (**above**) at the realtive times shown in the plot (**below**). As can be seen from the plot there was not a significant difference in NO_2 either for the different points or the same points at different times. The small fluctuations in concentration over time are caused by slight fluctuations in total flow through the duct.

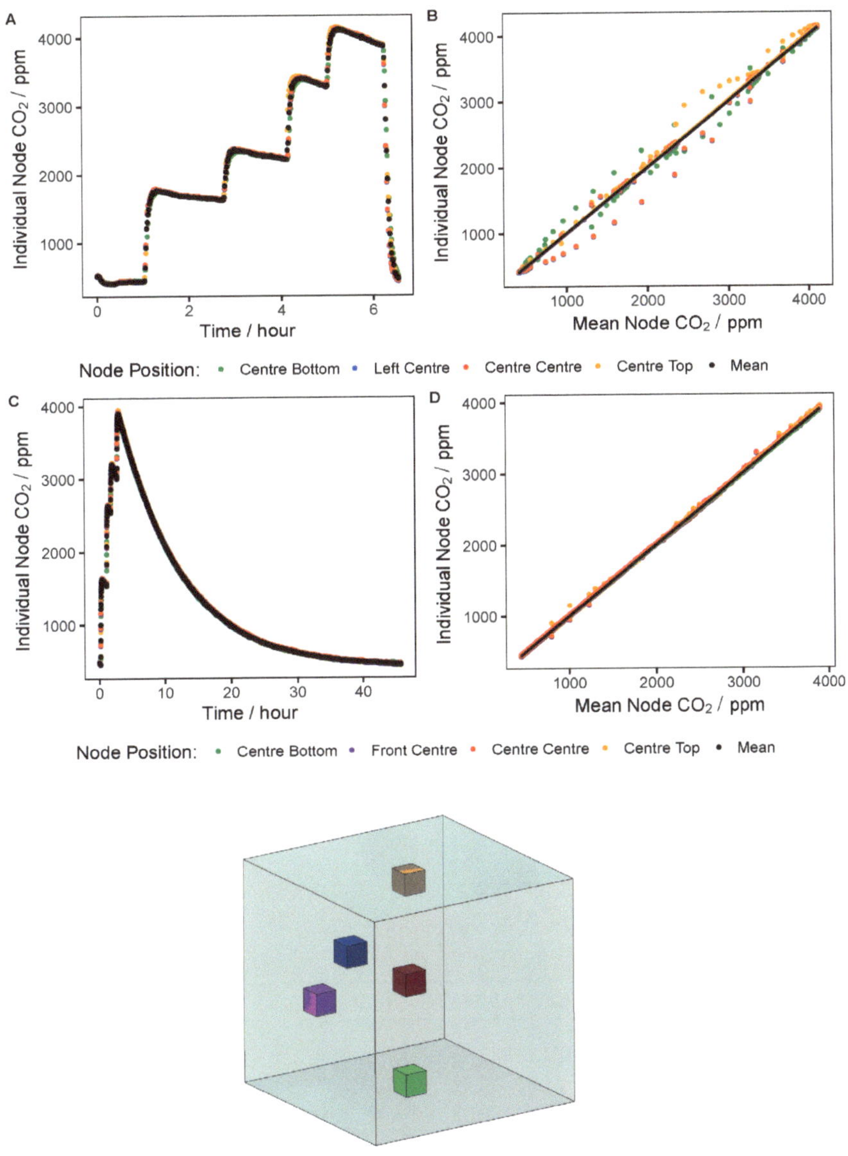

Figure A4. Figure displaying the results of chamber mixing tests, including: schematic showing 5 different node placements in the chamber, in their corresponding colours (**below**), and time-series (plots A and C) and scatter-plots (plots B and D) of two different mixing tests, where the CO_2 concentration was incrementally increased before rapidly venting the chamber (**A**,**B**) and slowly increased before leaving to steadily decay (**C**,**D**).

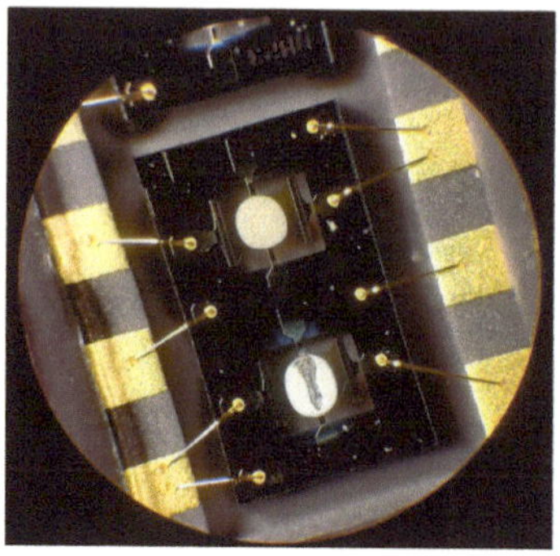

Figure A5. Example MOx sensor surfaces (4× magnification), bottom sensor is damaged.

Table A2. Sensitivity and zero currents for OX-B431 cells, provided by Alphasense.

Node	OX-B431 Cell Serial	WE Zero Current/nA	AE Zero Current/nA	Sensitivity/ nA ppm^{-1}
ANG500012	204050221	25.22	9.77	−656.87
ANG500149	204071554	37.20	24.59	−548.26
ANG500151	204071553	39.09	21.12	−586.25
ANG500174	204070150	47.29	17.66	−480.64
ANG500194	204751304	−29.01	11.67	−599.18
ANG500208	204070439	34.36	11.98	−543.37
ANG500218	204070440	33.42	7.25	−585.94
ANG500219	204070441	44.45	17.97	−489.46
ANG500224	204070407	20.49	14.19	−627.87
ANG500225	204070406	37.52	10.72	−570.02
ANG500245	204851926	33.73	16.71	−584.20
ANG500252	204851922	39.41	20.18	−613.37
ANG500254	204851921	43.51	19.86	−613.05
ANG500255	204851916	36.89	17.34	−626.76
ANG500259	204851918	35.63	19.86	−559.77

Table A3. Field offsets applied to the Laboratory calibration of the EC nodes, including the minimum and maximum offset applied for each pollutant, and the difference between them.

Node	Offset Applied NO_2/ppb	Offset Applied O_3/ppb
ANG500012	24.80	52.20
ANG500149	55.28	52.50
ANG500151	58.55	−62.56
ANG500174	69.03	−11.60
ANG500194	37.55	43.46
ANG500208	76.79	−55.55
ANG500218	−21.66	46.76
ANG500219	44.61	−127.34
ANG500224	−20.53	−10.62
ANG500225	36.25	−90.69
ANG500245	−7.93	122.58
ANG500252	18.78	125.58
ANG500254	45.48	−7.23
ANG500255	105.74	−89.34
ANG500259	101.61	−175.24

Table A3. *Cont.*

Node	Offset Applied NO_2/ppb	Offset Applied O_3/ppb
Max	105.74	125.58
Min	−21.66	−175.24
Difference	127.41	300.82

Table A4. Summary of calibration period dates. * EASE separated into 2 batches, ** EASE separated into 7 batches, tested alongside other nodes.

Node	Field	EASE	Laboratory
MOx	22nd December 2020 : 3rd February 2021	* 21st February 2021 : 26th February 2021	12th March 2021 : 15th March 2021
EC	9th December 2021 : 23rd December 2021	** 1st September 2021 : 5th October 2021	Prior to dispatch (then sealed)

Table A5. Evaluation statistics for training of the MOx nodes with all of the calibration methods. All values are means over all MOx nodes with their corresponding standard deviation shown in brackets.

Pollutant	Method	R^2	Slope	Intercept/ppb	RMSE/ppb	MBE/ppb	MAE/ppb
NO_2	Laboratory	0.90 (0.047)	1.0 (0.0)	0.0 (0.0)	9.9 (2.7)	0.0 (0.0)	6.6 (1.7)
	EASE	0.96 (0.014)	1.0 (0.0)	0.0 (0.0)	3.7 (0.67)	0.0 (0.0)	2.6 (0.32)
	Field	0.92 (0.040)	1.0 (0.0)	0.0 (0.0)	3.6 (0.86)	0.0 (0.0)	2.6 (0.67)
O_3	Laboratory	0.99 (0.011)	1.0 (0.0)	0.0 (0.0)	3.3 (1.2)	0.0 (0.0)	2.2 (0.86)
	EASE	0.97 (0.013)	1.0 (0.0)	0.0 (0.0)	2.7 (0.72)	0.0 (0.0)	1.5 (0.51)
	Field	0.96 (0.040)	1.0 (0.0)	0.0 (0.0)	1.2 (0.71)	0.0 (0.0)	0.90 (0.57)

Table A6. Evaluation statistics for training of the EC nodes with all of the calibration methods. Laboratory coefficients were not available from the calibration performed by the manufacturer. Note that the Field training period encompassed low NO_2 concentrations, near the LOD, resulting in a low R^2 value. All values are means over all EC nodes with their corresponding standard deviation shown in brackets.

Pollutant	Method	R^2	Slope	Intercept/ppb	RMSE/ppb	MBE/ppb	MAE/ppb
NO_2	EASE	0.94 (0.040)	1.0 (0.0)	0.0 (0.0)	5.2 (1.6)	0.0 (0.0)	4.0 (1.4)
	Field	0.49 (0.082)	1.0 (0.0)	0.0 (0.0)	3.3 (0.22)	0.0 (0.0)	2.5 (0.19)
O_3	EASE	0.96 (0.026)	1.0 (0.0)	0.0 (0.0)	4.6 (1.1)	0.0 (0.0)	2.9 (1.2)
	Field	0.84 (0.047)	1.0 (0.0)	0.0 (0.0)	2.8 (0.38)	0.0 (0.0)	2.2 (0.29)

References

1. Landrigan, P.J.; Fuller, R.; Acosta, N.J.R.; Adeyi, O.; Arnold, R.; Basu, N.N.; Baldé, A.B.; Bertollini, R.; Bose-O'Reilly, S.; Boufford, J.I.; et al. The Lancet Commission on Pollution and Health. *Lancet* **2018**, *391*, 462–512. [CrossRef]
2. Forouzanfar, M.H.; Afshin, A.; Alexander, L.T.; Anderson, H.R.; Bhutta, Z.A.; Biryukov, S.; Brauer, M.; Burnett, R.; Cercy, K.; Charlson, F.J.; et al.Global, regional, and national comparative risk assessment of 79 behavioural, environmental and occupational, and metabolic risks or clusters of risks, 1990–2015: A Syst. Anal. Global Burden Disease Study 2015.*Lancet* **2016**, *388*, 1659–1724. [CrossRef]
3. Vohra, K.; Vodonos, A.; Schwartz, J.; Marais, E.A.; Sulprizio, M.P.; Mickley, L.J. Global Mortality from Outdoor Fine Particle Pollution Generated by Fossil Fuel Combustion: Results from GEOS-Chem. *Environ. Res.* **2021**, *195*, 110754. [CrossRef]
4. Fuller, R.; Landrigan, P.J.; Balakrishnan, K.; Bathan, G.; Bose-O'Reilly, S.; Brauer, M.; Caravanos, J.; Chiles, T.; Cohen, A.; Corra, L.; et al. Pollution and Health: A Progress Update. *Lancet Planet. Health* **2022**, *6*, e535–e547. [CrossRef]
5. Thurston, G.D.; Kipen, H.; Annesi-Maesano, I.; Balmes, J.; Brook, R.D.; Cromar, K.; De Matteis, S.; Forastiere, F.; Forsberg, B.; Frampton, M.W.; et al. A Joint ERS/ATS Policy Statement: What Constitutes an Adverse Health Effect of Air Pollution? An Analytical Framework. *Eur. Respir. J.* **2017**, *49*, 1600419. [CrossRef] [PubMed]
6. Rundle, A.; Hoepner, L.; Hassoun, A.; Oberfield, S.; Freyer, G.; Holmes, D.; Reyes, M.; Quinn, J.; Camann, D.; Perera, F.; et al. Association of Childhood Obesity With Maternal Exposure to Ambient Air Polycyclic Aromatic Hydrocarbons during Pregnancy. *Am. J. Epidemiol.* **2012**, *175*, 1163–1172. [CrossRef]

7. Shehab, M.A.; Pope, F.D. Effects of Short-Term Exposure to Particulate Matter Air Pollution on Cognitive Performance. *Sci. Rep.* **2019**, *9*, 8237. [CrossRef] [PubMed]
8. Schraufnagel, D.E.; Balmes, J.R.; Cowl, C.T.; Matteis, S.D.; Jung, S.H.; Mortimer, K.; Perez-Padilla, R.; Rice, M.B.; Riojas-Rodriguez, H.; Sood, A.; et al. Air Pollution and Noncommunicable Diseases: A Review by the Forum of International Respiratory Societies' Environmental Committee, Part 1: The Damaging Effects of Air Pollution. *Chest* **2019**, *155*, 409–416. [CrossRef] [PubMed]
9. World Health Organization. *WHO Global Air Quality Guidelines: Particulate Matter (PM2.5 and PM10), Ozone, Nitrogen Dioxide, Sulfur Dioxide and Carbon Monoxide*; World Health Organization: Geneva, Switzerland, 2021.
10. Apte, J.S.; Messier, K.P.; Gani, S.; Brauer, M.; Kirchstetter, T.W.; Lunden, M.M.; Marshall, J.D.; Portier, C.J.; Vermeulen, R.C.; Hamburg, S.P. High-Resolution Air Pollution Mapping with Google Street View Cars: Exploiting Big Data. *Environ. Sci. Technol.* **2017**, *51*, 6999–7008. [CrossRef] [PubMed]
11. Goldberg, M.S. On the Interpretation of Epidemiological Studies of Ambient Air Pollution. *J. Expo. Sci. Environ. Epidemiol.* **2007**, *17* (Suppl. S2), S66–S70. [CrossRef] [PubMed]
12. Hernández-Gordillo, A.; Ruiz-Correa, S.; Robledo-Valero, V.; Hernández-Rosales, C.; Arriaga, S. Recent Advancements in Low-Cost Portable Sensors for Urban and Indoor Air Quality Monitoring. *Air Qual. Atmos. Health* **2021**, *14*, 1931–1951. [CrossRef]
13. *BS EN 14211:2012*; Ambient Air. Standard Method for the Measurement of the Concentration of Nitrogen Dioxide and Nitrogen Monoxide by Chemiluminescence. Technical Report BS EN 14211:2012. Publications Office of the European Union: Luxembourg, 2012.
14. *BS EN 14625:2012*; Ambient Air. Standard Method for the Measurement of the Concentration of Ozone by Ultraviolet Photometry. Technical Report BS EN 14625:2012. Publications Office of the European Union: Luxembourg, 2012.
15. Hertel, O.; Ellermann, T.; Palmgren, F.; Berkowicz, R.; Løfstrøm, P.; Frohn, L.M.; Geels, C.; Skjøth, C.A.; Brandt, J.; Christensen, J.; et al. Integrated Air-Quality Monitoring – Combined Use of Measurements and Models in Monitoring Programmes. *Environ. Chem.* **2007**, *4*, 65–74. [CrossRef]
16. Morawska, L.; Thai, P.K.; Liu, X.; Asumadu-Sakyi, A.; Ayoko, G.; Bartonova, A.; Bedini, A.; Chai, F.; Christensen, B.; Dunbabin, M.; et al. Applications of Low-Cost Sensing Technologies for Air Quality Monitoring and Exposure Assessment: How Far Have They Gone? *Environ. Int.* **2018**, *116*, 286–299. [CrossRef]
17. Frederickson, L.B.; Petersen-Sonn, E.A.; Shen, Y.; Hertel, O.; Hong, Y.; Schmidt, J.; Johnson, M.S. Low-cost sensors for indoor and outdoor pollution. In *Air Pollution Sources, Statistics and Health Effects*; Springer: Berlin/Heidelberg, Germany, 2021; pp. 423–453.
18. Maag, B.; Zhou, Z.; Thiele, L. A Survey on Sensor Calibration in Air Pollution Monitoring Deployments. *IEEE Internet Things J.* **2018**, *5*, 4857–4870. [CrossRef]
19. Tonne, C. A Call for Epidemiology Where the Air Pollution Is. *Lancet Planet. Health* **2017**, *1*, e355–e356. [CrossRef]
20. Lewis, A.; Edwards, P. Validate Personal Air-Pollution Sensors. *Nat. News* **2016**, *535*, 29. [CrossRef] [PubMed]
21. Castell, N.; Dauge, F.R.; Schneider, P.; Vogt, M.; Lerner, U.; Fishbain, B.; Broday, D.; Bartonova, A. Can Commercial Low-Cost Sensor Platforms Contribute to Air Quality Monitoring and Exposure Estimates? *Environ. Int.* **2017**, *99*, 293–302. [CrossRef] [PubMed]
22. Karagulian, F.; Barbiere, M.; Kotsev, A.; Spinelle, L.; Gerboles, M.; Lagler, F.; Redon, N.; Crunaire, S.; Borowiak, A. Review of the Performance of Low-Cost Sensors for Air Quality Monitoring. *Atmosphere* **2019**, *10*, 506. [CrossRef]
23. Lewis, A.C.; Schneidemesser, E.V.; Peltier, R.E. *Low-Cost Sensors for the Measurement of Atmospheric Composition: Overview of Topic and Future Applications*; Technical Report; World Meteorological Organization: Geneva, Switzerland, 2018.
24. Rai, A.C.; Kumar, P.; Pilla, F.; Skouloudis, A.N.; Di Sabatino, S.; Ratti, C.; Yasar, A.; Rickerby, D. End-User Perspective of Low-Cost Sensors for Outdoor Air Pollution Monitoring. *Sci. Total. Environ.* **2017**, *607–608*, 691–705. [CrossRef] [PubMed]
25. Peterson, P.J.D.; Aujla, A.; Grant, K.H.; Brundle, A.G.; Thompson, M.R.; Hey, J.V.; Leigh, R.J. Practical Use of Metal Oxide Semiconductor Gas Sensors for Measuring Nitrogen Dioxide and Ozone in Urban Environments. *Sensors* **2017**, *17*, 1653. [CrossRef] [PubMed]
26. Baron, R.; Saffell, J. Amperometric Gas Sensors as a Low Cost Emerging Technology Platform for Air Quality Monitoring Applications: A Review. *ACS Sens.* **2017**, *2*, 1553–1566. [CrossRef] [PubMed]
27. Korotcenkov, G.; Cho, B.K. Instability of Metal Oxide-Based Conductometric Gas Sensors and Approaches to Stability Improvement (Short Survey). *Sens. Actuators B Chem.* **2011**, *156*, 527–538. [CrossRef]
28. Li, J.; Hauryliuk, A.; Malings, C.; Eilenberg, S.R.; Subramanian, R.; Presto, A.A. Characterizing the Aging of Alphasense NO_2 Sensors in Long-Term Field Deployments. *ACS Sens.* **2021**, *6*, 2952–2959. [CrossRef] [PubMed]
29. Castell, N.; Viana, M.; Minguillón, M.; Guerreiro, C.; Querol, X. *Real-World Application of New Sensor Technologies for Air Quality Monitoring*; Technical Report; European Topic Centre on Air Pollution and Climate Change Mitigation (ETC/ACM): Bilthoven, The Netherlands, 2013.
30. Topalović, D.B.; Davidović, M.D.; Jovanović, M.; Bartonova, A.; Ristovski, Z.; Jovašević-Stojanović, M. In Search of an Optimal In-Field Calibration Method of Low-Cost Gas Sensors for Ambient Air Pollutants: Comparison of Linear, Multilinear and Artificial Neural Network Approaches. *Atmos. Environ.* **2019**, *213*, 640–658. [CrossRef]
31. Piedrahita, R.; Xiang, Y.; Masson, N.; Ortega, J.; Collier, A.; Jiang, Y.; Li, K.; Dick, R.; Lv, Q.; Hannigan, M.; et al. The next Generation of Low-Cost Personal Air Quality Sensors for Quantitative Exposure Monitoring. *Atmos. Meas. Tech. Discuss.* **2014**, *7*, 2425–2457. [CrossRef]

32. Cantrell, C. Review of methods for linear least-squares fitting of data and application to atmospheric chemistry problems. *Atmos. Chem. Phys.* **2008**, *8*, 5477–5487. [CrossRef]
33. Casey, J.G.; Hannigan, M.P. Testing the Performance of Field Calibration Techniques for Low-Cost Gas Sensors in New Deployment Locations: Across a County Line and across Colorado. *Atmos. Meas. Tech.* **2018**, *11*, 6351–6378. [CrossRef]
34. Levy Zamora, M.; Buehler, C.; Datta, A.; Gentner, D.R.; Koehler, K. Optimizing Co-Location Calibration Periods for Low-Cost Sensors. *EGUsphere* **2022**, 1–20. [CrossRef]
35. De Vito, S.; Esposito, E.; Castell, N.; Schneider, P.; Bartonova, A. On the Robustness of Field Calibration for Smart Air Quality Monitors. *Sens. Actuators B Chem.* **2020**, *310*, 127869. [CrossRef]
36. Tancev, G.; Pascale, C. The Relocation Problem of Field Calibrated Low-Cost Sensor Systems in Air Quality Monitoring: A Sampling Bias. *Sensors* **2020**, *20*, 6198. [CrossRef] [PubMed]
37. Vikram, S.; Collier-Oxandale, A.; Ostertag, M.H.; Menarini, M.; Chermak, C.; Dasgupta, S.; Rosing, T.; Hannigan, M.; Griswold, W.G. Evaluating and Improving the Reliability of Gas-Phase Sensor System Calibrations across New Locations for Ambient Measurements and Personal Exposure Monitoring. *Atmos. Meas. Tech.* **2019**, *12*, 4211–4239. [CrossRef]
38. Papapostolou, V.; Zhang, H.; Feenstra, B.J.; Polidori, A. Development of an Environmental Chamber for Evaluating the Performance of Low-Cost Air Quality Sensors under Controlled Conditions. *Atmos. Environ.* **2017**, *171*, 82–90. [CrossRef]
39. Aleixandre, M.; Gerboles, M.; Spinelle, L. *Protocol of Evaluation and Calibration of Low-Cost Gas Sensors for the Monitoring of Air Pollution;* Technical Report; Publications Office of the European Union, Institute for Environment and Sustainability (Joint Research Centre): Luxembourg, 2013.
40. RC Team. *R: A Language and Environment for Statistical Computing;* R Foundation for Statistical Computing: Vienna, Austria, 2020.
41. Hadley. *Ggplot2: Elegant Graphics for Data Analysis;* Springer: New York, NY, USA, 2016.
42. Carslaw, D.C.; Ropkins, K. Openair—An R Package for Air Quality Data Analysis. *Environ. Model. Softw.* **2012**, *27–28*, 52–61. [CrossRef]
43. Sensortech, S. *MiCS-6814 Datasheet 1143 Rev 8;* Technical Report; SGX, Sensortech, Courtils 1: Corcelles-Cormondrèche, Switzerland, 2015.
44. Dey, A. Semiconductor Metal Oxide Gas Sensors: A Review. *Mater. Sci. Eng. B* **2018**, *229*, 206–217. [CrossRef]
45. Korotcenkov, G. The Role of Morphology and Crystallographic Structure of Metal Oxides in Response of Conductometric-Type Gas Sensors. *Mater. Sci. Eng. R Rep.* **2008**, *61*, 1–39. [CrossRef]
46. Fine, G.F.; Cavanagh, L.M.; Afonja, A.; Binions, R. Metal Oxide Semi-Conductor Gas Sensors in Environmental Monitoring. *Sensors* **2010**, *10*, 5469–5502. [CrossRef]
47. Schütze, A.; Baur, T.; Leidinger, M.; Reimringer, W.; Jung, R.; Conrad, T.; Sauerwald, T. Highly Sensitive and Selective VOC Sensor Systems Based on Semiconductor Gas Sensors: How to? *Environments* **2017**, *4*, 20. [CrossRef]
48. Viricelle, J.; Pauly, A.; Mazet, L.; Brunet, J.; Bouvet, M.; Varenne, C.; Pijolat, C. Selectivity Improvement of Semi-Conducting Gas Sensors by Selective Filter for Atmospheric Pollutants Detection. *Mater. Sci. Eng. C* **2006**, *26*, 186–195. [CrossRef]
49. Alphasense. *Nitrogen Dioxide Sensors;* Technical Report; Alphasense: Braintree, UK, 2022.
50. Alphasense. *Ozone Sensors;* Technical Report; Alphasense: Braintree, UK, 2022.
51. Bulot, F.M.J.; Russell, H.S.; Rezaei, M.; Johnson, M.S.; Ossont, S.J.J.; Morris, A.K.R.; Basford, P.J.; Easton, N.H.C.; Foster, G.L.; Loxham, M.; et al. Laboratory Comparison of Low-Cost Particulate Matter Sensors to Measure Transient Events of Pollution. *Sensors* **2020**, *20*, 2219. [CrossRef]
52. Ellermann, T.; Nygaard, J.; Nøjgaard, J.K.; Nordstrøm, C.; Brandt, J.; Christensen, J.; Ketzel, M.; Massling, A.; Bossi, R.; Frohn, L.M.; et al. *The Danish Air Quality Monitoring Programme—Annual Summary for 2018;* Technical Report No. 218; DCE—Danish Centre for Environment and Energy: Roskilde, Denmark, 2020.
53. Alphasense. *AAN 803-05 Correcting For Background Currents In Four Electrode Toxic Gas Sensors;* Technical Report AAN 803-05; Alphasense: Braintree, UK, 2019.
54. Chatzidiakou, L.; Krause, A.; Popoola, O.A.M.; Di Antonio, A.; Kellaway, M.; Han, Y.; Squires, F.A.; Wang, T.; Zhang, H.; Wang, Q.; et al. Characterising Low-Cost Sensors in Highly Portable Platforms to Quantify Personal Exposure in Diverse Environments. *Atmos. Meas. Tech.* **2019**, *12*, 4643–4657. [CrossRef]
55. Alexander, D.L.J.; Tropsha, A.; Winkler, D.A. Beware of R(2): Simple, Unambiguous Assessment of the Prediction Accuracy of QSAR and QSPR Models. *J. Chem. Inf. Model.* **2015**, *55*, 1316–1322. [CrossRef] [PubMed]
56. Spinelle, L.; Gerboles, M.; Villani, M.G.; Aleixandre, M.; Bonavitacola, F. Field Calibration of a Cluster of Low-Cost Available Sensors for Air Quality Monitoring. Part A: Ozone and Nitrogen Dioxide. *Sens. Actuators B Chem.* **2015**, *215*, 249–257. [CrossRef]

sensors

MDPI

Article

Graphene Oxide Thin Films for Detection and Quantification of Industrially Relevant Alcohols and Acetic Acid

Pedro Catalão Moura *, Thais Priscilla Pivetta, Valentina Vassilenko, Paulo António Ribeiro and Maria Raposo *

Laboratory for Instrumentation, Biomedical Engineering and Radiation Physics (LIBPhys-NOVA), Department of Physics, NOVA School of Science and Technology, NOVA University of Lisbon, Campus FCT-NOVA, 2829-516 Caparica, Portugal

* Correspondence: pr.moura@campus.fct.unl.pt (P.C.M.); mfr@fct.unl.pt (M.R.)

Abstract: Industrial environments are frequently composed of potentially toxic and hazardous compounds. Volatile organic compounds (VOCs) are one of the most concerning categories of analytes commonly existent in the indoor air of factories' facilities. The sources of VOCs in the industrial context are abundant and a vast range of human health conditions and pathologies are known to be caused by both short- and long-term exposures. Hence, accurate and rapid detection, identification, and quantification of VOCs in industrial environments are mandatory issues. This work demonstrates that graphene oxide (GO) thin films can be used to distinguish acetic acid, ethanol, isopropanol, and methanol, major analytes for the field of industrial air quality, using the electronic nose concept based on impedance spectra measurements. The data were treated by principal component analysis. The sensor consists of polyethyleneimine (PEI) and GO layer-by-layer films deposited on ceramic supports coated with gold interdigitated electrodes. The electrical characterization of this sensor in the presence of the VOCs allows the identification of acetic acid in the concentration range from 24 to 120 ppm, and of ethanol, isopropanol, and methanol in a concentration range from 18 to 90 ppm, respectively. Moreover, the results allows the quantification of acetic acid, ethanol, and isopropanol concentrations with sensitivity values of $(3.03 \pm 0.12) * 10^4$, $(-1.15 \pm 0.19) * 10^4$, and $(-1.1 \pm 0.50) * 10^4$ mL^{-1}, respectively. The resolution of this sensor to detect the different analytes is lower than 0.04 ppm, which means it is an interesting sensor for use as an electronic nose for the detection of VOCs.

Keywords: volatile organic compounds; VOC; industrial environment; indoor air; air quality; acetic acid; ethanol; methanol; isopropanol; electronic nose; impedance spectroscopy; layer-by-layer films

1. Introduction

The indoor environment of industrial facilities, particularly in production lines and warehouses, is commonly populated by a large variety of potentially polluting and hazardous compounds [1–3]. Their presence in the air arises from the multitude of emitting sources existent in these kinds of scenarios and leads to well-known and worthy-of-attention consequences for the employee's health [4,5]. Volatile organic compounds (VOCs) are among the most concerning of these potentially hazardous analytes.

Volatile organic compounds correspond to organic compounds whose vapor pressure, at 293.15 K, equals or exceeds 10 Pa, i.e., they are volatile at room temperature [6]. Considering their nature, VOC-emitting sources can be arranged into two distinct categories, natural sources and anthropogenic sources. Natural sources include fauna and flora emissions. Smoking, cooking, or cleaning, and all the products related to these activities, such as tobacco, food, perfumes, personal care creams, and detergents, for example, are among the main anthropogenic sources of VOCs. A vast range of daily-use objects, namely clothes, furniture, building materials, paints, fuels, sprays, pesticides, glues, writing materials, and copying devices, are equally relevant anthropogenic sources of VOCs [2,7,8].

At an industrial level, VOCs are a common element of the indoor air of the facilities since the activities that are usually undertaken in such locations are conducive to the emission of these kinds of analytes. For instance, in coating industries and facilities with painting, printing, or similar activities, it is rather common for the detection of relevant amounts of alcohol-based VOCs in the atmosphere due to the frequent use of solvents, paints, and other coating solutions [9–11]. Automotive, electronics, and comparable assembly lines are equally replete with numerous sources of VOCs, namely, the chemicals, solvents, or rubbers, and the welding, drying, heating, and coating processes often employed during the production [3,12,13]. The manufacturing facilities of personal care and cleaning products, due to the intense utilization of VOCs-based chemicals in their formulas, are major contributors to the presence of VOCs in both indoor and outdoor air [14–16]. Summarizing, independently of the undertaken activities, industrial facilities are often crowded by sources of VOCs such as acetic acid, ethanol, isopropanol, methanol, and many others.

Exposure to VOCs in both short- and long-term scenarios is known for causing an extensive list of pathologies and health conditions that ranges from harmless biological reactions to health-threatening diseases [17]. In simpler cases, exposure to VOCs leads to allergic or inflammatory reactions in the respiratory tract, and cutaneous and ocular tissues. Headaches, nausea, dizziness, visual disorders, memory impairment, emesis, epistaxis, and fatigue are equally ordinary and well-known reactions of the human organism to the presence of VOCs [18,19]. A cause–consequence relation between continued exposure to VOCs and the development of dangerous forms of cancer has equally been studied. Lung, oral, and even breast cancer are examples of VOC-related carcinogenic pathologies [20,21].

Due to all the aforementioned facts, it is mandatory to study, develop, and implement analytical tools that enable the accurate and rapid detection, identification, and quantification of the presence of VOCs in the indoor air of industrial facilities and, consequently, the prevention of potential hazards to both the environment and employees' health.

Several techniques have been scientifically addressed regarding their suitability for the assessment of VOCs. These techniques include both multisensor array-based procedures and analytical techniques such as chromatographic and spectrometric approaches. Independently of the designation or the nature of the system, their core purpose is the detection of specific and potentially dangerous analytes.

Regarding chromatographic and spectrometric systems, their suitability for the detection of VOCs has been largely investigated in a substantial amount of practical applications that include air quality assessment, safety conditions, food characterization, drug detection, clinical scenarios, and many others [1,22–24]. The main advantages of analytical techniques such as liquid chromatography (LC), gas chromatography (GC), mass spectrometry (MS), or ion mobility spectrometry (IMS) include their high levels of sensitivity and precision, wide dynamic concentration ranges, analytical flexibility, and almost real-time monitoring capability. Notwithstanding these advantages, they also have some limitations, namely, their lack of portability and high costs, the necessity for sample preparation, and the requirement for qualified personnel [25–27].

Aiming to circumvent the mentioned limitations, the development of sensor array-based systems dedicated to specific scenarios has gained relevance as a cheaper and simpler solution [28,29]. These systems enable an accurate and rapid qualification and quantification of VOCs through the interactions that occur on the surface of the sensor when they experience contact with the analyte [30,31]. Due to their flexibility and scientifically relevant results, sensor array-based systems have proved their suitability for the assessment of common VOCs such as acetone [32], ethanol [33], butanol [34], formaldehyde [35], triethylamine [36], methanol [37], isopropanol [38], ethyl acetate [39], benzene [40], or acetic acid [41], among many others.

The development of electronic tongues [42–44] and noses [30] based on featured organic thin films of graphene as sensing units have been widely addressed over the past years; an approach that can also be used for the development of VOC-dedicated electronic noses and tongues. For these cases, a sensor array is normally required. Graphene molecules and derivatives were shown to be suitable to be used to detect a panoply of molecules, macromolecules, and even viruses; thus, making them an invaluable tool in many fields (e.g., medicine, industry, genetics, criminology) [45]. Moreover, graphene oxide (GO) can be used in both electrical and optical devices [46]. Graphene oxide and poly (allylamine hydrochloride) (PAH) layer-by-layer thin films (LBL) have been characterized with respect to their growth with the number of bilayers, morphology, and electrical properties [47]. The electrical characterization of these PAH/GO films revealed a semiconductor behaviour that makes these films interesting for the development of sensors by probing their electrical property changes when submitted to different environments [47]. This work proposes the development of a graphene oxide thin-film-based sensor using the layer-by-layer technique, towards the detection, identification, and quantification of industrially relevant VOCs, namely acetic acid, ethanol, isopropanol, and methanol. Impedance spectroscopy was used as a probe of the sensor response in terms of the analyte and concentration and data were processed through principal component analysis (PCA). The achieved results revealed the potential of the sensor being used not only to discriminate these compounds in a complex mixture but also to quantify them, which could be a factor that adds value towards its use for air quality control and public health.

2. Materials and Methods

2.1. Materials

Polyethyleneimine (PEI) and graphene oxide (GO), utilized for the preparation of thin films, were purchased from Sigma-Aldrich. Standards of acetone (C_3H_6O; 58.08 gmol^{-1}; 99.0%) and isopropanol (C_3H_8O; 60.10 gmol^{-1}; 98.0%) were purchased from Laborspirit-Labchem. Standards of ethanol (C_2H_6O; 46.07 gmol^{-1}; 99.8%) and methanol (CH_4O; 32.04 gmol^{-1}; 99.8%) were obtained from Honeywell. Standard of acetic acid ($C_2H_4O_2$; 60.05 gmol^{-1}; 99.8%) was purchased from Fisher Scientific. The ceramic-based sensor supports with deposited gold interdigitated electrodes (IDE) were acquired from Metrohm DropSens (length: 22.8 mm; width: 7.6 mm; thickness: 1 mm; electrodes width: 200 μm; distance between electrodes: 200 μm).

2.2. Preparation of Sensor

Thin films of PEI and GO polyelectrolytes were adsorbed by layer-by-layer (LbL) technique on ceramic-based sensor supports with deposited gold IDE. This technique basically settles down on electrostatic forces of the polyelectrolytes that enable the application of alternated electrically charged polyelectrolytes [48]. To do so, the supports were alternatively immersed in the PEI and GO aqueous solutions, positively and negatively charged polyelectrolyte solutions with concentrations of 2.5×10^{-1} and 3.2×10^{-1} mg/mL, respectively. Between each immersion, a wash procedure, consisting of the immersion of the support in ultrapure water to remove eventual excesses of polyelectrolyte, was undertaken. Once completing each immersion sequence, the support was dried with a gentle nitrogen air blasting. The described procedure corresponds to the deposition of a bilayer and was repeated 15 times, leading to the deposition of the multilayer thin films of polyelectrolyte in the surface of the ceramic-based support, and forming the sensors denominated as $(PEI/GO)_{15}$. Detailed information regarding the procedure was previously described in the literature [49,50].

2.3. Impedance Spectroscopy

Impedance spectroscopy was the analytical technique selected for the characterization of sensor units' response to the industrial solvents mentioned above. Detailed information on experimental setup has been described by Magro et al. [30]. Basically, a custom-made chamber with a volume of 58 L was employed to create a controlled atmosphere and assess the electrical impedance response of the thin films. Initially, the chamber was evacuated to pressure of 10^{-3} mbar and the previously calibrated sample of the target VOCs was introduced into a round-bottom glass flask. The VOC sample was then volatilized and purged through compressed synthetic air (ALPHAGAZ^{TM} 1 AR—Air Liquide). The sensor units were previously placed in the respective sample holder in the chamber and, as next step, their electrical response to the VOCs was measured with an impedance analyzer (Solartron 1260 Impedance/Gain-Phase Analyzer coupled with a 1296A Dielectric Interface-AMETEK Scientific Instruments) and assessed with a dedicated software (SMaRT Impedance Measurement Software, version 3.3.1-AMETEK Scientific Instruments). A frequency range of 1 to 10^6 Hz and an AC voltage of 25 mV were applied during the impedance assessment. To ensure that the signal is representative of the sensor's response, these measurements were performed in triplicate.

2.4. Data Treatment

Principal component analysis (PCA) was used to reduce the data size and to obtain a new space of orthogonal components aiming to distinguish the different samples and respective concentrations [51]. For this analysis, the electrical impedance and impedance angle spectra were considered. It is relevant to emphasize that these values were both collected in a frequency range of 1 Hz to 1 MHz for each sample at different concentrations. Since three replicas were registered, the spectra used in this analysis correspond to the average of those three measurements.

3. Results

Impedance Results

Figure 1 illustrates the impedance (a) and impedance angle (b) spectra of $(PEI/GO)_{15}$ films deposited on the surface of gold IDEs when submitted to an atmosphere of different concentrations of acetic acid (I), ethanol (II), methanol (III), and isopropanol (IV), represented by the VOCs' evaporated volume. It should be mentioned that, for better clarity and interpretation of the plots present in the figure, the error bars measured for these spectra were not included. Nonetheless, it is relevant to state that both these frequency-dependent measurements presented error values lower than 1% among the three replicas.

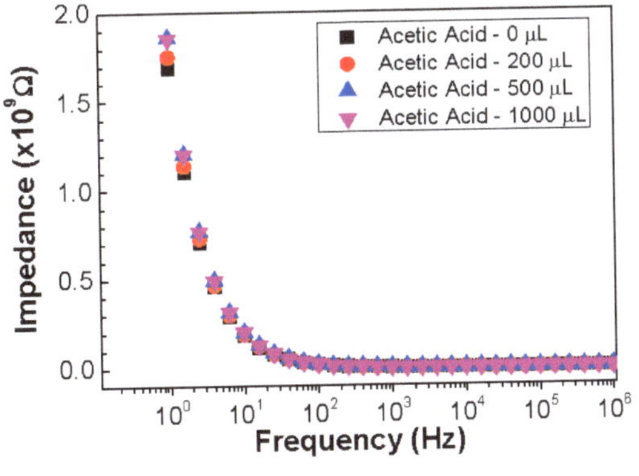

(I-a)

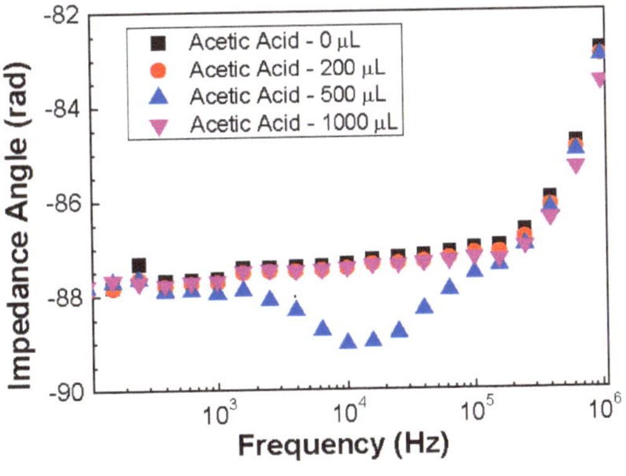

(I-b)

Figure 1. *Cont.*

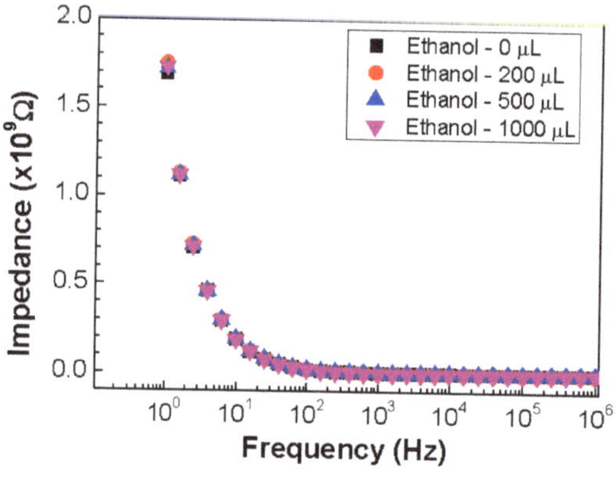

(II-a)

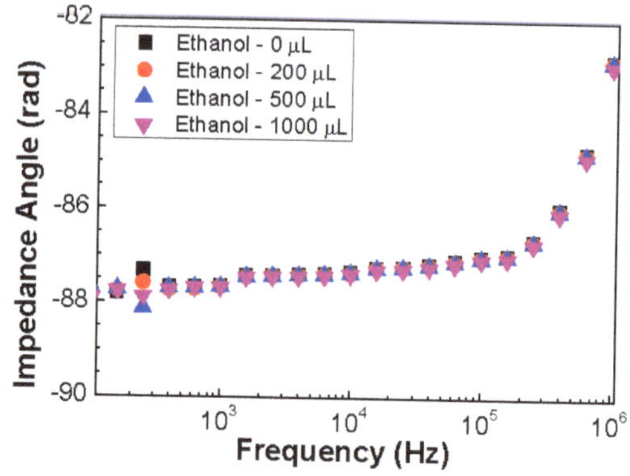

(II-b)

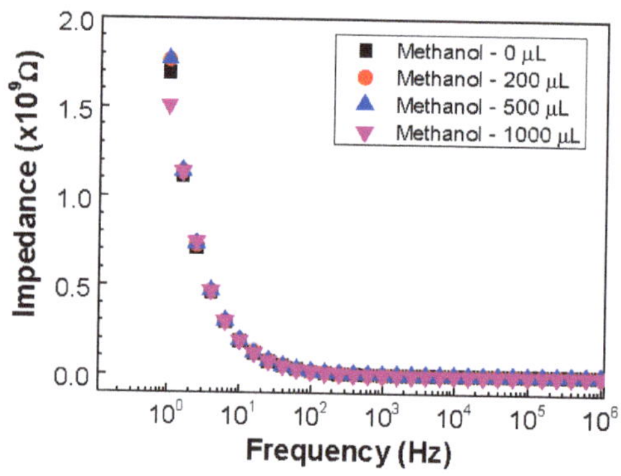

(III-a)

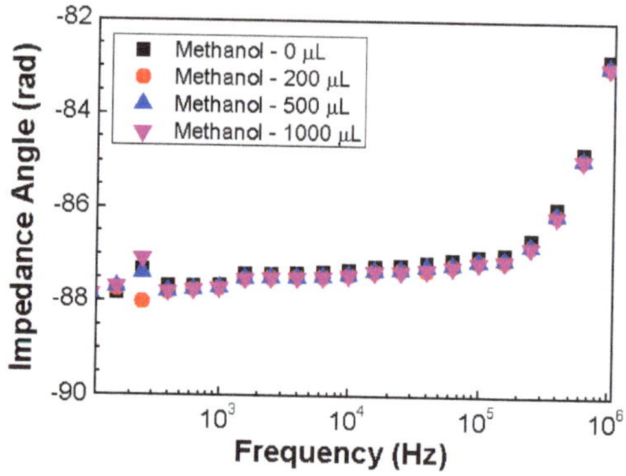

(III-b)

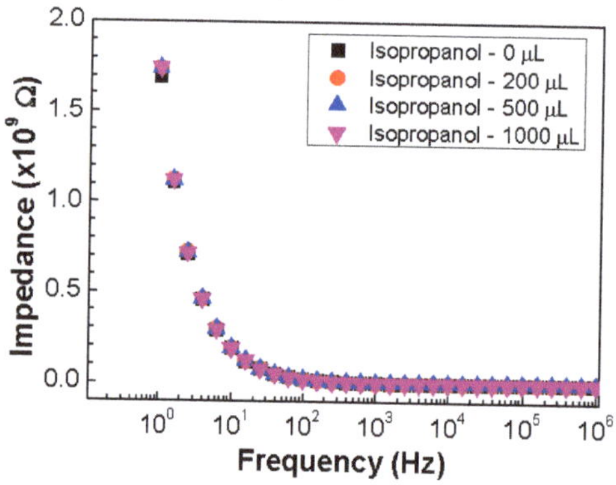

(IV-a)

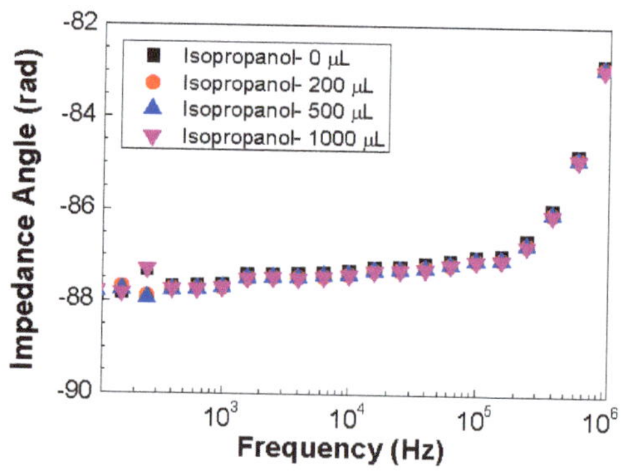

(IV-b)

Figure 1. Impedance (**a**) and impedance angle (**b**) spectra of the sensor devices when exposed to atmospheres with different concentrations of acetic acid (**I**), ethanol (**II**), methanol (**III**), and isopropanol (**IV**).

To verify if the PEI/GO thin films can distinguish the different concentrations of the measured VOCs, the impedance magnitude and impedance angle at fixed frequencies were analyzed for the different VOCs' concentrations. The evaporated volume can be directly related with the concentration levels through the ratio between the mass of the analyte and the mass of the air in the chamber. The mass of each analyte was calculated for each volume through their density. The same approach was employed to calculate the mass of the air existent in the interior of the 58 L volume chamber. Once both the analyte mass for each volume and the total mass in the interior of the chamber were calculated, the corresponding concentrations were estimated by the mentioned ratio. Table 1 summarizes the evaporated volumes and respective concentration levels for the four analysed analytes. The concentration levels were converted to ppm scale for easier comprehension.

Table 1. Evaporated volumes (μL) and respective concentration levels (ppm) for the four considered VOCs; acetic acid, ethanol, methanol, and isopropanol.

Volume (μL)	Concentration (ppm)			
	Acetic Acid	Ethanol	Methanol	Isopropanol
200	24	18	18	18
500	60	45	45	45
1000	120	90	90	90

Figure 2a–d show both the impedance magnitude and impedance angle at 10^4 Hz, plotted as a function of the evaporated volume and, consequently, the concentration of acetic acid, ethanol, methanol, and isopropanol, respectively. These graphs clearly demonstrate that the electrical measurements can distinguish between the different concentrations since both the magnitude and angle vary with the concentration if only a VOC type is considered. However, when analyzing the measured values of magnitude and angle at this chosen frequency, one cannot distinguish between the different alcohol VOCs, meaning that is necessary to analyze the data achieved for all the frequencies with mathematical methods such as the PCA method. This analysis allows conclusions to be made as to whether the impedance magnitude and impedance angle spectra depend on the different samples and allows discrimination between the VOCs and their respective concentrations using a single sensor.

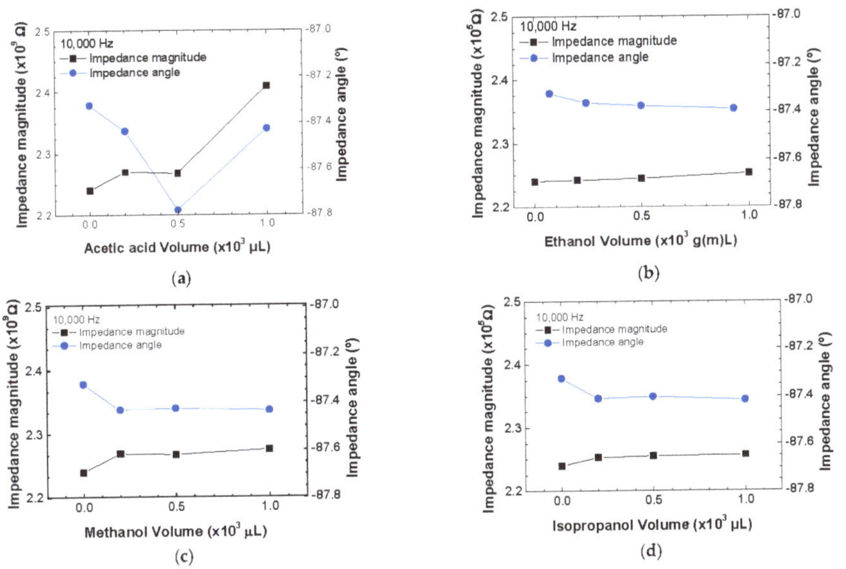

Figure 2. Impedance magnitude and impedance angle at a fixed frequency of 10^4 Hz for different concentrations of acetic acid (**a**), ethanol (**b**), methanol (**c**), and isopropanol (**d**) in air. The lines between the experimental points are guidelines.

4. Discussion

As one intends to distinguish between different VOCs using a single sensor, the PCA method was applied to both the impedance magnitude and impedance angle spectra data measured for the different concentrations of acetic acid, ethanol, methanol, and isopropanol. The PCA score plots of all the measured data for all the four target analytes in air at different concentrations are shown in Figure 3a. By analyzing this figure, one can observe that well defined PCA score regions can be defined for each type of measured VOC, allowing discrimination between the samples in the concentrations measured using a single sensor.

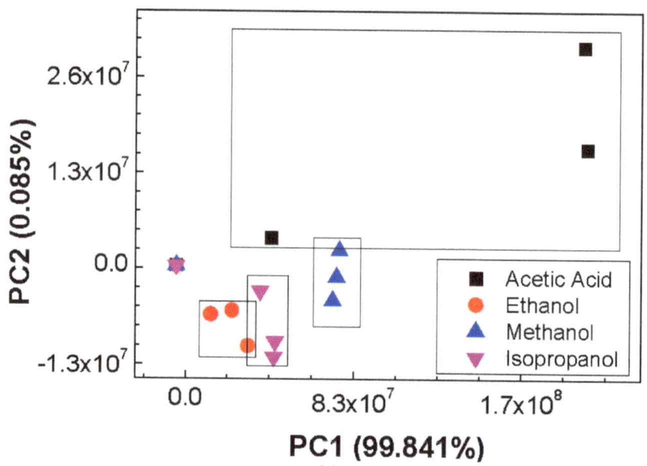

(a)

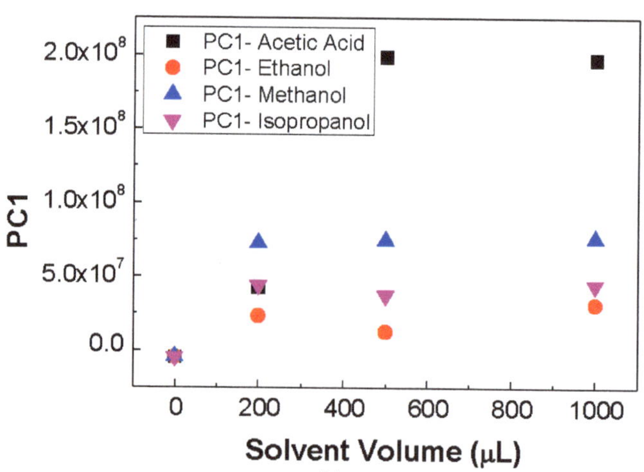

(b)

Figure 3. *Cont.*

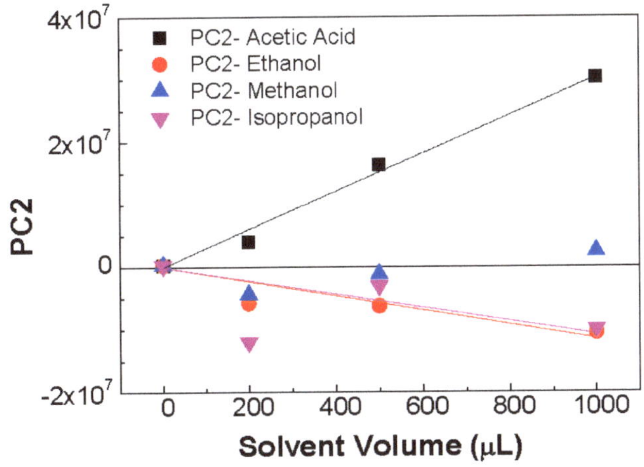

(c)

Figure 3. (**a**) PCA score plot after analyzing all the measured data for the detection of acetic acid, ethanol, methanol, and isopropanol in air at different concentrations; (**b**) evolution of PC1 components as a function of solvent volume; (**c**) evolution of PC2 components as a function of solvent volume.

To analyze if the calculated principal components PC1 and PC2 are concentration-dependent, the achieved values of PC1 and PC2 were plotted as a function of the VOC's volume in Figure 3b,c, respectively. Interestingly, the evolution of PC1 with the solvent volume shows that, except for the PC1 value associated to samples without VOCs and for the acetic acid sample with a volume of 200 μL, one can calculate a PC1 average value for each type of sample. The calculated values and respective error bars are present in Table 2. Therefore, it is possible to state that the PC1 value can distinguish the ethanol, methanol, and isopropanol in air samples and the acid acetic for higher concentrations. On the other hand, Figure 3c clearly demonstrates that the PC2 components can discriminate the VOCs' concentrations, with methanol being the exception. The sensitivity of the sensor when submitted to the different VOCs was estimated by fitting the PC2 data versus concentration with a straight line. The fitting parameters are listed in Table 2, where the sensitivity values correspond to the slope of the PC2 parameters versus the concentration. Sensitivity values of 30,300 ± 1200, −11,500 ± 1900, and −11,000 ± 5000 mL^{-1} were calculated for acetic acid, ethanol, and isopropanol, respectively. Since the slope calculated for the case of methanol PC2 data is very low, one decided to calculate the PC2 average. All these values are displayed in Table 2. From these results, one can conclude that this unique sensor can distinguish the different target VOCs and, except for methanol, the VOCs' concentration in the analyzed range.

Table 2. Summarization of the values achieved from Figure 3b,c. The columns PC1 Average and PC2 Average represent the mean value calculated from data of Figure 3b,c for the case of methanol, not considering the null concentration data. The sensitivity corresponds to the slope of the straight lines fitting PC2 data displayed in Figure 3c.

VOC	PC1 Average	Sensitivity (mL^{-1})	PC2 Average
Acetic acid	$(1.98 \pm 0.01) \times 10^8$ *	$(3.03 \pm 0.12) \times 10^4$	-
Ethanol	$(2.23 \pm 0.54) \times 10^7$	$(-1.15 \pm 0.19) \times 10^4$	-
Methanol	$(7.4 \pm 0.1) \times 10^7$	-	$(-0.066 \pm 2.81) \times 10^6$
Isopropanol	$(4.19 \pm 0.23) \times 10^7$	$(-1.1 \pm 0.5) \times 10^4$	-

* For concentrations higher than or equal to 60 ppm.

To compare the achieved data with the results of other sensors existent in the literature, one presents Table 3 where the values of the resolution and range values of the developed sensor are compared with the values of different sensors. To calculate the sensor resolution for this work, the minimum measurable values were considered, as described elsewhere [52]. As in the present results, the principal component 2 (*PC2*) values are linearly dependent on the concentration in ppm (*C*). The sensitivity (*S*) was calculated by the slope of the straight line, $\Delta PC2/\Delta C$, used to fit the data. Therefore, $\Delta PC2/\Delta C = S \pm u(S)$, with *u(S)* being the uncertainty of sensitivity given also by the fitting. The resolution corresponds to the calculated value of ΔC in which $\Delta C = u(S)/S$. This procedure enabled the estimation of the resolution values of 0.005, 0.015, and 0.04 ppm for acetic acid, ethanol, and isopropanol, respectively. As aforementioned, due to the very low slope calculated for methanol, one opted to not estimate the sensitivity and, consequently, the resolution values of this case. From the comparison included in Table 3, one can conclude that the methodology described in this work leads to limited resolution even though the studied range is of an intermediate level. The achieved values of resolution indicate that this sensor can be used in the development of an electronic nose for the detection of VOCs.

Table 3. Comparison of the achieved sensors with others available in literature.

Sensor	Resolution (ppm)	Range (ppm)
Acetic acid [53]	1.2	1–13
Acetic acid [54]	1	10–100
Acetic acid [55]	0.5	0.5–2000
Acetic acid [56]	0.73	1–15
Acetic acid (this work)	0.005	24–240
Ethanol [57]	0.05	1–200
Ethanol [58]	3	30–145
Ethanol [59]	1	1–200
Ethanol [60]	0.15	0.15–5
Ethanol (this work)	0.015	18–180
Methanol [61]	0.015	1.14–11.36
Methanol [62]	10	100–300
Methanol [63]	0.5	0.5–700
Methanol [64]	10	100–500
Methanol (this work)	-	18–180
Isopropanol [65]	2	2–100
Isopropanol [66]	1	1–100
Isopropanol [67]	1	1–1000
Isopropanol [68]	1	5–1000
Isopropanol (this work)	0.04	18–180

5. Conclusions

A unique sensor based on GO oxide thin films was used to simultaneously detect four industrially relevant VOCs, acetic acid, ethanol, methanol, and isopropanol, by measuring the impedance magnitude and impedance angle spectra responses in terms of concentrations and processing the data through PCA. The results lead to the conclusion that the impedance data allow both the different VOCs samples and their concentrations to be distinguished in the range of hundreds of ppm. From the PCA results, one can conclude that the principal component PC1 values can distinguish the ethanol, methanol, and isopropanol in air samples and also the acid acetic for higher concentrations, while from principal component PC2, one can discriminate the VOCs' concentrations with the exception of methanol. The sensitivities of the sensor are 30,300 $\pm$ 1200, $-$11,500 $\pm$ 1900, and $-$11,000 $\pm$ 5000 mL^{-1} for the acetic acid, ethanol, and isopropanol, respectively. The resolution values for this sensor are lower than 0.04 ppm, which proves the relevancy of using this sensor in the sensor array of an electronic nose for the qualification and quantification of VOCs.

Author Contributions: Conceptualization, P.A.R. and M.R.; methodology, T.P.P. and P.A.R.; software, P.A.R. and M.R.; validation, P.A.R. and M.R.; formal analysis, P.C.M., T.P.P. and M.R.; investigation, T.P.P.; resources, V.V., P.A.R. and M.R.; data curation, P.C.M., T.P.P. and M.R.; writing—original draft preparation, P.C.M., T.P.P. and M.R.; writing—review and editing, V.V., P.A.R. and M.R.; supervision, V.V., P.A.R. and M.R.; funding acquisition, V.V., P.A.R. and M.R. All authors have read and agreed to the published version of the manuscript.

Funding: This research was funded by Fundação para a Ciência e Tecnologia (FCT -MCTES, Portugal) through the grants PTDC/FIS-NAN/0909/2014, UIDB/04559/2020 (LIBPhys), and UIDP/04559/2020 (LIBPhys), and through the grant PD/BDE/150627/2020 in collaboration with Volkswagen Autoeuropa.

Institutional Review Board Statement: Not applicable.

Informed Consent Statement: Not applicable.

Data Availability Statement: Datasets are available from the corresponding authors upon reasonable request.

Acknowledgments: P.C.M. acknowledges Volkswagen Autoeuropa for his fellowship, PD/BDE/150627/2020.

Conflicts of Interest: The authors declare no conflict of interest.

References

1. Kim, M.-J.; Seo, Y.-K.; Kim, J.-H.; Baek, S.-O. Impact of industrial activities on atmospheric volatile organic compounds in Sihwa-Banwol, the largest industrial area in South Korea. *Environ. Sci. Pollut. Res.* **2020**, *27*, 28912–28930. [CrossRef] [PubMed]
2. Baek, S.-O.; Suvarapu, L.N.; Seo, Y.-K. Occurrence and Concentrations of Toxic VOCs in the Ambient Air of Gumi, an Electronics-Industrial City in Korea. *Sensors* **2015**, *15*, 19102–19123. [CrossRef] [PubMed]
3. Liu, Z.; Yan, Y.; Lv, T.; Huang, Z.; Liu, T.; Huang, Q.; Yang, J.; Chen, Y.; Zhao, Y.; Zhou, T. Comprehensive understanding the emission characteristics and kinetics of VOCs from automotive waste paint sludge in an environmental test chamber. *J. Hazard. Mater.* **2022**, *429*, 128387. [CrossRef] [PubMed]
4. Stute, N.L.; Stickford, J.L.; Augenreich, M.A.; Kimball, K.C.; Cope, J.M.; Bennett, C.; Grosicki, G.J.; Ratchford, S.M. Arterial stiffness and carotid distensibility following acute formaldehyde exposure in female adults. *Toxicol. Ind. Health* **2021**, *37*, 535–546. [CrossRef] [PubMed]
5. Sadeghi-Yarandi, M.; Karimi, A.; Ahmadi, V.; Sajedian, A.A.; Soltanzadeh, A.; Golbabaei, F. Cancer and non-cancer health risk assessment of occupational exposure to 1,3-butadiene in a petrochemical plant in Iran. *Toxicol. Ind. Health* **2020**, *36*, 960–970. [CrossRef]
6. Koppmann, R. *Volatile Organic Compounds in the Atmosphere*, 1st ed.; Blackwell Publishing Ltd.: Oxford, UK, 2007.
7. Montero-Montoya, R.; López-Vargas, R.; Arellano-Aguilar, O. Volatile Organic Compounds in Air: Sources, Distribution, Exposure and Associated Illness in Children. *Ann. Glob. Health* **2018**, *84*, 225–238. [CrossRef]
8. Cheung, P.K.; Jim, C.Y.; Siu, C.T. Air quality impacts of open-plan cooking in tiny substandard homes in Hong Kong. *Air Qual. Atmos. Health* **2019**, *12*, 865–878. [CrossRef]
9. Baumann, W.; Dinglreiter, U. Method for reprocessing and recycling of aqueous rinsing liquids from car painting with water-based paints in automobile industry. *Heat Mass Transf.* **2011**, *47*, 1043–1049. [CrossRef]
10. Zhong, Z.; Sha, Q.; Zheng, J.; Yuan, Z.; Gao, Z.; Ou, J.; Zheng, Z.; Li, C.; Huang, Z. Sector-based VOCs emission factors and source profiles for the surface coating industry in the Pearl River Delta region of China. *Sci. Total. Environ.* **2017**, *583*, 19–28. [CrossRef]
11. Tiessen, M.; Stock, N.L.; Stotesbury, T. Untargeted SPME-GC-MS Characterization of VOCs Released from Spray Paint. *J. Chromatogr. Sci.* **2020**, *59*, 103–111. [CrossRef]
12. Chang, C.-T.; Lee, C.-H.; Wu, Y.-P.; Jeng, F.-T. Assessment of the strategies for reducing volatile organic compounds emissions in the automotive industry in Taiwan. *Resour. Conserv. Recycl.* **2002**, *34*, 117–128. [CrossRef]
13. Lv, Z.; Liu, X.; Wang, G.; Shao, X.; Li, Z.; Nie, L.; Li, G. Sector-based volatile organic compounds emission characteristics from the electronics manufacturing industry in China. *Atmos. Pollut. Res.* **2021**, *12*, 101097. [CrossRef]
14. Nematollahi, N.; Kolev, S.D.; Steinemann, A. Volatile chemical emissions from 134 common consumer products. *Air Qual. Atmos. Health* **2019**, *12*, 1259–1265. [CrossRef]
15. Palmisani, J.; Di Gilio, A.; Cisternino, E.; Tutino, M.; de Gennaro, G. Volatile Organic Compound (VOC) Emissions from a Personal Care Polymer-Based Item: Simulation of the Inhalation Exposure Scenario Indoors under Actual Conditions of Use. *Sustainability* **2020**, *12*, 2577. [CrossRef]

16. Heeley-Hill, A.C.; Grange, S.K.; Ward, M.W.; Lewis, A.C.; Owen, N.; Jordan, C.; Hodgson, G.; Adamson, G. Frequency of use of household products containing VOCs and indoor atmospheric concentrations in homes. *Environ. Sci. Process. Impacts* **2021**, *23*, 699–713. [CrossRef]
17. Rouf, Z.; Dar, I.Y.; Javaid, M.; Dar, M.Y.; Jehangir, A. Volatile Organic Compounds Emission from Building Sector and Its Adverse Effects on Human Health. In *Ecological and Health Effects of Building Materials*, 1st ed.; Malik, J.A., Marathe, S., Eds.; Springer: Cham, Switzerland, 2022; pp. 67–86. [CrossRef]
18. Guo, H.; Lee, S.; Chan, L.; Li, W. Risk assessment of exposure to volatile organic compounds in different indoor environments. *Environ. Res.* **2004**, *94*, 57–66. [CrossRef]
19. Soni, V.; Singh, P.; Shree, V.; Goel, V. Effects of VOCs on Human Health. In *Air Pollution and Control. Energy, Environment, and Sustainability*, 1st ed.; Sharma, N., Agarwal, A.K., Eastwood, P., Gupta, T., Singh, A.P., Eds.; Springer: Singapore, 2018; pp. 119–142.
20. Mokammel, A.; Rostami, R.; Niazi, S.; Asgari, A.; Fazlzadeh, M. BTEX levels in rural households: Heating system, building characteristic impacts and lifetime excess cancer risk assessment. *Environ. Pollut.* **2022**, *298*, 118845. [CrossRef]
21. Lim, S.K.; Shin, H.S.; Yoon, K.S.; Kwack, S.J.; Um, Y.M.; Hyeon, J.H.; Kwak, H.M.; Kim, J.Y.; Kim, T.H.; Kim, Y.J.; et al. Risk Assessment of Volatile Organic Compounds Benzene, Toluene, Ethylbenzene, and Xylene (BTEX) in Consumer Products. *J. Toxicol. Environ. Health Part A* **2014**, *77*, 1502–1521. [CrossRef]
22. DePaula, J.; Cunha, S.C.; Cruz, A.; Sales, A.L.; Revi, I.; Fernandes, J.; Ferreira, I.M.P.L.V.O.; Miguel, M.A.L.; Farah, A. Volatile Fingerprinting and Sensory Profiles of Coffee Cascara Teas Produced in Latin American Countries. *Foods* **2022**, *11*, 3144. [CrossRef]
23. Jońca, J.; Pawnuk, M.; Arsen, A.; Sówka, I. Electronic Noses and Their Applications for Sensory and Analytical Measurements in the Waste Management Plants—A Review. *Sensors* **2022**, *22*, 1510. [CrossRef]
24. Armenta, S.; Alcala, M.; Blanco, M. A review of recent, unconventional applications of ion mobility spectrometry (IMS). *Anal. Chim. Acta* **2011**, *703*, 114–123. [CrossRef]
25. Moura, P.C.; Vassilenko, V. Gas Chromatography—Ion Mobility Spectrometry as a tool for quick detection of hazardous volatile organic compounds in indoor and ambient air: A university campus case study. *Eur. J. Mass Spectrom.* **2022**, *28*, 113–126. [CrossRef]
26. Moura, P.C.; Vassilenko, V.; Fernandes, J.M.; Santos, P.H. Indoor and Outdoor Air Profiling with GC-IMS. In *Technological Innovation for Life Improvement, IFIP Advances in Information and Communication Technology, DoCEIS 2020, Caparica, Portugal, 20 July 2020*; Springer: Cham, Switzerland, 2020.
27. Günzler, H.; Williams, A. *Handbook of Analytical Techniques*, 1st ed.; Wiley-VCH Verlag GmbH & Co.: Weinheim, Germany, 2001.
28. Hübschmann, H.J. *Handbook of GC-MS: Fundamentals and Applications*, 1st ed.; Wiley-VCH Verlag GmbH & Co.: Weinheim, Germany, 2015.
29. Lundanes, E.; Reubsaet, L.; Greibrokk, T. *Chromatography—Basic Principles, Sample Preparations and Related Methods*, 1st ed.; Wiley-VCH Verlag GmbH & Co.: Weinheim, Germany, 2014.
30. Magro, C.; Gonçalves, O.C.; Morais, M.; Ribeiro, P.A.; Sério, S.; Vieira, P.; Raposo, M. Volatile Organic Compound Monitoring during Extreme Wildfires: Assessing the Potential of Sensors Based on LbL and Sputtering Films. *Sensors* **2022**, *22*, 6677. [CrossRef]
31. Shinkai, T.; Masumoto, K.; Iwai, M.; Inomata, Y.; Kida, T. Study on Sensing Mechanism of Volatile Organic Compounds Using Pt-Loaded ZnO Nanocrystals. *Sensors* **2022**, *22*, 6277. [CrossRef]
32. Szkudlarek, A.; Kollbek, K.; Klejna, S.; Rydosz, A. Electronic sensitization of CuO thin films by Cr-doping for enhanced gas sensor response at low detection limit. *Mater. Res. Express* **2018**, *5*, 126406. [CrossRef]
33. Mirzaei, A.; Park, S.; Sun, G.-J.; Kheel, H.; Lee, C.; Lee, S. Fe2O3/Co3O4 composite nanoparticle ethanol sensor. *J. Korean Phys. Soc.* **2016**, *69*, 373–380. [CrossRef]
34. Yang, B.; Liu, J.; Qin, H.; Liu, Q.; Jing, X.; Zhang, H.; Li, R.; Huang, G.; Wang, J. PtO_2-nanoparticles functionalized CuO polyhedrons for n-butanol gas sensor application. *Ceram. Int.* **2018**, *44*, 10426–10432. [CrossRef]
35. Chen, H.I.; Hsiao, C.Y.; Chen, W.C.; Chang, C.H.; Liu, I.P.; Chou, T.C.; Liu, W.C. Formaldehyde Sensing Characteristics of a NiO-Based Sensor Decorated with Pd Nanoparticles and a Pd Thin Film. *IEEE Trans. Electron Devices* **2018**, *65*, 1956–1961. [CrossRef]
36. Shang, Y.; Shi, W.; Zhao, R.; Ahmed, M.; Li, J.; Du, J. Simple self-assembly of 3D laminated CuO/SnO_2 hybrid for the detection of triethylamine. *Chin. Chem. Lett.* **2020**, *31*, 2055–2058. [CrossRef]
37. Lee, J.; Kim, D.; Kim, T. Synthesis of Vapochromic Dyes Having Sensing Properties for Vapor Phase of Organic Solvents Used in Semiconductor Manufacturing Processes and Their Application to Textile-Based Sensors. *Sensors* **2022**, *22*, 4487. [CrossRef]
38. Chien, P.-J.; Suzuki, T.; Ye, M.; Toma, K.; Arakawa, T.; Iwasaki, Y.; Mitsubayashi, K. Ultra-Sensitive Isopropanol Biochemical Gas Sensor (Bio-Sniffer) for Monitoring of Human Volatiles. *Sensors* **2020**, *20*, 6827. [CrossRef]
39. Rodríguez-Torres, M.; Altuzar, V.; Mendoza-Barrera, C.; Beltrán-Pérez, G.; Castillo-Mixcóatl, J.; Muñoz-Aguirre, S. Discrimination Improvement of a Gas Sensors' Array Using High-Frequency Quartz Crystal Microbalance Coated with Polymeric Films. *Sensors* **2020**, *20*, 6972. [CrossRef] [PubMed]
40. Marikutsa, A.; Khmelevsky, N.; Rumyantseva, M. Synergistic Effect of Surface Acidity and PtO_x Catalyst on the Sensitivity of Nanosized Metal–Oxide Semiconductors to Benzene. *Sensors* **2022**, *22*, 6520. [CrossRef]
41. Genzardi, D.; Greco, G.; Núñez-Carmona, E.; Sberveglieri, V. Real Time Monitoring of Wine Vinegar Supply Chain through MOX Sensors. *Sensors* **2022**, *22*, 6247. [CrossRef]

42. Raposo, M.; Ribeiro, P.A.; Bari, N.E.; Bouchikhi, B. Sensing of Component Traces in Complex Systems. In *Electrokinetics Across Disciplines and Continents*, 1st ed.; Ribeiro, A.B., Mateus, E.P., Couto, N., Eds.; Springer: Cham, Switzerland, 2016; pp. 401–426.
43. Magro, C.; Mateus, E.P.; Raposo, M.; Ribeiro, A. Overview of electronic tongue sensing in environmental aqueous matrices: Potential for monitoring emerging organic contaminants. *Environ. Rev.* **2019**, *27*, 202–214. [CrossRef]
44. Magro, C.; Zagalo, P.; Pereira-Da-Silva, J.; Mateus, E.P.; Ribeiro, A.B.; Ribeiro, P.; Raposo, M. Polyelectrolyte Based Sensors as Key to Achieve Quantitative Electronic Tongues: Detection of Triclosan on Aqueous Environmental Matrices. *Nanomaterials* **2020**, *10*, 640. [CrossRef] [PubMed]
45. Zagalo, P.; Sério, S.; Ribeiro, P.A.; Raposo, M. Graphene-based biosensors. In *Recent Advances in Graphene and Graphene-Based Technologies*, 1st ed.; IOP Publishing Ltd. of Temple Circus: Bristol, UK, 2023; *in press*.
46. Monteiro, C.; Raposo, M.; Ribeiro, P.; Silva, S.; Frazão, O. Acoustic Optical Fiber Sensor Based on Graphene Oxide Membrane. *Sensors* **2021**, *21*, 2336. [CrossRef]
47. Assunção, I.C.C.; Sério, S.; Ferreira, Q.; Jones, N.C.; Hoffmann, S.V.; Ribeiro, P.A.; Raposo, M. Graphene Oxide Layer-by-Layer Films for Sensors and Devices. *Nanomaterials* **2021**, *11*, 1556. [CrossRef]
48. Decher, G.; Schmitt, J. Fine-Tuning of the film thickness of ultrathin multilayer films composed of consecutively alternating layers of anionic and cationic polyelectrolytes. In *Trends in Colloid and Interface Science VI*, 1st ed.; Helm, C., Lösche, M., Möhwald, H., Eds.; Springer: New York, NY, USA, 1992; pp. 160–164.
49. Zagalo, P.M.; Ribeiro, P.A.; Raposo, M. Effect of Applied Electrical Stimuli to Interdigitated Electrode Sensors While Detecting 17α-Ethinylestradiol in Water Samples. *Chemosensors* **2022**, *10*, 114. [CrossRef]
50. Zagalo, P.M.; Ribeiro, A.P.; Raposo, M. Detecting Traces of 17α-Ethinylestradiol in Complex Water Matrices. *Sensors* **2020**, *20*, 7324. [CrossRef]
51. Jackson, J.E. *A User's Guide to Principal Components*, 1st ed.; John Wiley & Sons: Hoboken, NJ, USA, 2003.
52. Marques, I.; Magalhâes-Mota, G.; Pires, F.; Sério, S.; Ribeiro, P.A.; Raposo, M. Detection of traces of triclosan in water. *Appl. Surf. Sci.* **2017**, *421*, 142–147. [CrossRef]
53. Turemis, M.; Zappi, D.; Giardi, M.T.; Basile, G.; Ramanaviciene, A.; Kapralovs, A.; Ramanavicius, A.; Viter, R. ZnO/polyaniline composite based photoluminescence sensor for the determination of acetic acid vapor. *Talanta* **2020**, *211*, 120658. [CrossRef]
54. Panigrahi, S.; Sankaran, S.; Mallik, S.; Gaddam, B.; Hanson, A.A. Olfactory receptor-based polypeptide sensor for acetic acid VOC detection. *Mater. Sci. Eng. C* **2012**, *32*, 1307–1313. [CrossRef]
55. Wang, Y.-C.; Sun, Z.-S.; Wang, S.-Z.; Wang, S.-Y.; Cai, S.-X.; Huang, X.-Y.; Li, K.; Chi, Z.-T.; Pan, S.-D.; Xie, W.-F. Sub-ppm acetic acid gas sensor based on In2O3 nanofibers. *J. Mater. Sci.* **2019**, *54*, 14055–14063. [CrossRef]
56. Cai, J.; Yan, Y.; Wang, W.; Ma, Y.; Cai, L.; Wu, L.; Zhou, H. Detection of formic acid and acetic acid gases by a QCM sensor coated with an acidified multi-walled carbon nanotube membrane. *Environ. Technol.* **2021**, 1–11. [CrossRef]
57. Ma, X.; Yuan, Y.; Peng, J.; Sun, M.; Chen, Z.; Yin, R.; Su, P.; Wang, X.; Wang, S. High-performance gas sensor based on GO/In2O3 nanocomposite for ethanol detection. *J. Mater. Sci. Mater. Electron.* **2022**, *33*, 15460–15472. [CrossRef]
58. Pienutsa, N.; Roongruangsree, P.; Seedokbuab, V.; Yannawibut, K.; Phatoomvijitwong, C.; Srinives, S. SnO_2-graphene composite gas sensor for a room temperature detection of ethanol. *Nanotechnology* **2021**, *32*, 115502. [CrossRef] [PubMed]
59. Liu, T.; Wang, T.; Li, H.; Su, J.; Hao, X.; Liu, F.; Liu, F.; Liang, X. Ethanol sensor using gadolinia-doped ceria solid electrolyte and double perovskite structure sensing material. *Sens. Actuators B Chem.* **2021**, *349*, 130771. [CrossRef]
60. Zhu, Z.; Kao, C.T.; Wu, R.J. A highly sensitive ethanol sensor based on Ag@TiO_2 nanoparticles at room temperature. *Appl. Surf. Sci.* **2014**, *320*, 348–355. [CrossRef]
61. Phasuksom, K.; Prissanaroon-Ouajai, W.; Sirivat, A. A highly responsive methanol sensor based on graphene oxide/polyindole composites. *RSC Adv.* **2020**, *10*, 15206–15220. [CrossRef]
62. Singh, S.K.; Azad, P.; Akhtar, M.J.; Kar, K.K. Improved Methanol Detection Using Carbon Nanotube-Coated Carbon Fibers Integrated with a Split-Ring Resonator-Based Microwave Sensor. *ACS Appl. Nano Mater.* **2018**, *1*, 4746–4755. [CrossRef]
63. Acharyya, D.; Huang, K.Y.; Chattopadhyay, P.P.; Ho, M.S.; Fecht, H.-J.; Bhattacharyya, P. Hybrid 3D structures of ZnO nanoflowers and PdO nanoparticles as a highly selective methanol sensor. *Analyst* **2016**, *141*, 2977–2989. [CrossRef]
64. Sahay, P.; Nath, R. Al-doped ZnO thin films as methanol sensors. *Sens. Actuators B Chem.* **2008**, *134*, 654–659. [CrossRef]
65. Xu, H.; Xu, J.; Wei, J.; Zhang, Y. Fast Response Isopropanol Sensing Properties with Sintered $BiFeO_3$ Nanocrystals. *Materials* **2020**, *13*, 3829. [CrossRef] [PubMed]
66. Jayababu, N.; Poloju, M.; Shruthi, J.; Reddy, M.R. NiO decorated CeO_2 nanostructures as room temperature isopropanol gas sensors. *RSC Adv.* **2019**, *9*, 13765–13775. [PubMed]
67. Wang, S.-C.; Wang, X.-H.; Qiao, G.-Q.; Chen, X.-Y.; Wang, X.-Z.; Wu, N.-N.; Tian, J.; Cui, H.-Z. NiO nanoparticles-decorated ZnO hierarchical structures for isopropanol gas sensing. *Rare Met.* **2022**, *41*, 960–971. [CrossRef]
68. Jiang, L.; Wang, C.; Wang, J.; Liu, F.; Lv, S.; You, R.; Yao, D.; Yan, X.; Wang, C.; Sun, P.; et al. Ultrathin BiVO4 nanosheets sensing electrode for isopropanol sensor based on pyrochlore-Gd2Zr2O7 solid state electrolyte. *Sens. Actuators B Chem.* **2020**, *321*, 128478. [CrossRef]

sensors

MDPI

Article

Humidity Effect on Low-Temperature NH_3 Sensing Behavior of In_2O_3/rGO Composites under UV Activation

Abulkosim Nasriddinov [1,*], Tatiana Shatalova [1], Sergey Maksimov [1], Xiaogan Li [2] and Marina Rumyantseva [1,*]

1 Chemistry Department, Moscow State University, Moscow 119991, Russia
2 Key Lab of Liaoning for Integrated Circuits Technology, School of Microelectronics, Dalian University of Technology, Dalian 116024, China
* Correspondence: a.f.nasriddinov@gmail.com (A.N.); roum@inorg.chem.msu.ru (M.R.)

Abstract: The nature of the constituent components of composite materials can significantly affect the character of their interaction with the gas phase. In this work, nanocrystalline In_2O_3 was synthesized by the chemical precipitation method and was modified using reduced graphene oxide (rGO). The obtained composites were characterized by several analysis techniques—XRD, TEM, SEM, FTIR and Raman spectroscopy, XPS, TGA, and DRIFTS. The XPS and FTIR and Raman spectroscopy results suggested the formation of interfacial contact between In_2O_3 and rGO. The results of the gas sensor's properties showed that additional UV illumination led to a decrease in resistance and an increase in sensor response at room temperature. However, the presence of humidity at room temperature led to the disappearance of the response for pure In_2O_3, while for the composites, an inversion of the sensor response toward ammonia was observed. The main reason may have been the formation of NH_4NO_3 intermediates with further hydrolysis and decomposition under light illumination with the formation of nitrite and nitrate species. The presence of these species was verified by in situ DRIFT spectroscopy. Their strong electron-accepting properties lead to an increase in resistance, which possibly affected the sensor signal's inversion.

Keywords: metal oxide gas sensor; nanocrystalline indium oxide; reduced graphene oxide; UV activation; humidity effect; low temperature detection; NH_3 sensor

1. Introduction

Resistive-type gas sensors based on wide-gap semiconductor oxides are widely used in practical applications. A variety of commercial types are available on the market for the detection of toxic, polluting, and explosive gases [1,2]. However, the main effort of researchers is aimed at eliminating and minimizing the existing disadvantages, such as the thermal degradation of the sensitive layer, high power consumption, negative effects of humidity, and low selectivity [3–8]. The commonly used metal oxide semiconductors (MOSs), such as SnO_2, In_2O_3, ZnO, WO_3, TiO_2, etc., show significant sensitivity toward different pollutant gases, but their high electrical resistance limits their gas-sensing performances at low operating temperatures. Many recent studies highlighted the use of perspective strategies to improve sensor performance: the use of modifiers depending on the nature of the analyte [9,10], applying filters [11], the use of photoactivation [12–15], and heterostructure creation [16–18].

In this regard, van der Waals two-dimensional (2D) materials, heterostructures, and devices, such as graphene and transition metal dichalcogenides, due to their unusual electronic and optical characteristics, can be quite effective [19–22]. The main advantages are their flexibility to change their electronic properties, in particular, the high electron mobility at room temperature for graphene, which allows the creation of p–n junctions, as well as the dependence of the band gap in such materials on the number of layers, which allows the control of the spectral characteristics of the resulting materials [23,24]. Previously, it was shown that the formation of a heterostructure between MOS and graphene can lead to efficient charge transfer, leading to an improvement in photocatalytic activity [25–29].

It should be noted that composite heterostructures of MOS with 2D materials, including graphene and its derivatives, have also shown promise in the field of gas sensors [30,31]. Hence, Quang et al. showed that tuning the Schottky barrier height and barrier width at the tiny area of contact between graphene and a SnO_2 nanowire through the adsorption/desorption of gas molecules led to outstanding gas-sensing characteristics [32]. Shekhirev et al. studied CVD-grown graphene nanoribbon films that could reliably recognize VOCs from different chemical classes [33]. Abideen et al. [34] developed rGO nanosheet-loaded ZnO nanofibers with significantly higher responses toward different oxidizing and reducing gases than pure ZnO. This enhancement was proposed to be due to the creation of local p–n heterojunctions. Similarly, Tammanoon et al. [35] achieved a sensitive and selective NO_2 sensor made from an electrolytically exfoliated graphene/flame-spray-made SnO_2 composite operated at low temperatures. In combination with In_2O_3, which is highly chemically stable and has a large number of free charge carriers in the conduction band, surface oxygen vacancies, and active chemisorbed oxygen, such composites are able to effectively detect gases at room temperature [36–39]. On one hand, ammonia is widely used in various fields, including the agricultural, medical, and chemical industries. On the other hand, the production of large volumes leads to an increase in its concentration in the environment, which negatively affects human health. Thus, according to the NIOSH (National Institute for Occupational Safety and Health) for NH_3, the TWA (time-weighted average concentration for up to a 10 h workday during a 40 h workweek) is 25 ppm and the ST (short-term exposure limit) is 35 ppm [40].

Nevertheless, it is also important to study the influence of light irradiation on the processes occurring at the solid–gas interface. Light radiation with the corresponding emission energy can lead to an increase in the concentration of charge carriers in a semiconductor matrix due to the generation of an electron–hole pair, to a change in the type and concentration of surface adsorption centers, and also to the formation of highly active radical particles that contribute to the oxidation of analyte molecules. All of the above can contribute to an increase in the sensitivity of sensors by accelerating ongoing processes. In particular, photogenerated holes can recombine with electrons localized on chemisorbed particles and lead to a decrease in the recovery time. In addition, for future practical applications, it is important to know the nature and character of the interaction of gas molecules with the material of the sensitive layer under different operating conditions, including high humidity values.

Herein, in this work, we obtained In_2O_3 and In_2O_3/rGO nanocomposites with varying rGO contents and conducted a systematic analysis of the chemical composition and interfacial interactions between components. The research explored the simultaneous influence of light irradiation and relative humidity, which can simulate the environmental conditions outside. The experimental results show that UV activation can enhance the sensing response toward NH_3 at room temperature. The effect of humidity appears as an inversion of the sensor signal. A possible reason may be the photochemical conversion of surface NH_4NO_3 to nitrite and nitrate species. This study provides useful information for further understanding the influence of complex conditions on the sensing behavior of composite materials at room temperature.

2. Materials and Methods

2.1. Material Synthesis

2.1.1. Synthesis of Nanocrystalline In_2O_3

The synthesis of the In_2O_3 nanocrystalline semiconductor oxide was performed via the chemical precipitation method. The synthesis procedure consisted of the precipitation of indium (III) hydroxide from an $In(NO_3)_3$ aqueous solution of its salt (5.00 g of $In(NO_3)_3 \times 4.5\ H_2O$ in 50 mL of deionized water). Precipitation was carried out by adding 13.3 M ammonia solution dropwise until pH ~ 6 was reached. The obtained dense gel was repeatedly washed several times and precipitated by centrifugation (8000× g rpm for 1 min) with 25 mM NH_4NO_3 solution and finally with deionized water to remove side products. The quality of purification from nitrate groups was checked using the Quantofix indicator (Macherey-Nagel, Düren, Germany). The resulting product was dried at T = 50 °C and then annealed at T = 300 °C for 24 h in air.

2.1.2. Synthesis of Composite Materials

Commercial rGO from "Lition Company" (Dubna, Russia) was used to obtain the composite materials. rGO was synthesized by a modified Hummer's method, then dispersed and stored in ethylene glycol to maintain the degree of reduction. A pre-weighed amount of the rGO suspension was added to the In_2O_3 powder and further diluted with 1 mL of citric acid (6.5 mM). The resulting mixture was stored for 20 min in an ultrasonic bath at a temperature of 50 °C to obtain a uniform distribution for the constituent components. The powder was dried at T = 50 °C and then annealed at T = 150 °C in air for 10 h. As a result, In_2O_3/rGO composites with a rGO content of 0.5 and 1.0 wt.% were obtained.

2.2. Materials Characterization

The phase composition of the as-prepared samples was characterized by X-ray powder diffraction (XRD) using a DRON-4 X-ray diffractometer (Burevestnik, Moscow, Russia) with Cu Ka radiation (λ = 1.5418 Å) and Raman spectroscopy using an i-Raman Plus spectrometer (BW Tek, Plainsboro, NJ, USA) equipped with a BAC 151C microscope and a laser with an excitation wavelength of 532 nm. The morphology and composition of the samples were analyzed by scanning electron microscopy (SEM) using a Carl Zeiss SUPRA 40 FE-SEM instrument (Carl Zeiss AG, Jena, Germany) with an Inlens SE detector. Transmission electron microscopy (TEM) studies were performed using the high-resolution electronic transmission microscope JEOL JEM-2100F at 200 kV.

Thermogravimetric analysis with mass spectral analysis of gaseous products (TG-MS) was carried out using a NETZSCH STA 449 instrument combined with a QMS-409 mass spectrometer (Netzsch-Gerätebau GmbH, Selb, Germany). The samples were heated in airflow (30 mL/min) at a rate of 10 °C/min in the temperature range of 25–700 °C.

Infrared spectra were recorded on a Spectrum One Fourier-transform infrared spectrometer (Perkin Elmer Inc., Waltham, MA, USA) in the transmission mode and the wavenumber range of 400–4000 cm^{-1} with a step of 1 cm^{-1}. Sample preparation consisted of pressing the samples (about 5 mg) with KBr (50 mg) to obtain tablets (6 mm diameter and 0.5 mm thickness). Diffuse reflectance infrared Fourier-transform spectroscopy (DRIFTS) was also performed on a Spectrum One Fourier-transform infrared spectrometer (Perkin Elmer Inc.) with the DiffusIR annex and flow chamber HC900 (Pike Technologies, Fitchburg, WI, USA) in the range of 4000–1000 cm^{-1} with a resolution of 4 cm^{-1} and accumulation of 30 scans. The samples (35 mg) were placed in ceramic crucibles (5.0 mm diameter, 3.5 mm depth) and placed into the heating chamber. The samples were preheated to 150 °C for 1 h to remove weakly adsorbed species and then cooled down to room temperature. The spectra were recorded at room temperature under a controlled flow rate of 100 mL/min in a gas mixture containing 100 ppm NH_3 in dry air or with a relative humidity (RH) of 70%.

X-ray photoelectron spectroscopy (XPS) studies were carried out on an OMICRON ESCA+ spectrometer (Scienta Omicron, Uppsala, Sweden) with an aluminum anode equipped with an AlKα XM1000 monochromatic X-ray source (radiation energy 1486.6 eV and power 252 W). To eliminate the local charge on the analyzed surface, a CN-10 charge neutralizer with an emission current of 6 μA and beam energy of 1 eV was used. The analyzer transmission energy was 20 eV. The spectrometer was calibrated toward the Au4f 7/2 line at 84.1 eV. The pressure in the analyzer chamber did not exceed 10^{-9} mbar. All spectra were accumulated at least three times. The fluctuation in the position of the peaks did not exceed ±0.1 eV.

Specially designed micro-hotplates were used for resistance measurements. The micro-hotplates consisted of a dielectric substrate (0.9 mm × 0.9 mm × 0.15 mm) made from Al_2O_3. There were platinum electrodes for heating on the back side of the substrates, and the electrodes on the top side were intended to measure the resistance. The electrodes and heater were made using Pt-based paste by the screen-printing method. The synthesized samples in the form of powders were pre-dispersed with ethanol and deposited as a thick film on the surface of the dielectric substrate to cover the electrodes. The films were sintered at T = 150 °C in air for 5 h.

Gas sensor measurements were performed using laboratory-made equipment with a flow cell. DC measurements were carried out in the temperature range of 150–25 °C with a step of 25 °C. During the tests, the concentration of the analyte gas was controlled by mass flow controllers by mixing a certified gas mixture with purified air. Gas was supplied through the cell with a flow rate of 100 mL/min. The measurements were carried out with a cyclic change in the atmosphere's composition: air with analyte for 15 min and pure air for 30 min. Preliminarily, for 100 min, the chamber with sensors was purged with purified air at a temperature of 150 °C. The sensor signal was calculated as:

$$S = \frac{R(air) - R(gas)}{R(gas)} \times 100\%,$$

where $R(air)$ is the sensor's resistance in air and $R(gas)$ is the sensor's resistance under a gas-containing atmosphere (NH_3).

The relative humidity in the DRIFTS and sensor measurements was controlled by regulating the ratio of dry and wet stream flows and was registered by a hygrometer IVTM-7 (Practic-NC, Zelenograd, Russia). A UV light-emitting diode (LED, λ_{max} = 365 nm, P = 3.5 mW/cm^2) was used for irradiation.

3. Results and Discussion

3.1. Characteristics of Composite Materials

The phase composition of the obtained samples was characterized using ICDD PDF-2 for In_2O_3 (6-416) (Figure 1). The X-ray diffraction patterns corresponded to the In_2O_3 phase with the cubic bixbyite structure, which indicated that the synthesized sample was single-phase. The X-ray patterns of the composite materials did not have significant differences, suggesting no influences on the structural characteristics after modification. The particle sizes of indium oxide in the pure sample and composites, estimated by the Scherrer formula from the most intense peaks, were approximately equal and corresponded to 7–8 nm.

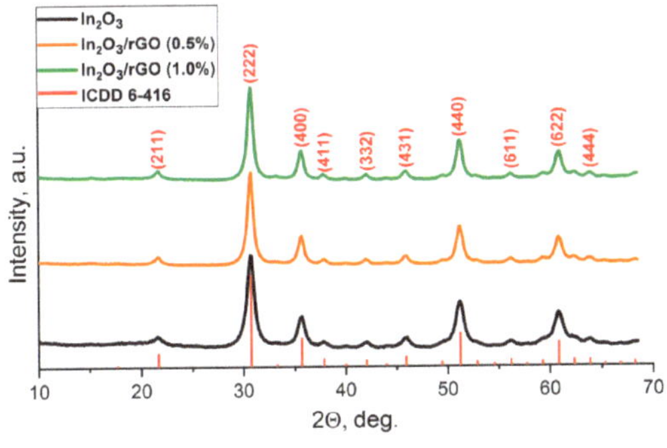

Figure 1. XRD patterns of the In_2O_3 and In_2O_3/rGO composites.

According to the TEM images (Figure 2a), indium oxide nanoparticles had a shape close to spherical with a size in the range of 7–10 nm, which was in good agreement with the values obtained from XRD data. In addition, crystalline particles with interplanar spacings d = 0.29 nm corresponding to the In_2O_3 (222) plane were found.

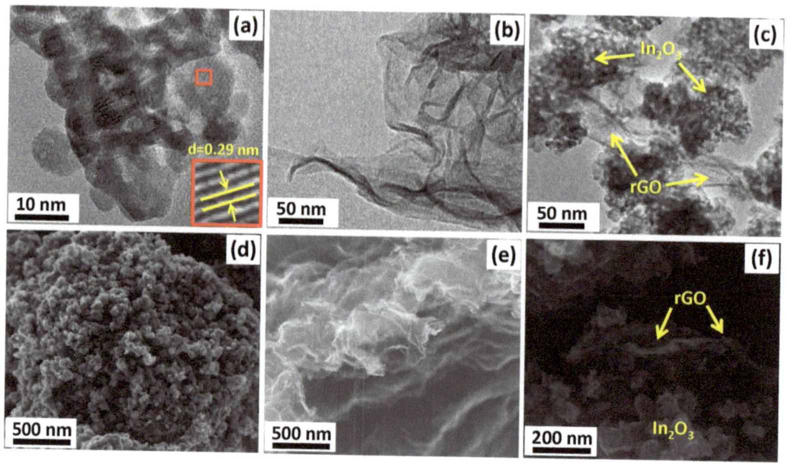

Figure 2. TEM images (**a–c**) and SEM images (**d–f**) of the In_2O_3 (**a**,**d**), rGO (**b**,**e**), and In_2O_3/rGO composite (**c**,**f**).

The rGO nanosheets (Figure 2b,e) had a typical 2D morphology and visually appeared to be crumpled due to tightly attached layers. According to the SEM, images the nanocrystalline In_2O_3 had a three-dimensional and porous structure with sintered grains (Figure 2d). The obtained composites consisted of agglomerated In_2O_3 nanoparticles attached by two-dimensional graphene flakes in the form of bridges (Figure 2c,f).

The structure of the semiconductor oxide and composite materials was also studied by Raman spectroscopy. Figure 3a shows the Raman spectra of nanocrystalline In_2O_3, rGO, and the composite materials in the frequency range of 90–2500 cm^{-1}. Characteristic Raman modes corresponding to the body-centered cubic lattice of In_2O_3 were observed at 122.5, 298.5, 357.2, 487.6, and 621.5 cm^{-1}. The vibration of the In-O bond of the [InO_6] structural units was observed at 122.5 cm^{-1}. The vibrational mode at 298.5 cm^{-1} was associated with a bending vibration of the [InO_6] octahedra, while the modes at 487.6 and 621.5 cm^{-1} were attributed to the stretching vibrations of the [InO_6] octahedra. The band at 357.2 cm^{-1} corresponded to the stretching vibrations of the In-O-In bonds [41–43]. The broad band at 447.3 cm^{-1} corresponded to surface structural defects due to the small particle size of nanocrystalline In_2O_3 [44].

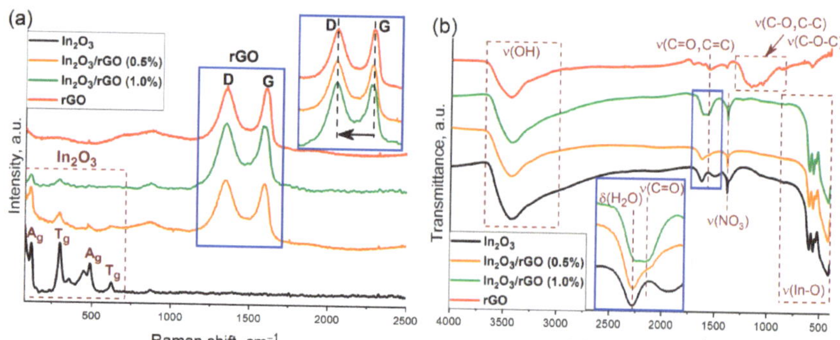

Figure 3. Raman spectra (**a**) and FTIR spectra (**b**) of the In_2O_3, rGO, and In_2O_3/rGO composites.

Reduced graphene oxide has two characteristic vibrational modes at 1346.3 and 1599.8 cm^{-1}, which are designated as D and G modes, respectively. The G-mode (E_{2g} symmetry) is due to the stretching vibrations of carbon atoms (C-C) in the plane. This peak can be observed in the Raman spectra for all carbon structures containing sp^2 hybridized bonds, while the D-mode (A_{1g} symmetry) becomes active in the presence of any defect in the ideal structure. Therefore, it can be seen as a band caused by disturbances and it can be described as a "breathing oscillation" of hexagonal aromatic carbon rings [45]. For the In_2O_3/rGO (0.5%) and In_2O_3/rGO (1.0%) composite materials, the intensity of the A_g and T_g vibrational modes corresponding to indium oxide decreased with increasing rGO content. This may have been due to the fact that rGO, which had a black color, absorbed green laser radiation to a greater extent; therefore, scattering in it will occur with greater probability. The presence of D and G bands indicated the preservation of the rGO structure after synthesis and thermal treatment. Moreover, a red shift in the position of the G-band occurred with increasing rGO content in composites from 1599.8 cm^{-1} (for rGO) to 1592.6 cm^{-1} (for In_2O_3/rGO (0.5%)) and to 1582.8 cm^{-1} (for In_2O_3/rGO (1%)). This gradual shift may indicate a charge transfer as a result of chemical bond formation with the surface of In_2O_3 [46].

The level of disorder in graphene oxide can be qualitatively estimated from the ratio of the intensities of the D and G bands (I_D/I_G). From pure rGO (I_D/I_G = 0.98) to composites (I_D/I_G = 1.02 and 1.04), an increase in this ratio was observed, which may have indicated an increase in the surface oxygen-containing functional groups that formed upon bonding with a semiconductor oxide [47,48]. The increase in the I_D/I_G ratio may have also indicated a decrease in the number of graphene layers [27].

Figure 3b shows the results of the study of samples by FTIR spectroscopy. The peaks of adsorbed water (1628 cm^{-1}) and hydroxyl groups (3000–3670 cm^{-1}) can be observed in the spectra. In the low-frequency region, In_2O_3 and composite materials exhibited characteristic peaks corresponding to the vibrations of the In-O bonds in the crystal lattice [49]. The presence of vibration bands corresponding to nitrate ions (1385 cm^{-1}) may have been due to residual impurities from the precursor that could not be washed during the synthesis procedure.

The spectrum of rGO is characterized by the presence of vibration bands with weak intensities, which most likely belong to different surface functional groups: hydroxyl (C–OH), epoxy (C–O–C), carbonyl (C=O), or carboxyl (COOH). Thus, a wide band in the region of 910–1320 cm^{-1} corresponds to the superposition of stretching vibrations of the (C-O), (C-C), and (C-O-C) bonds. The bands at 580, 1400, 1565, and 1720 cm^{-1} were associated with vibrations of the (C-C), (-COOH), (C=C), and (C=O) bonds, respectively [49–51].

At the same time, the FTIR spectra of the composites were characterized by the appearance of a new peak at 1576 cm^{-1} (C=O or C=C). The intensity of this peak increased with an increase in the rGO content. The obtained results indicated the efficient immobilization of rGO with the semiconductor oxide. Such contact can provide an electronic interaction between In_2O_3 and rGO, which can lead to better charge separation.

XP spectra of the samples are shown in Figure 4. For the rGO sample, both the C 1s and O 1s spectra differed from those of pure In_2O_3 and the In_2O_3/rGO (1%) composite. The deconvolution of the C 1s spectrum showed that it consisted of sp^2 bonding C=C, sp^3 bonding C-C, and C=O (or O-C=O) bonds at 284.2, 285.7, and 288.2 eV, respectively [52]. The O 1s spectrum consisted of the following three components: a C=O (carbonyl and carboxyl) bond at 531.1 eV, a C-O (epoxy) bond at 531.9 eV, and an O-H (hydroxyl) bond at 533 eV [53]. The C1s spectrum of the pure In_2O_3 sample consisted of several components of adventitious carbon. Thus, an intense peak at 285.2 eV corresponded to sp^3 bonding C-C, the broad peak at 286.4 eV was due to C-OH and C-O-C bonds, while the peak at 289 eV corresponded to the O-C=O bond or carbonate contamination CO_3^{2-} [54]. For the In_2O_3/rGO (1%) composite sample, a shift and an increase in the quantitative ratio of these components could be observed, which may have indicated an interaction between In_2O_3 and rGO. An additional shift of 0.3 eV for the In 3d spectrum toward higher binding energies also indicated charge redistribution and transfer at the interface.

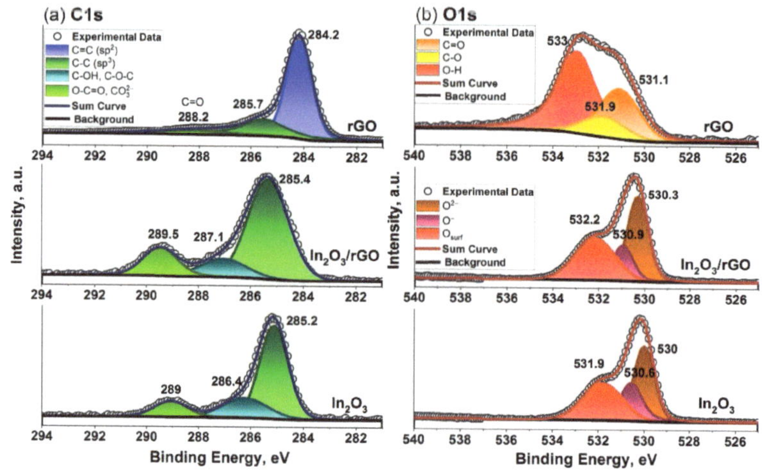

Figure 4. C 1s (**a**) and O 1s (**b**) X-ray photoelectron spectra of the In_2O_3 and In_2O_3/rGO (1%) composite.

The O1s spectra of pure In_2O_3 could be fitted using three peaks with maxima at binding energies of 530, 530.6, and 531.9 eV. The lowest binding energy region was associated with lattice oxygen (O^{2-}), while the middle region could be assigned to oxygen ions with lower electron density (O^-) in the subsurface. The coordination number of the oxygen ions in these sites was lower and it could indicate the defective structure of the metal oxide [55]. The appearance of the peaks in the higher-binding-energy region was due to chemically adsorbed oxygen-containing species on the sample's surface [56,57]. These peaks shifted by 0.3 eV to the higher binding energy for the In_2O_3/rGO (1%) composite. Moreover, the ratio of their content (O^{2-}:O^-:O_{surf}) changed between pure In_2O_3 (40:24:36) and the composite (41:20:39), which indicated the compensation of oxygen deficiencies leading to an increase in surface species, including chemisorbed oxygen ions.

Figure 5 shows the TG curve and temperature dependencies of the ionic currents corresponding to gaseous products released from the reduced graphene oxide. The analysis showed that, in the temperature range of 50–200 °C, there was a slight increase in the mass of the sample, which was accompanied by a decrease in the ion currents for $m/z = 18$ (H_2O). Most likely, rGO was partially oxidized at this stage. At T = 450 °C, the destruction of the rGO skeleton began: C–C bonds were broken, resulting in the formation of CO_2 ($m/z = 44$). It should be noted that, in this temperature range, the mass loss was almost 100%.

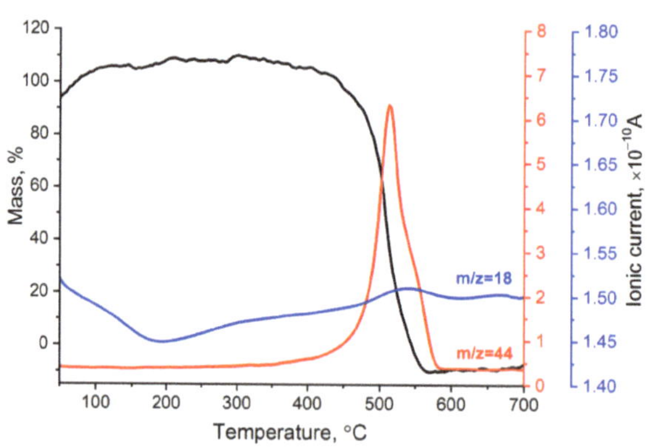

Figure 5. TG curve and temperature dependencies of ionic currents corresponding to H_2O ($m/z = 18$) and CO_2 ($m/z = 44$) during rGO fragmentation and oxidation.

3.2. Gas Sensor Measurements

Figure 6 shows the change in the resistance of the obtained sensors depending on the gas phase composition under different conditions. It is noteworthy that the sensors did not exhibit baseline drift across the entire measurement range, even under a humid atmosphere. It can be seen that, under an atmosphere of dry air (Figure 6a,b), all samples behaved like n-type semiconductors: when 20 ppm of ammonia (reducing gas) was added, the resistance decreased due to reaction (1), and under an atmosphere of purified air, the resistance increased again as reaction (2) could proceed.

$$2NH_{3(gas)} + \frac{3}{\beta}O^{\alpha-}_{\beta(ads)} \rightarrow N_{2(gas)} + 3H_2O_{(gas)} + \frac{3\alpha}{\beta}e^-, \quad (1)$$

$$\frac{\beta}{2}O_{2(gas)} + \alpha e^- \rightarrow O^{\alpha-}_{\beta(ads)} \quad (2)$$

where $NH_{3(gas)}$ represents the ammonia molecules in the gas phase and $O^{-\alpha}_{\beta(ads)}$ represents chemisorbed oxygen species (α = 1 and 2 for once- and twice-charged particles, respectively; β = 1 and 2 for atomic and molecular forms, respectively).

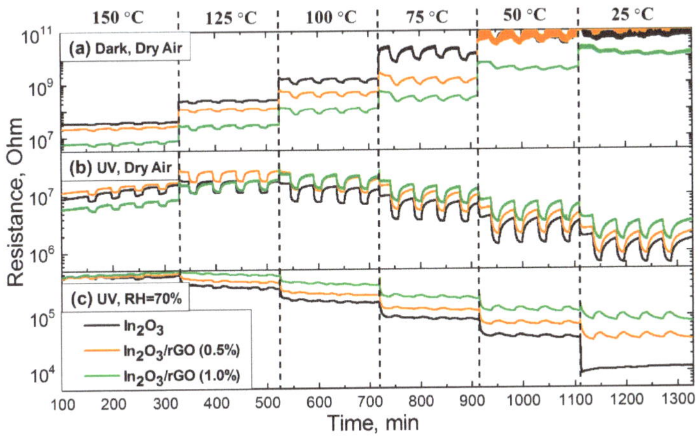

Figure 6. Dynamic changes in sensors' resistance with periodic changes in the gas phase composition (20 ppm NH_3–purified air) in the temperature range of 150–25 °C under various experimental conditions: (**a**) under dark conditions with dry air; (**b**) under UV illumination with dry air; and (**c**) under UV illumination with a relative humidity of 70% (RH = 70%).

A high electrical conductivity of the rGO compared with MOS led to a decrease in the baseline resistance for composites under dark conditions (Figure 6a) [22]. As the resistance of the semiconductor materials increased with decreasing temperature, at low operating temperatures (25 and 50 °C), the resistance value reached above the detection limit of the device (R > 10^{10} Ohm), so the data were noisy and illegible, which led to measurement and calculation difficulties.

As illustrated in Figure 7a, the addition of p-type rGO to the n-type In_2O_3 semiconductor matrix formed a p-to-n transition at the interface of the obtained composite. Electrons would be transferred from In_2O_3 with a lower work function (4.3 eV) to rGO with a higher work function (4.7 eV), resulting in depletion layer formation until the equilibrium of the Fermi level was reached [58–61]. As a result, a potential barrier would be created at the heterojunction. The formation of p–n heterojunction led to an increase in the resistance value of the composites, which could be observed based on the dependences of the resistance over time (Figure 6b,c). In addition, it could significantly reduce the rate of recombination of electron–hole pairs and promote their separation and migration to the semiconductor surface under the action of internal electric fields. Figure 7b represents the response and recovery time for the In_2O_3/rGO (1%) composite to 20 ppm NH_3 under UV illumination. The results show that increasing the operating temperature led to a decrease in both the response and recovery times. At the same time, comparing with Figure 8, one can notice that the slower the reaction preceded, the higher the sensor signal. This was achieved in the low temperature range, while at higher temperatures, partial desorption of oxygen from the surface may have occurred.

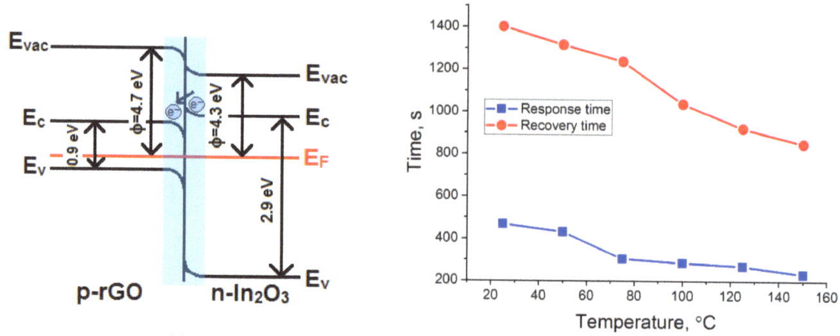

(a) (b)

Figure 7. A schematic illustration of the p–n heterojunction formation in the rGO–In_2O_3 interface (**a**); response and recovery times of the In_2O_3/rGO (1%) composite under UV illumination at different operating temperatures (**b**).

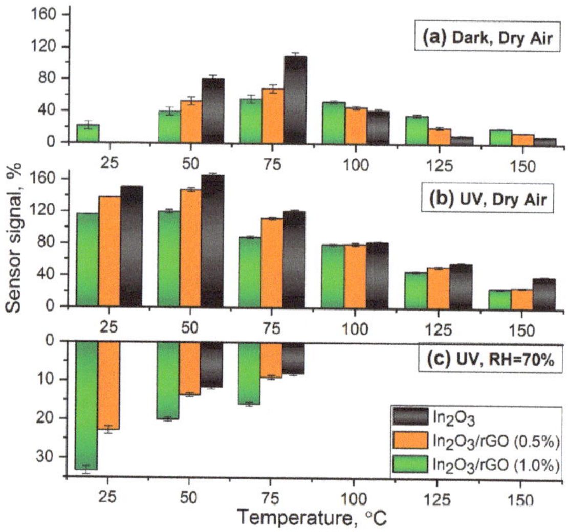

Figure 8. Temperature dependence of the sensor signal of the materials upon the detection of 20 ppm NH_3 under various experimental conditions: (**a**) under dark conditions with dry air; (**b**) under UV illumination with dry air; and (**c**) under UV illumination with a relative humidity of 70% (RH = 70%).

In the case of UV illumination (Figure 6b), a decrease in resistance of all sensors was observed, thereby achieving a reproducible response to NH_3, even at room temperature. When In_2O_3 semiconductor particles were exposed to UV irradiation, photogenerated electrons and holes were formed (3). An increase in the concentration of photoelectrons in the conduction band led to a decrease in the base resistance of the In_2O_3 sample, particularly compared with composites. In the latter case, rGO flakes could cover the surface of the semiconductor, thereby preventing the interaction of UV radiation with the semiconductor's solid surface. An increase in the concentration of charge carriers in the conduction band could enhance the adsorption of oxygen from the atmosphere (reaction 2), which, in turn, could stimulate reaction (1). Acting as an electron acceptor, rGO could prevent the rapid recombination of electrons and holes [62].

$$In_2O_3 \overset{h\nu}{\rightarrow} e^- + h^+ \quad (3)$$

Nevertheless, in the first two cases, the unmodified In_2O_3 exhibited a higher sensor signal compared with the composite materials (Figure 8a,b). However, it is worth noting that UV illumination significantly improved the response to NH_3, especially in the low-temperature region.

An interesting case is the third one (Figures 6c and 8c), when the measurements were carried out under an atmosphere with RH = 70%. First, it can be noticed that the baseline resistance of the sensors decreased by more than one order of magnitude compared with the measurements under a dry atmosphere. It is mentioned in the literature that this phenomenon may have been associated with an increase in the electron concentration in the conduction band due to the following reactions [63]:

$$H_2O_{(gas)} + In_{(lat)} + O_{(lat)} = [In_{(lat)} - OH] + [O_{(lat)}H]^{\cdot\cdot} + e^- \quad (4)$$

$$H_2O_{(gas)} + 2In_{(lat)} + O_{(lat)} = 2[In_{(lat)} - OH] + V_O^{\cdot\cdot} + 2e^- \quad (5)$$

$$2H_2O_{(gas)} + 4In_{(lat)} + O_2{}^-{}_{(ads)} = 4[In_{(lat)} - OH] + e^- \quad (6)$$

In the temperature range of 100–150 °C, the resistance of the sensors fluctuated at the noise level and there were no significant changes, indicating the negative effect of water vapor on the detection of ammonia and the lack of charge transfer occurring. In the low temperature range (25–75 °C), the effect of inversion of the sensor signal was observed, and the lower the measurement temperature, the better this effect revealed itself. This situation implies that, under an atmosphere containing ammonia, the resistance increased, and under an atmosphere of purified air, it decreased.

The minimum detectable NH_3 concentration was calculated by plotting calibration curves, which had a good linear relationship with the ammonia concentration (Figure 9). The minimum measurable sensor response was estimated using the ratio of $R(av)/(R(av) - 3\sigma)$, where $R(av)$ is the average resistance in pure air and σ is the standard deviation of resistance in pure air. The noise level of the sensors was calculated as the changes in the relative response of the sensor over the baseline or the root-mean-square deviation (RMS_{noise}). Sensitivity was determined as $\Delta R/\Delta c$. The obtained results are presented in Table 1. It can be observed that the In_2O_3/rGO (1%) composite demonstrated the lowest value of the minimum detectable NH_3 concentration, RMS_{noise}, and sensitivity at T = 50 °C. However, it should be noted that the obtained values were quite close.

Table 1. Minimum detectable NH_3 concentration c_{min}, RMS_{noise}, and sensitivity of the sensors measured at 25 °C and 50 °C under dry air conditions and UV illumination.

Sample	T = 25 °C			T = 50 °C		
	C_{min}, ppm	**RMS_{noise}, $\times 10^{-3}$**	**Sensitivity, ppb^{-1}**	**C_{min}, ppm**	**RMS_{noise}, $\times 10^{-3}$**	**Sensitivity, ppb^{-1}**
In_2O_3	1.71	18.02	31.5	1.1	14.12	38.5
In_2O_3/rGO (0.5%)	1.78	20.65	34.7	1.04	13.5	38.8
In_2O_3/rGO (1%)	1.88	21.12	33.6	1.0	12.53	37.4

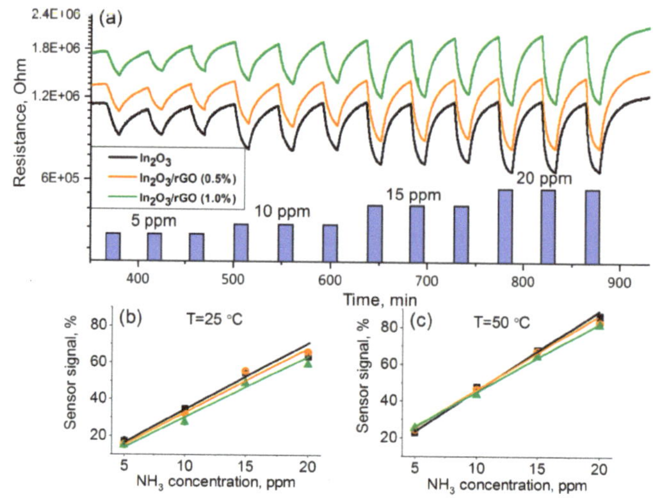

Figure 9. Change in the samples' resistances depending on the NH_3 concentration (5–10–15–20 ppm) (**a**); calibration curves at T = 25 °C (**b**) and T = 50 °C (**c**).

3.3. In Situ DRIFTS Analysis

In order to analyze in more detail the effect of humidity on the nature of the change in resistance, a study entailing DRIFT spectroscopy was carried out. The spectra were recorded at room temperature in dry air and humid air (RH = 70%). The samples were preliminarily kept under a flow of purified air at T = 150 °C for 40 min. The results are shown in Figure 10.

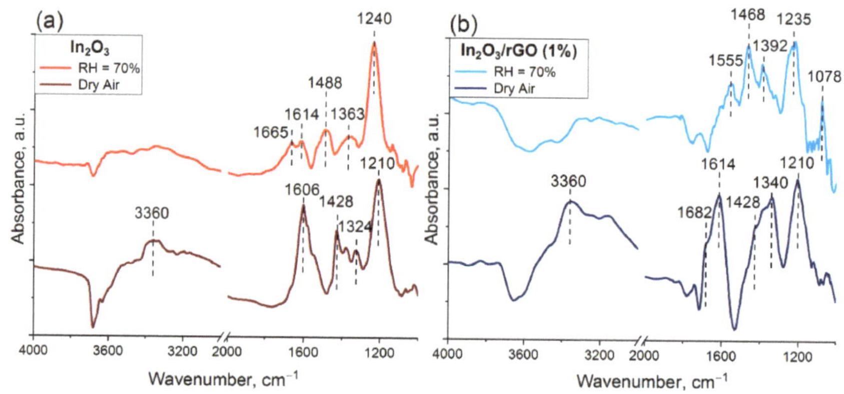

Figure 10. In situ DRIFT spectra of the In_2O_3 (**a**) and In_2O_3/rGO (1%) (**b**) samples after 100 ppm NH_3 adsorption for 50 min at room temperature in dry air and humid air (RH = 70%).

After 5 min of NH_3 exposure, narrow peaks in the range of 1210–1240 cm^{-1} were immediately detected. These bands were assigned to NH_3^+ species adsorbed on Lewis acid sites. An additional confirmation may be offered by the appearance of N-H stretching vibrations at the wavenumber of 3360 cm^{-1} related to NH_3^+ species under an atmosphere of dry air. The appearance of bands in the range of 1428–1488 cm^{-1} was associated with NH_4^+ species as a result of ammonia adsorption on Brønsted acid sites, including terminal In-OH groups. Such adsorption was accompanied by a decrease in the intensity of the bands of OH groups [49,64].

Intense peaks corresponding to molecularly adsorbed NH_3 on Lewis acid sites in the range of 1606–1682 cm^{-1} appeared only under the dry atmosphere, while under humid conditions, they disappeared or decreased in intensity [65]. This may indicate, by implication, the predominant coverage of the surface and, accordingly, the occupation of active sites by water molecules. However, the IR band corresponding to NH_3^+ did not change in intensity both under dry and humid air, which could be due to stronger interaction with Lewis acid sites compared with hydroxyl groups.

It can be noticed from Figure 3b that residual nitrate groups remained in the composition of the samples after synthesis due to the precursor. In the presence of ammonia, they could react with the formation of a surface intermediate, likely NH_4^+–NO_3^- species. This was confirmed by the appearance of absorption bands in the region of 1324–1392 cm^{-1}. Previously, the same surface species with similar vibration frequencies were also detected on different catalysts: during the reaction between NO_2 and a NH_3-pre-adsorbed Cu-exchanged SAPO-34 catalyst [65], a V_2O_5–WO_3/TiO_2 catalyst after exposure to NH_3 and NO_2 [66], and an Fe-zeolite-based catalyst after exposure to NO_2 over a NH_3-pre-adsorbed sample [67]. The main characteristic infrared vibrational frequencies, which were found according to the results of DRIFTS analysis, are shown in Table 2 [49,64,68,69].

Table 2. Assignments of IR absorption bands (cm^{-1}) that appeared in the DRIFT spectra on the surface of In_2O_3 and In_2O_3/rGO (1%) samples under different conditions.

Functional Groups	In_2O_3		In_2O_3/rGO (1%)	
	Dry Air	RH = 70%	Dry Air	RH = 70%
NO_2^-, monodentate nitrite	-	-	-	1078
NH_3^+ on Lewis acid site	1210	1240	1210	1235
ν(NO_3) in NH_4NO_3 species	1324, 1378	1363	1340, 1380	1392
NH_4^+ on Brønsted acid site	1428	1488	1428	1468
NO_3^-, chelating bidentate nitrate	-	-	-	1555
NH_3, molecularly adsorbed on Lewis acid sites	1606	1614, 1665	1614, 1682	-
ν(N-H) in NH_3	3360	-	3360	-
ν(OH)	3530–3715	3600–3715	3450–3740	3320–3820

From Figure 10b, it is clear that, after NH_3 was introduced to the chamber with 70% background humidity, new bands at 1078 and 1555 cm^{-1} appeared. These bands could be attributed to monodentate nitrite and chelating bidentate nitrate species. However, the exact identification of these bands is an ambiguous task due to the overlapping of the absorption regions of various structural fragments, including nitrates and nitrites (both monodentate and bidentate species) [68,69].

It can be observed from Figure 6c that the effect of sensor signal inversion was observed, even for pure In_2O_3 at T = 50 and 75 °C under a humid atmosphere, while at T = 25 °C, it was observed exclusively for composites. Hence, it can be assumed that the main factor affecting the signal inversion was the indium oxide matrix. From the DRIFTS results, it was found that the interaction with ammonia resulted in the formation of NH_4NO_3 species. It could be assumed that further interaction of this fragment with water molecules in a sufficiently high humid atmosphere (RH = 70%) led to the formation of nitric acid. The reaction pathway might have involved NH_4NO_3 hydrolysis on the surface of particles. The formed intermediate nitric acid may have been decomposed under light illumination to produce NO_2 [70–74]. In the case of the In_2O_3/rGO composites, rGO flakes could serve as an additional path for charge transfer due to heterocontact and side reactions on the surface.

Recently, Ma et al. comprehensively investigated the photolysis of various nitrates on different mineral oxides [75]. It was found that NH_4NO_3 had the highest rate of NO_2 production, even at room temperature, compared with the other studied nitrates. Moreover, this rate was higher under a humid atmosphere compared with dry air. The authors used UV irradiation with 365 nm wavelength (as in this work) and assumed that photoinduced electrons and holes (reaction (3)) could promote the photolysis of NH_4NO_3 on the metal oxide's surface (reactions (7) and (8)). The photochemical reaction between the water and photogenerated hole could enhance the surface acidity and facilitate NO_2 production.

$$NO_3^- + h^+ \rightarrow NO_3 \quad (7)$$

$$2NO_3 \xrightarrow{h\nu} 2NO_2 + O_2 \quad (8)$$

Nitrogen dioxide, being a strong electron acceptor, can attract electrons from the conduction band, thereby leading to an increase in the resistance of the sensors and consequently inverting the signal (reaction (9)). The electron affinity of NO_2 (2.27 eV, [76]) is greater than that of O_2 (0.44 eV, [77]). Therefore, in the subsequent competing process between NO_2 and O_2 (according to the products of reaction (5)), nitrogen dioxide will predominate. It is also worth noting that the signal inversion is reproducible (Figure 6c), which may indicate the regeneration of NH_4NO_3. The formation of nitrate species can proceed via reaction (10) and, according to the DRIFTS results, further interaction with ammonia can lead to NH_4NO_3 formation again.

$$NO_2 + e^- \rightarrow NO_2^- \quad (9)$$

$$2NO_2 + O_2 + e^- \rightarrow 2NO_3^- \quad (10)$$

4. Conclusions

Composite materials based on nanocrystalline In_2O_3 and rGO were synthesized and investigated. The obtained Raman and FTIR spectroscopy results indicate the efficient immobilization of rGO with a semiconductor oxide matrix. The influence of UV activation and humidity on the gas-sensing behavior of In_2O_3/rGO composites was also studied.

When ammonia interacted with composite materials, the main adsorption sites were provided by the porous surface of In_2O_3. Additional surface modification with rGO, which consisted of flakes with large lateral sizes, could limit the access to and interaction with the analyte gas molecules, resulting in a reduced sensor signal in dry air. However, UV illumination led to the generation of electron–hole pairs in the In_2O_3 structure, as this energy was comparable to its band gap. An increase in the electron concentration in the conduction band promoted greater oxygen adsorption and, accordingly, more efficient interaction with ammonia. At the same time, under an atmosphere with high relative humidity, the predominant active centers, including chemisorbed oxygen, were occupied or replaced by adsorbed water. On one hand, this led to an increase in conductivity, and on the other hand, the decrease in the concentration of chemisorbed oxygen limited the oxidation of ammonia and further charge transfer. As a result, the sensor signal for pure In_2O_3 was noticeably reduced.

Based on the in situ DRIFTS analysis, it is proposed that residual nitrate groups can react with ammonia, resulting in the formation of surface intermediates, likely NH_4NO_3 species. The combined influence of humidity and UV illumination could lead to the hydrolysis of NH_4NO_3 on the In_2O_3 surface, followed by photolysis, or immediately undergo a photochemical reaction. As a result, nitrite and nitrate species were formed. Due to their electron-accepting nature, they led to a decrease in conductivity, resulting in an inversion of the sensor signal when detecting ammonia at low temperatures. The appearance of these groups was proven by DRIFT spectroscopy. The effect of signal inversion was most clearly expressed for a composite with a rGO content of 1% at room temperature in RH = 70%. In this case, rGO flakes could serve as an additional path for charge transfer due to heterocontact and side reactions on the surface.

Author Contributions: Conceptualization, X.L. and M.R.; methodology, M.R., X.L. and A.N.; formal analysis, A.N.; investigation, A.N., T.S. and S.M.; data curation, A.N. and T.S.; writing—original draft preparation, A.N. and M.R.; writing—review and editing, A.N., X.L. and M.R. All authors have read and agreed to the published version of the manuscript.

Funding: This research was funded by the Russian Foundation for Basic Research, grant number 21-53-53018, and the National Natural Science Foundation of China, grant number 62111530055.

Institutional Review Board Statement: Not applicable.

Informed Consent Statement: Not applicable.

Data Availability Statement: The data that support the findings of this study are available from the corresponding author upon reasonable request.

Acknowledgments: The spectral research and thermal analysis were carried out using the equipment purchased using funds from the Lomonosov Moscow State University Program of Development. The TEM experiments were conducted using equipment from the "Nanochemistry and Nanomaterials" center supported by the Program of Development of Lomonosov Moscow State University.

Conflicts of Interest: The authors declare no conflict of interest.

References

1. Gas Sensor Market Size, Share & Trends Analysis Report by Product, by Type, by Technology, by End Use, by Region, and Segment Forecasts, 2022–2030. Available online: https://www.grandviewresearch.com/industry-analysis/gas-sensors-market (accessed on 5 January 2023).
2. Gas Sensor Market by Gas Type. Available online: https://www.marketsandmarkets.com/Market-Reports/gas-sensor-market-245141093.html (accessed on 16 January 2023).
3. Park, S.Y.; Kim, Y.; Kim, T.; Eom, T.H.; Kim, S.Y.; Jang, H.W. Chemoresistive materials for electronic nose: Progress, perspectives, and challenges. *InfoMat* **2019**, *1*, 289–316. [CrossRef]
4. Saruhan, B.; Lontio Fomekong, R.; Nahirniak, S. Review: Influences of Semiconductor Metal Oxide Properties on Gas Sensing Characteristics. *Front. Sens.* **2021**, *2*, 657931. [CrossRef]
5. Kim, S.; Brady, J.; Al-Badani, F.; Yu, S.; Hart, J.; Jung, S.; Tran, T.T.; Myung, N.V. Nanoengineering Approaches Toward Artificial Nose. *Front. Chem.* **2021**, *9*, 629329. [CrossRef] [PubMed]
6. Jeong, S.Y.; Kim, J.S.; Lee, J.H. Rational Design of Semiconductor-Based Chemiresistors and their Libraries for Next-Generation Artificial Olfaction. *Adv. Mater.* **2020**, *32*, 2002075. [CrossRef] [PubMed]
7. Reddy, B.K.S.; Borse, P.H. Review—Recent Material Advances and Their Mechanistic Approaches for Room Temperature Chemiresistive Gas Sensors. *J. Electrochem. Soc.* **2021**, *168*, 057521. [CrossRef]
8. Majhi, S.M.; Mirzaei, A.; Kim, H.W.; Kim, S.S.; Kim, T.W. Recent advances in energy-saving chemiresistive gas sensors: A review. *Nano Energy* **2021**, *79*, 105369. [CrossRef] [PubMed]
9. Krivetskiy, V.V.; Rumyantseva, M.N.; Gaskov, A.M. Chemical modification of nanocrystalline tin dioxide for selective gas sensors. *Russ. Chem. Rev.* **2013**, *82*, 917–941. [CrossRef]
10. Rumyantseva, M.N.; Gas'Kov, A.M. Chemical modification of nanocrystalline metal oxides: Effect of the real structure and surface chemistry on the sensor properties. *Russ. Chem. Bull.* **2008**, *57*, 1106–1125. [CrossRef]
11. Van Den Broek, J.; Weber, I.C.; Güntner, A.T.; Pratsinis, S.E. Highly selective gas sensing enabled by filters. *Mater. Horiz.* **2021**, *8*, 661–684. [CrossRef]
12. Chizhov, A.; Rumyantseva, M.; Gaskov, A. Light Activation of Nanocrystalline Metal Oxides. *Nanomaterials* **2021**, *11*, 892. [CrossRef]
13. Nasriddinov, A.; Rumyantseva, M.; Shatalova, T.; Tokarev, S.; Yaltseva, P.; Fedorova, O.; Khmelevsky, N.; Gaskov, A. Organic-inorganic hybrid materials for room temperature light-activated sub-ppm no detection. *Nanomaterials* **2020**, *10*, 70. [CrossRef] [PubMed]
14. Suh, J.M.; Eom, T.H.; Cho, S.H.; Kim, T.; Jang, H.W. Light-activated gas sensing: A perspective of integration with micro-LEDs and plasmonic nanoparticles. *Mater. Adv.* **2021**, *2*, 827–844. [CrossRef]
15. Kumar, R.; Liu, X.; Zhang, J.; Kumar, M. *Room-Temperature Gas Sensors under Photoactivation: From Metal Oxides to 2D Materials*; Springer: Singapore, 2020; Volume 12, ISBN 0123456789.
16. Sowmya, B.; John, A.; Panda, P.K. A review on metal-oxide based p-n and n-n heterostructured nano-materials for gas sensing applications. *Sens. Int.* **2021**, *2*, 100085. [CrossRef]
17. Walker, J.M.; Akbar, S.A.; Morris, P.A. Synergistic effects in gas sensing semiconducting oxide nano-heterostructures: A review. *Sens. Actuators B Chem.* **2019**, *286*, 624–640. [CrossRef]
18. Rumyantseva, M.N.; Vladimirova, S.A.; Vorobyeva, N.A.; Giebelhaus, I.; Mathur, S.; Chizhov, A.S.; Khmelevsky, N.O.; Aksenenko, A.Y.; Kozlovsky, V.F.; Karakulina, O.M.; et al. p-CoOx/n-SnO_2 nanostructures: New highly selective materials for H_2S detection. *Sens. Actuators B Chem.* **2018**, *255*, 564–571. [CrossRef]
19. Novoselov, K.S.; Mishchenko, A.; Carvalho, A.; Castro Neto, A.H. 2D materials and van der Waals heterostructures. *Science* **2016**, *353*, 1–25. [CrossRef]

20. Avsar, A.; Tan, J.Y.; Taychatanapat, T.; Balakrishnan, J.; Koon, G.K.W.; Yeo, Y.; Lahiri, J.; Carvalho, A.; Rodin, A.S.; O'Farrell, E.C.T.; et al. Spin-orbit proximity effect in graphene. *Nat. Commun.* **2014**, *5*, 1–6. [CrossRef] [PubMed]
21. Benítez, L.A.; Savero Torres, W.; Sierra, J.F.; Timmermans, M.; Garcia, J.H.; Roche, S.; Costache, M.V.; Valenzuela, S.O. Tunable room-temperature spin galvanic and spin Hall effects in van der Waals heterostructures. *Nat. Mater.* **2020**, *19*, 170–175. [CrossRef]
22. Liu, Y.; Weiss, N.O.; Duan, X.; Cheng, H.C.; Huang, Y.; Duan, X. Van der Waals heterostructures and devices. *Nat. Rev. Mater.* **2016**, *1*, 16042. [CrossRef]
23. Novoselov, K.S.; Geim, A.K.; Morozov, S.V.; Jiang, D.; Zhang, Y.; Dubonos, S.V.; Grigorieva, I.V.; Firsov, A.A. Electric Field Effect in Atomically Thin Carbon Films. *Mater. Sci.* **2004**, *306*, 666–669. [CrossRef]
24. Mak, K.F.; Lee, C.; Hone, J.; Shan, J.; Heinz, T.F. Atomically thin MoS_2: A new direct-gap semiconductor. *Phys. Rev. Lett.* **2010**, *105*, 136805. [CrossRef] [PubMed]
25. Huang, H.; Yue, Z.; Li, G.; Wang, X.; Huang, J.; Du, Y.; Yang, P. Heterostructured composites consisting of In_2O_3 nanorods and reduced graphene oxide with enhanced interfacial electron transfer and photocatalytic performance. *J. Mater. Chem. A* **2014**, *2*, 20118–20125. [CrossRef]
26. An, X.; Yu, J.C.; Wang, Y.; Hu, Y.; Yu, X.; Zhang, G. WO 3 nanorods/graphene nanocomposites for high-efficiency visible-light-driven photocatalysis and NO 2 gas sensing. *J. Mater. Chem.* **2012**, *22*, 8525–8531. [CrossRef]
27. Perera, S.D.; Mariano, R.G.; Vu, K.; Nour, N.; Seitz, O.; Chabal, Y.; Balkus, K.J. Hydrothermal Synthesis of Graphene-TiO_2 Nanotube Composites with Enhanced Photocatalytic Activity. *ACS Catal.* **2012**, *2*, 949–956. [CrossRef]
28. Zhang, H.; Lv, X.; Li, Y.; Wang, Y.; Li, J. P25-Graphene Composite as a High Performance Photocatalyst. *ACS Nano* **2010**, *4*, 380–386. [CrossRef]
29. Devi, P.; Singh, J.P. Visible light induced selective photocatalytic reduction of CO_2 to CH_4on In2O_3-rGO nanocomposites. *J. CO2 Util.* **2021**, *43*, 101376. [CrossRef]
30. Joshi, N.; Hayasaka, T.; Liu, Y.; Liu, H.; Oliveira, O.N.; Lin, L. A review on chemiresistive room temperature gas sensors based on metal oxide nanostructures, graphene and 2D transition metal dichalcogenides. *Microchim. Acta* **2018**, *185*, 213. [CrossRef]
31. Sun, D.; Luo, Y.; Debliquy, M.; Zhang, C. Graphene-enhanced metal oxide gas sensors at room temperature: A review. *Beilstein J. Nanotechnol.* **2018**, *9*, 2832–2844. [CrossRef]
32. Van Quang, V.; Van Dung, N.; Sy Trong, N.; Duc Hoa, N.; Van Duy, N.; Van Hieu, N. Outstanding gas-sensing performance of graphene/SnO_2 nanowire Schottky junctions. *Appl. Phys. Lett.* **2014**, *105*, 1–5. [CrossRef]
33. Shekhirev, M.; Lipatov, A.; Torres, A.; Vorobeva, N.S.; Harkleroad, A.; Lashkov, A.; Sysoev, V.; Sinitskii, A. Highly Selective Gas Sensors Based on Graphene Nanoribbons Grown by Chemical Vapor Deposition. *ACS Appl. Mater. Interfaces* **2020**, *12*, 7392–7402. [CrossRef]
34. Abideen, Z.U.; Katoch, A.; Kim, J.H.; Kwon, Y.J.; Kim, H.W.; Kim, S.S. Excellent gas detection of ZnO nanofibers by loading with reduced graphene oxide nanosheets. *Sens. Actuators B Chem.* **2015**, *221*, 1499–1507. [CrossRef]
35. Tammanoon, N.; Wisitsoraat, A.; Sriprachuabwong, C.; Phokharatkul, D.; Tuantranont, A.; Phanichphant, S.; Liewhiran, C. Ultrasensitive NO_2 Sensor Based on Ohmic Metal-Semiconductor Interfaces of Electrolytically Exfoliated Graphene/Flame-Spray-Made SnO_2 Nanoparticles Composite Operating at Low Temperatures. *ACS Appl. Mater. Interfaces* **2015**, *7*, 24338–24352. [CrossRef] [PubMed]
36. Gu, F.; Nie, R.; Han, D.; Wang, Z. In2O3-graphene nanocomposite based gas sensor for selective detection of NO_2 at room temperature. *Sens. Actuators B Chem.* **2015**, *219*, 94–99. [CrossRef]
37. Andre, R.S.; Mercante, L.A.; Facure, M.H.M.; Mattoso, L.H.C.; Correa, D.S. Enhanced and selective ammonia detection using In_2O_3/reduced graphene oxide hybrid nanofibers. *Appl. Surf. Sci.* **2019**, *473*, 133–140. [CrossRef]
38. Tian, Z.; Song, P.; Yang, Z.; Wang, Q. Reduced graphene oxide-porous In_2O_3 nanocubes hybrid nanocomposites for room-temperature NH_3 sensing. *Chin. Chem. Lett.* **2020**, *31*, 2067–2070. [CrossRef]
39. Fang, W.; Yang, Y.; Yu, H.; Dong, X.; Wang, R.; Wang, T.; Wang, J.; Liu, Z.; Zhao, B.; Wang, X. An In_2O_3 nanorod-decorated reduced graphene oxide composite as a high-response NOx gas sensor at room temperature. *New J. Chem.* **2017**, *41*, 7517–7523. [CrossRef]
40. The National Institute for Occupational Safety and Health (NIOSH). *NIOSH Pocket Guide to Chemical Hazards, September 2007*; NIOSH Publication No. 2005–149; NIOSH: Cincinnnati, OH, USA, 2007; Volume 15.
41. Kranert, C.; Schmidt-Grund, R.; Grundmann, M. Raman active phonon modes of cubic In_2O_3. *Phys. Status Solidi-Rapid Res. Lett.* **2014**, *8*, 554–559. [CrossRef]
42. Garcia-Domene, B.; Ortiz, H.M.; Gomis, O.; Sans, J.A.; Manjón, F.J.; Muñoz, A.; Rodríguez-Hernández, P.; Achary, S.N.; Errandonea, D.; Martínez-García, D.; et al. High-pressure lattice dynamical study of bulk and nanocrystalline In $2O_3$. *J. Appl. Phys.* **2012**, *112*, 123511. [CrossRef]
43. Nasriddinov, A.; Tokarev, S.; Fedorova, O.; Bozhev, I.; Rumyantseva, M. In_2O_3 Based Hybrid Materials : Interplay between Microstructure, Photoelectrical and Light Activated NO_2 Sensor Properties. *Chemosensors* **2022**, *10*, 135. [CrossRef]
44. Kim, W.J.; Pradhan, D.; Sohn, Y. Fundamental nature and CO oxidation activities of indium oxide nanostructures: 1D-wires, 2D-plates, and 3D-cubes and donuts. *J. Mater. Chem. A* **2013**, *1*, 10193–10202. [CrossRef]
45. Ferrari, A.C. Interpretation of Raman spectra of disordered and amorphous carbon. *Schweiz. Z. Hydrol.* **1969**, *31*, 632–645. [CrossRef]

46. Yadav, R.; Joshi, P.; Hara, M.; Yoshimura, M. In situ electrochemical Raman investigation of charge storage in rGO and N-doped rGO. *Phys. Chem. Chem. Phys.* **2021**, *23*, 11789–11796. [CrossRef] [PubMed]
47. Gangwar, P.; Singh, S.; Khare, N. Study of optical properties of graphene oxide and its derivatives using spectroscopic ellipsometry. *Appl. Phys. A* **2018**, *124*, 620. [CrossRef]
48. Childres, I.; Jauregui, L.A.; Park, W.; Caoa, H.; Chena, Y.P. Raman spectroscopy of graphene and related materials. *New Dev. Photon Mater. Res.* **2013**, *1*, 1–20.
49. Socrates, G. *Infrared and Raman Characteristic Group Frequencies: Tables and Charts*; Wiley: Hoboken, NJ, USA, 1981; Volume 5, ISBN 0471852988.
50. Nakamoto, K. *Infrared and Raman Spectra of Inorganic and Coordination Compounds. Part A: Theory and Applications in Inorganic Chemistry*, 6th ed.; John Wiley & Sons, Inc.: Hoboken, NJ, USA, 2009.
51. Sheka, E.F. Digital Twins Solve the Mystery of Raman Spectra of Parental and Reduced Graphene Oxides. *Nanomaterials* **2022**, *12*, 4209. [CrossRef]
52. Chen, X.; Wang, X.; Fang, D. A review on C1s XPS-spectra for some kinds of carbon materials. *Fullerenes Nanotub. Carbon Nanostruct.* **2020**, *28*, 1048–1058. [CrossRef]
53. Kwan, Y.C.G.; Ng, G.M.; Huan, C.H.A. Identification of functional groups and determination of carboxyl formation temperature in graphene oxide using the XPS O 1s spectrum. *Thin Solid Films* **2015**, *590*, 40–48. [CrossRef]
54. Biesinger, M.C. Accessing the robustness of adventitious carbon for charge referencing (correction) purposes in XPS analysis: Insights from a multi-user facility data review. *Appl. Surf. Sci.* **2022**, *597*, 153681. [CrossRef]
55. Dupin, J.C.; Gonbeau, D.; Vinatier, P.; Levasseur, A. Systematic XPS studies of metal oxides, hydroxides and peroxides. *Phys. Chem. Chem. Phys.* **2000**, *2*, 1319–1324. [CrossRef]
56. Zhou, S.; Chen, M.; Lu, Q.; Zhang, Y.; Zhang, J.; Li, B.; Wei, H.; Hu, J.; Wang, H.; Liu, Q. Ag Nanoparticles Sensitized In_2O_3 Nanograin for the Ultrasensitive HCHO Detection at Room Temperature. *Nanoscale Res. Lett.* **2019**, *14*, 365. [CrossRef]
57. Wu, L.Q.; Li, Y.C.; Li, S.Q.; Li, Z.Z.; Tang, G.D.; Qi, W.H.; Xue, L.C.; Ge, X.S.; Ding, L.L. Method for estimating ionicities of oxides using O1s photoelectron spectra. *AIP Adv.* **2015**, *5*, 097210. [CrossRef]
58. Wang, Y.; Liu, L.; Sun, F.; Li, T.; Zhang, T.; Qin, S. Humidity-Insensitive NO_2 Sensors Based on SnO_2/rGO Composites. *Front. Chem.* **2021**, *9*, 681313. [CrossRef]
59. Zhang, X.; Sun, J.; Tang, K.; Wang, H.; Chen, T.; Jiang, K.; Zhou, T.; Quan, H.; Guo, R. Ultralow detection limit and ultrafast response/recovery of the H_2 gas sensor based on Pd-doped rGO/ZnO-SnO_2 from hydrothermal synthesis. *Microsyst. Nanoeng.* **2022**, *8*, 67. [CrossRef] [PubMed]
60. Jin, Y.; Zheng, Y.; Podkolzin, S.G.; Lee, W. Band gap of reduced graphene oxide tuned by controlling functional groups. *J. Mater. Chem. C* **2020**, *8*, 4885–4894. [CrossRef]
61. Lang, O.; Pettenkofer, C.; Sánchez-Royo, J.F.; Segura, A.; Klein, A.; Jaegermann, W. Thin film growth and band lineup of In_2O_3 on the layered semiconductor InSe. *J. Appl. Phys.* **1999**, *86*, 5687–5691. [CrossRef]
62. Gillespie, P.N.O.; Martsinovich, N. Origin of Charge Trapping in TiO_2/Reduced Graphene Oxide Photocatalytic Composites: Insights from Theory. *ACS Appl. Mater. Interfaces* **2019**, *11*, 31909–31922. [CrossRef]
63. Can, I.; Weimar, U.; Barsan, N. Operando Investigations of Differently Prepared In_2O_3-Gas Sensors. *Proceedings* **2017**, *1*, 432. [CrossRef]
64. Zhou, G.; Zhong, B.; Wang, W.; Guan, X.; Huang, B.; Ye, D.; Wu, H. In situ DRIFTS study of NO reduction by NH_3 over Fe-Ce-Mn/ZSM-5 catalysts. *Catal. Today* **2011**, *175*, 157–163. [CrossRef]
65. Luo, J.Y.; Oh, H.; Henry, C.; Epling, W. In Situ-DRIFTS Study of Selective Catalytic Reduction of NOx by NH3 over Cu-Exchanged SAPO-34. *Appl. Catal. B Environ.* **2012**, *123–124*, 296–305. [CrossRef]
66. Nova, I.; Ciardelli, C.; Tronconi, E.; Chatterjee, D.; Bandl-Konrad, B. NH_3-NO/NO_2 chemistry over V-based catalysts and its role in the mechanism of the Fast SCR reaction. *Catal. Today* **2006**, *114*, 3–12. [CrossRef]
67. Malpartida, I.; Marie, O.; Bazin, P.; Daturi, M.; Jeandel, X. The NO/NO x ratio effect on the NH 3-SCR efficiency of a commercial automotive Fe-zeolite catalyst studied by operando IR-MS. *Appl. Catal. B Environ.* **2012**, *113–114*, 52–60. [CrossRef]
68. Leblanc, E.; Perier-Camby, L.; Thomas, G.; Gibert, R.; Primet, M.; Gelin, P. NOx adsorption onto dehydroxylated or hydroxylated tin dioxide surface. Application to SnO_2-based sensors. *Sens. Actuators B Chem.* **2000**, *62*, 67–72. [CrossRef]
69. Hadjiivanov, K.I. Identification of neutral and charged NxOy surface species by IR spectroscopy. *Catal. Rev.-Sci. Eng.* **2000**, *42*, 71–144. [CrossRef]
70. Baergen, A.M.; Donaldson, D.J. Photochemical renoxification of nitric acid on real urban grime. *Environ. Sci. Technol.* **2013**, *47*, 815–820. [CrossRef]
71. Ye, C.; Gao, H.; Zhang, N.; Zhou, X. Photolysis of Nitric Acid and Nitrate on Natural and Artificial Surfaces. *Environ. Sci. Technol.* **2016**, *50*, 3530–3536. [CrossRef]
72. Schuttlefield, J.; Rubasinghege, G.; El-Maazawi, M.; Bone, J.; Grassian, V.H. Photochemistry of adsorbed nitrate. *J. Am. Chem. Soc.* **2008**, *130*, 12210–12211. [CrossRef]
73. Nanayakkara, C.E.; Jayaweera, P.M.; Rubasinghege, G.; Baltrusaitis, J.; Grassian, V.H. Surface photochemistry of adsorbed nitrate: The role of adsorbed water in the formation of reduced nitrogen species on α-Fe_2O_3 particle surfaces. *J. Phys. Chem. A* **2014**, *118*, 158–166. [CrossRef]

74. Ye, C.; Zhang, N.; Gao, H.; Zhou, X. Photolysis of particulate nitrate as a source of HONO and NOx. *Environ. Sci. Technol.* **2017**, *51*, 6849–6856. [CrossRef]
75. Ma, Q.; Zhong, C.; Ma, J.; Ye, C.; Zhao, Y.; Liu, Y.; Zhang, P.; Chen, T.; Liu, C.; Chu, B.; et al. Comprehensive Study about the Photolysis of Nitrates on Mineral Oxides. *Environ. Sci. Technol.* **2021**, *55*, 8604–8612. [CrossRef]
76. Zhou, Z.; Gao, H.; Liu, R.; Du, B. Study on the structure and property for the NO_2 + NO_2- electron transfer system. *J. Mol. Struct. THEOCHEM* **2001**, *545*, 179–186. [CrossRef]
77. Protocol, M.; Cao, C.; Chen, Y.; Wu, Y.; Deumens, E. Electron Affinity of the O_2 Molecule. *Int. J. Quantum Chem.* **2011**, *111*, 4020–4029. [CrossRef]

sensors

MDPI

Article

Laboratory Comparison of Low-Cost Particulate Matter Sensors to Measure Transient Events of Pollution—Part B—Particle Number Concentrations

Florentin Michel Jacques Bulot [1,2,*], Hugo Savill Russell [3,4,5,6], Mohsen Rezaei [6], Matthew Stanley Johnson [4,6], Steven James Ossont [7], Andrew Kevin Richard Morris [8], Philip James Basford [1], Natasha Hazel Celeste Easton [2,9], Hazel Louise Mitchell [1], Gavin Lee Foster [9], Matthew Loxham [2,10,11,12] and Simon James Cox [1,2]

1 Faculty of Engineering and Physical Sciences, University of Southampton, Southampton SO17 1BJ, UK; p.j.basford@soton.ac.uk (P.J.B.); hlm1g16@soton.ac.uk (H.L.M.); s.j.cox@soton.ac.uk (S.J.C.)
2 Southampton Marine and Maritime Institute, University of Southampton, Southampton SO16 7QF, UK; nhcs1g13@soton.ac.uk (N.H.C.E.); m.loxham@soton.ac.uk (M.L.)
3 Danish Big Data Centre for Environment and Health (BERTHA), Aarhus University, DK-4000 Roskilde, Denmark; hugo.russell@envs.au.dk
4 AirScape UK, London W1U 6TQ, UK; matthew.johnson@airscape.ai or msj@chem.ku.dk
5 Department of Environmental Science, Atmospheric Measurement, Aarhus University, DK-4000 Roskilde, Denmark
6 Department of Chemistry, University of Copenhagen, DK-2100 Copenhagen, Denmark; mohsen@chem.ku.dk
7 BizData, Melbourne, VIC 3000, Australia; steven.ossont@bizdata.co.nz
8 National Oceanography Centre, Southampton SO14 3ZH, UK; andmor@noc.ac.uk
9 School of Ocean and Earth Science, National Oceanography Centre, University of Southampton, Southampton SO14 3ZH, UK; gavin.foster@noc.soton.ac.uk
10 Faculty of Medicine, University of Southampton, Southampton SO17 1BJ, UK
11 National Institute for Health Research, Southampton Biomedical Research Centre, Southampton SO16 6YD, UK
12 Institute for Life Sciences, University of Southampton, Southampton SO17 1BJ, UK
* Correspondence: florentin.bulot@centraliens.net

Abstract: Low-cost Particulate Matter (PM) sensors offer an excellent opportunity to improve our knowledge about this type of pollution. Their size and cost, which support multi-node network deployment, along with their temporal resolution, enable them to report fine spatio-temporal resolution for a given area. These sensors have known issues across performance metrics. Generally, the literature focuses on the PM mass concentration reported by these sensors, but some models of sensors also report Particle Number Concentrations (PNCs) segregated into different PM size ranges. In this study, eight units each of Alphasense OPC-R1, Plantower PMS5003 and Sensirion SPS30 have been exposed, under controlled conditions, to short-lived peaks of PM generated using two different combustion sources of PM, exposing the sensors' to different particle size distributions to quantify and better understand the low-cost sensors performance across a range of relevant environmental ranges. The PNCs reported by the sensors were analysed to characterise sensor-reported particle size distribution, to determine whether sensor-reported PNCs can follow the transient variations of PM observed by the reference instruments and to determine the relative impact of different variables on the performances of the sensors. This study shows that the Alphasense OPC-R1 reported at least five size ranges independently from each other, that the Sensirion SPS30 reported two size ranges independently from each other and that all the size ranges reported by the Plantower PMS5003 were not independent of each other. It demonstrates that all sensors tested here could track the fine temporal variation of PNCs, that the Alphasense OPC-R1 could closely follow the variations of size distribution between the two sources of PM, and it shows that particle size distribution and composition are more impactful on sensor measurements than relative humidity.

Keywords: low-cost sensors; particle number concentration; laboratory study; fine particles; particulate matter; air pollution

1. Introduction

Exposure to air pollution is a major cause of environmental morbidity and mortality in the world at present, with Particulate Matter (PM) air pollution being associated with 8.9 million premature deaths per year [1,2]. PM air pollution varies with fine spatio-temporal granularity and can have heterogeneous composition and concentration over a specific area [3]. Current regulatory monitoring networks are based on cumbersome and expensive apparatus that means monitoring with the spatial coverage required to comprehensively understand the spread of air pollution is not feasible. Given the recently substantially reduced WHO exposure limits [4], down to 5 $\mu g/m^3$ as an annual mean for $PM_{2.5}$, there is an increased need for monitoring. At this lower threshold, local sources can often be the factor causing exceedance, which makes information concerning local levels and sources more important than they have been in the past [5]. The EU is moving towards adopting the more stringent WHO standard [6] and voices in the community are saying that the only way to ensure compliance is by using dense networks of low-cost sensors in populated areas [7].

Low-cost PM sensors have been used in the literature and in various projects around the world to determine PM mass concentrations, especially when deployed as networks of sensors to improve the limited spatio-temporal coverage of existing monitoring networks [5]. Considerable research has been conducted to reach a known level of precision and accuracy with some studies achieving the data quality objectives of reference-grade instruments with the proper calibration methods and frequencies [8], at high temporal resolution, providing data that was not previously available to determine population exposure to PM air pollution at a finer level. However, some of these sensors provide not only PM concentrations but also Particle Number Concentrations (PNCs) for different size ranges, for example by giving PNC in the range 0.3–1 µm and PNC in the range 1–2.5 µm. They are based on light scattering and generally claim to measure particles of diameters 0.3–10 µm, and it is important to note that the scattering efficiency decreases as the diameter is close to or lower than 0.3 µm therefore implying a lower performance in the response to the lower size ranges of particles. Standard metrics for PM have evolved through the last few decades [9]. For example, in the US, the 1971 National Ambient Air Quality Standards was set for PM as total suspended particles. Later in 1987, following new evidence on the health effects of PM, the standards were revised to focus on PM_{10}. In 1997, the first standards for $PM_{2.5}$ were issued to account for the health impact of this size fraction. Although current legal limits are based on PM mass concentration, not all PM is equally harmful and other properties of the particles may be significant in terms of health impact, such as their composition, shape, size, etc. [10]. Therefore, size distribution of PM could be a promising metrics to better capture the health impact of PM.

There is variation in what is reported with some sensors giving a detailed size distribution and others only outputting PNC with a restricted number of size ranges. For instance, the Plantower PMS5003 outputs six different size ranges and the Alphasense OPC-R1 outputs 13 size ranges. The ability of the sensors to report PNCs of different size fractions can be used to identify sources of pollution. Indeed, in Delhi, India, Hagan et al. [11] used the first three size ranges of an Alphasense OPC-N2, in conjunction with data on other air pollutants (CO, NO_2, SO_2, O_3) to successfully identify sources of pollution using positive matrix factorisation. Additionally, we previously demonstrated reference-grade improvements to the performances of Plantower PMS5003 and Sensirion SPS30 through calibration methods based on the PNCs reported by these sensors [12]. Similarly, Wallace et al. [13,14] developed a calibration method using the PNCs reported by the Plantower PMS5003 that outperformed calibration methods based on mass concentrations. This improvement in performances was confirmed by further long-term studies [15,16].

There are broadly two types of low-cost PM sensors [17]: (1) volume scattering, or integrating nephelometers, that measure the light scattered by an ensemble of particles; and (2) single particle counters which count individual particles. The two types have different sensitivities to aerosol parameters and environmental factors [18]. However, there is disagreement in the literature about which sensor belongs to which type. There is also concern about whether these low-cost PM sensors can accurately segregate PNCs into different size ranges [17–21]. Recently, Ouimette et al. [22] conducted a detailed study of the PurpleAir sensors (PurpleAir, Draper, UT, USA) (which use two Plantower PMS5003s (Plantower, Nanchang City, China)) comparing them to a research-grade integrating nephelometer and developed a physical model that showed that the Plantower PMS5003 is a cell-reciprocal nephelometer providing a reliable measurement of the aerosol scattering coefficients for particles in the range 0.26–0.46 µm. Ouimette et al. [22] is one of the rare studies that focused on sensor-reported PNCs. A few laboratory studies have been conducted regarding the size segregation capacity of the sensors [19,21,23,24]. Three of these studies have focused only on sensor-reported mass concentrations, while one has also studied sensor-reported PNCs. All of the above studies examined sensor performances with stable concentrations of PM over periods ranging from 5 min to 1 h, depending on the study. They exposed the sensors to PM of a variety of sources and sizes. Several studies highlighted that low-cost PM sensors are susceptible to a range of environmental factors, namely particle composition, size distribution and Relative Humidity (RH). However, different studies obtained contrasting results concerning RH which suggests that other factors may be at play that are not accounted for.

Feature selection methods quantify the contribution of individual features (here environmental factors) to the variability of an output variable (here sensor-reported PNCs) [25]. They are one of the most popular techniques to improve the explainability of machine learning models [26], which are often used to correct measurements from low-cost sensors. Feature selection methods are divided into three sub-categories of method: filter-based, wrapper-based and embedded methods [27]. Filter-based methods class the variables using different metrics such as the Pearson coefficient or the Akaike information criterion (AIC). They do not account for possible correlation between variables and are prone to missing patterns [28]. Wrapper-based methods iteratively use supervised learning techniques (e.g., linear model, support vector machine) to classify the variables. They apply algorithms such as recursive feature selection and greedy forward selection. They are generally more accurate than filter-based methods but risk over-fitting and are more computationally intensive [28]. Finally, embedded methods have the reduction of the number of variables embedded in their algorithms, such as Lasso regression, elastic net regression or random forest. They constitute a trade-off between filter and wrapper-based methods [29]. Nonetheless, the features selected are dependent on the methods chosen and the best practice is to use different methods concomitantly and to compare their results [25].

The current study is the second part of a comprehensive experiment that aimed to characterise the response of a range of low-cost PM sensors to transient events of PM pollution. The first part of this study [30] focused on sensor-reported mass concentrations while this current contribution focuses on sensor-reported PNCs. Sensors measuring at a 10 s temporal resolution were exposed to short-lived peaks of PM pollution (≈1 min) generated by lighting candles and incense sticks at different RH levels. Using two combustion sources, we can assess the performance of the sensors across different size distributions. Understanding the response of these sensors to short-lived events of PM pollution is important especially if these sensors are to be used indoors, where polluting activities may last only for a few minutes [31], or used outdoors as a network for tracking events of PM pollution as they spread through an area [32]. These data could also be integrated into models to further their granularity through data fusion techniques [33]. Here, we compare eight units of each sensor model, Sensirion SPS30, Plantower PMS5003 and Alphasense OPC-R1, at a total of 24 low-cost PM sensors, at 10 s resolution. A TSI OPS 3330 (TSI Inc., Shoreview, MN, USA)is used as a reference instrument. The aim is to characterise sensor-reported particle size distribution, to determine whether sensor-reported PNCs can follow the transient variations of PM observed by the reference instruments and to determine the relative impact of different variables on the performances of the sensors.

The objectives of this study are:

1. To determine whether the sensors can be used at a high temporal resolution to follow trends of PNCs.
2. To determine whether the different sensor-reported PNCs are independent from each other and characterise their accuracy.
3. To characterise the capacity of these sensors to capture the size distribution of PM.
4. To determine environmental factors impacting the response of the sensors.

2. Materials and Methods

2.1. Low-Cost PM Sensors

The low-cost sensors were mounted in the air quality monitors developed in Johnston et al. [34], without their environmental enclosure as can be seen in Figure 1. The absence of enclosure helps reduce potential residual heat build-up in the vicinity of the sensors. The low-cost sensors studied here are the Plantower PMS5003, the Sensirion SPS30 and the Alphasense OPC-R1. These sensors were chosen because they output PNCs of the PM measured for different size fractions. Table 1 presents the different size ranges of these three models of sensors. The Honeywell HPMA115S0 and the Novafitness SDS018 were also measuring during the experiment but the data they produced were not used in this study as they only report PM mass concentrations. All the low-cost sensors tested here are optical measurement devices based on Mie light-scattering. Four air quality monitors were used concomitantly, each containing two of each of the sensor models mentioned above, at a total of eight sensors of each model. In each air quality monitor, the sensors were plugged in via USB to a Raspberry Pi, powered through Power Over Ethernet (PoE) and controlled using Python 3.6 libraries [35–37]. The data recorded by all the sensors were averaged over 10 s for cross-comparison purposes. Relative humidity and temperature were measured by each of the four air quality monitors using a Sensirion SHT35 [38] ($\pm$1.5% RH and $\pm$0.1 °C).

The Plantower PMS5003 reports six size ranges called gr03um, gr05um, gr10um, gr25um, gr50um and gr100um, which represent, respectively, PNCs of particles >0.3 µm, >0.5 µm, >1 µm, >2.5 µm, >5 µm and >10 µm. The PNCs are reported as particles per 0.1 L of air. According to Sayahi et al. [39], the Plantower PMS5003 has a flow rate of ≈0.1 L/min and a wavelength of 640 $\pm$ 10 nm with light polarised at 90° [22]. The size ranges of the Plantower PMS5003 have been recalculated to obtain distinct size ranges similarly to Wallace et al. [14]. The size ranges obtained are as follows: n03_05, n05_10, n10_25, n25_50 and n50_100 which represent, respectively, PNCs of particles 0.3–0.5 µm, 0.5–1.0 µm, 1.0–2.5 µm, 2.5–5.0 µm and 5.0–10.0 µm.

The Sensirion SPS30 reports size ranges called n05, n1, n25, n4 and n10, which represent, respectively, PNCs in the range 0.3–0.5 µm, in the range 0.3–1 µm, in the range 0.3–2.5 µm, in the range 0.3–4 µm and in the range 0.3–10 µm. It utilises a laser beam of 660 nm wavelength and reports PNCs as particles/cm^3. The Sensirion SPS30s are calibrated by their manufacturer against a TSI OPS 3330 or a TSI DustTrak DRX 8533. The accuracy of the calibration is then verified by the manufacturer using an atomized potassium chloride solution [40]. For the Sensirion SPS30, according to the manufacturer, particles above 4 µm are not directly measured but determined from the other size ranges using a particle distribution profile. The Sensirion SPS30 is certified for UK indicative monitoring and, although the sensors used in this study were acquired prior to the certification, private communication with the manufacturer confirmed that there had been no significant changes between the sensors used here and the sensors used for the certification. The size ranges of the Sensirion SPS30 were also recalculated to obtain distinct size ranges. The size ranges obtained are as follows: n03_05, n05_1, n1_25, n25_4 and n4_10 which represent, respectively, PNCs in the range 0.3–0.5 µm, in the range 0.5–1 µm, in the range 1–2.5 µm, in the range 2.5–4 µm and in the range 4–10 µm.

(a) 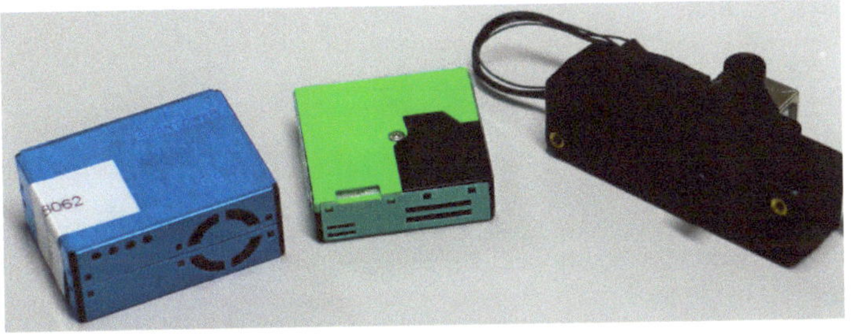

(b) 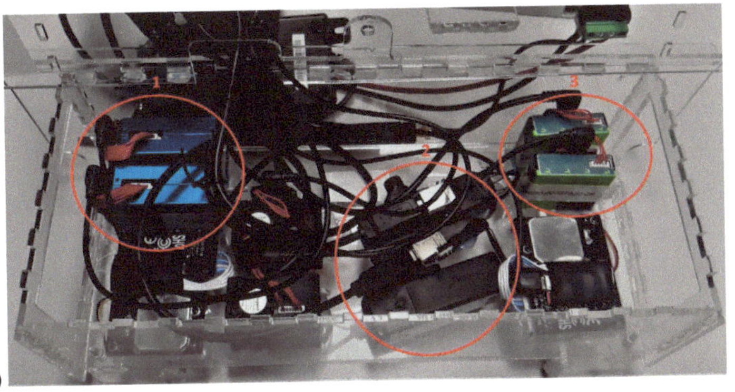

Figure 1. (**a**) Sensors tested from left to right: Plantower PMS5003, Sensirion SPS030, Alphasense OPC-R1, adapted from Bulot et al. [30]. (**b**) Position of the sensors tested within each air quality monitor. From left to right, top to bottom: two Plantower PMS5003s (red circle 1), one Novafitness SDS018, two Honeywell HPMA115S0s, two Alphasense OPC-R1 (red circle 2), two Sensirion SPS30s (red cicle 3) and one Novafitness SDS018. All the inlets are facing down. Reprinted/adapted with permission from Bulot et al. [30]. 2020, by the authors.

The Alphasense OPC-R1 is a single particle counter, which utilises a laser beam at a 639 nm wavelength, which can theoretically count up to 10,000 particles/s or 2500 particles/cm^3 with a maximum coincidence probability of 0.7% at 1000 particles/cm^3. The PNCs are output as particles/cm^3 into 16 different size ranges (see Table 1) and can measure particles in the range 0.35–12.4 µm. It has a flow rate of 0.24 L/min. The Alphasense OPC-R1 was calibrated by its manufacturer using monodisperse Polystyrene Sphere (PLS) particles against an Alphasense OPC-R1, which itself had previously been calibrated against a TSI OPS 3330 [41].

Table 1. Size ranges of the Plantower PMS5003, Sensirion SPS30, Alphasense OPC-R1 and TSI OPS 3330.

Sensor	Size Ranges (µm)
PMS5003	>0.3; >0.5; >1; >2.5; >5; >10
SPS30	0.3–0.5; 0.3–1; 0.3–2.5; 0.3–4; 0.3–10
OPC-R1	0.4–0.7; 0.7–1.1; 1.1–1.5; 1.5–1.9; 1.9–2.4; 2.4–3; 3–4; 4–5; 5–6; 6–7; 7–8; 8–9; 9–10; 10–11; 11–12; 12–12.4
OPS 3330	0.3–0.4; 0.4–0.5; 0.5–0.6; 0.6–0.7; 0.7–0.9; 0.9–1.1; 1.1–1.4; 1.4–1.7; 1.7–2.2; 2.2–2.7; 2.7–3.3; 3.3–4.2; 4.2–5.2; 5.2–6.5; 6.5–8.0; 8.0–10

2.2. Reference Instruments

2.2.1. TSI OPS 3330

The TSI Optical Particle Sizer (OPS) 3330 (TSI Inc., Shoreview, MN, USA) is an optical instrument based on light scattering which reports PNCs divided into 16 size ranges (see Table 1), reported in particles/cm^3, in the range 0.3–10 µm. It uses a laser beam at 660 nm and a flow rate of 1 L/min. It is calibrated by its manufacturer for size using Polystyrene Sphere (PLS) [42]. In this study, the TSI OPS 3330 was used as a reference instrument and set to measure every 10 s. The size ranges of the TSI OPS 3330 were recalculated to match the size ranges of the different models of sensors to enable comparison. The method for redistributing the size ranges is described below in the Section 2.4.

2.2.2. Aerasense Nanotracer

The Aerasense Nanotracer (Oxility BV, Venray, The Netherlands) counts particles in the range 10–300 nm based on diffusion charging. It is measuring below the advertised cut-off size of the low-cost sensors. It was calibrated by its manufacturer using KNO_3 polydisperse particles and has a flow rate of 0.3–0.4 L/min. It is used because some of the sensors may be able to measure below 0.3 µm.

2.3. Experimental Conditions

The experimental conditions and set-up are the same as in Bulot et al. [30], the relevant elements are summarised here and the experimental set-up is described in Figure 2. The test chamber was placed in an environmental chamber where the temperature was controlled and set to 23 °C. The temperature inside the test chamber varied in the range 26–29 °C throughout the experiments. Candle and incense smoke were used as the two different combustion sources, enabling testing at different particle size distributions. Candle smoke, here produced by smouldering, contains mostly particles in the range 0.02–0.1 µm with particle size peaks in the range 0.03–0.05 µm in terms of PNC [43,44]. For incense smoke, particles are mostly within the range 0.05–0.7 µm with a peak at 0.2 µm, in terms of PNC [45]. Five sets of experiments were conducted. For each set of experiments, several peaks of candle smoke were generated, followed by a longer concentration of candle smoke, then the air was cleaned of particles using the Electrostatic Precipitator (ESP), then a series of peaks of incense smoke were generated followed by a stable concentration of incense smoke. The peaks of PM lasted around 1 min and had a targeted concentration of 20–50 µg/m^3 as measured in real-time by a DustTrak DRX 8533 Desktop (TSI Inc., Shoreview, RC, USA). During each experiment, the RH was set at different targets: 54, 69, 72, 76 and 79% RH. RH was controlled by a mist generator. RH was measured by the Sensirion SHT35 RH and temperature sensors built into each of the air quality monitors. RH was taken as the median reading of the four SHT35 sensors (as the distribution of RH recorded by the sensors was not normal). The experimental conditions are further described in Bulot et al. [30], along with some statistical analysis of the RH measured by the SHT35 sensors. Targeted RHs were set higher but could not be achieved with this experimental set-up. The range of RH attained in this study may not be sufficient to capture the impact of this environmental factor and the behaviour of the sensors may change for higher levels of RH.

The residual heat generated by the electronics and the sensors inside the air quality monitors means that the RH and temperature condition inside the air quality monitor are higher than the conditions surrounding the air quality monitors. For instance, in one of our previous studies, we noticed the RH was 15% lower inside the air quality monitor than outside [12]. It is not clear whether the particles have sufficient time to adjust to these conditions before being measured by the sensors. To avoid any impact of the phenomenon described above, the enclosure of the air quality monitors was removed during the experiment as can be seen in Figure 2.

The particle size distribution from the TSI OPS 3330 is available in Figure A13 and shows that, for candle-generated PM, there is less than 10 particles/cm^3 above 5 µm and for incense-generated PM above 2.5 µm. Given the low values of PNC above 2.5 µm, only the size ranges of the sensors having a lower cut size < 2.5 μm will be considered during this study.

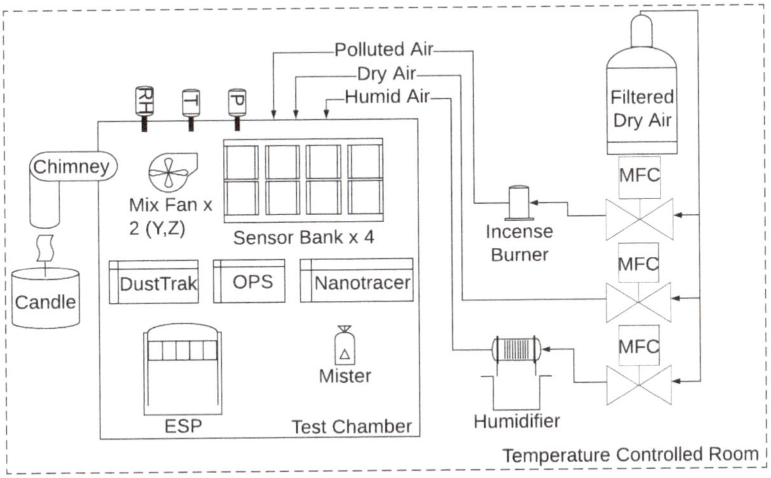

(a)

(b)

Figure 2. (**a**) Schematic showing the arrangement of the test chamber and supporting equipment, adapted from Bulot et al. [30]. (**b**) Image showing the air quality boxes located in the test chamber. Reprinted/adapted with permission from Bulot et al. [30]. 2020, by the authors. Mass Flow Controllers (MFCs).

2.4. Data Analysis

2.4.1. Feature Selection Methods

As detailed in the introduction of this paper, the outcome of feature selection methods depends on the methods chosen, and it is best practice to use different methods concomitantly and to compare their results. In this paper, we tested three methods: Ridge regression,

Boruta and Recursive Feature Elimination (RFE) with Support Vector Machines (SVM). Ridge regression feature selection is an extension of the linear model with a penalisation term on the residual of the sums of squares using the L_2 norm [46]. Ridge regression is an embedded method, as the selection of features is part of its algorithms. Boruta and RFE are both wrapper methods. Boruta is a wrapper based on random forest; it starts by adding shuffled copies of the existing variables to the dataset, called shadow variables. It then trains a random forest on this dataset and measures the variable importance (using the variable importance measure built into the random forest) and compares the importance of the initial variables to the importance of the shadow variables. Variables that obtained a significantly lower score than the higher score of the shadow variables are removed. It then reiterates the process [47]. RFE was initially developed to enable SVM to perform feature selection [48]. It trains a SVM model, computes a ranking criterion for all the variables considered (the weights of the SVM) and then removes the feature with the smallest ranking criterion. Filter-based methods have not been tested here as they generally do not allow for complex interactions between the variables considered. It is important to note that the scores obtained across the different methods cannot be compared; only the relative scores of each variable within a method can be compared.

In this study, the following variables are used for feature selection to predict each size range of each sensor: the source of PM (candle or incense), the number of particles <0.3 µm (measured by the Nanotracer), the number of particles in the range 0.3–0.8 µm (measured by the TSI OPS 3330), the number of particles in the range 0.3–10 µm (measured by the TSI OPS 3330) and RH. The feature selection methods are performed on the data from the five different levels of RH and using both sources of PM, aggregated by model of sensor.

2.4.2. Lognormal Size Distribution

Particle size distribution is presented using lognormal distributions of the normalised concentrations calculated using the following formula, for each size range of the instrument considered:

$$\frac{dN}{dlog(D_p)} = \frac{dN}{log(D_{p,u}) - log(D_{p,l})} \tag{1}$$

with dN the PNC, $D_{p,u}$ the diameter of the upper boundary of the size range and $D_{p,l}$ the diameter of the lower boundary of the size range.

2.4.3. Redistribution of OPS Size Ranges

The sensors and the TSI OPS 3330 measure the particle distribution using different numbers of size ranges or different cut-off diameters. In this study, we recalculate the size ranges of the TSI OPS 3330 to match the size ranges of each sensor tested. Overlapping size range fractions are computed with the formulas used by Di Antonio et al. [49] and shown in Figure 3 for a simple example. For this example, the equivalent TSI OPS 3330 size range b_{eq}^{ops} is defined by:

$$b_{eq}^{ops} = b_0^{ops} \times f_{low} + b_1^{ops} + b_2^{ops} \times f_{upp} \tag{2}$$

with

$$f_{low} = \frac{b_{0,upp}^{ops} - b_{low}}{b_{0,upp}^{ops} - b_{0,low}^{ops}} \tag{3}$$

and

$$f_{upp} = \frac{b_{upp} - b_{2,low}^{ops}}{b_{2,upp}^{ops} - b_{2,low}^{ops}} \tag{4}$$

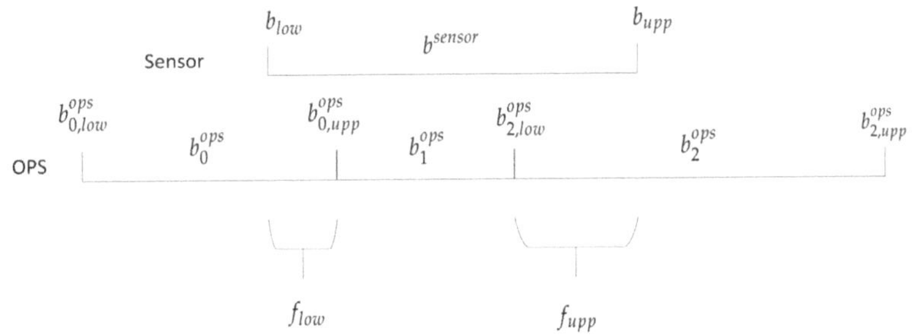

Figure 3. Principle of the redistribution of the size ranges of the TSI OPS 3330 to match the size ranges of each model of sensor with b^{sensor} the sensor size range, b_{low} and b_{upp} the lower and upper cut size of the sensor size range, b_i^{ops} the i corresponding size ranges of the TSI OPS 3330, $b_{i,low}^{ops}$ and $b_{i,upp}^{ops}$ the lower and upper cut sizes of the ith size range of the TSI OPS 3330, and f_{low} and f_{upp} the lower and upper fractions of the size ranges corresponding to b_{low} and b_{upp}.

2.4.4. Software and Data

The data were analysed using R 4.2.2 (R Foundation for Statistical Computing, Vienna, Austria) [50]. The underlying dataset is openly available at https://doi.org/10.5281/zenodo.7808620), and the code used for the analysis and to generate the tables and graphs of this study is openly available at https://doi.org/10.5281/zenodo.7808794. Boruta feature selection was conducted using the Boruta package [47]. RFE-SVM was conducted using the Caret package [51] using a 10 times repeated cross-validation, based on SVM radial [52]. Ridge was performed using the packages caret and glmnet [53] with a 10 times repeated cross-validation, and a grid search to optimise the penalisation coefficient λ between 0 and 1. The in-between variability between sensors of the same models was characterised by some of our previous works [12,30] which showed that this variability was relatively low. Therefore, the results presented here were aggregated over all the sensors for a given model, apart from the time series presented later for which a single sensor of each model was randomly selected. The size distributions of each individual sensor are presented in Appendix B.

3. Results

3.1. Correlation Between the Different Size Ranges

Tables 2–4 present the correlations obtained by the different models of sensors across the different size ranges they report. For comparison, the same has been done for the equivalent size ranges calculated from the TSI OPS 3330 readings and these are presented between brackets in the same tables. This gives a baseline for the levels of correlation to expect in the actual size distribution of the particles measured. If the difference between correlation between the sensor size ranges and the correlation of the TSI OPS 3330 is >0.15, we consider that the two considered size ranges of the sensors are not truly independent. If the difference in correlation is <0.15, the size ranges of the sensors are considered independent.

For the Plantower PMS5003, the three size ranges were not independent from each other. For the Alphasense OPC-R1, all the size ranges were independent from each other. For the Sensirion SPS30, n03_05 and n05_1 were not independent from each other although to a lower extent than what was observed for the first size range of the Plantower PMS5003. n03_05 and n1_25 were independent from each other, and n05_1 and n1_25 were also independent from each other.

Table 2. Correlation and linear model between the different Particle Number Concentration (PNC) size ranges reported by the Plantower PMS5003 during the period of the study. The numbers between brackets represent the correlation between equivalent size ranges of the TSI OPS 3330. Cells are shaded in red if the difference between the correlation of the sensor and the TSI OPS 3330 is greater than 0.15. If the sensor size ranges accurately measured the size distribution, they should obtain a similar correlation to the TSI OPS 3330.

PMS5003		**Range**	**n05_10 0.5–1 μm**	**n10_25 1–2.5 μm**
R^2	n03_05	0.3–0.5 μm	0.99 (0.53)	0.86 (0.14)
R^2	n05_10	0.5–1 μm		0.80(0.64)

Table 3. Correlation and linear model between the different Particle Number Concentration (PNC) size ranges reported by the Sensirion SPS30 during the period of the study. The numbers between brackets represent the correlation between equivalent size ranges of the TSI OPS 3330. Cells are shaded in red if the difference between the correlation of the sensor and the TSI OPS 3330 is greater than 0.15. If the sensor size ranges accurately measured the size distribution, they should obtain a similar correlation to the TSI OPS 3330.

SPS30		**Range**	**n05_1 0.5–1 μm**	**n1_25 1–2.5 μm**
R^2	n03_05	0.3–0.5 μm	0.73(0.53)	0.18 (0.14)
R^2	n05_1	0.5–1 μm		0.70 (0.64)

Table 4. Correlation and linear model between the different Particle Number Concentration (PNC) size ranges reported by the Alphasense OPC-R1 during the period of the study. The numbers between brackets represent the correlation between equivalent size ranges of the TSI OPS 3330. Cells are shaded in red if the difference between the correlation of the sensor and the TSI OPS 3330 is greater than 0.15. If the sensor size ranges accurately measured the size distribution, they should obtain a similar correlation to the TSI OPS 3330.

OPCR1		**Range**	**Bin1 0.7–1.1 μm**	**Bin2 1.1–1.5 μm**	**Bin3 1.5–1.9 μm**	**Bin4 1.9–2.4 μm**
R^2	Bin0	0.35–0.7 μm	0.31 (0.43)	0.19 (0.17)	0.14 (0.09)	0.12 (0.06)
R^2	Bin1	0.7–1.1 μm		0.78 (0.80)	0.61 (0.64)	0.53 (0.53)
R^2	Bin2	1.1–1.5 μm			0.93 (0.96)	0.88 (0.89)
R^2	Bin3	1.5–1.9 μm				0.96 (0.98)

3.2. Time Series of the Experiments

The time series presented in Figures 4–6 are focused on the experiment performed at 69% RH for brevity and the time series for the other experiments are available in Figures A1–A12. They show similar results to Experiment 2. The first seven peaks correspond to the generation of peaks of candle-generated PM followed by stable concentrations of candle-generated PM, then a series of six peaks of incense-generated PM and a stable concentration of incense-generated PM. For this section, the size ranges of the OPS have been converted to the size ranges of each individual sensor. The y-axis follows a logarithmic scale.

For the three models of sensors, the time series of the different size ranges closely followed the variation of the size ranges of the TSI OPS 3330, for both sources of particles. For the Plantower PMS5003, for candle-generated PM, the magnitude of sensor-reported PNC was 10 times lower than the magnitude of the TSI OPS 3330, for all the size ranges of this sensor model. For incense-generated PM, the first size range (0.3–0.5 μm) was also 10 times lower than the magnitude of the TSI OPS 3330, but the second size range (0.5–1 μm) obtained the same magnitude as the TSI OPS 3330 for peaks but underestimated

PNC for the stable concentration of incense-generated PM. This size range also presented a downward slope for stable concentrations of incense-generated PM, which did not match the TSI OPS 3330 measurements. The third size range (1–2.5 µm) for incense-generated PM overestimated PNC for peak concentrations but, for the stable concentration, it started by over-reporting before then under-reporting. For the Sensirion SPS30, the magnitude was about 100 times lower for all the size ranges. For its first size range (0.3–0.5 µm) little difference was observed between candle- and incense-generated PM; for its second size range (0.5–1 µm), while the TSI OPS 3330 recorded lower PNC for incense-generated PM than for candle-generated PM, the Sensirion SPS30 recorded similar levels of PNC for the two sources. For the third size range (1–2.5 µm), the sensor under-reported more for incense than for candle-generated PM. For the Alphasense OPC-R1, the magnitude of the first size range (from 0.35–0.7 µm) was about 10 times lower; this size range also presented a lot of variability that was not present in the measurement made by the TSI OPS 3330. For the other size ranges of this sensor, the magnitude was similar to the TSI OPS 3330: for the size range 0.7–1.1 µm, the Alphasense OPC-R1 slightly over-reported PNC while, for the three remaining size ranges, it slightly under-reported PNC compared to the TSI OPS 3330.

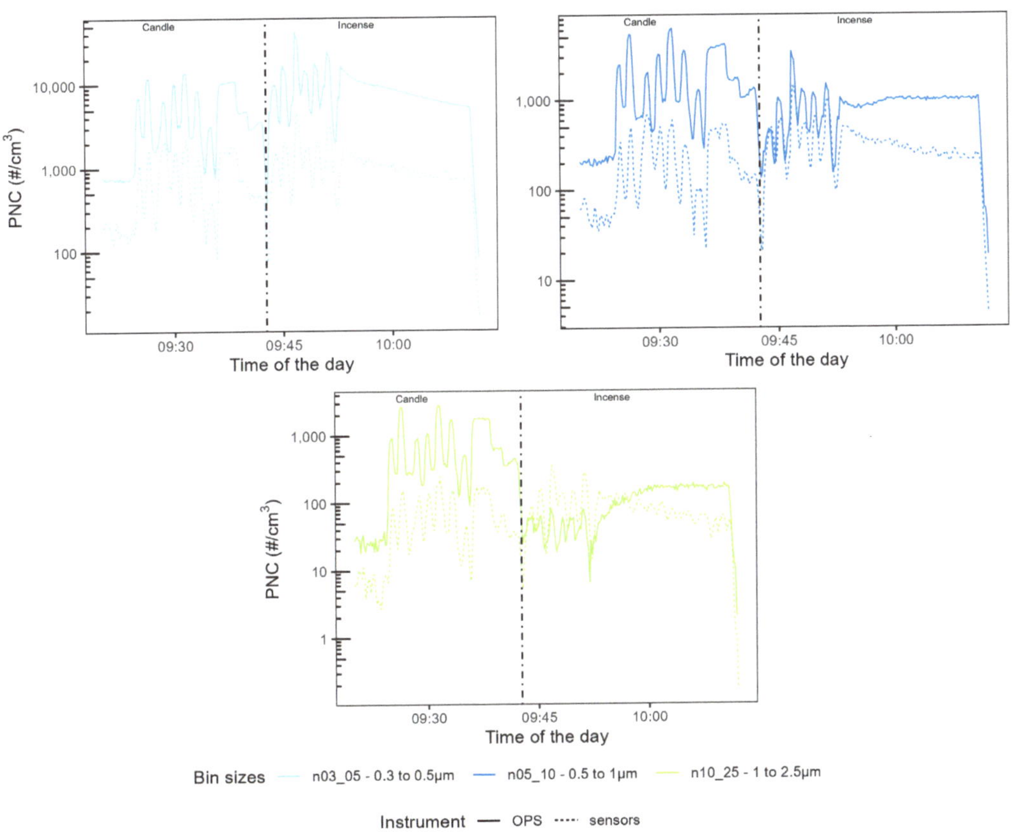

Figure 4. Time series of the Particle Number Concentration (PNC) size ranges of the Plantower PMS5003 compared with the size ranges computed from the OPS size ranges. On the left of the dotted line, PM was generated using a candle, on the right using incense. The different categories are, from left to right and top to bottom, n03_05 (in the range 0.3–0.5 µm), n05_10um (in the range 0.5–1 µm), and n10_25um (in the range 1–2.5 µm).

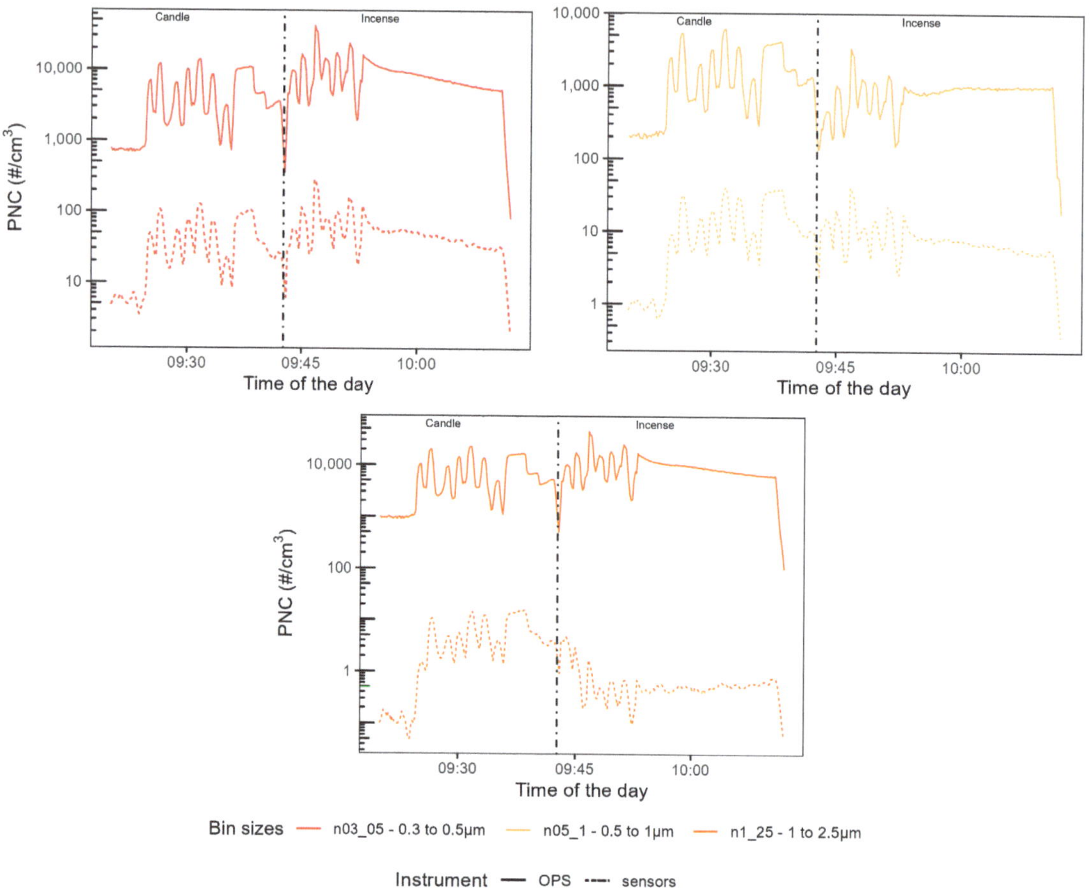

Figure 5. Time series of the Particle Number Concentration (PNC) size ranges of the Sensirion SPS30 compared with the size ranges computed from the OPS size ranges. On the left of the dotted line, PM was generated using a candle, on the right using incense. The different categories are, from left to right and top to bottom, n03_05 (in the range 0.3–0.5 µm), n05_1 (in the range 0.5–1 µm), and n1_25 (in the range 1–2.5 µm).

3.3. Feature Selection

Tables 5–7 present the importance of the variables (source; PNC 0.01–0.3 µm; PNC 0.3–0.8 µm; PNC 0.3–10 µm; RH) for the different size ranges of the sensors, computed by using the three methods described in the Section 2.4.1: Boruta, Ridge and RFE-SVM. High scores denote the relevance of the variable to explain the size range considered. Each method computes their score differently and the values obtained should not be compared between the methods.

For the Plantower PMS5003, RH was consistently given a score of zero or close-to-zero. For the two first size ranges, the PNC for particles 0.01–0.3 µm was given the highest scores for the Boruta and Ridge method and a high score for the RFE-SVM method. The third size range was given lower scores for this variable for the three methods. The source of the PM was given relatively low scores for the first two size ranges and scores close to zero for the third size range. PNCs 0.3–0.8 µm and 0.3–10 µm were given high scores for all methods and all size ranges.

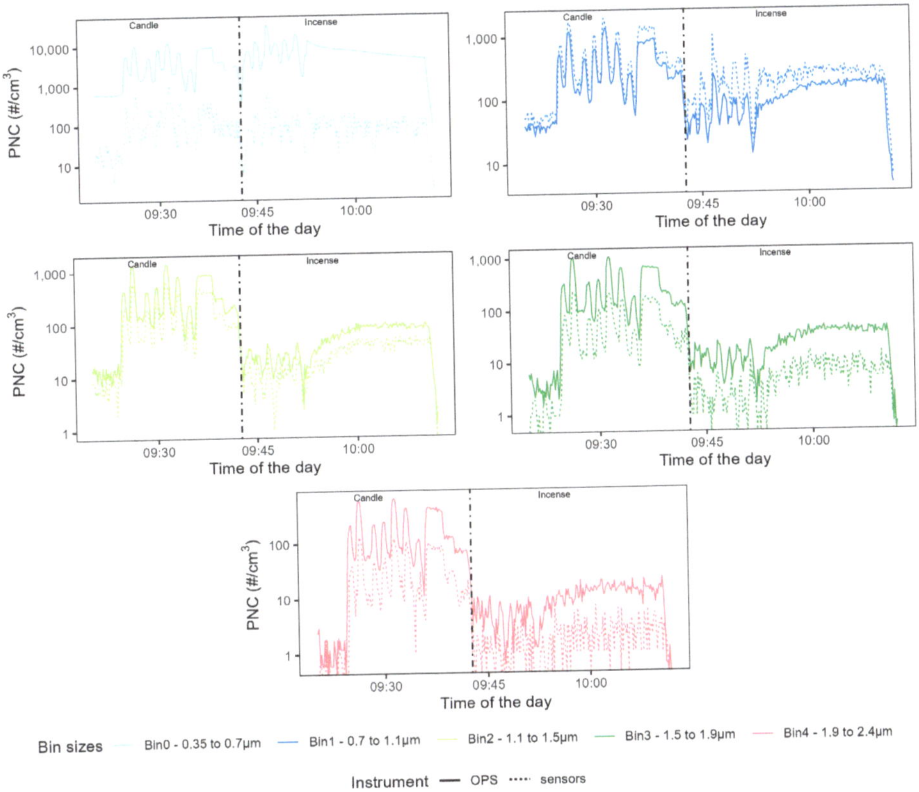

Figure 6. Time series of the Particle Number Concentration (PNC) size ranges of the Alphasense OPC-R1 compared with the size ranges computed from the OPS size ranges. On the left of the dotted line, PM was generated using a candle, on the right using incense. The different categories are, from left to right and top to bottom, Bin0 (in the range 0.4–0.7 µm), Bin1 (in the range 0.7–1.1 µm), Bin2 (in the range 1.1–1.5 µm), Bin3 (in the range 1.5–1.9 µm) and Bin4 (in the range 1.9–2.4 µm).

For the Sensirion SPS30, similarly, RH was given low scores for its three size ranges for all three methods. Source was given a low score for the first size range but obtained high scores for the second size range for Boruta and Ridge, and the highest scores for the third size range for Ridge and RFE-SVM. PNC for particles 0.01–0.3 µm was given relatively low scores on the three size ranges.

PNCs 0.3–0.8 µm and 0.3–10 µm were given the highest scores for all methods for the first size range and relatively high scores for the second size range. For the third size range, PNC 0.3–0.8 µm was given low scores for Ridge and RFE-SVM but a high score for Boruta, and PNC 0.3–10 µm was given high scores for Boruta and Ridge but a low score for RFE-SVM.

For the Alphasense OPC-R1, RH was given low scores, of zero or close-to-zero, for most size ranges and most methods except for Bin1 for Boruta and Ridge for which it was given moderate scores. Source was given the highest score for Bin2 to Bin4 for the three methods, the highest score for Bin1 for Boruta and Ridge, and a moderate score for RFE-SVM. The first size range was given lower scores for source. PNC for particles 0.01–0.3 µm was given relatively low scores for all methods for all size ranges. PNCs 0.3–0.8 µm and 0.3–10 µm were given moderate to high scores on all size ranges for the three methods considered.

Table 5. Scores of the feature selection methods for the different size ranges of the Plantower PMS5003 computed using (a) Boruta, (b) Ridge, (c) RFE-SVM. High scores denote the relevance of the variable to explain the size range. Each method computes the score differently and the values obtained should not be compared between the methods. PNC 0.01–0.3 µm is measured by the Nanotracer and the other two size fractions are measured by the TSI OPS 3330.

PMS5003	**n03_05 0.3–0.5 µm**			**n05_1 0.5–1 µm**			**n10_25 1–2.5 µm**		
Method	**(a)**	**(b)**	**(c)**	**(a)**	**(b)**	**(c)**	**(a)**	**(b)**	**(c)**
Source	18	31	13	22	38	17	6	1	0
PNC 0.01–0.3 µm	49	100	50	52	100	52	16	30	24
PNC 0.3–0.8 µm	29	75	63	27	67	59	30	95	80
PNC 0.3–10 µm	28	65	60	26	51	55	33	100	81
RH	9	0	0	10	0	0	7	0	0

Table 6. Scores of the feature selection methods for the different size ranges of the Sensirion SPS30 computed using (a) Boruta, (b) Ridge, (c) RFE-SVM. High scores denote the relevance of the variable to explain the size range. Each method computes the score differently and the values obtained should not be compared between the methods.

SPS30	**n03_05 0.3–0.5 µm**			**n05_1 0.5–1 µm**			**n1_25 1–2.5 µm**		
Method	**(a)**	**(b)**	**(c)**	**(a)**	**(b)**	**(c)**	**(a)**	**(b)**	**(c)**
Source	6	0	4	32	68	5	6	100	23
PNC 0.01–0.3 µm	17	26	22	16	0	6	17	0	5
PNC 0.3–0.8 µm	31	100	79	26	36	35	31	11	3
PNC 0.3–10 µm	33	84	79	27	100	40	33	76	5
RH	9	8	0	16	28	2	7	16	3

Table 7. Scores of the feature selection methods for the different size ranges of the Alphasense OPC-R1 computed using (a) Boruta, (b) Ridge, (c) RFE-SVM. High scores denote the relevance of the variable to explain the size range. Each method computes the score differently and the values obtained should not be compared between the methods.

OPC-R1	**Bin0 0.35–0.7 µm**			**Bin1 0.7–1.1 µm**			**Bin2 1.1–1.5 µm**			**Bin3 1.5–1.9 µm**			**Bin4 1.9–2.4 µm**		
Method	**(a)**	**(b)**	**(c)**	**(a)**	**(b)**	**(c)**	**(a)**	**(b)**	**(c)**	**(a)**	**(b)**	**(c)**	**(a)**	**(b)**	**(c)**
Source	17	22	10	35	100	16	74	100	35	69	100	36	69	100	35
PNC 0.01–0.3 µm	12	3	23	10	1	3	14	15	5	13	12	5	13	10	5
PNC 0.3–0.8 µm	35	100	98	20	0	32	23	33	15	23	34	13	21	35	13
PNC 0.3–10 µm	29	85	96	21	81	34	33	95	16	32	95	15	32	96	14
RH	5	0	0	15	46	7	8	0	0	5	0	0	4	0	0

3.4. Particle Size Distribution

Figure 7 presents the size distribution measured by the sensors and the TSI OPS 3330 during stable concentrations of candle- and incense-generated PM. The data are averaged per sensor model and are average over the five sets of experiments. The different size distributions recorded by the TSI OPS 3330 in each experiment are presented in Figure A13. To facilitate the visualisation, the data from the Sensirion SPS30 has been multiplied by 100 and the data from the Plantower PMS5003 by 10.

For the TSI OPS 3330, the incense-generated PM showed a relatively steeper decrease of PNC with increasing sizes and higher PNC with candle-generated PM for particles >0.5 µm than for incense-generated PM.

The Plantower PMS5003 reported almost the same distribution for candle- and incense-generated PM and surprisingly reported higher numbers for incense-generated PM than for candle-generated PM. For candle-generated PM, the magnitude was off by a factor 10. For incense-generated PM, the Plantower PMS5003 over-reported PNC for particles about <0.6 µm and under-reported for particles of wider diameter. The Sensirion SPS30 presented the same steeper decrease than the TSI OPS 3330 between candle- and incense-generated PM and detected fewer particles >0.7 µm for incense-generated PM than for candle-generated PM. As in the time series, the magnitude of this sensor differed by a factor of 100. For the Alphasense OPC-R1, the steeper decrease between incense- and candle-generated PM was also present. The magnitude of the first size range was much lower for the sensor than for the TSI OPS 3330 for both sources of PM but the other sizes followed each other with almost similar magnitude.

The readings of the individual sensors are available in Appendix B Figures A14–A16 and limited variability was observed between units of the same model of sensor for each size range although the Alphasense OPC-R1 demonstrated a higher variability for its first size range.

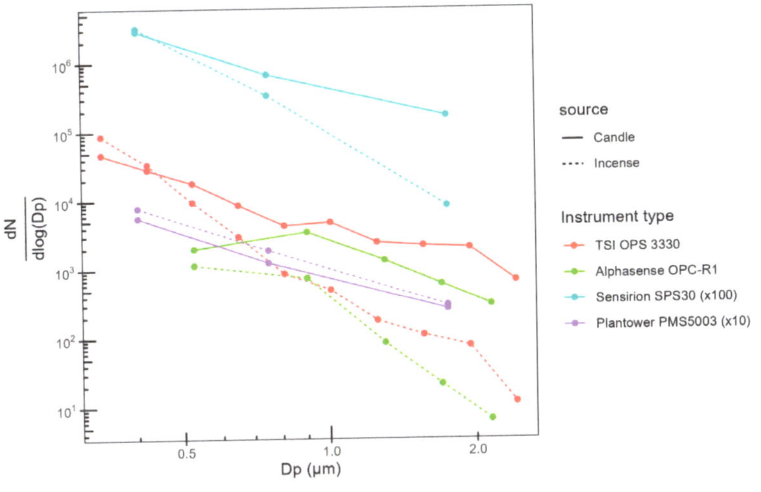

Figure 7. Particle size distribution reported by the sensors and the OPS TSI for stable concentrations of incense- and candle-generated PM, aggregated over the five experiments and per sensor model. For each sensor model, the data presented is the average per size range. The dots are on the midpoint of the size ranges. To facilitate the visualisation, the data from the Sensirion SPS30 has been multiplied by 100 and the data from the Plantower PMS5003 by 10. The dots represent the actual datapoints used to construct the plot.

4. Discussion

The particle size distribution showed that Plantower PMS5003 did not capture the difference in size distribution between candle and incense smoke. Incense smoke had clearly fewer particles >0.5 µm than candle smoke, according to the TSI OPS 3330; however, while this was not captured by the Plantower PMS5003, it was captured by the Sensirion SPS30 to a certain extent and more clearly by the Alphasense OPC-R1. The size distribution captured by the latter sensor was close to the size distribution measured by the TSI OPS 3330, except for its first size range (0.35–0.7 µm). Ouimette et al. [22] found that the Plantower PMS5003 behaved as an integrating nephelometer, reporting correctly the aerosol scattering coefficient for particles in the range 0.26–0.46 µm. Using the global database of PurpleAir sensors, they also showed that the Plantower PMS5003 obtained similar shape of size distribution in different sites outdoors around the world, this being partly attributed to the fact that aerosol scattering coefficient of outdoor PM is generally constant. Similarly,

He et al. [21] showed that the different size ranges of the Plantower PMS5003 had cut-off diameters in the range 0.1–0.7 µm, when testing the sensors with ammonium sulfate and sodium chloride. Together, these support the claim that the size distribution is computed by an algorithm rather than actually measured for each size ranges. Tryner et al. [19] studied the size segregation of the mass concentration reported by the Plantower PMS5003 and the Sensirion SPS30 in an environmental chamber, and exposed eight of each of these sensor models to stable concentrations of PM in the range 10–1000 µg/m^3 lasting around 45 min each and generated using ammonium sulfate, Arizona road dust, NIST urban PM, wood smoke and oil mist with PLS of different diameters (0.1, 0.27, 0.72 and 2 µm). They found that the Plantower PMS5003 obtained similar shape of size distribution for all the diameters of PLS and for the different sources of pollution. However, Kuula et al. [23] exposed Plantower PMS5003 and Sensirion SPS30 sensors, amongst other sensor models, to monodisperse particles of diameters in the range 0.45–9.8 µm and they showed that, while the Plantower PMS5003 misclassified the size of the particles, it was still producing two different signals, one for particles 0.3–2.5 µm and one for particles 2.5–10 µm. Similarly, Zamora et al. [24] exposed the sensor to PLS of 0.081, 0.3, 0.8, 1.1, 2.5 and 4.8 µm and while the Plantower PMS5003 misclassified and misreported the size distribution of the particle measured, it showed some differences in the size distribution it reported between the different diameters tested. These suggest that the algorithm used by the Plantower PMS5003 may include a second measurement to compute the size distribution it reports. This is further supported by: (1) the between size range correlations obtained here by the sensor; and (2) the Plantower PMS5003 differences between the scores obtained for PNC <0.3 µm for its first two size ranges and the third size range.

The Sensirion SPS30 captured some of the variations in the size distribution between incense and candle-generated PM. Tryner et al. [19] obtained two different sizes, using sensor-reported PM mass concentrations, for PLS particles of 0.1 and 0.27 µm, and of 0.72 and 2 µm, in contrast to the Plantower PMS5003. Nonetheless, in their study, the Sensirion SPS30 did not agree with the Aerodynamic Particle Sizer Spectrometer that was used as a reference instrument. Kuula et al. [23], again using sensor-reported PM mass concentrations, suggested that this sensor was able to differentiate two different size ranges, 0.3–0.9 µm and 0.7–1.3 µm, with a valid detection range for PM_1 mass concentration. The correlations obtained here between the different size ranges of the sensors support the fact that this sensor is able to differentiate two size ranges: 0.3–0.5 µm and 1–2.5 µm.

The Alphasense OPC-R1 showed clear differences between its size ranges, which generally followed the correlation obtained by the TSI OPS 3330, for both PM sources. In our 2020 study [30], which analysed the same set of experiments but focused on sensor-reported mass concentrations, the Alphasense OPC-R1 obtained lower correlation coefficients between the mass concentration of the sensors and a DustTrak DRX 8533 for incense- than for candle-generated PM. This could have several explanations: (1) the algorithm used by the Alphasense OPC-R1 to convert PNC to mass concentration is based on factors, which differ according to specific properties of the particles measured; (2) there are some differences in the measurement taken by the TSI OPS 3330 and the DustTrak DRX 8533, which is unlikely as they are based on the same technology and are made by the same manufacturer.

For the three models of sensors, the variables that impacted their readings the most were the variables linked to the particle size distribution, followed by source, with RH having less impact on the readings. This corroborates the suggestion in our earlier paper [30] that the performances of the sensors are primarily impacted by the size distribution of the particles and secondarily the source of those particles. This also means that, at RH < 79%, this variable does not need to be corrected. This would need to be verified on different sources of PM, especially with sources having a different size distribution and refractive index than candle and incense. Jayaratne et al. [54] showed that the Plantower PMS1003 and a DustTrak DRX8530 started to over-report PM mass concentration for RH > 75%. Tryner et al. [19] also found an impact of RH on the readings of the Plantower PMS5003 and the Sensirion SPS30 for RH > 75–80%. Conversely, Holder et al. [55] and Liang et al. [56], who both studied the response of the sensors to wild fire smoke, showed that it was not necessary to include RH in the correction models. Similarly, in a year-long outdoor study in a port city, Bulot et al. [12] showed only marginal improvements in models including RH compared to models not including it. Taken together, these suggest that the impact of RH is highly dependent on the composition of the aerosol being measured. This would explain why different studies conducted in different settings, in different places or countries, obtain conflicting results with regard to RH.

We found here that the accuracy of the readings of the first two size ranges of the Plantower PMS5003 were impacted by PNC <0.3 µm. This is similar to the results obtained by both He et al. [21], who developed a transfer-function based model that predicted that the sensor would output a signal for particles with diameter <0.3 µm, and by Ouimette et al. [22] whose physical model of the Plantower PMS5003 as an integrating nephelometer, based on the Mie theory, also predicted that the sensor would be able to measure particles <0.3 µm, in direct proportion to their contribution to the aerosol scattering coefficient. The Alphasense OPC-R1 and the Sensirion SPS30 readings were not impacted by the PNC in that size range. While it is quite clear that the Alphasense OPC-R1 is an Optical Particle Counter (OPC), this, along with the differences observed earlier on the sensitivity to particle size distribution between the Plantower PMS5003 and the Sensirion SPS30, may suggest that the Sensirion SPS30 is not an integrated nephelometer and/or measures and interprets the PNC differently from the Plantower PMS5003. Tryner et al. [19] also suggested that these two sensors had a different method for measuring or interpreting light-scattering data.

This experimental set-up has some limitations that must be accounted for before extrapolating its results. The sensors were only tested against two combustion sources of pollution, and environmental particles will have different physical and chemical properties (i.e., hygroscopicity, refractive index and particle size). Additionally, the range of RH was limited and the temperature was set in the range 26–29 °C which does not reflect the range of environmental conditions to which the sensors would be exposed in outdoor conditions.

As these sensors are expected to be used in large monitoring networks, for extended periods of time, it is critical to characterise their potential drift with time. This is not possible with the data collected during this study. There is conflicting evidence in the literature on this subject. Tryner et al. [19], in their laboratory experiment, simulated the ageing of the Sensirion SPS30 and the Plantower PMS5003 by exposing them to mass concentrations in the range 7300–33,000 µg/m^3 for 18 h to emulate the outdoor concentrations these sensors would encounter over a year, and showed that some performance degradation occurred for three out of eight of the Plantower PMS5003s tested but not for the Sensirion SPS30. However, it is likely that the really high concentrations to which the sensors were exposed may over-estimate their time drift. Indeed, in our previous outdoor study [57], the performances of different low-cost sensors, including the Plantower PMS5003, were evaluated over a year outdoor, and no performance degradation was found. Several studies lasting from a couple of months to a year or more found no drift over time of the sensors [24,58–60]. Wallace et al. [16] studied eight Purple Air sensors for 1.5 to 3 years indoors and outdoors, and found little evidence to support a temporal drift. Similarly, Collier-Oxandale et al. [61] conducted a three year long analysis of the performances of a network of 400 Purple Air II sensors (which use two Plantower PMS5003 each) deployed across 14 communities in the USA, and found that the performances of the sensors were mainly explained by seasonal variations but found little evidence of a temporal drift.

Although the three models of sensors were able to capture the temporal variations of the PNC, as reported by the TSI OPS 3330, the only sensor that reliably reported the particle size distribution of the aerosols was the Alphasense OPC-R1.

5. Conclusions

In this study, eight sensors of each of three models, Alphasense OPC-R1, Plantower PMS5003 and Sensirion SPS30, at a total of 24 sensors, were studied at a 10 s resolution and exposed to short-lived events of PM pollution, generated from two combustion sources having a different size distribution profile, at varying levels of RH.

The time series obtained revealed that the sensors were able to closely follow PNC variation measured by the reference TSI OPS 3330, for both sources of PM, but for the Plantower PMS5003 and the Sensirion SPS30 PNC measurements recorded were, respectively, 10 and 100 times lower than the measurements of the TSI OPS 3330. For the Sensirion SPS30, the second and third size ranges under-reported more than the first size range for incense-generated PM than for candle-generated PM. The magnitude was correct for the Alphasense OPC-R1, except for its first size range, 0.35–0.7 µm.

Regarding the independence of the size range reported by the sensors and their accuracy, the Plantower PMS5003 reported two independent signals; the Alphasense OPC-R1 reports an independent signal for each of its size ranges; the Sensirion SPS30 reported two independent signals for 0.3–0.5 µm and 1–2.5 µm. The analysis conducted suggested that the Plantower PMS5003 and the Sensirion SPS30 had a different method for measuring or interpreting light-scattering data, and reporting the determined PNC.

For capturing the particle size distribution, the Plantower PMS5003 showed no difference between incense- and candle-generated PM, while the two other sensors recorded differences that were also recorded by the TSI OPS 3330. The Alphasense OPC-R1 measured values that were close to the data reported by the TSI OPS 3330, except for its first size range (0.35–0.7 µm).

The analysis of the feature selection revealed that the sensors were more susceptible to the composition of the particles and their size distribution than to RH at the levels of humidity considered. We therefore recommend that a RH correction is not required below 75–79%. PNC in the range 0.01–0.3 µm impacted the first size range of the Plantower PMS5003 supporting the fact that this sensor is an integrating nephelometer.

For studies requiring more detailed knowledge of the particle size distribution of the aerosol measured, the Alphasense OPC-R1 should be preferred, although our previous study [30] also showed that this sensor was less suited to report PM mass concentration for the two sources of PM used here. If a more general image of the particle size distribution is sufficient, the Sensirion SPS30 should be considered. It is not clear from the results of this study whether the Plantower PMS5003 can be used in this scenario but it can be used to measure the general trends of PNC. This work shows that there is added value in directly using the PNC instead of PM mass concentration, as the size ranges provide some level of information about the particle size distribution, especially in the case of the Alphasense OPC-R1. This differential information collected by the PNC size ranges can be used to improve the calibration models developed to calibrate the sensors to standard performances and provide extra granularity with regard to source profiling.

Author Contributions: Conceptualisation, F.M.J.B., H.S.R., M.R., M.S.J., S.J.O., A.K.R.M., P.J.B., N.H.C.E., H.L.M., G.L.F., M.L. and S.J.C.; Data curation, F.M.J.B.; Formal analysis, F.M.J.B. and M.L.; Funding acquisition, F.M.J.B., M.S.J., S.J.O., A.K.R.M., G.L.F., M.L. and S.J.C.; Investigation, F.M.J.B., H.S.R., M.R. and M.S.J.; Methodology, F.M.J.B., H.S.R., M.R., M.S.J., S.J.O., A.K.R.M., P.J.B., N.H.C.E., H.L.M., G.L.F., M.L. and S.J.C.; Project administration, M.S.J.; Resources, F.M.J.B., H.S.R., M.R., M.S.J. and A.K.R.M.; Software, F.M.J.B., S.J.O. and P.J.B.; Supervision, M.S.J., S.J.O., A.K.R.M., G.L.F., M.L. and S.J.C.; Visualisation, F.M.J.B.; Writing—original draft, F.M.J.B.; Writing—review and editing, F.M.J.B., H.S.R., M.R., M.S.J., S.J.O., A.K.R.M., P.J.B., N.H.C.E., H.L.M., G.L.F., M.L. and S.J.C. All authors have read and agreed to the published version of the manuscript.

Funding: This research was funded by the Next Generation of Unmanned Systems Centre for Doctoral Training supported by the Natural Environmental Research Council grant number [NE/N012070/1]; the Leverhulme Trust through the Southampton Marine and Maritime Institute; Engineering and Physical Sciences Research Council UK grant [EP/T517859/1]. Matthew Loxham is supported by a BBSRC David Phillips Fellowship [BB/V004573/1] and a NIHR Southampton Biomedical Research Centre Senior Fellowship. Hugo S. Russell was supported by Airscape, Aarhus University Graduate School of Science and Technology (GSST) and BERTHA—the Danish Big Data Centre for Environment and Health funded by the Novo Nordisk Foundation Challenge Programme (grant

NNF17OC0027864). The test chamber at the University of Copenhagen is supported by ACTRIS-DK. The APC was funded by the Engineering and Physical Sciences Research Council.

Institutional Review Board Statement: Not applicable.

Informed Consent Statement: Not applicable.

Data Availability Statement: The underlying dataset is openly available at https://doi.org/10.5281/zenodo.7808620, and the code used for the analysis and to generate the tables and graphs of this study is openly available at https://doi.org/10.5281/zenodo.7808794.

Conflicts of Interest: M.S.J. and H.S.R. receive part of their salary from Airscape, London, UK, which builds and sells low cost sensor nodes. No Airscape equipment was used in this study.

Appendix A. Time Series

Appendix A.1. RH = 54%

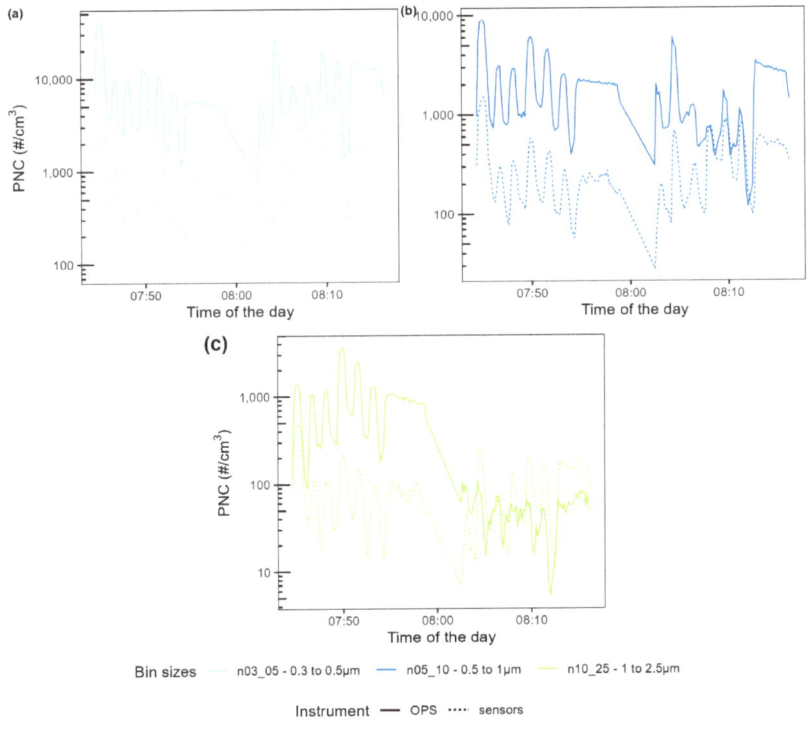

Figure A1. Time series of the Particle Number Concentration (PNC) size ranges of the Plantower PMS5003 compared with the size ranges computed from the OPS size ranges for RH = 54%. The different categories are (**a**) n03_05 (in the range 0.3–0.5 μm), (**b**) n05_10 (in the range 0.5–1 μm) and (**c**) n10_25 (in the range 1–2.5 μm).

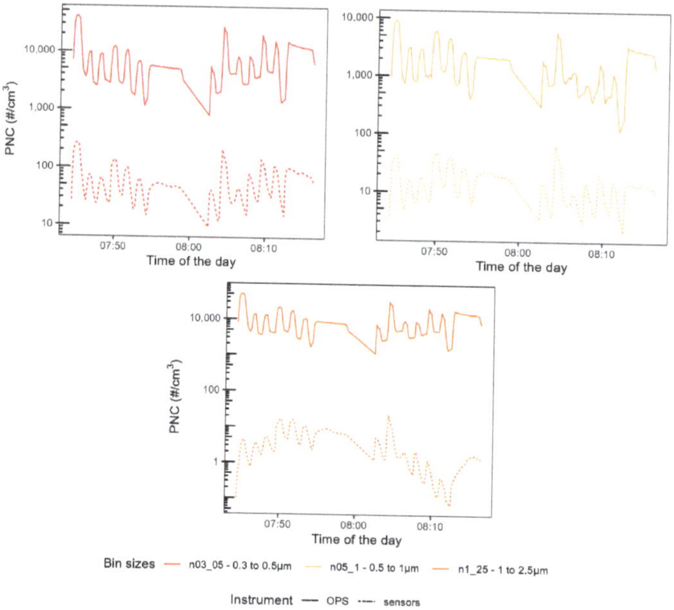

Figure A2. Time series of the Particle Number Concentration (PNC) size ranges of the Sensirion SPS30 compared with the size ranges computed from the OPS size ranges for RH = 54%. The different categories are n03_05 (in the range 0.3–0.5 µm), n05_1 (in the range 0.5–1 µm) and n1_25 (in the range 1–2.5 µm).

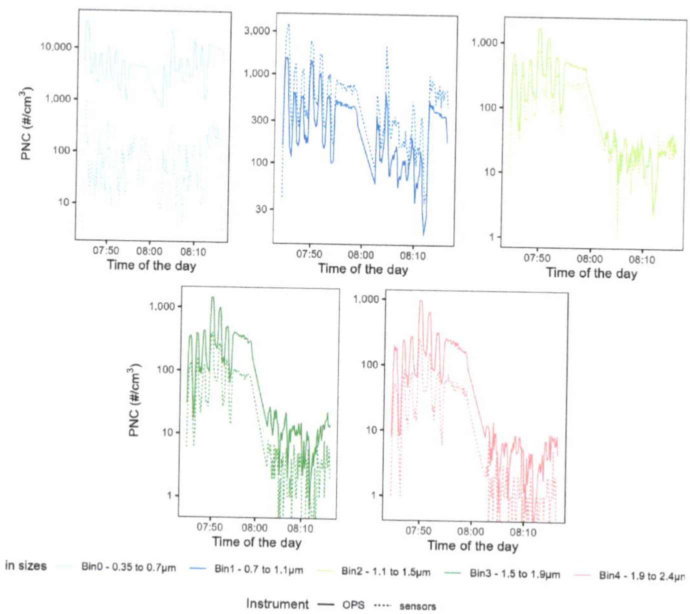

Figure A3. Time series of the Particle Number Concentration (PNC) size ranges of the Alphasense OPC-R1 compared with the size ranges computed from the OPS size ranges for RH = 54%. The different categories are Bin0 (in the range 0.4–0.7 µm), Bin1 (in the range 0.7–1.1 µm), Bin2 (in the range 1.1–1.5 µm), Bin3 (in the range 1.5–1.9 µm and Bin4 (in the range 1.9–2.4 µm).

Appendix A.2. RH = 72%

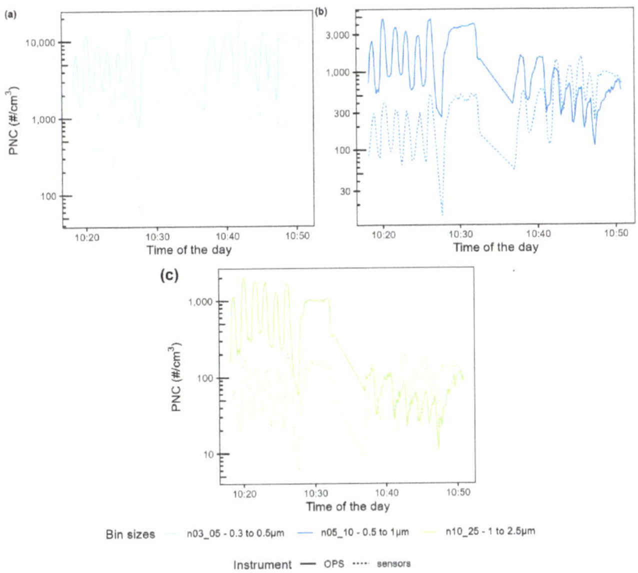

Figure A4. Time series of the Particle Number Concentration (PNC) size ranges of the Plantower PMS5003 compared with the size ranges computed from the OPS size ranges for RH = 72%. The different categories are (**a**) n03_05 (in the range 0.3–0.5 µm), (**b**) n05_10 (in the range 0.5–1 µm) and (**c**) n10_25 (in the range 1–2.5 µm).

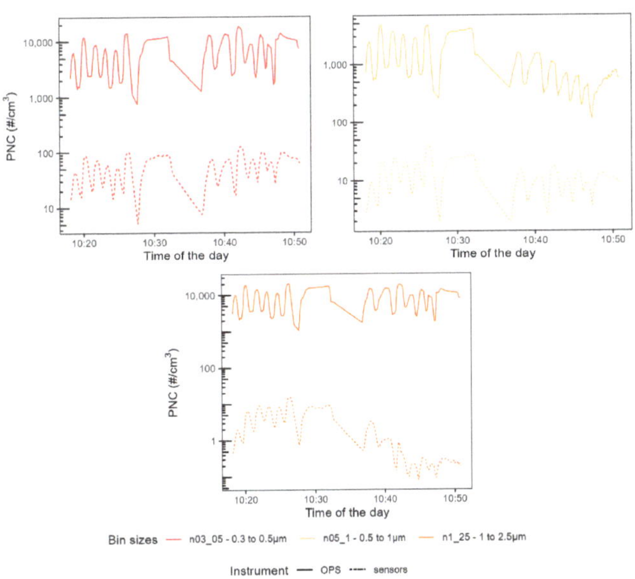

Figure A5. Time series of the Particle Number Concentration (PNC) size ranges of the Sensirion SPS30 compared with the size ranges computed from the OPS size ranges for RH = 72%. The different categories are n03_05 (in the range 0.3–0.5 µm), n05_1 (in the range 0.5–1 µm) and n1_25 (in the range 1–2.5 µm).

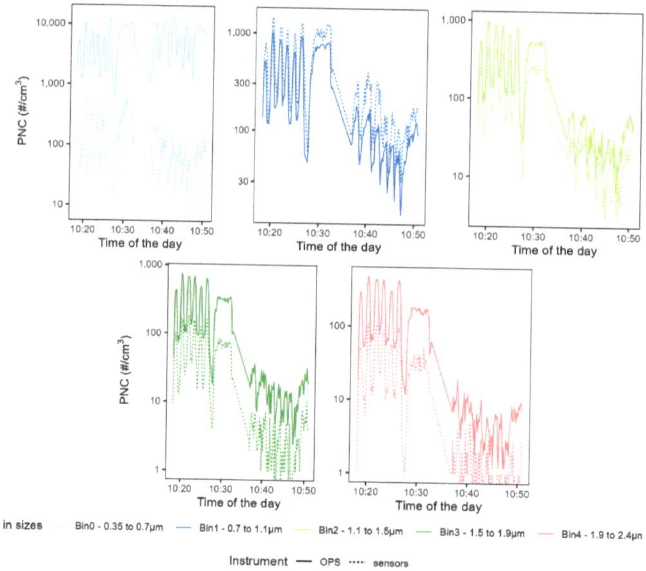

Figure A6. Time series of the Particle Number Concentration (PNC) size ranges of the Alphasense OPC-R1 compared with the size ranges computed from the OPS size ranges for RH = 72%. The different categories are Bin0 (in the range 0.4–0.7 µm), Bin1 (in the range 0.7–1.1 µm), Bin2 (in the range 1.1–1.5 µm), Bin3 (in the range 1.5–1.9 µm and Bin4 (in the range 1.9–2.4 µm).

Appendix A.3. RH = 76%

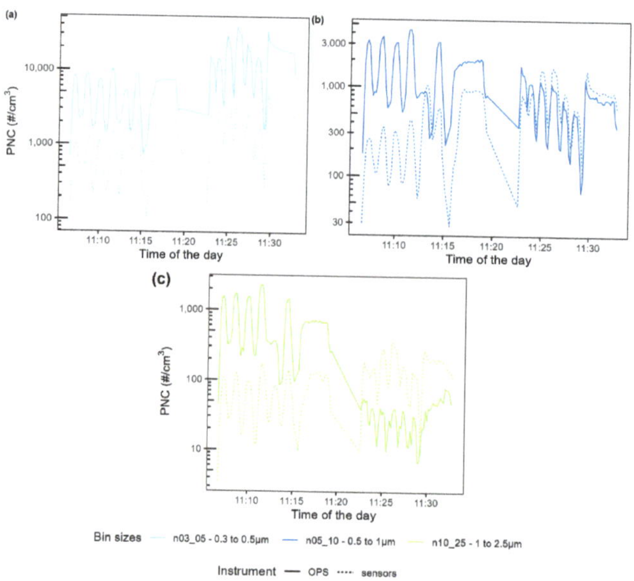

Figure A7. Time series of the Particle Number Concentration (PNC) size ranges of the Plantower PMS5003 compared with the size ranges computed from the OPS size ranges for RH = 76%. The different categories are (**a**) n03_05 (in the range 0.3–0.5 µm), (**b**) n05_10 (in the range 0.5–1 µm) and (**c**) n10_25 (in the range 1–2.5 µm).

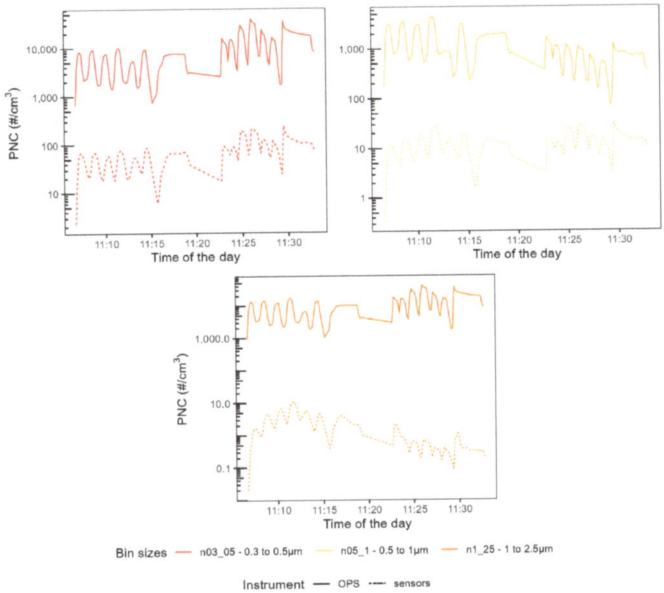

Figure A8. Time series of the Particle Number Concentration (PNC) size ranges of the Sensirion SPS30 compared with the size ranges computed from the OPS size ranges for RH = 76%. The different categories are n03_05 (in the range 0.3–0.5 µm), n05_1 (in the range 0.5–1 µm) and n1_25 (in the range 1–2.5 µm).

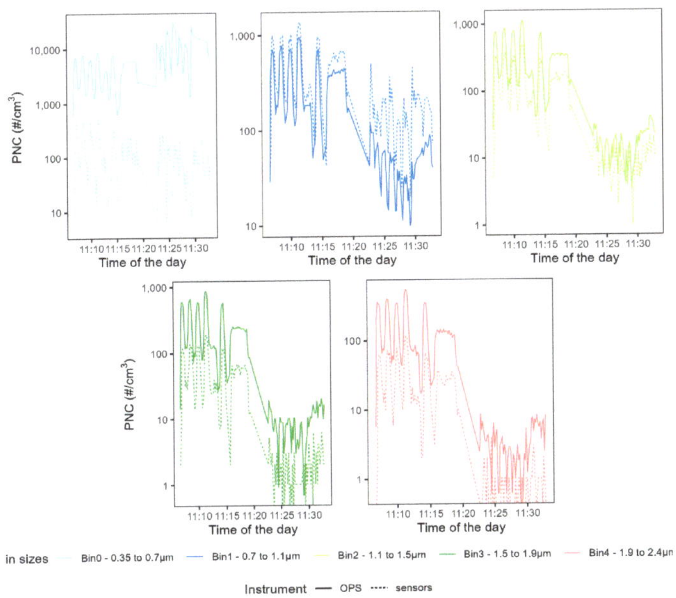

Figure A9. Time series of the Particle Number Concentration (PNC) size ranges of the Alphasense OPC-R1 compared with the size ranges computed from the OPS size ranges for RH = 76%. The different categories are Bin0 (in the range 0.4–0.7 µm), Bin1 (in the range 0.7–1.1 µm), Bin2 (in the range 1.1–1.5 µm), Bin3 (in the range 1.5–1.9 µm and Bin4 (in the range 1.9–2.4 µm).

Appendix A.4. RH = 79%

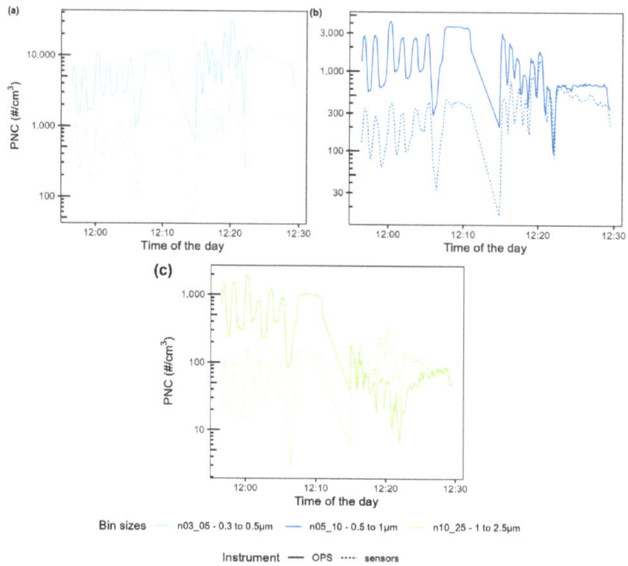

Figure A10. Time series of the Particle Number Concentration (PNC) size ranges of the Plantower PMS5003 compared with the size ranges computed from the OPS size ranges for RH = 79%. The different categories are (**a**) n03_05 (in the range 0.3–0.5 µm), (**b**) n05_10 (in the range 0.5–1 µm) and (**c**) n10_25 (in the range 1–2.5 µm).

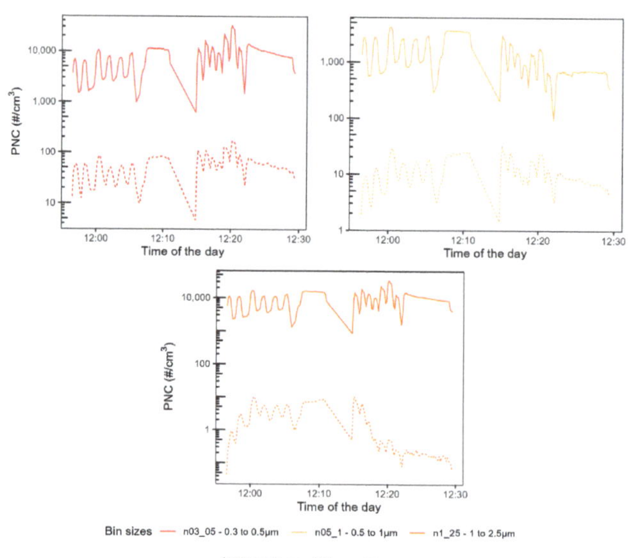

Figure A11. Time series of the Particle Number Concentration (PNC) size ranges of the Sensirion SPS30 compared with the size ranges computed from the OPS size ranges for RH = 79%. The different categories are n03_05 (in the range 0.3–0.5 µm), n05_1 (in the range 0.5–1 µm) and n1_25 (in the range 1–2.5 µm).

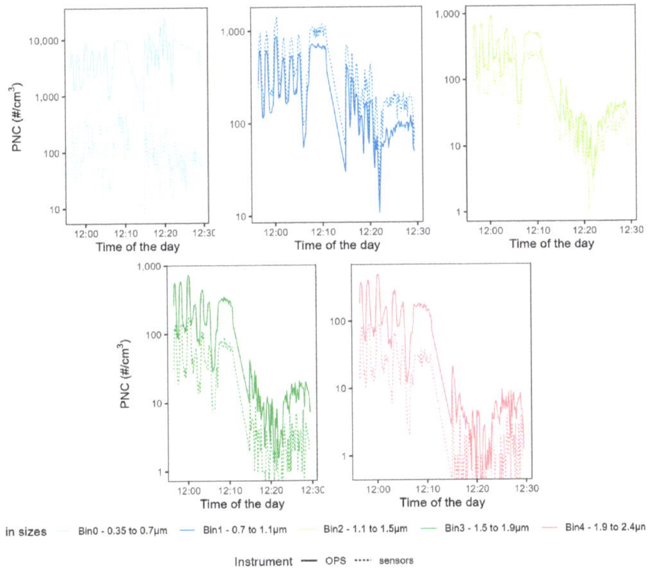

Figure A12. Time series of the Particle Number Concentration (PNC) size ranges of the Alphasense OPC-R1 compared with the size ranges computed from the OPS size ranges for RH = 79%. The different categories are Bin0 (in the range 0.4–0.7 µm), Bin1 (in the range 0.7–1.1 µm), Bin2 (in the range 1.1–1.5 µm), Bin3 (in the range 1.5–1.9 µm and Bin4 (in the range 1.9–2.4 µm).

Appendix B. Particle Size Distribution

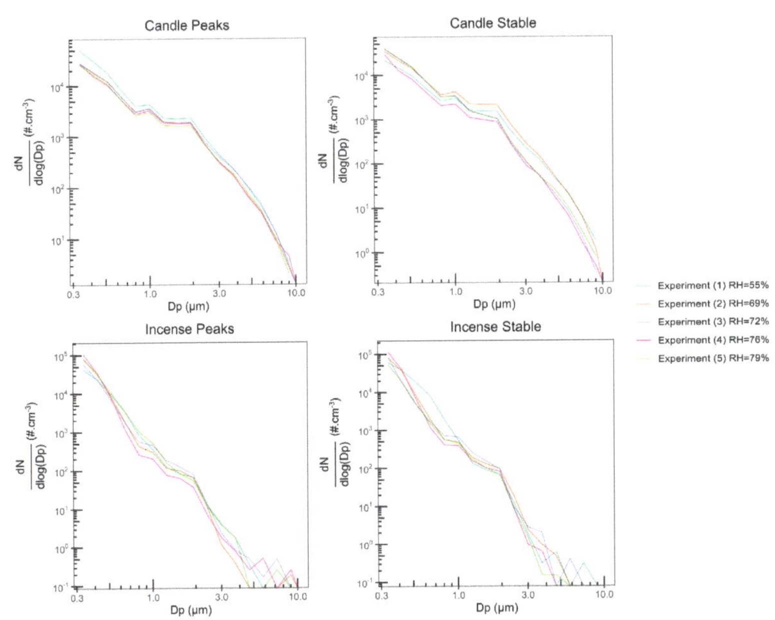

Figure A13. Size distribution measured from the OPS during the different experiments for peaks and stable concentrations of incense and candle smoke. The axes are in a logarithmic scale. N is the number of particles in # 1/cm; D_p is the mean diameter of the particles. Adapted from Bulot et al. [30].

Particle Size Distribution per Sensor

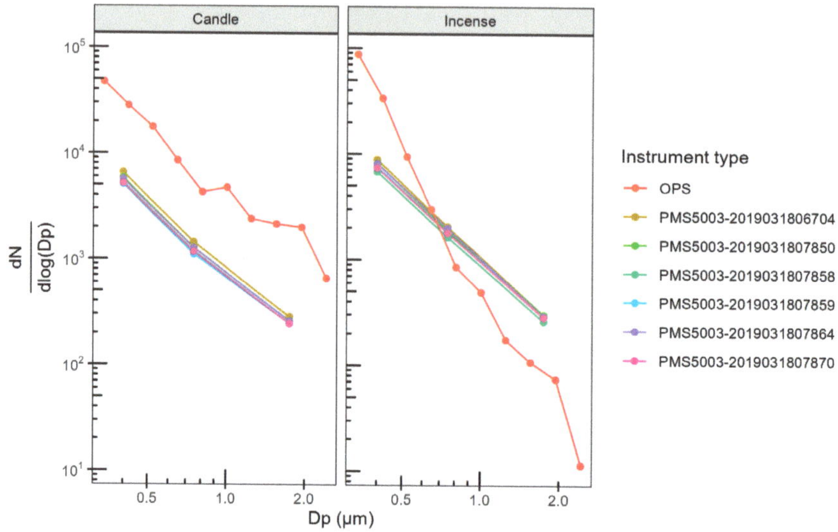

Figure A14. Particle size distribution reported by the PMS5003 and the OPS TSI for stable concentrations of incense and candle. The dots represent the actual datapoints used. To facilitate the visualisation, the data from the Plantower PMS5003 were multiplied by a factor of 10.

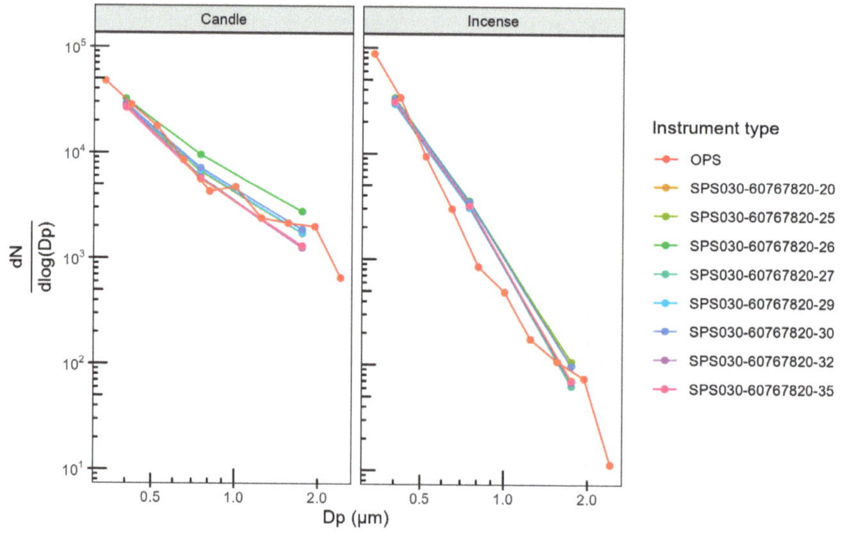

Figure A15. Particle size distribution reported by the SPS30 and the OPS TSI for stable concentrations of incense and candle. The dots represent the actual datapoints used. To facilitate the visualisation, the data from the Sensirion SPS30 were multiplied by a factor of 100.

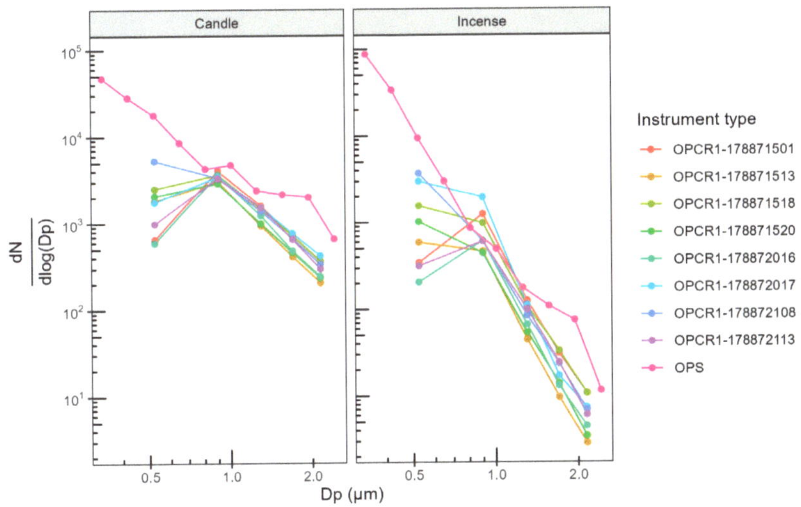

Figure A16. Particle size distribution reported by the OPCR1 and the OPS TSI for stable concentrations of incense and candle. The dots represent the actual datapoints used.

References

1. Landrigan, P.J.; Fuller, R.; Acosta, N.J.; Adeyi, O.; Arnold, R.; Basu, N.; Baldé, A.B.; Bertollini, R.; Bose-O'Reilly, S.; Boufford, J.I.; et al. The Lancet Commission on pollution and health. *Lancet* **2017**, *391*, 462–512. [CrossRef]
2. Burnett, R.; Chen, H.; Fann, N.; Hubbell, B.; Pope, C.A.; Frostad, J.; Lim, S.S.; Kan, H.; Walker, K.D.; Thurston, G.D.; et al. Global estimates of mortality associated with long- term exposure to outdoor fine particulate matter. *Proc. Natl. Acad. Sci. USA* **2018**, *115*, 9592–9597. [CrossRef] [PubMed]
3. Kim, E.; Hopke, P.K.; Pinto, J.P.; Wilson, W.E. Spatial variability of fine particle mass, components, and source contributions during the Regional Air Pollution Study in St. Louis. *Environ. Sci. Technol.* **2005**, *39*, 4172–4179. [CrossRef]
4. WHO. *WHO Global Air Quality Guidelines: Particulate Matter (PM2.5 and PM10), Ozone, Nitrogen Dioxide, Sulfur Dioxide and Carbon Monoxide*; World Health Organization: Geneva, Switzerland, 2021.
5. Frederickson, L.B.; Sidaraviciute, R.; Schmidt, J.A.; Hertel, O.; Johnson, M.S. Are dense networks of low-cost nodes really useful for monitoring air pollution? A case study in Staffordshire. *Atmos. Chem. Phys.* **2022**, *22*, 13949–13965. [CrossRef]
6. Questions and Answers on New Air Quality Rules. Available online: https://ec.europa.eu/commission/presscorner/detail/en/qanda_22_6348 (accessed on 5 April 2023).
7. Kuula, J.; Timonen, H.; Niemi, J.V.; Manninen, H.E.; Rönkkö, T.; Hussein, T.; Fung, P.L.; Tarkoma, S.; Laakso, M.; Saukko, E.; et al. Opinion: Insights into updating Ambient Air Quality Directive 2008/50/EC. *Atmos. Chem. Phys.* **2022**, *22*, 4801–4808. [CrossRef]
8. Bulot, F. Systematic Studies of Commodity Particulate Matter Air Pollution Sensors. Ph.D. Thesis, University of Southampton, Southampton, UK, 2022.
9. US EPA. How Science Is Protecting the Air We Breathe. 2020. Available online: https://www.epa.gov/sciencematters/how-science-protecting-air-we-breathe (accessed on 22 July 2023).
10. Kim, K.H.; Kabir, E.; Kabir, S. A review on the human health impact of airborne particulate matter. *Environ. Int.* **2015**, *74*, 136–143. [CrossRef] [PubMed]
11. Hagan, D.H.; Gani, S.; Bhandari, S.; Patel, K.; Habib, G.; Apte, J.S.; Hildebrandt Ruiz, L.; Kroll, J.H. Inferring Aerosol Sources from Low-Cost Air Quality Sensor Measurements: A Case Study in Delhi, India. *Environ. Sci. Technol. Lett.* **2019**, *6*, 467–472. [CrossRef]
12. Bulot, F.M.J.; Ossont, S.J.; Morris, A.K.R.; Basford, P.J.; Easton, N.H.C.; Mitchell, H.L.; Foster, G.L.; Cox, S.J.; Loxham, M. Characterisation and calibration of low-cost PM sensors at high temporal resolution to reference-grade performance. *Heliyon* **2023**, *9*, e15943. [CrossRef]
13. Wallace, L.; Bi, J.; Ott, W.R.; Sarnat, J.; Liu, Y. Calibration of low-cost PurpleAir outdoor monitors using an improved method of calculating PM2.5. *Atmos. Environ.* **2021**, *256*, 118432. [CrossRef]
14. Wallace, L.; Ott, W.; Zhao, T.; Cheng, K.C.; Hildemann, L. Secondhand exposure from vaping marijuana: Concentrations, emissions, and exposures determined using both research-grade and low-cost monitors. *Atmos. Environ. X* **2020**, *8*, 100093. [CrossRef]
15. Wallace, L.; Zhao, T.; Klepeis, N.E. Calibration of PurpleAir PA-I and PA-II Monitors Using Daily Mean PM2.5 Concentrations Measured in California, Washington, and Oregon from 2017 to 2021. *Sensors* **2022**, *22*, 4741. [CrossRef]

16. Wallace, L. Intercomparison of PurpleAir Sensor Performance over Three Years Indoors and Outdoors at a Home: Bias, Precision, and Limit of Detection Using an Improved Algorithm for Calculating PM2.5. *Sensors* **2022**, *22*, 2755. [CrossRef] [PubMed]
17. Morawska, L.; Thai, P.K.; Liu, X.; Asumadu-Sakyi, A.; Ayoko, G.; Bartonova, A.; Bedini, A.; Chai, F.; Christensen, B.; Dunbabin, M.; et al. Applications of low-cost sensing technologies for air quality monitoring and exposure assessment: How far have they gone? *Environ. Int.* **2018**, *116*, 286–299. [CrossRef]
18. Hagan, D.H.; Kroll, J.H. Assessing the accuracy of low-cost optical particle sensors using a physics-based approach. *Atmos. Meas. Tech.* **2020**, *13*, 6343–6355. [CrossRef]
19. Tryner, J.; Mehaffy, J.; Miller-Lionberg, D.; Volckens, J. Effects of aerosol type and simulated aging on performance of low-cost PM sensors. *J. Aerosol Sci.* **2020**, *150*, 105654. [CrossRef]
20. Alfano, B.; Barretta, L.; Giudice, A.D.; De Vito, S.; Francia, G.D.; Esposito, E.; Formisano, F.; Massera, E.; Miglietta, M.L.; Polichetti, T. A review of low-cost particulate matter sensors from the developers' perspectives. *Sensors* **2020**, *20*, 6819. [CrossRef] [PubMed]
21. He, M.; Kuerbanjiang, N.; Dhaniyala, S. Performance characteristics of the low-cost Plantower PMS optical sensor. *Aerosol Sci. Technol.* **2020**, *54*, 232–241. [CrossRef]
22. Ouimette, J.R.; Malm, W.C.; Schichtel, B.A.; Sheridan, P.J.; Andrews, E.; Ogren, J.A.; Arnott, W.P. Evaluating the PurpleAir monitor as an aerosol light scattering instrument. *Atmos. Meas. Tech.* **2022**, *15*, 655–676. [CrossRef]
23. Kuula, J.; Mäkelä, T.; Aurela, M.; Teinilä, K.; Varjonen, S.; González, Ó.; Timonen, H. Laboratory evaluation of particle-size selectivity of optical low-cost particulate matter sensors. *Atmos. Meas. Tech.* **2020**, *13*, 2413–2423. [CrossRef]
24. Levy Zamora, M.; Xiong, F.; Gentner, D.; Kerkez, B.; Kohrman-Glaser, J.; Koehler, K. Field and Laboratory Evaluations of the Low-Cost Plantower Particulate Matter Sensor. *Environ. Sci. Technol.* **2019**, *53*, 838–849. [CrossRef]
25. Saarela, M.; Jauhiainen, S. Comparison of feature importance measures as explanations for classification models. *SN Appl. Sci.* **2021**, *3*, 272. [CrossRef]
26. Bhatt, U.; Xiang, A.; Sharma, S.; Weller, A.; Taly, A.; Jia, Y.; Ghosh, J.; Puri, R.; Moura, J.M.F.; Eckersley, P. Explainable Machine Learning in Deployment. *arXiv* **2020**. arXiv:1909.06342. [CrossRef]
27. Khaire, U.M.; Dhanalakshmi, R. Stability of feature selection algorithm: A review. *J. King Saud Univ.-Comput. Inf. Sci.* **2022**, *34*, 1060–1073. [CrossRef]
28. Venkatesh, B.; Anuradha, J. A Review of Feature Selection and Its Methods. *Cybern. Inf. Technol.* **2019**, *19*, 3–26. [CrossRef]
29. Li, J.; Cheng, K.; Wang, S.; Morstatter, F.; Trevino, R.P.; Tang, J.; Liu, H. Feature Selection: A Data Perspective. *ACM Comput. Surv.* **2018**, *50*, 1–45. [CrossRef]
30. Bulot, F.M.J.; Russell, H.S.; Rezaei, M.; Johnson, M.S.; Johnston Ossont, S.J.; Morris, A.K.R.; Basford, P.J.; Easton, N.H.C.; Foster, G.L.; Loxham, M.; et al. Laboratory comparison of low-cost particulate matter sensors to measure transient events of pollution. *Sensors* **2020**, *20*, 2219. [CrossRef] [PubMed]
31. Cowell, N.; Chapman, L.; Bloss, W.; Srivastava, D.; Bartington, S.; Singh, A. Particulate matter in a lockdown home: Evaluation, calibration, results and health risk from an IoT enabled low-cost sensor network for residential air quality monitoring. *Environ. Sci. Atmos.* **2023**, *3*, 65–84. [CrossRef]
32. Gupta, P.; Doraiswamy, P.; Levy, R.; Pikelnaya, O.; Maibach, J.; Feenstra, B.; Polidori, A.; Kiros, F.; Mills, K.C. Impact of California Fires on Local and Regional Air Quality: The Role of a Low-Cost Sensor Network and Satellite Observations. *GeoHealth* **2018**, *2*, 172–181. [CrossRef]
33. Gressent, A.; Malherbe, L.; Colette, A.; Rollin, H.; Scimia, R. Data fusion for air quality mapping using low-cost sensor observations: Feasibility and added-value. *Environ. Int.* **2020**, *143*, 105965. [CrossRef]
34. Johnston, S.J.; Basford, P.J.; Bulot, F.M.J.; Apetroaie-Cristea, M.; Easton, N.H.C.; Davenport, C.; Foster, G.L.; Loxham, M.; Morris, A.K.R.; Cox, S.J. City scale particulate matter monitoring using LoRaWAN based air quality IoT devices. *Sensors* **2019**, *19*, 209. [CrossRef]
35. Bulot, F.M.J.; Basford, P.J.; Johnston, S.J.; Cox, S.J. FEEprojects/sensirion-sps030: Fix and Improve Test.py. 2019. [CrossRef]
36. Bulot, F.M.J.; Basford, P.J.; Johnston, S.J.; Cox, S.J. FEEprojects/Plantower: Timestamp Now Included. 2019. [CrossRef]
37. Hagan, D.H.; Bulot, F.; Tolmie, A.; Rach8cww; Will; Badger, T.G.; Trochim, J.; Smith, A. FlorentinBulotUoS/py-opc: Implementation of the OPC-R1. 2019. [CrossRef]
38. Sensirion. Datasheet SHT3x-DIS Humidity and Temperature Sensor. Available online: https://sensirion.com/media/documents/213E6A3B/63A5A569/Datasheet_SHT3x_DIS.pdf (accessed on 2 February 2023).
39. Sayahi, T.; Butterfield, A.; Kelly, K.E. Long-term field evaluation of the Plantower PMS low-cost particulate matter sensors. *Environ. Pollut.* **2019**, *245*, 932–940. [CrossRef]
40. Sensirion. Datasheet SPS30—Particulate Matter Sensor for Air Quality Monitoring and Control. Available online: https://sensirion.com/media/documents/73EED62F/61812A35/SE_Flyer_SPS3x_EN_Web_210621.pdf (accessed on 2 February 2023).
41. Alphasense User Manual OPC-R1 Optical Particle Counter. 2019. Available online: https://kolegite.com/EE_library/datasheets_and_manuals/sensors/OPC/072-0500_OPC-R1_manual_issue_1_250219.pdf (accessed on 2 February 2023).
42. Han, H.S.; Sreenath, A.; Birkeland, N.T.; Chancellor, G.J. Performance of a High Resolution Optical Particle Spectrometer. In Proceedings of the EAC2011, Manchester, UK, 4–9 September 2011.
43. Klosterköther, A.; Kurtenbach, R.; Wiesen, P.; Kleffmann, J. Determination of the emission indices for NO, NO2, HONO, HCHO, CO, and particles emitted from candles. *Indoor Air* **2021**, *31*, 116–127. [CrossRef]

44. Zai, S.; Zhen, H.; Jia-song, W. Studies on the size distribution, number and mass emission factors of candle particles characterized by modes of burning. *J. Aerosol Sci.* **2006**, *37*, 1484–1496. [CrossRef]
45. Li, W.; Hopke, P.K. Initial size distributions and hygroscopicity of indoor combustion aerosol particles. *Aerosol Sci. Technol.* **1993**, *19*, 305–316. [CrossRef]
46. Hastie, T.; Tibshirani, R.; Friedman, J. *The Elements of Statistical Learning Data Mining, Inference, and Prediction*, 2nd ed.; Springer Series in Statistics; Springer: New York, NY, USA, 2009.
47. Kursa, M.B.; Rudnicki, W.R. Feature Selection with the Boruta Package. *J. Stat. Softw.* **2010**, *36*, 1–13. [CrossRef]
48. Guyon, I.; Weston, J.; Barnhill, S.; Vapnik, V. Gene Selection for Cancer Classification using Support Vector Machines. *Mach. Learn.* **2002**, *46*, 389–422. [CrossRef]
49. Di Antonio, A.; Popoola, O.A.; Ouyang, B.; Saffell, J.; Jones, R.L. Developing a relative humidity correction for low-cost sensors measuring ambient particulate matter. *Sensors* **2018**, *18*, 2790. [CrossRef]
50. R Core Team. *R: A Language and Environment for Statistical Computing*; R Foundation for Statistical Computing: Vienna, Austria, 2021.
51. Kuhn, M. Caret: Classification and Regression Training; R Package Version 6.0-93. Available online: https://CRAN.R-project.org/package=caret (accessed on 22 July 2023)
52. Karatzoglou, A.; Smola, A.; Hornik, K.; Zeileis, A. kernlab—An S4 Package for Kernel Methods in R. *J. Stat. Softw.* **2004**, *11*, 1–20. [CrossRef]
53. Friedman, J.; Hastie, T.; Tibshirani, R. Regularization Paths for Generalized Linear Models via Coordinate Descent. *J. Stat. Softw.* **2010**, *33*, 1–22. [CrossRef]
54. Jayaratne, R.; Liu, X.; Thai, P.; Dunbabin, M.; Morawska, L. The influence of humidity on the performance of a low-cost air particle mass sensor and the effect of atmospheric fog. *Atmos. Meas. Tech.* **2018**, *11*, 4883–4890. [CrossRef]
55. Holder, A.L.; Mebust, A.K.; Maghran, L.A.; McGown, M.R.; Stewart, K.E.; Vallano, D.M.; Elleman, R.A.; Baker, K.R. Field Evaluation of Low-Cost Particulate Matter Sensors for Measuring Wildfire Smoke. *Sensors* **2020**, *20*, 4796. [CrossRef]
56. Liang, Y.; Sengupta, D.; Campmier, M.J.; Lunderberg, D.M.; Apte, J.S.; Goldstein, A.H. Wildfire smoke impacts on indoor air quality assessed using crowdsourced data in California. *Proc. Natl. Acad. Sci. USA* **2021**, *118*, e2106478118. [CrossRef]
57. Bulot, F.M.J.; Johnston, S.J.; Basford, P.J.; Easton, N.H.C.; Cox, S.J.; Foster, G.L.; Morris, A.K.R.; Apetroia-Cristea, M.; Loxham, M. Long-term field comparison of the performances of multiple low-cost particulate matter sensors in an urban area. *Sci. Rep.* **2019**, *9*, 7497. [CrossRef] [PubMed]
58. Zusman, M.; Schumacher, C.S.; Gassett, A.J.; Spalt, E.W.; Austin, E.; Larson, T.V.; Carvlin, G.; Seto, E.; Kaufman, J.D.; Sheppard, L. Calibration of low-cost particulate matter sensors: Model development for a multi-city epidemiological study. *Environ. Int.* **2020**, *134*, 105329. [CrossRef] [PubMed]
59. Crilley, L.R.; Shaw, M.; Pound, R.; Kramer, L.J.; Price, R.; Young, S.; Lewis, A.C.; Pope, F.D. Evaluation of a low-cost optical particle counter (Alphasense OPC-N2) for ambient air monitoring. *Atmos. Meas. Tech.* **2018**, *11*, 709–720. [CrossRef]
60. Masic, A.; Bibic, D.; Pikula, B.; Blazevic, A.; Huremovic, J.; Zero, S. Evaluation of optical particulate matter sensors under realistic conditions of strong and mild urban pollution. *Atmos. Meas. Tech.* **2020**, *13*, 6427–6443. [CrossRef]
61. Collier-Oxandale, A.; Feenstra, B.; Papapostolou, V.; Polidori, A. AirSensor v1.0: Enhancements to the open-source R package to enable deep understanding of the long-term performance and reliability of PurpleAir sensors. *Environ. Model. Softw.* **2021**, *148*, 105256. [CrossRef]

sensors

MDPI

Article

ArtEMon: Artificial Intelligence and Internet of Things Powered Greenhouse Gas Sensing for Real-Time Emissions Monitoring

Ali Yavari [1,2,*], Irfan Baig Mirza [1], Hamid Bagha [3], Harindu Korala [4], Hussein Dia [5], Paul Scifleet [6], Jason Sargent [6], Caroline Tjung [7] and Mahnaz Shafiei [1,*]

1 School of Science, Computing and Engineering Technologies, Swinburne University of Technology, Melbourne, VIC 3122, Australia; imirza@swin.edu.au
2 6G Research and Innovation Lab, Swinburne University of Technology, Melbourne, VIC 3122, Australia
3 Department of Infrastructure Engineering, University of Melbourne, Melbourne, VIC 3010, Australia; baghah@unimelb.edu.au
4 Institute of Railway Technology, Monash University, Melbourne, VIC 3800, Australia; harindu.korala@monash.edu
5 Department of Civil and Construction Engineering, Swinburne University of Technology, Melbourne, VIC 3122, Australia; hdia@swin.edu.au
6 School of Business, Law and Entrepreneurship, Swinburne University of Technology, Melbourne, VIC 3122, Australia; jpsargent@swin.edu.au (J.S.)
7 School of Design and Architecture, Swinburne University of Technology, Melbourne, VIC 3122, Australia; ctjung@swin.edu.au
* Correspondence: mail@aliyavari.com (A.Y.); mshafiei@swin.edu.au (M.S.)

Abstract: Greenhouse gas (GHG) emissions reporting and sustainability are increasingly important for businesses around the world. Yet the lack of a single standardised method of measurement, when coupled with an inability to understand the true state of emissions in complex logistics activities, presents enormous barriers for businesses to understanding the extent of their emissions footprint. One of the traditional approaches to accurately capturing and monitoring gas emissions in logistics is through using gas sensors. However, connecting, maintaining, and operating gas sensors on moving vehicles in different road and weather conditions is a large and costly challenge. This paper presents the development and evaluation of a reliable and accurate sensing technique for GHG emissions collection (or monitoring) in real-time, employing the Internet of Things (IoT) and Artificial Intelligence (AI) to eliminate or reduce the usage of gas sensors, using reliable and cost-effective solutions.

Keywords: IoT; greenhouse gas; sustainable logistics; emissions; supply chain; AI

1. Introduction

Transport activity accounts for around one fifth of global carbon dioxide (CO_2) emissions. Freight road activity accounts for around 29.4% of all transport emissions [1]. Demand for freight is expected to triple by 2050 compared to 2015, according to the International Transport Forum [2], fuelled by global supply chains, burgeoning economies in the developing world, and a rise in e-commerce activities.

In Australia, transport is the third-largest source of local greenhouse gases, accounting for 18.7% of all national emissions in 2022 [3]. The transport sector is also one of the strongest contributors to emissions growth in Australia. Emissions from transport have increased by nearly 60% since 1990—more than any other sector in the economy [3]. Within transport, road freight and supply chains present challenges to successful emissions reduction. Freight transport is a significant contributor to the sector's emissions, and it is expected to grow as a proportion of total emissions [3].

Transport is facing growing pressure from regulators, financiers, and consumers to accelerate the transition to zero emissions. To enable the required reductions in the use of fossil fuels in the transportation sector, the sector will need to undergo a profound transformation in its energy use and the ability to accurately measure emissions from fossil fuels. There is a growing focus on the logistics emissions reporting function, as companies seek to uncover one of the least-known yet potentially highly carbon-intensive parts of their organisation's operations. The granularity and accuracy of reporting is critical for the advancement of future states of logistics emissions management. Whilst many companies are still grappling with the backward-looking reporting task, to make a meaningful impact in logistics emissions, there needs to be a transition to forward-looking and performance monitoring of logistics emissions. Transport sector emissions analysis has been the focal point of numerous researches for many years. The studies show that there are multiple factors—including weather, road condition, vehicle condition, vehicle weight, and driving behaviour—that have significant impacts on vehicle emissions [4–10].

Analysis of vehicle emissions is normally conducted in laboratory environments, on a chassis dynamometer [11,12]. However, the studies show that there is a significant difference between data captured in a laboratory environment and data captured on roads [12,13].

Although laboratory analysis can capture useful data regarding vehicle emissions, it is almost impractical to evaluate all the real-world metrics that impact vehicle emissions in a laboratory environment. Laboratory emissions analysis has been found to underestimate emissions rates by 10–20% [11]. As a result, many researchers tend to install portable exhaust gas analysis devices on vehicles, to capture on-road data. However, installing and capturing on-road data can pose significant challenges.

The accuracy of these devices for long-term data collection is one of the challenges, because they need calibration before each measurement. Exhaust gas analysers are analytical devices that measure the concentration of a specific gas within a mixture of other gases. These devices typically comprise different gas sensors, using industry-standard non-dispersive infrared (NDIR) and chemical sensor technologies. NDIR technology measures the concentration of a gas, by determining how much infrared energy is absorbed at a select wavelength band that corresponds to a resonant mode spectrum of the analysed gas molecule. The chemical sensors technologies used in the analyser are usually based on electrochemical and metal-oxide semiconducting sensors, which suffer from long-term stability, selectivity, and fast response and recovery. The electrochemical sensors require regular replacement of electrolytes; therefore, the short life of these chemical sensors would add to the cost of experiments.

We previously conducted research in developing and installing an Internet of Things (IoT)-based platform, called ParcEMon [14], to capture live data from a delivery van, so as to analyse parcel-level emissions. During this research, multiple challenges were faced, in retrieving continuous data from the delivery van by using a gas analyser comprising different gas sensors, so as to measure the concentration of carbon monoxide (CO), carbon dioxide (CO_2), hydrocarbons (HC), oxygen (O_2), and nitrogen oxide (NO_x) gases, as shown in Table 1.

One of the main challenges was installation of the gas analyser in a secure location, to be able to collect and analyse the gas flow without being impacted by vehicle movement and road condition in a real world context.

Since the gas analyser control unit was required to be close and connected to the gas flow, it could not be installed inside the cabin, as it could pose safety issues to the driver and other passengers. Therefore, the gas analyser was secured outside the car. However, there are multiple factors that can impact the possibility of installing the gas analyser and related accessories outside the car.

Table 1. Gas analyser specifications.

Gases	CO (carbon monoxide) HC (hydrocarbons—hexane (gasoline), propane (LPG), or methane (CNG or LNG)) CO_2 (carbon dioxide) O_2 (oxygen) NO (NOx, nitric oxide)
Analysis Method	CO, HC, CO_2: NDIR (non-dispersive infra-red) O_2, NO: electro-chemical sensor
Ranges	CO: 0–10.00% HC (hexane and propane): 0–9999 ppm HC (methane) 0.000–5.000% CO_2: 0–20.0% O_2: 0.0–25.0% NO: 0–5000 ppm
Gas Sample Rate	350 mL/min typical. (flow-control pneumatic system).

The first issue is that a vehicle's sudden movements and vibrations can loosen the different components that transfer the gas flow to the gas analyser, including the exhaust nuzzle and gas pipe. As a result, all these components must be firmly installed, to withstand the vehicle movements and vibrations. In addition, a change in road conditions can also impact the gas analyser functionality. Bumpy roads can cause road surface contact with the devices and stop or suspend the sensors' functionality. Moreover, dirt roads can cause gravel to hit the analyser or can cause mud splash on the analyser, which can cause full or partial blockage of gas flow, thus impacting the data quality. Moreover, rainy weather can also cause water splash on the analyser, which can damage the sensors and impact their functionality. Furthermore, installing the analyser in a sealed enclosure is also not practical, because there must be airflow, so as to avoid concentrated gas around the gas analyser, which can result in false data readings. Another issue regarding analyser installation is the high degree of heat emitted from the exhaust. This can damage the analyser pipe and sensors and cause data loss. A further challenge, regarding utilisation of a gas analyser in live data collection, is that the sensors normally require periodic checks, to ensure data quality and the functionality of the sensors. For example, one of the components of such gas analysers, which is connected to the exhaust nuzzle, is a water filter. This filter is required to reduce the moisture entering the gas analyser unit. After some hours of driving, and also depending on the weather conditions, this filter must be unplugged, to empty the water accumulated inside it, because the moisture that enters the gas analyser can impact the sensors' accuracy and data quality. Moreover, the sensors require to be periodically restarted (i.e., zeroing) and calibrated, to improve measurement accuracy. Such periodic check-up requirements hinder the usage of such sensors for long-term data collection in a real world context.

Having discussed the challenges of deployment of a gas analyser to collect live data from on-road vehicles, investigation of alternative solutions to analysing vehicle emissions without employing gas sensors was a potential research area. These challenges could be addressed by using artificial intelligence (AI) techniques to estimate vehicle emissions based on vehicle data, such as speed, revolutions per minute (RPM), acceleration, and deceleration. By employing the dataset that has been collected using a gas analyser, the correlation between driver behaviour and vehicle emissions could be determined. The obtained dataset could then be used as a training set, to predict vehicle emissions in the long run, based on driver behaviour without a gas analyser. Such an approach could help to eliminate the challenge of installing a gas analyser and, at the same time, estimating the impact of driving behaviour on vehicle emissions.

In this research, the dataset collected, using a physical gas analyser and ParcEMon platform [14], was used to develop the training set, as well as to analyse the accuracy and validity of the machine learning technique, in predicting vehicle emissions.

This paper presents an AI-enabled emissions monitoring technique (referred as ArtEMon) that aims to utilise AI to replace a physical gas analyser attached to the exhausts and to produce accurate emissions reporting. The main contributions of this paper are as follows:

- An effective technique to align and merge the measurements from various IoT data sources, through a series of methods, from processing raw data to predictive emissions modelling using information from on-board diagnostics (OBDs), temperature, and humidity sensors.
- A weighted ensemble model that assigns weights based on the performance of the tree-based and stacked models, to estimate CO_2 emissions.
- A feasibility study of our solution, through an experimental approach to evaluating how we predict emissions from the IoT data, without using a gas analyser.

The remainder of this paper is organised as follows. Section 2 describes three emissions reporting frameworks, as well as their comparison. Section 3 describes the architecture and methodology of the ArtEMon. Sections 4 and 5 describe the IoT data collection, preprocessing, and the performance of our emissions prediction. Finally, Sections 6 and 7 describe the discussion, future work, and conclusion of this paper.

2. Greenhouse Gas Emissions Reporting Frameworks

Corporate monitoring and reporting of Scope 3 GHG emissions [15] from the 'last-mile' in the up-and-downstream transportation of goods services has been prioritised by governments and companies since the Paris Agreement of 2015 [16], with international and regional frameworks developed on the foundations of the Greenhouse Gas Protocol [17] initiative now in use.

The GHG Protocol Corporate Standard groups a company's GHG emissions into three categories, from Scope 1 to Scope 3: Scope 1 refers to direct emissions; Scope 2 refers to indirect emissions and represents the emissions resulting from the generation of purchased energy (e.g., grid-supplied electricity); Scope 3 covers all the other indirect emissions in the company's value chain that are not included in Scope 2 [17,18].

Scope 3 emissions can constitute a significant part of a company's overall GHG footprint [19]. With estimates placed as high as 75% of all emissions for many companies [20], it is not surprising that the value of modelling, working with, and reporting Scope 3 emissions data is being recognised. Emissions data can be used to support investment, procurement, and sales strategies, by assessing the impact of various freight scenarios and predicting carbon return on investments. It provides a metric that allows companies to identify inefficiencies in their logistics network and work towards improved efficiencies. Companies can track progress towards climate goals and demonstrate corporate social responsibility, including positive climate and health impacts [21].

However, managing the collection and analysis of GHG emissions data remains a complex and frustrating challenge for companies. Current, local, and real-time data that account for the emissions produced in the delivery of domestic goods and services (whether by truck or freight consignment, courier, express post and parcel, or public transport networks) are not readily available data that are accessible for use. Current reporting is based on applying one of a range of available calculators, to estimate GHG emissions by measuring the distance a parcel travels and the fuel expended over the distance. In such approaches, there are many variables that cannot be easily accounted for, such as speed, road incline, and driver behaviour.

For companies to report Scope 3 emissions, they must apply the methodology from one of several key frameworks available for use, developed by different industry consortia in different regions and contexts. These frameworks include the Carbon Disclosure Project (CDP) [22], the EconTransistIT methodology [23], the GHG Protocol [17], the Global

Logistics Emissions Council (GLEC) European framework [24], the Taskforce on climate-related financial disclosures [25], the USA's Environmental Protection Agency's Smartway Program [26], and the Australian National Greenhouse Energy Reporting Scheme [3]. More recently the International Standards Organisation, have published ISO 14083 [27], "Greenhouse gas management and related activities—Quantification and reporting of greenhouse gas emissions of transport operations", with the aim of providing a common methodology for the quantification and reporting of greenhouse gas emissions; however, its use remains voluntary.

For this project, the research team evaluated three of the leading regional frameworks, to develop an understanding of their use, the types of data collected, approaches to data collection, methodologies for calculating Scope 3 emissions, and granularity of reporting. These frameworks include GLEC [24], Smartway [26], and NGERS [28]. The objectives and approaches of the three frameworks are informed by and inform one other, but there are significant differences in their methodologies. Each framework varies in its approach to the measurement of GHG emissions and gases prioritisation (see Figure 1). While NGERS [28] establishes goals and specifies gases for emissions reporting, it does not present a methodology for Scope 3 data collection at all. Therefore, further evaluation has focused on a comparison of the GLEC [24] and Smartway [26] frameworks, including consideration of how each works with different calculation tools, different decision trees, and sets of historic data made available (See Figure 2). Each framework applies different assumptions about routes and different approaches to the measurement of fuel consumption per gram (FPG), and each varies in the range of gases reported [24,26,28].

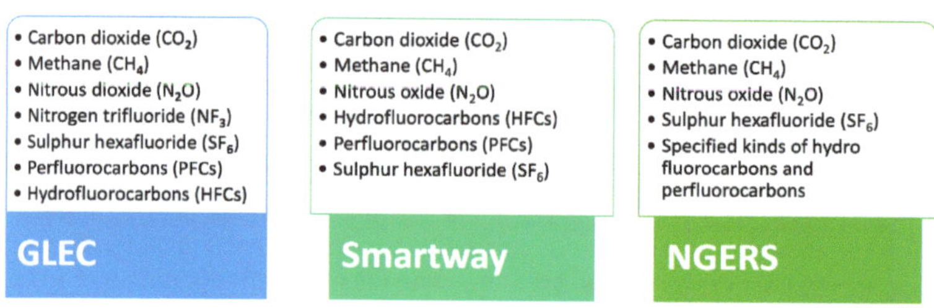

Figure 1. Gases included in the compared frameworks [24,26,28].

For Scope 3 emissions, the absence of a single standardised method for measuring and modelling emissions across complex logistics activities, compounded by significant issues in sourcing and collecting accurate and timely data for reporting, presents enormous barriers for companies. There is an uncertainty about the accuracy of reported data, and about their ability to pursue the most cost-effective carbon mitigation strategies. This is exacerbated in Australia, where—under the National Greenhouse and Energy Reporting Act [29]—GHG emissions reporting is not yet mandatory [3], and the sheer complexity of data collection and reporting means that many who could participate voluntarily and benefit from more effective and efficient close-to-real-time reporting, are simply not participating.

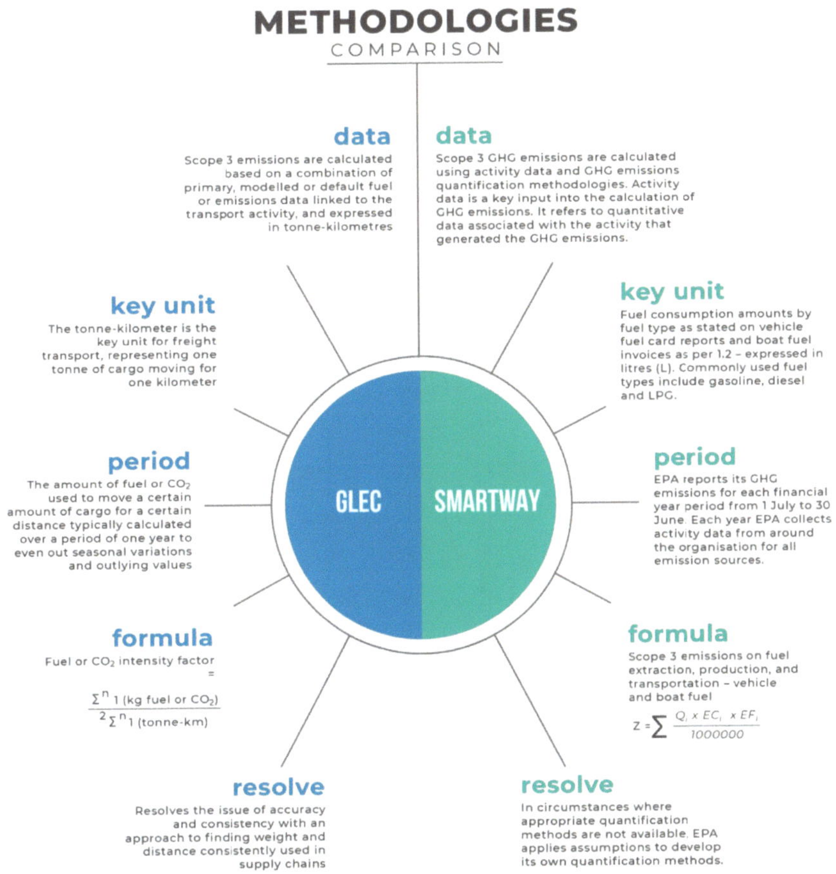

Figure 2. GHG reporting frameworks comparison.

3. Architecture and Methodology

The main objective of the research reported in this paper was to develop a complete end-to-end IoT-and-AI-enabled solution for monitoring GHG emissions from combustion engine vehicles, without relying on exhaust gas analysers. The IoT is defined as connecting a countless number of sensors and various devices (referred to as 'things') through the internet, paving the way for innovative services and products [30].

Figure 3 illustrates the testbed architecture used to collect the raw data employed in this research. The development of this testbed was presented in [14]. The sensor data were collected from four different types of sensors, as follows:

- Gas Analyser (5 Gases): The analyser measured the concentration of carbon monoxide (CO), carbon dioxide (CO_2), fuel-dependant hydrocarbons (HC), oxygen (O_2), and nitrogen oxide (NO_x) gases emitted from the exhaust of a delivery van. The data collected from the gas analyser were used to train and validate the AI models.
- OBDs: vehicle on-board diagnostics data were collected from the instrumented vehicle, to be used by the AI models. An OBD2 scanner provided comprehensive access to live and recorded data from the vehicle, including vehicle speed, acceleration, deceleration, and engine and fuel system operating conditions. These data were required, to correlate driving conditions to the emissions recorded by the gas analyser. OBDs sensory data are easy to collect, and are available from the majority of new cars produced over the past several years.

- Temperature and humidity sensor: the data related to the ambient condition of the delivery van were collected, using a temperature and humidity sensor located underneath the van.

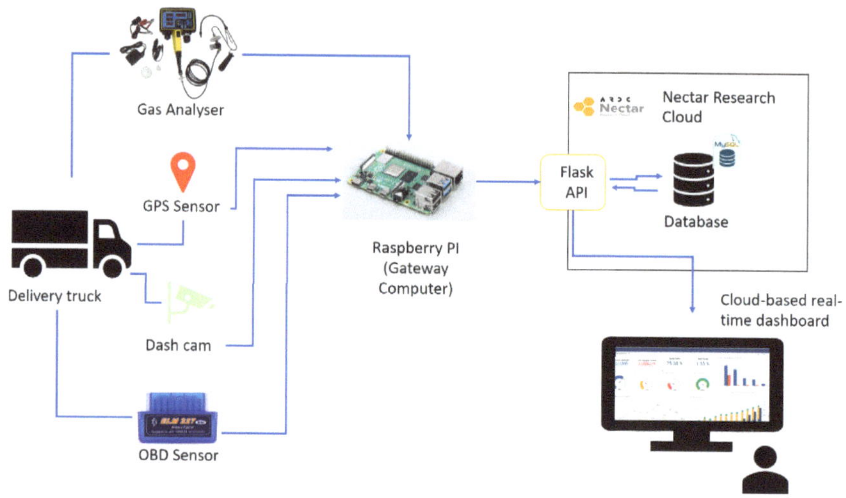

Figure 3. IoT -based testbed architecture.

The IoT sensor node included a small single-board computer (i.e., a Raspberry Pi) that was connected to the reporting gas analyser and OBD2 port through Bluetooth. The gas analyser collected the emissions data in real time and sent it to the computer through a serial cable. The data streams were annotated and synchronised in the computer and sent to the cloud through a broadband internet connection. This IoT sensor node was also connected to a dashcam and GPS module, to track the location of the vehicle and its speed and acceleration. A power bank was included, to provide the required power for running the IoT sensor node. In addition, the implemented enclosure included an automobile auxiliary power outlet for charging the power bank. Finally, a cooling system and environmental monitoring sensors (i.e., temperature and humidity) were deployed, to monitor the ambient temperature and humidity and to maintain the temperature inside the IoT sensor node enclosure. A detailed description of the testbed is described in [14].

Field testing was undertaken with a diesel delivery van in Vermont South in Victoria, Australia. The field trial was conducted under typical 'everyday' or naturalistic driving conditions, employing a driver who regularly drove the vehicle. The vehicle was monitored while making its usual deliveries under normal driving conditions on urban roads, major arterial roads, and freeways. This included regular stop-and-go traffic, acceleration, deceleration, and smooth uniform driving during the trial.

The gas analyser was installed in the carriage, close to the exhaust, without interfering with driving conditions or safety, allowing for the collection of data under the ordinary driving conditions experienced by the driver, as illustrated in Figure 4.

The primary sensors used for the data collection included a gas analyser, vehicle onboard diagnostics (OBDs), and environmental condition sensors (i.e., humidity and temperature sensors). The data collected from these sensors were heterogeneous and contained inherent noise. These heterogeneous data were then processed, to remove noise and to identify relevant features. The data collected from the sensors were stored on the computer, using the JSON format. The first step involved reading the JSON files containing the sensor measurements and converting the raw timestamp (i.e., UNIX time) to a valid date–time format, for time-based data reporting. The timestamp was used as the main parameter, to perform the fusion of the heterogeneous data collected from various different sensors.

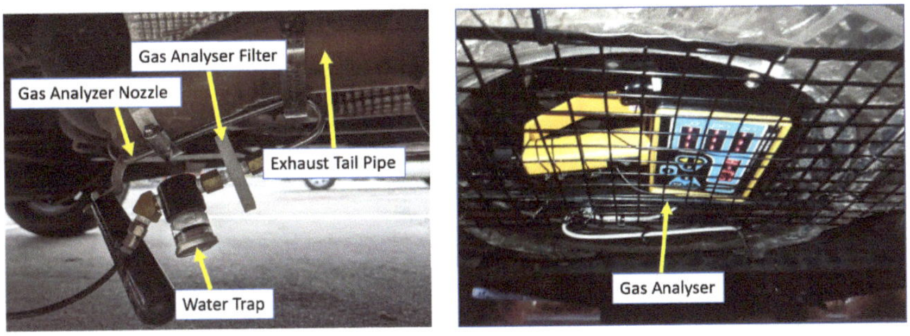

Figure 4. Gas analyser installed under the van.

4. IoT Sensory Data Preprocessing

This section presents a discussion on the processing of the IoT sensory data. The first step involved reading the sensor data measurements and converting the raw timestamp to a valid date–time format. This produced 43,782 (gas analyser, including five gases), 14,562 (OBDs, including RPM, speed, FPG, etc.), and 192,972 (temperature and humidity) data points on the first day of the experiment.

4.1. Noise Removal

The next part of the IoT data processing involved identifying and removing incorrect sensor measurements. Most of the incorrect measurements were from the gas analyser. These measurements contained incorrect characters or partial values. Such measurements were considered as noise and were dropped from the analysis. There were several reasons for these noises, such as data loss because of communication issues, as the gas analyser was using serial protocol to transfer the data, using a long cable.

After removing all the rows containing the incorrect values, a total of 40,894 valid data points from the gas sensor measurements were used in the analysis.

4.2. Zeroing Process

Gas analysers need to purge the gas and perform calibration, based on the reference gases in the ambient air, frequently, to provide accurate readings. The gas analyser we used was set to purge the gas and perform the calibration process every 10 min.

We flagged the readings between the start-zeroing and finish-zeroing intervals. For that duration, the gas reading were not valid and were removed from the dataset, accordingly.

4.3. Lag Identification

There is usually a delay associated with gas analyser measurements when compared to the OBDs. This is because the gas concentration is measured by the gas analyser at the end of the exhaust pipe. The gas must travel through the exhaust pipe to reach the gas analyser, where it is sensed, and this normally takes a few seconds. The challenge here was to identify the right amount of delay and then offset the gas analyser measurements, to adjust the delay and align them with measurements from the other sensors. All the sensor measurements were first indexed based on their timestamp, and the measurements were re-sampled to the nearest one second of the time series, because the sensors reported measurements at millisecond granularity. Re-sampling usually introduces *null* values at the timestamps that were not in the original time series. Imputation is an important data preprocessing step for handling incomplete data, as the machine learning models for analysing IoT data from heterogeneous sensors commonly assume that the sensor data is complete. Such modelling results, which include missing or incomplete data, may be inaccurate and unreliable. It was, therefore, important to interpolate the missing values, by using the mean value of the respective features. Then, the used RPM measurements from the OBDs were used as a reference timestamp and were compared to the CO_2 measurements from the gas analyser at various time intervals, to visually study the lag. As seen in Figure 5, the visual analysis of the time series helped to identify the delay, which was found to be approximately 6 s.

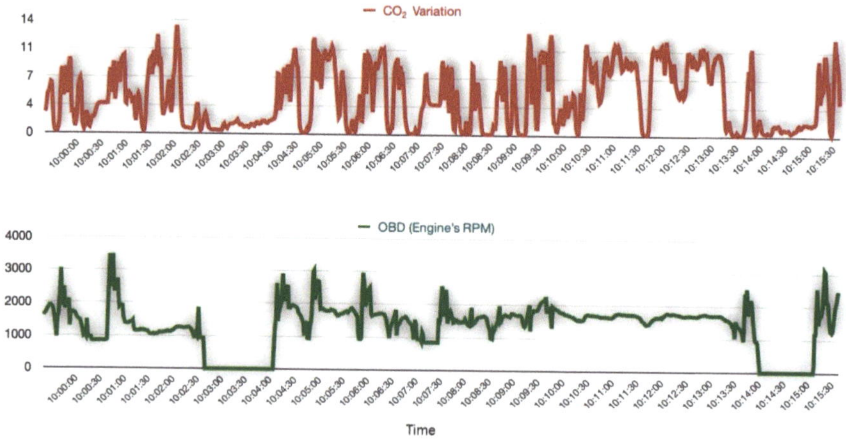

Figure 5. Identifying delay in gas analyser measurements.

4.4. Offsetting, Re-Sampling, and Interpolating Measurements

To align the gas analyser measurements with the rest of the sensor measurements, the gas analyser measurements were offset by around 6 s and were aligned with the rest of the sensor measurements. IoT sensors do not necessarily produce data at regular intervals, and the IoT time series is highly irregular, with regard to its sampling rate, which makes the processing and analysis of such data challenging—specifically, when the data are received independently from multiple sensors, while they are being processed and analysed together. This makes the re-sampling of such irregular and unevenly spaced time series to a more regular and consistent frequency an important part of the IoT data processing step. To obtain a refined set of measurements, all data received from the gas analyser, OBDs, and temperature sensors were interpolated with their mean values at every one-second interval.

After performing interpolation on the data with the mean values of individual features, to fill the time points where there were no measurements, the next step was to align all measurements that had a common start and end time, because the starting and ending times of measurements are different for all sensors. This was required, to eliminate any missing values for the time series that ended early.

Table 2 identifies the starting and ending times of all the IoT sensors for the first day of the experiment. After re-sampling for every one second, interpolating, and aligning the measurements, a total of 15,778 data points were obtained from the gas, OBDs, temperature, and humidity sensors on a single day. All measurements from the sensors where the corresponding speed was reported as zero were also excluded. This allowed us to have a uniform time series for all measurements.

Table 2. Starting and ending time of all IoT Sensors.

Sensor	Starting Time	Ending Time
OBDs	08:54:58	13:17:57
Gas Analyser	08:50:16	14:57:23
Temperature and humidity	08:52:26	14:58:34

4.5. Relevant Feature Selection

To estimate the CO_2 emissions, various features of the vehicles—such as speed, RPM, coolant temperature, acceleration, and throttle—were used [31–33]. Singh et al. [33] identified that speed and acceleration are highly correlated with CO_2 emissions; however, these features on their own provide limited support in estimating CO_2 emissions.

As seen in Algorithm 1, the merged measurements from all the sensors in this research (the OBDs sensors combined with the gas, temperature, and humidity sensors) were used to construct a dataset for understanding CO_2 emissions. The final feature set, presented in Table 3, included information on the RPM, FPG, speed, temperature, and humidity, and the target feature was CO_2.

Algorithm 1: IoT sensor data preprocessing.

```
Input : sensor measurements from gas analyser (A), OBDs (B), temperature and
  humidity (C)
Output : modelling dataset E
for each measurement in A do
  remove noise
  perform zeroing process
end for
for each measurement in A do
  offset ∀ measurement by 6 s
end for
for all measurements in A,B,C do
  re-sample ∀ 1 s
  interpolate with μ values
  align timestamps
end for
Construct modelling dataset E by merging A,B,C
```

Table 3. Feature set for estimating CO_2 emissions.

Feature	Units of Measurement	Range of Feature Values
engine RPM	revolutions per minute	0–6000
fuel consumption (FPG)	grams per second	0–300
vehicle speed	kilometres per hour (km/h)	0–120
air temperature	Celsius (°C)	0–30
humidity	relative humidity in percentage (%)	0–100

5. Predicting CO_2 Emissions

Ensemble learning is a class of machine learning technique that works by utilising multiple base learners that collectively improve the predictive performance. LightGBM [34], gradient boosting [35], and xGBoost [36] are some of the popular algorithms that have been used for regression and classification applications. In this paper, various gradient-boosting decision tree algorithms were used, including gradient boosting [35], xGBoost [36], and LightGBM [34]. Gradient-boosting decision trees are an ensemble of decision trees that use the best split points to learn the decision trees, make predictions on each decision tree, and then combine the individual predictions with those of other trees in the ensemble, to produce a strong prediction. Gradient boosting works by optimising the loss function; it sequentially tries to find new weak learners, based on the residuals from the previous weak learners in each iteration. During each iteration, gradient boosting attempts to minimise the loss function.

Although gradient boosting works well with both categorical and continuous data, it sometimes overemphasises outliers and is computationally expensive. The xGBoost works by growing trees horizontally and is robust enough in handling heterogeneous data types and distributions. Moreover, xGBoost can efficiently handle sparse data patterns and also helps in handling overfitting [36]. LightGBM is a greedy implementation of the gradient boosting framework that uses gradient-based one-side sampling to grow trees vertically, which enables better identification of relationships between input and target features. However, LightGBMs are sensitive to outliers and can lead to overfitting if they are not handled properly [34,37].

Various techniques based on deep learning [32–36,38], support vector machines [32], ensemble models [38], and xGBoost [36] have been used in the literature, for estimating CO_2 emissions from data gathered from vehicles.

In this paper, predicting CO_2 emissions was based on ensemble learning techniques [39] that combine multiple regression models [40], which have been reported, in previous studies, to improve the prediction accuracy of machine learning algorithms [39]. Further improvements have also been achieved by incorporating strategies that combine various ensemble formation designs, such as stacked generalisation [40] and blending [41]. Stacked generalisation, proposed by Wolpert [42], is a procedure [43] that consists of several base learners (first level) and a meta learner (second level), where the outputs from the first level serve as inputs to the second level meta learners. The first-level regressors are fitted to the same training set that is used to prepare the inputs for the second-level regressor, which may lead to overfitting. However, the techniques proposed in [44] extend the standard stacking algorithm, using out-of-fold predictions, and produce the data from the first-level regressor, which serves as input for the second-level regressor, by splitting the dataset into k folds and, in every successive k rounds, $k - 1$ folds are then utilised, to fit the first-level regressor. Furthermore, in each round, the first-level regressors are applied on the subset of the data that were initially excluded from the model fitting during each iteration. This allows the resulting predictions to be stacked and then forwarded as inputs to the second-level regressor, and then the first-level regressors are fitted on the entire dataset, to maximise the prediction accuracy. Blending is another ensemble approach that is derived from stacking [41], but differs by not using k-fold cross-validation to generate training data for the meta learner. However, it makes use of a one-holdout set, which allows a small portion of the data from the training set to be used for making predictions that can be used as inputs to the meta model.

In this paper, a weighted ensemble approach was used, which worked by assigning weights based on the individual model performance. Before the construction of the stacked model, the actual dataset was divided into training and test sets. The training set was then used to construct the stacked model, as shown in Figure 6. Following the identification of the best performing model from the individual baseline models during the stacking phase, weights were assigned to each of these models, with a higher weight to the stacked model followed by the best performing baseline model. The weighted predictions were then blended, to generate the final prediction. Using the model performances of the baseline models and the weighted ensemble (blended) model (Figure 7), weights of $w_s = 0.4$, $w_g = 0.3$, $w_x = 0.15$, and $w_l = 0.15$ were assigned, respectively, to the stacked, gradient boosting, xGBoost, and LightGBM models, following which, the weighted predictions were blended, to estimate the CO_2 emissions.

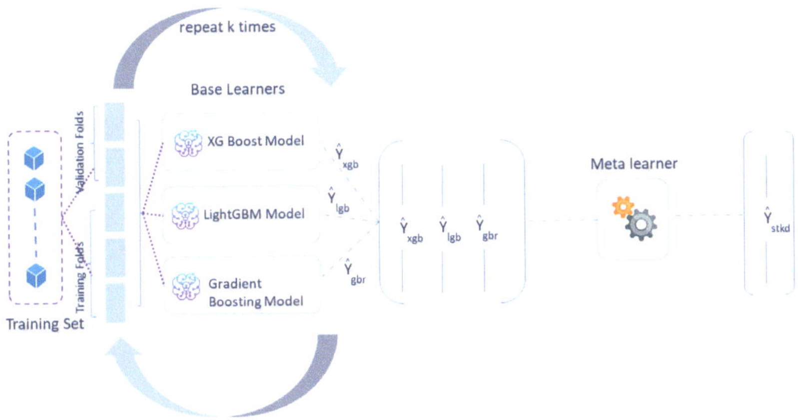

Figure 6. Stacked model, to estimate CO_2 emissions.

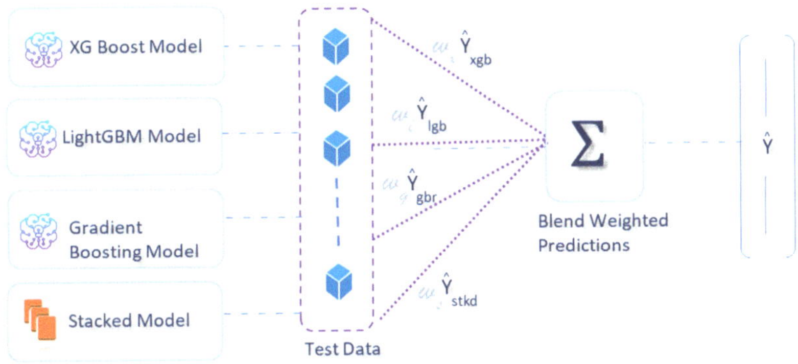

Figure 7. Weighted ensemble model, to estimate CO_2 emissions.

5.1. Model Performance

In the weighted blended approach, the dataset is divided into k = 10 consecutive folds, with the shuffle parameter set to *True,* to avoid non-random assignment of the data points in the training and test sets. Each fold is used once for validation, while the rest of the $k - 1$ folds then make up the training set. The root mean square error was used as a key performance measure for scoring. The mean square error returned by the cross-validation scoring function is always negative, and by selecting a negative mean squared error for scoring, the scoring function returns a positive score when the score has to be maximised and negates it when the score should be minimised. The baseline models were first trained individually, using the most common gradient-boosting tree-based machine learning algorithms in gradient boosting [35], xGBoost [36], and LightGBM [34], to identify the best-performing model baseline model. The stacked model was constructed using the best-performing baseline model as the meta model. The baseline models and the stacked model were used to construct the weighted ensemble model. The root mean square error (RMSE) metric was applied, to measure and compare the models' performance. The RMSE metric indicated how far the predicted CO_2 was from the average CO_2 emissions. As seen in Figure 8, the RMSE score on the test dataset, using a higher weight gradient-boosting regressor as a meta regressor in the weighted ensemble model, was 1.8625, which was marginally better than the baseline gradient boosting model.

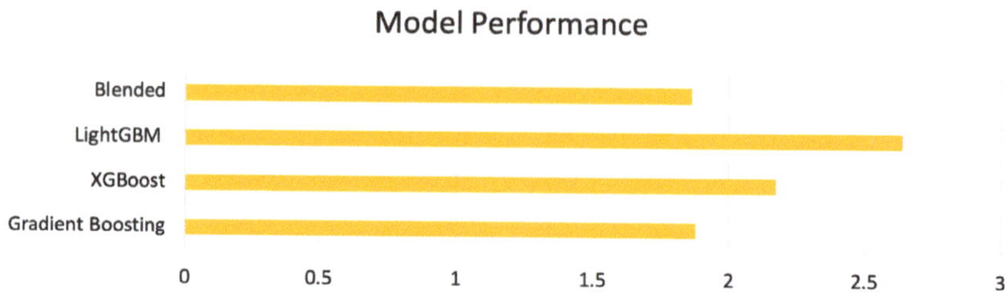

Figure 8. Model performance of blended, gradient boosting, xGBoost, and LightGBM baseline models.

Figure 9 illustrates a visual assessment of how well the predictions made by a weighted ensemble model align with actual CO_2 measurements. The plot has the model's predictions on the vertical (y) axis and the observed CO_2 values on the horizontal (x) axis. From this figure, it is evident that the data points cluster closely around a diagonal line, often referred to as the regression line. This line represents the scenario where predicted values perfectly match actual values.

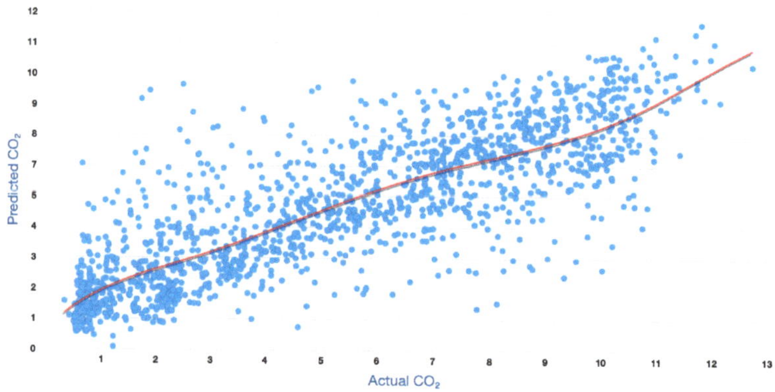

Figure 9. Actual vs. Predicted CO_2 Emissions of the Weighted ensemble model from Test Data.

In our context of predicting CO_2 emissions, this closeness indicates that the model's predictions are highly accurate. The tight distribution around the diagonal line suggests that the predicted CO_2 values closely resemble the actual CO_2 measurements. Furthermore, the vertical distance between each data point and the regression line, known as the prediction error or residual, appears small for most data points. Small prediction errors imply that the model's predictions are very close to the actual values.

It is worth mentioning that the absence of significant patterns or systematic deviations from the diagonal line indicates that the model does not exhibit bias. This uniformity in predictions across the entire range of observed values is desirable in various applications.

In summary, the proximity of predicted CO_2 values to the diagonal line reflects the excellent performance of the weighted ensemble model in providing accurate predictions. This outcome is highly favourable in data analysis, particularly in the context of IoT-based greenhouse gas sensing for real-time emissions monitoring applications, as it instils confidence in the model's reliability for making predictions and drawing inferences.

5.2. Explaining CO_2 Predictions

Lundberg et al. [45], have identified that interpretations and explanations of the predictions are as important as the accuracy of a predictive model.

When explaining CO_2 predictions from the machine learning models, the predicted CO_2, $f(x)$, can be represented as the sum of its computed SHAP values and a fixed base value:

$$f(x) = basevalue + sum(SHAP_{values}). \quad (1)$$

A SHAP (SHapley Additive exPlanations) [45] or Shapley value is the average marginal contribution of a feature value across all possible combinations, whereas the base value is the mean value of the target feature (CO_2) in the dataset. The distribution of Shapley values helps in understanding the impact of FPG, RPM, humidity, temperature, and speed on CO_2 predictions, both at local (individual) and global (population) levels. Local interoperability explains the individual predictions for each data instance, i.e., how the model arrived at a decision, in terms of the contributions of its input features, whereas global interoperability describes the expected behaviour of the machine learning model, with respect to the overall distribution of the values of all the input features. SHAP values are additive in nature, i.e., the SHAP values of FPG, RPM, humidity, temperature, and speed will always add up to the difference between the predicted CO_2 from the baseline model and the current CO_2 prediction that is being explained. Figure 10 provides information on the relationship between the features' actual values and Shap values, i.e., the colour bar represents the actual values of the features (FPG, RPM, humidity, temperature, and speed) for each instance. A red-coloured dot indicates a relatively high value of the feature, and vice versa with a blue-coloured dot. We can see that higher values of FPG have positive SHAP values, i.e., the dots extending towards the right, forming the horizontal (0-axis) line are red, which indicates that FPG leads to higher CO_2 predictions.

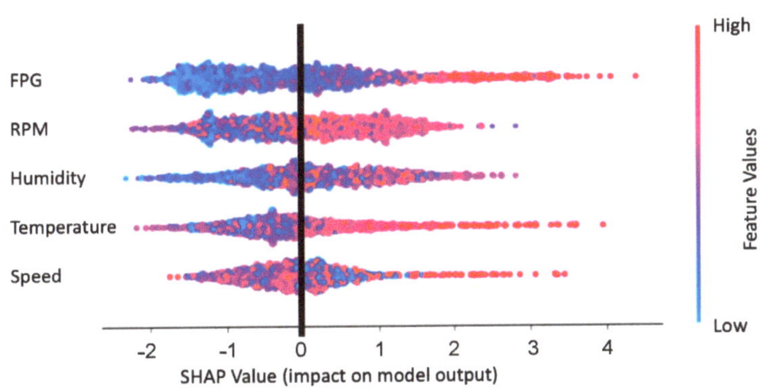

Figure 10. Relative feature importance and their relationship to the predicted CO_2.

6. Discussion and Future Research

This paper describes the weighted ensemble approach, and the modelling results of this technique demonstrate that CO_2 emissions can be predicted using data from onboard sensors from vehicles with considerably fewer features. The major contributions from this study can be summarised into (a) a weighted blended approach that uses data from onboard sensors to learn the CO_2 emissions characteristics, (b) visual analysis, to identify and offset the lag from gas analyser measurements, and (c) performance comparison of the weighted ensemble approach to other popular tree-based approaches, such as LightGBM, xGBoost, gradient boosting, and stacked modelling approaches. The weighted ensemble approach works by assigning weights to each of the individual baseline models and the stacked model for predicting CO_2 emissions.

In this approach, the highest weight was assigned to the stacked model, and then, among the baseline models, we assigned a higher weight to the best-performing baseline model, i.e., the gradient-boosting regressor, and then equally distributed the weight among the rest of the baseline models. However, it would be interesting to see the model performance on hyper-parameter tuned stacked and weighted ensemble models, instead of the baseline models, and also to compare the performance of the hyper-parameter tuned stacked model to the weighted ensemble model. Long et al. [46] have identified that geographical features (such as roads around mountainous regions), weather conditions (such as wind gusts, wind direction, and rainfall, etc.), and geographic terrains can impact CO_2 emissions from vehicles. Furthermore, as our model does not generalise well, to accurately predict CO_2 emissions at very low and high levels, it would be interesting to include contextual data in the model, describing the geographical features and traffic conditions, such as rush hours.

In Section 5.2, we also briefly explained the relative features importance and their relationships to the predicted CO_2 emissions. In this paper, we have only focused on global interpretability, when explaining CO_2 predictions [47]. However, in order to provide comprehensive explainability of the predicted CO_2 emissions and the feature values, it is important to (a) relate the CO_2 emissions with a variety of contextual information, such as geographical locations, and then explain the emissions, and (b) include local interpretability, when explaining the individual predictions.

Lastly, it is important to note that in application scenarios requiring substantial computational resources for real-time prediction of multiple transport vehicles, there are techniques like Map-Reduce [48,49], data approximation [50], contextualisation [51,52], and situation-aware computing [53] that can be employed, to achieve scalable real-time computation to meet the time-bound requirements [54,55].

7. Conclusions

Emissions reporting and sustainability have garnered significant global attention. However, a critical issue is the absence of a universally accepted method for quantifying emissions, especially in the case of Scope 3 emissions. This lack of standardisation, coupled with the complexities of logistics operations, poses substantial hurdles for businesses striving to comprehensively grasp and report their environmental impact. Traditionally, a common method for accurately capturing and monitoring gas emissions in logistics has centred on deploying gas sensors. Nonetheless, managing these sensors on moving vehicles, especially in the presence of varying road and weather conditions, poses a significant challenge. In this paper, we presented an innovative solution to addressing this challenge. Our approach leverages the IoT and AI, to eliminate reliance on gas sensors for real-time GHG emissions reporting, while accurately predicting CO_2 emissions.

Author Contributions: Conceptualisation, A.Y.; methodology, A.Y. and I.B.M.; software, A.Y., H.K. and I.B.M.; investigation, A.Y., H.D., P.S., J.S. and M.S.; writing—original draft preparation, A.Y., I.B.M. and H.B.; writing—review and editing, H.D., P.S., J.S., C.T. and M.S. All authors have read and agreed to the published version of the manuscript.

Funding: This research was co-funded by Scope 3 Pty Ltd. and Swinburne University of Technology's Research Office.

Institutional Review Board Statement: The study was ethics-approved at Swinburne University of Technology, No. 20204101-5592.

Acknowledgments: The authors acknowledge Michael Blake and Gerry Walsh from EmissionsIQ for their support. This research was supported by use of the Nectar Research Cloud, a collaborative Australian research platform supported by the Australian Research Data Commons (ARDC). The ARDC is funded by the National Collaborative Research Infrastructure Strategy (NCRIS).

Conflicts of Interest: The authors declare no conflict of interest.

References

1. Ritchie, H. Cars, Planes, Trains: Where Do CO_2 Emissions from Transport Come from? *Our World in Data*, 6 October 2020. Available online: https://ourworldindata.org/co2-emissions-from-transport (accessed on 16 June 2023).
2. International Transport Forum. *ITF Transport Outlook 2019*; OECD Publishing: Paris, France, 2019.

3. *Quarterly Update of Australia's National Greenhouse Gas Inventory;* Australian Government, Department of Climate Change, Energy, the Environment and Water: Canberra, Australia, 2022.
4. Jaworski, A.; Lejda, K.; Bilski, M. Effect of driving resistances on energy demand and exhaust emission in motor vehicles. *Combust. Engines* **2022**, *189*, 60–67. [CrossRef]
5. Greenwood, I.; Dunn, R.; Raine, R. Estimating the effects of traffic congestion on fuel consumption and vehicle emissions based on acceleration noise. *J. Transp. Eng.* **2007**, *133*, 96–104. [CrossRef]
6. Oak Ridge National Laboratory. Sensible Driving Saves More Gas Than Drivers Think. *ScienceDaily*, 28 September 2017.
7. Merkisz, J.; Andrzejewski, M.; Merkisz-Guranowska, A.; Jacyna-Gołda, I. The influence of the driving style on the CO_2 emissions from a passenger car. *J. KONES* **2014**, *21*, 219–226. [CrossRef]
8. Zhou, B.; He, L.; Zhang, S.; Wang, R.; Zhang, L.; Li, M.; Liu, Y.; Zhang, S.; Wu, Y.; Hao, J. Variability of fuel consumption and CO_2 emissions of a gasoline passenger car under multiple in-laboratory and on-road testing conditions. *J. Environ. Sci.* **2023**, *125*, 266–276. [CrossRef] [PubMed]
9. Davison, J.; Rose, R.A.; Farren, N.J.; Wagner, R.L.; Wilde, S.E.; Wareham, J.V.; Carslaw, D.C. Gasoline and diesel passenger car emissions deterioration using on-road emission measurements and measured mileage. *Atmos. Environ. X* **2022**, *14*, 100162. [CrossRef]
10. Tu, R.; Xu, J.; Li, T.; Chen, H. Effective and acceptable eco-driving guidance for human-driving vehicles: A review. *Int. J. Environ. Res. Public Health* **2022**, *19*, 7310. [CrossRef] [PubMed]
11. Pelkmans, L.; Debal, P. Comparison of on-road emissions with emissions measured on chassis dynamometer test cycles. *Transp. Res. Part D Transp. Environ.* **2006**, *11*, 233–241. [CrossRef]
12. Ježek, I.; Drinovec, L.; Ferrero, L.; Carriero, M.; Močnik, G. Determination of car on-road black carbon and particle number emission factors and comparison between mobile and stationary measurements. *Atmos. Meas. Tech.* **2015**, *8*, 43–55. [CrossRef]
13. Weiss, M.; Bonnel, P.; Hummel, R.; Manfredi, U.; Colombo, R.; Lanappe, G.; Le Lijour, P.; Sculati, M. Analyzing on-road emissions of light-duty vehicles with Portable Emission Measurement Systems (PEMS). *JRC Sci. Tech. Rep. Eur* **2011**, *24697*, 1–66.
14. Yavari, A.; Bagha, H.; Korala, H.; Mirza, I.; Dia, H.; Scifleet, P.; Sargent, J.; Shafiei, M. ParcEMon: IoT Platform for Real-Time Parcel Level Last-Mile Delivery Greenhouse Gas Emissions Reporting and Management. *Sensors* **2022**, *22*, 7380. [CrossRef]
15. Huang, Y.A.; Weber, C.L.; Matthews, H.S. Categorization of scope 3 emissions for streamlined enterprise carbon footprinting. *Environ. Sci. Technol.* **2009**, *43*, 8509–8515. [CrossRef] [PubMed]
16. Bodansky, D. The United Nations Framework Convention on Climate Change: A Commentary. *Yale J. Int. Law* **1993**, *18*, 451. Available online: https://heinonline.org/HOL/LandingPage?handle=hein.journals/yjil18&div=29&id=&page= (accessed on 8 September 2023).
17. World Resources Institute; World Business Council for Sustainable Development. *Technical Guidance for Calculating Scope 3 Emissions. Supplement to the Corporate Value Chain (Scope 3). Accounting & Reporting Standard;* World Resources Institute and World Business Council for Sustainable Development: Geneva, Switzerland, 2013. Available online: https://ghgprotocol.org/sites/default/files/standards/Scope3_Calculation_Guidance_0.pdf (accessed on 8 September 2023).
18. Wiedmann, T.; Chen, G.; Owen, A.; Lenzen, M.; Doust, M.; Barrett, J.; Steele, K. Three-scope carbon emission inventories of global cities. *J. Ind. Ecol.* **2021**, *25*, 735–750. [CrossRef]
19. Hertwich, E.G.; Wood, R. The growing importance of scope 3 greenhouse gas emissions from industry. *Environ. Res. Lett.* **2018**, *13*, 104013. [CrossRef]
20. Downie, J.; Stubbs, W. Evaluation of Australian companies' scope 3 greenhouse gas emissions assessments. *J. Clean. Prod.* **2013**, *56*, 156–163. [CrossRef]
21. Whitten, S.; Verikios, G.; Kitsios, V.; Mason-D'Croz, D.; Cook, S.; Holt, P. *Exploring Climate Risk in Australia;* CSIRO: Canberra, Australia, 2021.
22. Worldwide, C. 9. Carbon Disclosure Project. 2023. Available online: https://www.cdp.net (accessed on 27 August 2023).
23. IFEU Heidelberg; INFRAS Berne; IVE Hannover. *Ecological Transport Information Tool for Worldwide Transports;* EcoTransIT World: Paris, France, 2019. Available online: https://download.ecotransit.world/EcoTransIT_World_Methodology_Data_100521.pdf (accessed on 8 September 2023).
24. Greene, S.; Lewis, A. *GLEC Framework for Logistics Emissions Methodologies;* Smart Freight Centre: Amsterdam, The Netherlands, 2016.
25. Financial Stability Board. *Task Force on Climate-Related Financial Disclosures: 2019 Status Report;* FSB: Basel, Switzerland, 2019. Available online: https://www.fsb.org/2019/06/task-force-on-climate-related-financial-disclosures-2019-status-report/ (accessed on 8 September 2023).
26. U.S. Environmental Protection Agency. SmartWay. Available online: https://www.epa.gov/smartway (accessed on 27 August 2023).
27. *ISO/TR 14069:2013;* Greenhouse Gases-Quantification and Reporting of Greenhouse Gas Emissions for Organizations-Guidance for the Application of ISO 14064-1; International Organization for Standardization: Geneva, Switzerland, 2013.
28. NSW Minerals Council. *National Greenhouse and Energy Reporting (NGER): Important Things You Need to Know about NGER;* Fact Sheet; NSW Minerals Council: Sydney, Australia, 2009.
29. The Law Library. *National Greenhouse and Energy Reporting Act 2007;* 2007. Available online: https://www.legislation.gov.au/Details/C2023C00090 (accessed on 8 September 2023).
30. Yavari, A. Internet of Things Data Contextualisation for Scalable Information Processing, Security, and Privacy. Ph.D. Thesis, RMIT University, Melbourne, Australia, 2019.

31. Palconit, M.G.B.; Nuñez, W.A. Statistical analysis of CO_2 emission based on road grade, acceleration and vehicle specific power for public utility vehicles: An IoT application. In Proceedings of the 2018 IEEE 4th World Forum on Internet of Things (WF-IoT), Singapore, 5–8 February, 2018; pp. 245–250.
32. Zeng, W.; Miwa, T.; Morikawa, T. Prediction of vehicle CO_2 emission and its application to eco-routing navigation. *Transp. Res. Part C Emerg. Technol.* **2016**, *68*, 194–214. [CrossRef]
33. Singh, M.; Dubey, R. Deep learning model based CO_2 emissions prediction using vehicle telematics sensors data. *IEEE Trans. Intell. Veh.* **2021**, *8*, 768–777. [CrossRef]
34. Ke, G.; Meng, Q.; Finley, T.; Wang, T.; Chen, W.; Ma, W.; Ye, Q.; Liu, T.Y. Lightgbm: A highly efficient gradient boosting decision tree. In *Advances in Neural Information Processing Systems 30, Proceedings of the Annual Conference on Neural Information Processing Systems, Long Beach, CA, USA, 19 May 2017*; Curran Associates: New York, NY, USA, 2017; Volume 30, pp. 1–9.
35. Friedman, J.H. Greedy function approximation: A gradient boosting machine. *Ann. Stat.* **2001**, *29*, 1189–1232. [CrossRef]
36. Chen, T.; Guestrin, C. Xgboost: A scalable tree boosting system. In Proceedings of the 22nd ACM SIGKDD International Conference on Knowledge Discovery and Data Mining, San Francisco, CA, USA, 13–17 August 2016; pp. 785–794.
37. Mienye, I.D.; Sun, Y. A survey of ensemble learning: Concepts, algorithms, applications, and prospects. *IEEE Access* **2022**, *10*, 99129–99149. [CrossRef]
38. Shah, S.; Thakar, S.; Jain, K.; Shah, B.; Dhage, S. A Comparative Study of Machine Learning and Deep Learning Techniques for Prediction of CO_2 Emission in Cars. In Proceedings of the Third International Conference on Sustainable Expert Systems, Kathmandu, Nepal, 9–10 September 2013; Springer: Singapore, 2023; pp. 749–758.
39. Webb, G.I.; Zheng, Z. Multistrategy ensemble learning: Reducing error by combining ensemble learning techniques. *IEEE Trans. Knowl. Data Eng.* **2004**, *16*, 980–991. [CrossRef]
40. Breiman, L. Stacked regressions. *Mach. Learn.* **1996**, *24*, 49–64. [CrossRef]
41. Zhang, Y.; Liu, J.; Shen, W. A review of ensemble learning algorithms used in remote sensing applications. *Appl. Sci.* **2022**, *12*, 8654. [CrossRef]
42. Wolpert, D.H. Stacked generalization. *Neural Netw.* **1992**, *5*, 241–259. [CrossRef]
43. Martinez-Gil, J. A comprehensive review of stacking methods for semantic similarity measurement. *Mach. Learn. Appl.* **2022**, *10*, 100423. [CrossRef]
44. Raschka, S. MLxtend: Providing machine learning and data science utilities and extensions to Python's scientific computing stack. *J. Open Source Softw.* **2018**, *3*, 638. [CrossRef]
45. Lundberg, S.M.; Lee, S.I. A unified approach to interpreting model predictions. *Adv. Neural Inf. Process. Syst.* **2017**, *30*, 4768–4777.
46. Long, Y.; Yoshida, Y.; Li, Y.; Gasparatos, A. Spatial-temporal variation of CO_2 emissions from private vehicle use in Japan. *Environ. Res. Lett.* **2022**, *17*, 014042. [CrossRef]
47. Carvalho, D.V.; Pereira, E.M.; Cardoso, J.S. Machine learning interpretability: A survey on methods and metrics. *Electronics* **2019**, *8*, 832. [CrossRef]
48. Chu, C.T.; Kim, S.; Lin, Y.A.; Yu, Y.; Bradski, G.; Olukotun, K.; Ng, A. Map-reduce for machine learning on multicore. In Advances in Neural Information Processing Systems 19, Proceedings of the Annual Conference on Neural Information Processing Systems, Montreal, Canada, 4–7 December 2006; MIT Press: Cambridge, MA, USA, 2006; Volume 19, pp. 1–8.
49. Ghoting, A.; Krishnamurthy, R.; Pednault, E.; Reinwald, B.; Sindhwani, V.; Tatikonda, S.; Tian, Y.; Vaithyanathan, S. SystemML: Declarative machine learning on MapReduce. In Proceedings of the 2011 IEEE 27th International Conference on Data Engineering, Hannover, Germany, 11–16 April 2011; pp. 231–242.
50. Yavari, A.; Korala, H.; Georgakopoulos, D.; Kua, J.; Bagha, H. Sazgar IoT: A Device-Centric IoT Framework and Approximation Technique for Efficient and Scalable IoT Data Processing. *Sensors* **2023**, *23*, 5211. [CrossRef]
51. Yavari, A.; Jayaraman, P.P.; Georgakopoulos, D. Contextualised service delivery in the internet of things: Parking recommender for smart cities. In Proceedings of the 2016 IEEE 3rd World Forum on Internet of Things (WF-IoT), Reston, VA, USA, 12–14 December 2016; pp. 454–459.
52. Yavari, A.; Jayaraman, P.P.; Georgakopoulos, D.; Nepal, S. ConTaaS: An Approach to Internet-Scale Contextualisation for Developing Efficient Internet of Things Applications. In Proceedings of the Hawaii International Conference on System Sciences 2017 (HICSS-50), Waikoloa Village, HI, USA, 4–7 January 2017.
53. Mirza, I.B.; Georgakopoulos, D.; Yavari, A. Cyber-physical-social awareness platform for comprehensive situation awareness. *Sensors* **2023**, *23*, 822. [CrossRef]
54. Korala, H.; Georgakopoulos, D.; Jayaraman, P.P.; Yavari, A. A time-sensitive IoT data analysis framework. In Proceedings of the Hawaii International Conference on System Sciences 2021 (HICSS-54), Maui, HI, USA, 5–8 January 2021.
55. Korala, H.; Georgakopoulos, D.; Jayaraman, P.P.; Yavari, A. Managing time-sensitive iot applications via dynamic application task distribution and adaptation. *Remote Sens.* **2021**, *13*, 4148. [CrossRef]

sensors

Review

Electric Transmission and Distribution Network Air Pollution

Saverio De Vito, Antonio Del Giudice and Girolamo Di Francia *

ENEA—Italian National Agency for New Technologies, Energy and Sustainable Economic Development, P.le E. Fermi, 1, 80055 Napoli, Italy; saverio.devito@enea.it (S.D.V.); antonio.delgiudice@enea.it (A.D.G.)
* Correspondence: girolamo.difrancia@enea.it

Abstract: There is a consensus within the scientific community regarding the effects on the environment, health, and climate of the use of renewable energy sources, which is characterized by a rate of harmful polluting emissions that is significantly lower than that typical of fossil fuels. On the other hand, this transition towards the use of more sustainable energy sources will also be characterized by an increasingly widespread electrification rate. In this work, we want to discuss whether electricity distribution and transmission networks and their main components are characterized by emissions that are potentially harmful to the environment and human health during their operational life. We will see that the scientific literature on this issue is rather limited, at least until now. However, conditions are reported in which the network directly causes or at least promotes the emissions of polluting substances into the environment. For the most part, the emissions recorded, rather than their environmental or human health impacts, are studied as part of the implementation of techniques for the early determination of faults in the network. It is probable that with the increasing electrification of energy consumption, the problem reported here will become increasingly relevant.

Keywords: battery; electric distribution; electric transmission; pollution; power cable; sensor; switch; transformer

1. Introduction

Climate and environmental emergencies are increasingly drawing worldwide attention to the need to increase the rate of renewable energy sources (RES) in the global energy mix [1]. In 2022, electricity represented about 38% of the global energy mix, but since more than 60% of it was produced by fossil fuels, its real contribution in terms of an environmental and climate friendly source was around just 15% [2]. As a matter of fact, in the same year, the overall percentage of RES in terms of global world energy consumption was only just over 14%; it is worth mentioning here that this rate is expected to become higher than 60% in 2050 [3]. The growing use of renewable energy sources brings about environmental benefits that are effectively measurable in terms of a reduction in climate-changing gas emissions and, more generally, in a decrease in polluting substances [4]. For example, from 2005 to 2018, despite a global increase in primary energy consumption of around 44%, a threefold increase in the use of renewables made it possible to obtain a net decrease of almost 130 kt in the concentrations of all major pollutants, and the major impact on this effect can be attributed to photovoltaic and wind power diffusion [4,5].

Car electrification is expected to play a key role in this scenario. Although a large abatement of NO_x, CO, CO_2, and VOC exposure can be obtained by the introduction of electric cars at a fast pace, fine particulate exposure cannot be reduced comparatively due to non-exhaust emissions (brake, tyre, and road wear), which represent the most relevant particulate pollution sources [6]. Furthermore, tyre-related particulate emissions could increase due to a possible increase in electric car weight. Despite this, an overall gas pollutant exposure from private transport of close to zero can be foreseen [7,8].

The replacement of fossil fuels with renewable energy sources directly generating electricity calls for the substantial implementation and strengthening of the electric transmission and distribution network, as well as for the replacement of any machine or equipment based on fossil fuels with others based on electric propulsion. One of the results of such a transformation is that power lines, cables, transformers, substation equipment, and charge storage apparatus are expected to correspondingly increase both in number and in dimensions. How deeply this process would impact human lives and the quality of the environment in which we live is not completely known. Certainly, the use of lower temperature energy processes will bring benefits both in terms of the CO_2 footprint and a decrease in the concentrations of the pollutants specifically connected to the use of fossil fuels [9,10]. Nevertheless, the energy transition also involves aspects that, if not decidedly negative, are still questionable and whose overall environmental impact deserves to be investigated in greater detail. The first is related to the energy and environmental costs of production, recycling, and recovery of the raw materials that make it possible [11]. However, another aspect that needs to be considered is that the quantitative increase in transmission lines, substations, batteries, etc. will lead to a corresponding increase in sources that can modify the environment both in terms of physical and chemical polluting agents. With regard to the former, these effects are mainly related to electromagnetic pollution and acoustic noise. As far as the former is considered, from a strictly scientific point of view, the effect of electromagnetic fields on the health of living organisms may still be considered as an open research topic although low-frequency magnetic fields (ELF-MF), typical for example of transmission lines, have already been classified by the International Agency for Research on Cancer (IARC) as possibly carcinogenic, and recent studies have confirmed their possible negative role on health [12–14]. With regard to acoustic noise, intensity levels in the order of tens of decibels, perfectly detectable by those who live near electrical substations, were reported by Piana and Roozen and mainly attributed to mechanical effects [15]. Even more interesting is the case of the noise generated by high voltage lines that can reach several tens of decibels at the source, which is known to be correlated with the corona effect [16].

In this paper, we discuss a topic that has been less addressed, up to now, in the scientific and technical literature: electricity transmission and distribution networks as atmospheric chemical polluting agents. In Section 2, the network components, their malfunctions, and specific emissions are reported, and in Section 3 the commercial solutions available for the various network components are reported. Finally, a discussion of the results is presented in Section 4, and general conclusions are drawn in Section 5.

2. Network Components, Malfunctions, and Specific Emissions

In this section, the connections between emissions and normal or faulty operating conditions in different elements of the power transport, distribution, and use network are reviewed. Figure 1 schematically shows the main components of an electric transmission and distribution network: the HVPL cables carry the electricity from an electric power plant to a transformer, and through the use of proper switchgears, LVPLs eventually carry the energy underground to cities. In general, such a system is already equipped with various monitoring sensors [17]. A network of this type is normally assumed to be characterized by practically zero emissions in terms of air polluting substances during its operation, except when malfunctions are observed in one or more of the components that characterize the network itself. Consequently, chemical sensors are only referred to as some specific components. The scientific literature reporting how these components, during the operating lifetime and in case of malfunction, may emit one or more gaseous components will be reported and discussed with the aim of understanding if the emitted pollutant concentrations can be correlated with a specific failure.

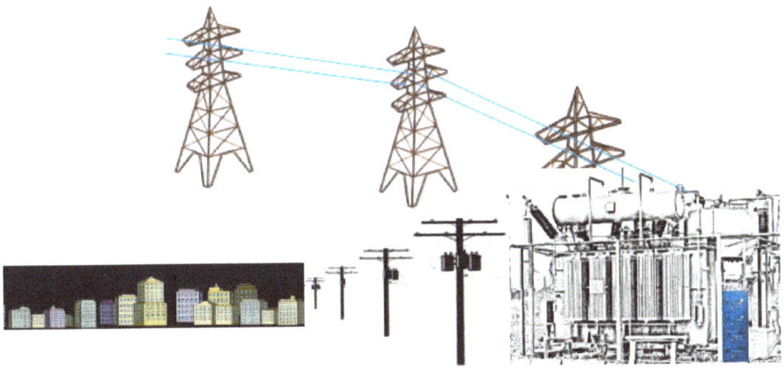

Figure 1. The general structure of an electric transmission and distribution network with its main components: the HVPL cables that carry the electricity from a power plant to a transformer, the switch cabinets, and the LVPLs carrying energy to the city.

In this scenario, the possibility of using arrays of solid-state sensors to monitor the operating status of the aforementioned components and prevent the onset of serious anomalies in the network is discussed. This paragraph is structured in sub-paragraphs, each of which focuses on a specific network component: Section 2.1 is related to transmission and distribution cables; Section 2.2 is related to transformation substations and distribution panels; Section 2.3 is related to switches; and Section 2.4 is related to batteries.

2.1. Transmission and Distribution Lines

During the operating lifetime of an electrical transmission and distribution power line, its components show signs of physical fatigue due to ageing wear in particular operating conditions or specific unexpected damage. At present, in the EU, the electric grid extends for more than 10^6 km and is expected to strongly increase in length because of the foreseen massive electrification. Due to such a large and pervasive diffusion, power lines and cables can be considered as the most pervasive source of air pollution in terms of toxic emissions into the atmosphere in cases of both overheating and fire. This last occurrence is, of course, the most dangerous. Burjupati and Arunjothi, for example, studied the toxic emissions associated with cable fires and concluded that even lethal concentrations of CO, CO_2, HCHO, HCl, and SO_2 can be reached in a burning event [18]. The main problem is obviously related to the combustion of the polymers and the various additives used for cable fabrication, with the burning of plastic components being well known in the literature as a source of toxic compounds at concentrations associated with this type of phenomenon, which are always very high. A detailed list of 40 pollutants released during such a burning event is reported by Ortner and Hensler [19]. It is worth noting here that such a problem is also correlated to any of the plastic-based components, such as the plastic cabinets used throughout the network. In case of cables overheating, dioctyl phthalate (DOP) and 2-ethylhexanol (2-EH) are normally released well before the burning event. Interestingly enough, an increased risk of overheating has also been reported for underground cables used for distribution networks if they partly come from overhead cables, especially in the case of solar exposition [20]. From the point of view of air pollution, both in case of overheating and in case of burning, the most appropriate equipment to detect specific air pollutants is FTIR, XPS, gas chromatography, and other similar techniques. These tools have the disadvantage of being expensive, complex, and unsuitable for continuous analysis, as Densley points out in Ref. [21]. However, in the case of overheating of electrical cables, solid-state monitoring-based solutions have recently been proposed that allow for the development of equipment suitable for more distributed monitoring. Han and coworkers, for example, proposed MOX sensors for the detection of DOP and 2-EH compounds generally present in the overheating phenomena of many electrical cables [22]. Although the work does not provide any evidence on the concentrations of the gases released during the overheating event and subsequently detected, the authors conclude that the solid-state sensors are suitable for detecting the presence of DOP and 2-EH well before the onset of fire phenomena. This technique can therefore be useful for preventing fire damage. Liu and coworkers compared both commercial and homemade sensor arrays for the same purpose to detect DOP, 2-EH, and benzene [23]. In addition, in this study, the concentrations of the target gases are not reported. The work concludes, similarly to Han and coworkers, that the investigated technique allows for the detection of DOP and 2EH well before the ignition of fire in electrical cables. A similar conclusion was reached by Knoblauch and co-authors [24], who studied a solid-state sensor array based on SnO_2 with different additives. The applied system is shown to be able to detect toxic components (CO, propylene) even before the cables show changes in color due to overheating. From the point of view of the concentrations emitted, one of the few relevant studies concerns the emission of toxic gases from the ageing of electrical cables used in nuclear power plants (and therefore subject to particularly restrictive regulations). This study shows how, in the case of twenty-year ageing structures, CO, HCl, and HBr concentrations that exceed the permitted limits are measured, whereas in the case of SO_2, a sensible increase in gas emission over time is observed. In conclusion, a significant increase in the overall toxicity index proposed for the system under examination can be measured in any case [25]. Transmission networks exhibit peculiar properties related to their HV operation. High-voltage power lines (HVPLs) may in fact behave as a source of charged aerosols and ions. Specifically, corona effects in HVDC lines usually generate atomic oxygen and other radicals, including OH or ions, mostly through the ionization of nitrogen molecules and their compounds. Deposition of particulate matter or dirt generally enhances the effect and hence ion production [26]. The production of ions and aerosols poses environmental and safety issues because it can add to inhalable pollutants in directly exposed populations, including HVPL maintenance workers [27,28]. Recently, Jung et al. implemented a measurement campaign in 2019 to establish correlations between aerosol concentrations and operative parameters near HVPLs. Their results confirmed that measured concentrations are correlated with operative conditions and specifically with current and magnetic fields, whereas a high correlation was found between fine particulate (equal or smaller than 10 um) concentrations and humidity [29]. It is worth noting that under extreme weather conditions, elastic extension or fatigue failures in conductors, poles, crossarms, or surrounding objects (trees) have been found to be responsible for wildfire ignition, generating volatiles whose detection can act as an alarm trigger [30]. Because of the corona effect, HVPLs are found to be sources of atmospheric pollutants such as ozone and nitrogen oxides [31]. Therefore, power transmission networks can be considered as a collection of connected linear sources of air pollution. It is extremely difficult to quantify the actual absolute (and hence relative) impact on the recorded pollution levels, which may vary locally in space and time due to accumulation and transport patterns depending on weather and topographic conditions [32]. Elansky et al. showed that ozone levels near 220 kV powerlines are 2 ppb higher than background recordings, whereas an average excess of 2.5 ppb to 4.6 ppb was recorded at multiple sites near 500 kV powerlines [33]. They also attempted an overall evaluation, revealing that emissions by high-voltage power transmission lines during the 1990s could have globally accounted for 400–600 $\times 10^3$ tons/year, at least 0.1% of ozone tropospheric formation by photochemical processes. Local exposure has also been investigated in other studies. Cociorva and co-authors quantified the potential additional intake of pollutants in the area of HVPLs through ad hoc measurement campaigns. Peak ozone and nitrogen oxide concentrations were found to exceed levels measured in the surrounding background areas by 13% and 30%, respectively [31]. Interestingly, hourly concentration patterns reported anomalous peaks during the night. Valuntaite and co-authors focused on ozone measurements near two 330 kV high-voltage transmission lines arranged parallel to each other and found an excess of 2% in terms of atmospheric ozone near the lines [34]. Dirty or faulty insulators are known cases of arcing issues that may cause increased emissions of ozone. Ionized ozone may in turn lower air insulation properties, causing direct phase-to-phase or phase-to-metal cage discharge with considerable risks for infrastructural faults targeting distribution nodes. Detection of high levels of ozone can be considered an alarming condition, signaling impending faults. Figure 2 provides a graphical representation of the most relevant polluting issues concerning power lines. On the left, the major pollutants observed in the case of cables overheating or the corona effect are shown, while the case of cables burning is shown on the right.

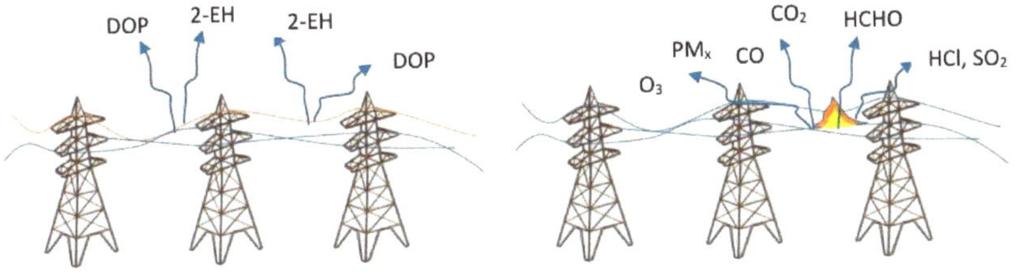

Figure 2. Polluting emissions in case of power lines overheating and the corona effect (**left**) and cables burning (**right**).

2.2. Switchgears

While the association between ozone and electrical discharge is well known in the electrical engineering maintenance field, there is limited use of volatiles or particulate monitoring tools in electrical power transmission and distribution components. A notable example is reported in the 2020 study by Kakar [35]. There, an IoT system based on ozone sensors was developed for switchgear monitoring purposes. Anomalous conditions are screened and detected for predictive maintenance applications. This capability can be exploited to activate and optimize maintenance actions on switchgear components. On the other hand, for its direct impact on fugitive currents in insulators (also captured in IEC68150 [36] design recommendations), air pollution analysis and, in particular, aerosol concentration measurements were also used in [37] to assess status and predict faults in HV disconnectors, which was reviewed in [38].

2.3. Power Transformers

Power transformers are critical nodes of the electric power grid. These generally highly reliable systems are expected to have an useful operational lifetime of more than 25 years when operating temperatures are maintained between 65 °C and 95 °C, but this can be extended to up to 60 years when these conditions are preserved and proper maintenance regularly conducted. Gas generation in transformers is a relevant process connected to regular operation. Therefore, it is an important indicator of its health status [39].

In fact, for oil- and silicone-immersed transformers, slight damage to different parts of this infrastructure node can lead to an increase in dissolved gases that can be detected well in advance of a critical malfunction arising [40,41]. During normal operations, several analytes and compounds are released in the oil, including hydrogen (H_2), methane (CH_4), acetylene (C_2H_2), ethylene (C_2H_4), ethane (C_2H_6), carbon monoxide (CO), and carbon dioxide (CO_2), most of which are found at low concentrations. Fast increases in their absolute or relative concentrations are anomalous events that can signal impending faults. Thermal failures, discharges, long-term exposure to electromagnetic fields, and the presence of water in the oil facilitate cracking and unpredicted chemical reactions, in turn generating H_2, CO, and VOCs that dissolve into the oil itself and whose concentrations are related to the temperature, as shown in Table 1 [42]. Hence, continuous monitoring of the mixture of dissolved gas by DGA can be a powerful tool for predictive maintenance applications for these components [43]. Chemical sensors, including gas sensors, have also been reported for use in continuous monitoring applications for power transformers, and IEEE standards based on fuzzy logic or machine learning have been proposed to recognize the type of faults and their origin, such as excessive heat, arcing, or discharge phenomena [44]. Under specific conditions, peculiar compounds are released in oil, which are consequences of specific types of faults. Furfural, for example, is released because of insulation paper (winding) ageing. Therefore, furfural content or its proxies (primarily CO and CO_2 and their ratio) can be exploited to obtain an estimate of the degree of polymerization in the winding insulation to predict the transformer's lifetime, as attempted in [45], or transformer status, as attempted in [46,47]. Insulation deterioration can also be highlighted by carbon monoxide emissions. Acetylene may be a proxy signal for overheating, partial discharge, or even arcing (which in turn can also induce overheating).

Table 1. H_2, CO, and VOCs observed during transformer operation, whose concentrations are related to the temperature and to the fault typology.

GAS	Fault Type	Concentration–Temp		
		<250 °C	250 °C < t < 350 °C	>350 °C
H_2	Discharge, arcing	Low	Average (growing)	High
C_2H_2	Temp fault	Absent	Low	High
CO	Temp fault, cellulose	Absent	Low	High
C_2H_2	Arcing	Absent	Absent	High
CH_4	Discharge	Low	High	Low
C_2H_6	Discharge	Absent	High	Low

Depending on the solubility of the target gas, fault detection can be best observed with gas sensors or by sampling of the oil reservoir, its headspace, or the transformer headspace itself. Hydrogen is, in fact, likely to easily escape the transformer and could be found at significant concentrations in the transformer headspace. This detection, however, will not provide information on the location of its production. Nitrogen, CO, and methane, which also share low solubility in the oil, are instead associated with the presence of both thermal and electrical faults [38], and their presence can be detected in the transformer headspace. Monitoring gas release in insulator fluids or headspace emissions is considered a critical issue for the preservation of these valuable assets. Regarding fugitive emissions, it has already been shown that distributed gas sensing may be an effective approach for their monitoring. In a study by Minglei et al., a complete solution following this approach was exploited and lab-tested for the monitoring of hydrogen emissions [48]. Figure 3 depicts the gas emissions that can be found in the transformer environment and whose relative concentrations can be correlated to specific faults.

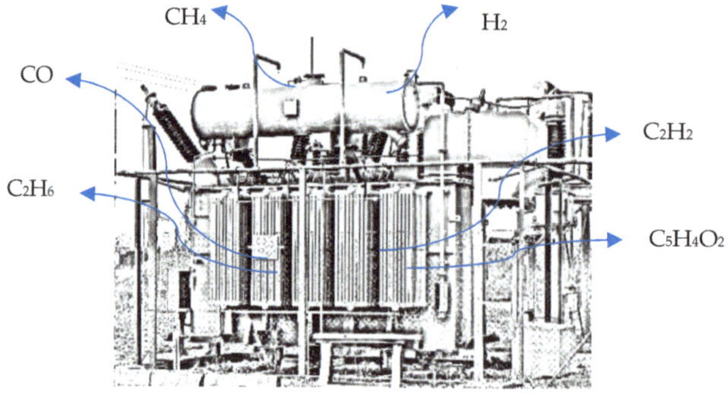

Figure 3. H_2, Methane, CO, and the various VOCs that are detectable in the electric transformer environment.

2.4. Batteries

In modern power grids, batteries are a basic facility that provides inertia for system balance in the presence of distributed and variable energy sources. Unfortunately, they are prone to catastrophic malfunctions leading to fires and, more rarely, explosions [17].

In the automotive industry, new regulations now oblige that dangers to car passengers be signaled minutes before a dangerous event may occur [49]. Once started, fires can be particularly difficult to extinguish when the storage system size is significant because of the large DC power arc. The most common causes are electrical or mechanical runaways caused by overcharging, undercharging, or even short circuits caused by connected devices, including inverters. Harmful or dangerous gases are emitted before the fire takes place or in its initial stages during thermal runaways. Among them are HF at high volumes, which could be fatal when emitted in confined environments, carbon monoxide, methane, and VOCs. Ethylene carbonate, ethyl-methyl carbonate, diethyl carbonate, dimethyl carbonate, and propylene carbonate can be found in Li-ion batteries off-gassing from electrolytes and are considered the most relevant [50,51]. Actual ignition may render most of these components oxidized to harmless combustion by-products; however, since ignition must be avoided by any means, ignition denial may cause them to reach concentrations that easily lead to fatal outcomes for inhaling humans [52]. Even before thermal runaway may start, it has been demonstrated that CO_2 emissions may signal this impending dangerous event, and chemical sensors may be used for alarming and for identifying the need for venting [53]. An increase in the concentration of hydrogen may also be considered as an early thermal runaway precursor due to unwanted electrolysis along with electrolyte vapor and gases produced by degassing of failing LIB batteries closer to thermal runaway events [54]. Because of the variety of emitted compounds and analytes, the use of a chemical multisensory device coupled with pattern recognition software for enhancing detection performance, rejecting false positives, and avoiding unwanted interference is recommended. In Table 2, the gaseous emissions so far reported by different authors and the main triggering events are summarized.

Table 2. Anomalous electric battery gas emissions so far reported.

Expected Gas/Volatile Emission	Relevant Event	Reference
HF (Fluoridic Acid)	Thermal runaway, suppressed fire and explosion	[52,53,55]
H_2 (Hydrogen)	Unwanted electrolysis, thermal runaway	[53,54,56–58]
CO (Carbon Monoxide)	Thermal runaway, fire and explosion	[52,53]
CO_2 (Carbon Dioxide)	Thermal runaway, fire and explosion	[53,56,58]
EMC (Ethyl Methyl Carbonate)	Electrolyte loss or vaporization, thermal runaway, suppressed fire and explosion, unsuppressed fire and explosion (traces)	[50–52,54–56,58,59]
DMC (Diethyl Methyl Carbonate)	Electrolyte loss or vaporization, thermal runaway, suppressed fire and explosion, unsuppressed fire and explosion (traces)	[50–52,54–56,58,59]
EC (Ethylene Carbonate)	Electrolyte loss or vaporization, thermal runaway, unsuppressed fire and explosion (traces), suppressed fire and explosion	[50–52,54–56,58,59]
DEC (Diethyl Carbonate)	Electrolyte loss or vaporization, unsuppressed fire and explosion (traces), suppressed fire and explosion	[50–52,54–56,58,59]
C_2H_4 (Ethylene)	Thermal runaway	[54,58]
CH_4 (Methane)	Thermal runaway	[54]
C_2H_6 (Ethane)	Thermal runaway	[54,56]
C_2H_2 (Acetylene)	Thermal runaway	[54]
C_4H_{10} (Butane)	Thermal runaway	[54]
C_6H_6 (Benzene)	Unsuppressed fire and explosion (traces), suppressed fire and explosion	[52–54]
$C_6H_5CH_3$ (Toluene)	Unsuppressed fire and explosion (traces), suppressed fire and explosion	[52]

Table 2. *Cont.*

Expected Gas/Volatile Emission	Relevant Event	Reference
$(C_6H_5)_2$ (Biphenyl)	Unsuppressed fire and explosion (traces), suppressed fire and explosion	[52]
C_3H_4O (Acrolein)	Unsuppressed fire and explosion (traces), suppressed fire and explosion	[52]

Finally, in Figure 4, a graphical representation of gas emissions in failing batteries is presented.

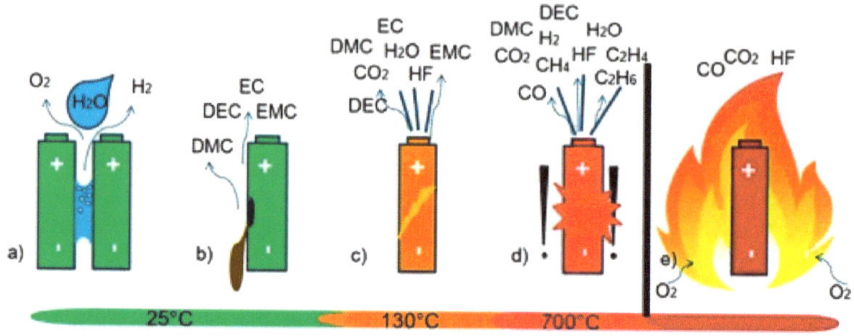

Figure 4. Gas emissions in failing batteries before unwanted electrolysis (**a**), in mechanical failure (**b**), and during thermal runaways (**c**,**d**) as conceptualized and depicted in [54]. The final ignition case is also considered (**e**). Refer to Table 2 for acronyms.

3. Commercial Systems

As mentioned above, during their operating lifetime, electric grid components may be sources of air polluting gases and aerosols. The effect is particularly severe up to the point of resulting in lethal concentrations in the case of malfunctions occurring in one or more equipment components. Continuous monitoring enables so-called data-based maintenance, a concept that assumes that component faults can be forecasted and that ageing processes can be monitored to properly schedule on-site intervention. Several companies are currently investing in research regarding methods for real-time monitoring of components of the electrical infrastructure to improve the reliability of power distribution systems. Honeywell (Charlotte, NC, USA) offers a large variety of gas sensors connected to the cloud that are directly remotely accessible for both reconfiguration and data transfer. The XNX, XCD, XRL, and S3000 series [60] all offer the possibility to monitor CO and CO_2. The XNX series also allows for the monitoring of H_2, while the XRL series can sample O_3. The mentioned quantities are important data in the case of transmission and distribution lines in the proximity of which these gases can be found, and the presence of these gases could reveal an upcoming fault. General Monitors (Irvine, CA, USA) has in its portfolio the toxic gas detector TS4000 [61] with replaceable electrochemical cells, and the system is reconfigurable to satisfy any specific requirements. This system can be used for transmission and distribution line monitoring. Honeywell series 700-AS switches are equipped with two types of sensors: catalytic sensors and electrochemical sensors [62]. WoMaster has in-catalog sensors for H_2, CO, CO_2, O_3, etc. They offer the possibility of putting together more sensors in a single case. Each sensor communicates with an RS485 bus connected to a gateway that exploits LORA communication or the Internet. Ad hoc WI-FI + LTE gateways and LORA NBIoT gateways to complete the network are also available [63]. The collected data can be sent to a proprietary server with no additional costs. These sensors are suitable for applications in the energy-transforming domain, as well as in transmission line applications or power station monitoring. Bosch proposes an all-in-one solution (BME688) that exploits artificial intelligence to obtain measurements of

VOCs, VSCs, H_2, CO, pressure, temperature, and humidity with a single sensor that can communicate the measurements to external systems [64]. This solution can be applied in the case of power transformers and battery monitoring. If a higher sampling frequency is required, Gas Sensing Solutions proposes the SprintIR®-R CO_2 Sensor (CO_2 Meter, Ormond Beach, FL, USA) for the measurement of carbon dioxide [65]. They claim that this sensor can take 50 measurements per second, ideal for high-speed monitoring of CO_2 or locations where gas concentration might change rapidly. Honeywell proposes the Li-ion Tamer Gen3 sensor, which detects the electrolytic vapors of lithium-ion batteries. It also provides monitoring of the temperature and humidity of the environment [66]. Readings of gas, temperature, and humidity are sent via CanBus to the hub for storage or resending. These last sensors are useful for battery condition monitoring. Honeywell series 700 RL sensors are methane sensors that can be used in battery status monitoring. Methane is in fact emitted before a fire occurs during thermal runaway. In addition, CO_2 sensors can be used for fire prevention purposes in the case of battery monitoring, and, in this case, the SprintIR®-R CO_2 Sensor mentioned before can be used to quickly sample the air surrounding the battery to capture the rapid increase in the concentration of CO_2. Schneider (Rueil-Malmaison, France) offers a closed solution for cable overheating, especially for power cabinets [67]. Vaisala (Vantaa, Finland) has been operating in the field of power measurement for many years. For power transformer condition monitoring there are several solutions, such as the OPT 100 [68], a DGA monitoring system claimed to require zero maintenance that is capable of auto-calibration.

4. Discussion

In this paper, a subset of research works highlighting the correlation between power grid electrical equipment operating conditions and specific gas emissions is reported and reviewed. It has been observed that in the case of transmission and distribution cables overheating, substantial concentrations of DOP and 2-EH are released into the air. Anomalous concentrations of ozone in the proximity of HV power lines, mainly due to the corona effect, have also been reported. However, only a limited subset of studies have analyzed such emissions in terms of their air pollution effect. Emitted ions are also known precursors of aerosol pollutants, but their concentration increase in the immediate surroundings is still disputed, although increased deposition due to HV has been clearly shown. To the best of our knowledge, the contribution of these emissions to air quality has not been considered in pollutant emissions inventories as they are expected to be, at least at present, too low to influence the balance. However, some authors have reported that the quantitative balance of ozone at the surface can be significantly affected near HVPLs. Ozone is present also in the case of the malfunction of switchgears and may itself contribute to further damages. Due to their extension, transmission and distribution power line cables should be carefully monitored in terms of their possible polluting emissions in case of cables burning when lethal limits for humans can be exceeded. During normal operation, oil-immersed power transformers release various gases both in the oil and eventually in air, including H_2, CH_4, C_2H_2, C_2H_4, CO, and CO_2. The concentration of these gases has been observed to increase rapidly in the case of equipment anomalies. Electric batteries carry with them an intrinsic danger of overheating and thermal runaway that can lead to fire. In the time directly preceding the fire, VOCs, CO_2, CH_4, and H_2 are released. Based on this evidence, some considerations can be made regarding the opportunity, perhaps the need, to enhance overall system reliability using condition monitoring, or in other words by electric system data-based maintenance operating by continuously monitoring the main asset of the system itself. Traditional approaches are based on redundancy, selectivity, and draw out technologies. Continuous condition monitoring allows system availability to be obtained at a reasonable cost. Installing IoT gas sensors along with power grid devices allows single components to communicate their status continuously. When a component reaches a predefined limit, it can be scheduled to be substituted at a time before the failure, thus avoiding environmental issues. In the case analyzed in this work, monitoring could be continuously performed on samples of the concentration of gases in proximity of the facility under observation. Several market solutions that allow for the design of an interconnected network of sensors are already available. Once all the useful information is collected, machine learning algorithms can be trained to recognize classified patterns of forthcoming failure. Furthermore, the grade of pollution can be estimated because, during normal operation, toxic gases are released.

5. Conclusions

The above-reviewed papers show that an operating electric transmission and distribution network may emit several atmospheric pollutants both in its normal everyday operation, such as in the case of HVPL cables because of the corona effect, and because of increasingly dangerous anomalies. Such emissions can, in general, be correlated to the specific fault a component is going to suffer. As a result, several IoT-based sensor systems that are suitable for implementing preventive fault detection are under investigation or even already on the market. It is worth noting that few papers have discussed the case of distribution networks, which is, on the contrary, quite relevant in terms of their possible effect on city environments. Moreover, quantitative investigations of the emitted gas concentrations are almost entirely absent. Future research should probably focus on these two aspects that are of relevance in terms of the expected massive electrification.

Author Contributions: Conceptualization, S.D.V., A.D.G. and G.D.F.; methodology, S.D.V., A.D.G. and G.D.F.; resources, G.D.F.; writing—original draft preparation, S.D.V., A.D.G. and G.D.F.; writing—review and editing, S.D.V., A.D.G. and G.D.F. All authors have read and agreed to the published version of the manuscript.

Funding: This research received partial funding from the EU Horizon 2020 Research and Innovation project VIDIS, grant agreement 952433, and from Ministero per lo Sviluppo Economico, Fondo per la Crescita Sostenibile, under the framework "Accordi per l'innovazione di cui al D.M. 31 Dicembre 2021 e DD 18 Marzo 2022" under project MARTA, n.: F/310193/01-02/X56.

Institutional Review Board Statement: Not applicable.

Informed Consent Statement: Not applicable.

Data Availability Statement: Not applicable.

Conflicts of Interest: The authors declare no conflict of interest.

References

1. World Climate Action Summit, Dubai, United Arab Emirates, 1–2 December 2023. Available online: https://www.cop28.com/ (accessed on 4 January 2024).
2. The Energy Institute. Statistical Review of World Energy 2023. Available online: https://www.energyinst.org/statistical-review (accessed on 28 December 2023).
3. Global Energy Transformation: A Roadmap to 2050—2019 Edition. Available online: https://www.irena.org/publications/2019/Apr/ (accessed on 28 December 2023).
4. Eionet Report-ETC/CME 7/2020-December 2020 Renewable Energy in Europe—2020 Recent Growth and Knock-On Effects. Available online: https://www.eionet.europa.eu/etcs/etc-cme/products/etc-cme-reports/etc-cme-report-7-2020-renewable-energy-in-europe-2020-recent-growth-and-knock-on-effects (accessed on 28 December 2023).
5. BP Statistical Review of World Energy June 2006. Available online: www.bp.com/statisticalreview (accessed on 28 December 2023).
6. Liu, Y.; Chen, H.; Gao, J.; Li, Y.; Dave, K.; Chen, J.; Federici, M.; Perricone, G. Comparative analysis of non-exhaust airborne particles from electric and internal combustion engine vehicles. *J. Hazard. Mater.* **2021**, *420*, 126626. [CrossRef] [PubMed]
7. Peters, D.R.; Schnell, J.L.; Kinney, P.L.; Naik, V.; Horton, D.E. Public health and climate benefits and trade-offs of US vehicle electrification. *GeoHealth* **2020**, *4*, e2020GH000275. [CrossRef]
8. Lin, W.Y.; Hsiao, M.C.; Wu, P.C.; Fu, J.S.; Lai, L.W.; Lai, H.C. Analysis of air quality and health co-benefits regarding electric vehicle promotion coupled with power plant emissions. *J. Clean. Prod.* **2020**, *247*, 119152. [CrossRef]
9. Bashir, M.A.; Dengfeng, Z.; Amin, F.; Mentel, G.; Ali Raza, S. Transition to greener electricity and resource use impact on environmental quality: Policy based study from OECD countries. *Util. Policy* **2023**, *81*, 101518. [CrossRef]
10. International Energy Agency. CO_2 Emissions from Fuel Combustion: Overview, Global CO_2 emissions by Sector. Available online: https://www.iea.org/reports/co2-emissions-in-2022 (accessed on 4 December 2023).

11. Penaherrera, F.; Pehlken, A. Limits of life cycle assessment in the context of the energy transition and its material basis. In *The Material Basis of Energy Transitions*; Bleicher, A., Pehlken, A., Eds.; Academic Press: London UK, 2020; Chapter 8; pp. 121–140.
12. Redlarski, G.; Lewczuk, B.; Zak, A.; Koncicki, A.; Krawczuk, M.; Piechocki, J.; Jakubiuk, K.; Tojza, P.; Jaworski, J.; Ambroziak, D.; et al. The influence of electromagnetic pollution on living organisms: Historical trends and forecasting changes. *BioMed Res. Int.* **2015**, *2015*, 234098. [CrossRef] [PubMed]
13. WHO; IARC. *Non-Ionizing Radiation, Part 1: Static and Extremely Low-Frequency (ELF) Electric and Magnetic Fields*; IARC Press: Lyon, France, 2002. Available online: https://www.ncbi.nlm.nih.gov/books/NBK390731/ (accessed on 28 December 2023).
14. Malagoli, C.; Malavolti, M.; Wise, L.A.; Balboni, E.; Fabbi, S.; Teggi, S.; Palazzi, G.; Cellini, M.; Poli, M.; Zanichelli, P.; et al. Residential exposure to magnetic fields from high-voltage power lines and risk of childhood leukemia. *Environ. Res.* **2023**, *232*, 116320. [CrossRef]
15. Piana, E.A.; Roozen, N.B. On the Control of Low-Frequency Audible Noise from Electrical Substations: A Case Study. *Appl. Sci.* **2020**, *10*, 637. [CrossRef]
16. Li, L.; Meng, W.; Li, Q.; Wang, Y.; Zheng, X.; Wang, H. Research Progress on Audible Noise Emitted from HVDC Transmission Lines. *Energies* **2023**, *16*, 4614. [CrossRef]
17. Swain, A.; Abdellatif, E.; Mousa, A.; Pong, P.W.T. Sensor Technologies for Transmission and Distribution Systems: A Review of the Latest Developments. *Energies* **2022**, *15*, 7339. [CrossRef]
18. Burjupati, R.; Arunjothi, R. Lethal combustion product evaluation of polymeric materials used in power cables. In Proceedings of the 9th International Conference on Insulated Power Cables, Paris, France, 21–25 June 2015.
19. von Kunststoffbränden, B. Bei Einer Störung des Bestimmungsgemäßen Betriebs Entstehende Stoffe nach den Anhängen II–IV der 12. BImSchV, Tab. 3.1 pag. 19. Available online: https://www.lfu.bayern.de/luft/doc/kunststoffbraende.pdf (accessed on 4 December 2023). (In German).
20. Czapp, S.; Ratkowski, F.; Szultka, S.; Tomaszewski, A. Overheating of Underground Power Cable Line Due to Its Partial Exposition to Solar Radiation. In Proceedings of the 2019 24th International Conference on Methods and Models in Automation and Robotics (MMAR), Miedzyzdroje, Poland, 26–29 August 2019; pp. 396–400. [CrossRef]
21. Densley, J. Ageing mechanisms and diagnostics for power cables—An overview. *IEEE Electr. Insul. Mag.* **2001**, *17*, 14–22. [CrossRef]
22. Han, J.; Chen, W.; Yu, A.; Yi, J. Detection of Semi-volatile Plasticizers as a Signature of Early Electrical Fire. *Front. Mater.* **2019**, *6*, 250. [CrossRef]
23. Liu, Y.; Geyik, U.; Kobald, A.; Yang, A.; Wang, X.; Weimar, U.; Rong, M.; Bârsan, N. Overheat diagnosis of power cable based on gas sensors: Device/material exploration. *Sens. Actuators B Chem.* **2022**, *350*, 130837. [CrossRef]
24. Knoblauch, J.; Illyaskutty, N.; Kohler, H. Early detection of fires in electrical installations by thermally modulated SnO2/additive-multi sensor arrays. *Sens. Actuators B Chem.* **2015**, *217*, 36–40. [CrossRef]
25. Seo, H.J.; Kim, N.K.; Lee, M.C.; Lee, S.K.; Moon, Y.S. Investigation into the toxicity of combustion products for CR/EPR cables based on aging period. *J. Mech. Sci. Technol.* **2020**, *34*, 1785–1794. [CrossRef]
26. Bailey, W.; Weil, D.; Stewart, J. *HVDC Power Transmission Environmental Issues Review*; Report ORNL/Sub-95SR893/2; Oak Ridge National Laboratory: Oak Ridge, TN, USA, 1997.
27. Bailey, W.; Johnson, G.B.; Bracken, T. *Dan, Charging and Transport of Aerosols near AC Transmission Lines: A Literature Review*; ESPRI Technical Report No. 1008148; National Grid Transco Plc.: London, UK, 2003; p. 1008148.
28. Fews, A.P.; Henshaw, D.L.; Keitch, P.; Close, J.J.; Wilding, R.J. Increased exposure to pollutant aerosols under high voltage power lines. *Int. J. Radiat. Biol.* **1999**, *75*, 1505–1521. [CrossRef]
29. Jung, J.S.; Lee, J.W.; Mailan Arachchige Don, R.K.; Park, D.S.; Hong, S.C. Characteristics and potential human health hazards of charged aerosols generated by high-voltage power lines. *Int. J. Occup. Saf. Ergon.* **2019**, *25*, 91–98. [CrossRef] [PubMed]
30. Mitchell, J.W. Power line failures and catastrophic wildfires under extreme weather conditions. *Eng. Fail. Anal.* **2013**, *35*, 726–735. [CrossRef]
31. Corciova, D.; Mitiu, M.A.; Mincu, N.; Raischi, M.S. Emissions of pollutants and air quality in the area of influence of high voltage overhead electrical lines. In Proceedings of the International Conference Air and Water Components of the Environment Conference, Cluj, Napoca, 17–19 March 2017; pp. 283–288.
32. Bekesiene, S.; Meidute-Kavaliauskiene, I.; Vasiliauskiene, V. Accurate Prediction of Concentration Changes in Ozone as an Air Pollutant by Multiple Linear Regression and Artificial Neural Networks. *Mathematics* **2021**, *9*, 356. [CrossRef]
33. Elansky, N.F.; Panin, L.V.; Belikov, I.B. Influence of High-Voltage Lines on the Surface Ozone Concnetration. *Atmos. Ocean. Phys.* **2001**, *37*, S10–S23.
34. Valuntaite, V.; Šerevičiene, V.; Girgždiene, R. Ozone concentration variations near high-voltage transmission lines. *J. Environ. Eng. Landsc. Manag.* **2009**, *17*, 28–35. [CrossRef]
35. Kakar, M. IOT Applications in Distribution Substations Review and New Approaches. Master's Thesis, Polytechnic of Milan, Milan, Italy, 2020. Available online: https://hdl.handle.net/10589/167163 (accessed on 6 January 2024).
36. *IEC TR 61850-1:2013*; Communication Networks and Systems for Power Utility Automation—Part 1: Introduction and Overview. IEC: Geneva, Switzerland, 2013.

37. Saraceni, R. HV Disconnector: Smart Integration and Real Time Monitoring of Pollution for Predictive Maintenance in APM Ecosystem. Master's Thesis, Polytechnic of Milan, Milan, Italy, 2021. Available online: https://thesis.unipd.it/handle/20.500.12608/28985 (accessed on 6 January 2024).
38. Alsumaidaee, Y.A.M.; Yaw, C.T.; Koh, S.P.; Tiong, S.K.; Chen, C.P.; Ali, K. Review of Medium-Voltage Switchgear Fault Detection in a Condition-Based Monitoring System by Using Deep Learning. *Energies* **2022**, *15*, 6762. [CrossRef]
39. Rutledge, C. Monitoring Gas Generation in Transformers. In Proceedings of the IEEE/PES Transmission and Distribution Conference and Ex-Position (T&D), Denver, CO, USA, 16–19 April 2018; pp. 1–9.
40. *IEEE Std C57.104-2019 (Revision of IEEE Std C57.104-2008)*; IEEE Guide for the Interpretation of Gases Generated in Mineral Oil-Immersed Transformers. IEEE: Piscataway, NJ, USA, 2019; pp. 1–98. [CrossRef]
41. Bustamante, S.; Mario, M.; Alberto, A.; Pablo, C.; Alberto, L.; Raquel, M. Dissolved Gas Analysis Equipment for Online Monitoring of Transformer Oil: A Review. *Sensors* **2019**, *19*, 4057. [CrossRef]
42. Patel, M.K.; Patel, A.M. Simulation and Analysis of DGA Analysis for Power Transformer Using Advanced Control Methods. *Asian J. Converg. Technol.* **2021**, *7*, 102–109. [CrossRef]
43. Sun, C.; Ohodnicki, P.R.; Stewart, E.M. Chemical Sensing Strategies for Real-Time Monitoring of Transformer Oil: A Review. *IEEE Sens. J.* **2017**, *17*, 5786–5806. [CrossRef]
44. Aburaghiega, E.; Emad Farrag, M.; Hepburn, D.; Haggag, A. Enhancement of Power Transformer State of Health Diagnostics Based on Fuzzy Logic System of DGA. In Proceedings of the 2018 Twentieth International Middle East Power Systems Conference (MEPCON), Cairo, Egypt, 18–20 December 2018; pp. 400–405. [CrossRef]
45. Elmashtoly, A.M.; Chang, C.-K. Prognostics health management system for power transformer with IEC61850 and Internet of Things. *J. Electr. Eng. Technol.* **2020**, *15*, 673–683. [CrossRef]
46. Cennamo, N.; De Maria, L.; D'Agostino, G.; Zeni, L.; Pesavento, M. Monitoring of Low Levels of Furfural in Power Transformer Oil with a Sensor System Based on a POF-MIP Platform. *Sensors* **2015**, *15*, 8499–8511. [CrossRef] [PubMed]
47. Prasojo, R.A.; Diwyacitta, K.; Suwarno; Gumilang, H. Transformer Paper Expected Life Estimation Using ANFIS Based on Oil Characteristics and Dissolved Gases (Case Study: Indonesian Transformers). *Energies* **2017**, *10*, 1135. [CrossRef]
48. Minglei, S.; Xiang, L.; Changping, Z.; Jiahua, Z. Gas concentration detection using ultrasonic based on wireless sensor networks. In Proceedings of the 2nd International Conference on Information Science and Engineering, Hangzhou, China, 4–6 December 2010; pp. 2101–2106.
49. *GB 38031-2020*; Electric Vehicles Traction Battery Safety Requirements. China National Standardization Administration Committee: Beijing, China, 2020.
50. Blum, A.F.; Long, R.T., Jr. *Fire Hazard Assessment of Lithium-Ion Battery Energy Storage Systems*; Springer: New York, NY, USA, 2016; Volume XXI, pp. 1–90.
51. Park, S.-H.; Lee, W.-J.; Lee, D.-J. On-Site Evaluation of BESS Fire Prevention System Considering the Environmental Factors. In Proceedings of the 24th International Conference on Electrical Machines and Systems (ICEMS), Gyeongju, Republic of Korea, 31 October–3 November 2021; pp. 2264–2269.
52. Nedjalkov, A.; Meyer, J.; Köhring, M.; Doering, A.; Angelmahr, M.; Dahle, S.; Sander, A.; Fischer, A.; Schade, W. Toxic Gas Emissions from Damaged Lithium Ion Batteries—Analysis and Safety Enhancement Solution. *Batteries* **2016**, *2*, 5. [CrossRef]
53. Cai, T.; Valecha, P.; Tran, E.; Engle, B.; Stefanopoulou, A.; Siegel, J. Detection of Li-ion battery failure and venting with Carbon Dioxide sensors. *eTransportation* **2021**, *7*, 100100. [CrossRef]
54. Essl, C.; Seifert, L.; Rabe, M.; Fuchs, A. Early Detection of Failing Automotive Batteries Using Gas Sensors. *Batterie* **2021**, *7*, 25. [CrossRef]
55. Larsson, F.; Andersson, P.; Blomqvist, P.; Mellander, B.-E. Toxic fluoride gas emissions from lithium-ion battery fires. *Sci. Rep.* **2017**, *7*, 10018. [CrossRef] [PubMed]
56. Essl, C.; Golubkov, A.W.; Gasser, E.; Nachtnebel, M.; Zankel, A.; Ewert, E.; Fuchs, A. Comprehensive hazard analysis of failing automotive Lithium-ion batteries in overtemperature experiments. *Batteries* **2020**, *6*, 30. [CrossRef]
57. Wang, M.; Wang, Z.; Gong, X.; Guo, Z. The intensification technologies to water electrolysis for hydrogen production—A review. *Renew. Sustain. Energy Rev.* **2014**, *29*, 573–588. [CrossRef]
58. Golubkov, A.W.; Fuchs, D.; Wagner, J.; Wiltsche, H.; Stangl, C.; Fauler, G.; Voitic, G.; Thaler, A.; Hacker, V. Thermal-runaway experiments on consumer Li-ion batteries with metal-oxide and olivin-type cathodes. *RSC Adv.* **2014**, *7*, 3633–3642. [CrossRef]
59. Golubkov, A.W. Thermal-runaway of commercial 18650 Li-ion batteries with LFP and NCA cathodes impact of state of charge and overcharge. *R. Soc. Chem.* **2015**, 57171–57186. [CrossRef]
60. Honeywell Product Guide. Available online: https://prod-edam.honeywell.com/content/dam/honeywell-edam/sps/his/it-it/products/gas-and-flame-detection/documents/ha_industrial_fixed_product_guide_1v0_2018.pdf?download=false (accessed on 6 December 2023).
61. General Monitors. Intelligent Sensors for Gas Detection. Available online: https://www.escventura.com/manuals/gm_h2sTS4000_um.pdf (accessed on 6 December 2023).
62. Honeywell Series 700 AS. Available online: https://prod-edam.honeywell.com/content/dam/honeywell-edam/hbt/en-us/documents/manuals-and-guides/user-manuals/notifier-it/HBT-Fire-202205041508-G700C-AS-H703-AS-sk-ita.pdf?download=false (accessed on 6 December 2023).

63. WoMaster Products. Available online: https://www.womaster.eu/products_list_Smart-City-%EF%BC%86-Industrial-IoT_IoT-Environmental-%7C-Water-%7C-Parking-%7C-Waste-Bin-Sensors.htm (accessed on 6 December 2023).
64. Bosch Gas Sensor BME 688. Available online: https://www.bosch-sensortec.com/products/environmental-sensors/gas-sensors/bme688/ (accessed on 6 December 2023).
65. GSS. SprintIR-R CO_2 Sensor. Available online: https://www.gassensing.co.uk/product/sprintir-r-co2-sensor/ (accessed on 6 December 2023).
66. Honeywell. Gas Sensors Li-ion. Available online: https://buildings.honeywell.com/it/it/products/by-category/fire-life-safety/sensors-and-detectors/gas-detectors/li-ion-tamer-gen-3-gas-sensors (accessed on 6 December 2023).
67. Schneider Electric. Power Logic Heat Tag. Available online: https://www.se.com/uk/en/product-range/38501657-powerlogic-heattag/ (accessed on 6 December 2023).
68. Vaisala Optimus DGA Monitor. Available online: https://www.vaisala.com/en/products/instruments-sensors-and-other-measurement-devices/instruments-industrial-measurements/opt100 (accessed on 6 December 2023).

sensors

MDPI

Article

Feasibility Study on the Use of NO_2 and $PM_{2.5}$ Sensors for Exposure Assessment and Indoor Source Apportionment at Fixed Locations

Miriam Chacón-Mateos [1,*], Erika Remy [2], Uta Liebers [3,4], Frank Heimann [5], Christian Witt [3] and Ulrich Vogt [1]

1 Department of Flue Gas Cleaning and Air Quality Control, University of Stuttgart, 70569 Stuttgart, Germany; ulrich.vogt@ifk.uni-stuttgart.de
2 Institute of Physics and Meteorology, University of Hohenheim, 70599 Stuttgart, Germany
3 Institute of Physiology, Charité—Universitätsmedizin Berlin, Corporate Member of Freie Universität Berlin and Humboldt-Universität zu Berlin, 10117 Berlin, Germany; christian.witt@charite.de (C.W.)
4 Department of Pneumology, Evangelische Lungenklinik Berlin Buch, 13125 Berlin, Germany
5 Ambulante Pneumologie mit Allergie Zentrum, 70178 Stuttgart, Germany
* Correspondence: miriam.chacon-mateos@ifk.uni-stuttgart.de

Abstract: Recent advances in sensor technology for air pollution monitoring open new possibilities in the field of environmental epidemiology. The low spatial resolution of fixed outdoor measurement stations and modelling uncertainties currently limit the understanding of personal exposure. In this context, air quality sensor systems (AQSSs) offer significant potential to enhance personal exposure assessment. A pilot study was conducted to investigate the feasibility of the NO_2 sensor model B43F and the particulate matter (PM) sensor model OPC-R1, both from Alphasense (UK), for use in epidemiological studies. Seven patients with chronic obstructive pulmonary disease (COPD) or asthma had built-for-purpose sensor systems placed inside and outside of their homes at fixed locations for one month. Participants documented their indoor activities, presence in the house, window status, and symptom severity and performed a peak expiratory flow test. The potential inhaled doses of $PM_{2.5}$ and NO_2 were calculated using different data sources such as outdoor data from air quality monitoring stations, indoor data from AQSSs, and generic inhalation rates (IR) or activity-specific IR. Moreover, the relation between indoor and outdoor air quality obtained with AQSSs, an indoor source apportionment study, and an evaluation of the suitability of the AQSS data for studying the relationship between air quality and health were investigated. The results highlight the value of the sensor data and the importance of monitoring indoor air quality and activity patterns to avoid exposure misclassification. The use of AQSSs at fixed locations shows promise for larger-scale and/or long-term epidemiological studies.

Keywords: air quality; low-cost sensors; indoor air; exposure assessment; source apportionment; I/O ratio

1. Introduction

Air pollution has long been known to affect health, and it contributes heavily to the Global Burden of Disease [1,2]. As stated in the United Nations Sustainable Development Goals, improving air quality is a pillar of improving global health and well-being (Goal 3), as well as creating safe and sustainable cities (Goal 11) [3]. However, as it stands now, approximately 6.7 million deaths can be attributed to air pollution each year [4]. The most recent WHO air quality guidelines describe air pollution as "the single biggest environmental threat to human health" [5].

The number of epidemiological studies showing the health effects of air pollution has been growing in recent years, but it is far from being complete [6]. With more research emerging, the effects of both long- and short-term exposure to pollutants are now better understood. It is also recognised that the level of exposure at which harmful long-term health problems can occur is significantly lower than was thought before [5,7].

The current standard to measure air quality relies on permanent outdoor monitoring stations. While these stations provide accurate and continuous measurements, both the equipment-purchase and maintenance costs are high. Due to their expense, their spatial distribution across the world is limited [8]. According to Fuller et al. [4], most urban areas in Europe and North America have at least one measurement station, which corresponds to one station for every 100,000 to 500,000 people, approximately. Due to the high spatial variability of pollutant concentrations, the existing distribution of monitoring stations is inadequate for accurately measuring air quality across different microenvironments [7,9].

Standard outdoor monitoring stations leave another gap in understanding air quality's relation to health, in that only ambient outdoor air is measured. The majority of the population in developed countries spends most of their time in indoor environments [10]. Bulot et al. [11] report that indoor pollution exposure causes around half of all pollutant-related deaths and that the associated risks "cannot be accurately studied by outdoor monitoring stations". Any individual's exposure to pollution cannot be fully described without understanding both the indoor and outdoor air quality [12].

As with many problems linked to climate change, there is a clear disparity in pollution exposure across the globe [13,14]. Low- and middle-income countries (LMICs) bear the brunt of deaths and economic loss due to air pollution [4]. Even as conditions improve in some wealthier countries, air quality continues to deteriorate in many LMICs [15]. According to the WHO, this is in part due to economic development in LMICs being reliant on fossil fuel burning [5]. Not only is there global inequality in exposure but also in resources available to effectively measure pollution levels [16,17]. Sub-Saharan Africa is reported by Fuller et al. [4] to have roughly one permanent measurement station per 15.9 million people. This extreme imbalance further emphasizes the need for alternative measurement methods [18].

Air quality sensor systems (AQSSs) for monitoring air quality have shown potential as practical solutions to these needs [19–22]. Current research in this field is focused on two main approaches: the dynamic or direct approach, which utilizes portable sensors, and the static or indirect approach, which relies on stationary sensors. The difference in cost and the portability of smaller sensors allow studies that track individual exposure, in contrast to the bulkier reference-grade instruments. For example, a study using wearable sensors on bicycles demonstrated that the health benefits associated with cycling could be partially offset by exposure to traffic-related air pollution along the route [23]. Moreover, multipollutant AQSSs have been developed and tested for estimating individual-level pollutant doses with promising results [24–26]. However, these studies also highlighted the limitations and significant resources required for evaluating in-transit or commuting data.

The widespread use of AQSSs has also enhanced research on exposure assessment across different microenvironments. A study conducted on a university campus in Beijing, China, used fixed AQSSs to investigate the impact of outdoor-origin $PM_{2.5}$ on potential inhaled doses indoors [27]. Amegah et al. [28] conducted a study at traffic hotspots in Accra (Ghana) using Purple Air PA-II monitors and self-reported health questionnaires and found consistent evidence that $PM_{2.5}$ exposure among street traders increases the occurrence of respiratory and cardiovascular symptoms.

The increased spatio-temporal resolution provided by AQSS networks has the potential to improve our understanding of individual exposure pathways [29]. The future of environmental health studies lies in the integration of big data from sensors and models. Research on data fusion has already explored both the dynamic and static approaches. For instance, recent studies have demonstrated that combining stationary AQSSs with dispersion models can enhance assessments of pollutant exposure, including human mobility patterns [30–32]. Other researchers have focused on integrating model data with information from wearable sensors [33,34], enabling more accurate exposure assessments and better health risk management.

Epidemiological studies will benefit immensely from the enhanced spatial resolution enabled by AQSSs. In particular, resource-limited regions can find in AQSSs a solution to close the data gap left by the scarce or non-existent monitoring and help address air quality management [35]. As stated by Vilcassim and Thurston [7], AQSSs will be a key part of "more democratized, high resolution, and inter-connected air (and health) monitoring, generating 'big data' for complex, but more inclusive, research".

In this study, a pilot project was carried out to evaluate the feasibility of using AQSSs at fixed locations for epidemiological investigations. This study aimed to evaluate the use of AQSSs over a longer timeframe, which so far has not been thoroughly studied [36,37]. To this end, custom-built stationary sensor systems for NO_2 and $PM_{2.5}$ were deployed for approximately 30 days both in- and outside the homes of seven individuals diagnosed with asthma or chronic obstructive pulmonary disease (COPD). Participants recorded their indoor daily activities as well as the window status and home presence on an hourly basis. Using this information alongside the AQSS data, we analysed the activity-specific indoor and outdoor NO_2 and $PM_{2.5}$ ratio (I/O ratio) and conducted indoor source apportionment. Additionally, indoor activities were classified into four intensity levels and the activity adjusted inhalation rates (IR) were calculated based on them. This information has been used to evaluate the exposure misclassification of the potential NO_2 and $PM_{2.5}$ dose. Comparisons of the calculated potential doses were made between using outdoor air quality monitoring stations or indoor AQSS data, and using generic IR versus activity adjusted IR. The results were compared with other studies, and a comprehensive discussion of the challenges encountered during the study was included as well. Finally, a qualitative symptomatology analysis was conducted to explore the suitability of the AQSS data for studying the relationship between air quality and health. The health data used in this analysis were self-reported by participants using a health questionnaire that we developed.

2. Methods

2.1. *Study Design and Population*

This study took place in Stuttgart, the sixth largest city in Germany with an estimated population of 630,000. Stuttgart has a unique and complicated meteorological and urban climate since the city centre is set in a basin surrounded by hills, and only opens to the northwest, where it meets the Neckar River Valley [38]. The geographical situation, combined with the fact that the busy main roads traverse the basin, often leads to temperature inversions during the cold months, which in turn worsens the air quality in the city centre [39].

The measurement campaign lasted approximately six months, spanning from winter to spring. The first measurements were taken on 19 December 2019 and the last measurements on 28 May 2020. The indoor and outdoor AQSSs were deployed in the patient's home for approximately one month.

The AQSSs were checked once in the middle of the measurement period to ensure that everything was working properly and that data were still being collected. The dates when AQSSs were deployed and checked in each patient's home are detailed in Table S1 in the Supplementary Materials.

Personnel of the pulmonology practice assisted in recruiting patients to participate in this study. The criteria for a patient to qualify for participation in this study included, among others, being aged 18 years and over, being diagnosed with COPD or asthma, not requiring supplemental oxygen therapy, and living near a busy road in Stuttgart. Seven adult subjects of varying ages and diagnoses participated in this study. The demographics of the patients are shown in Table S2. Patients provided written consent for their participation, and all collected data were kept pseudo-anonymous. Individuals are referred to by a patient identification number (ID). This number was assigned by the doctor to maintain patient anonymity. In this manuscript, the participants are referred to by numbers from 1 to 7.

2.2. Materials

Two AQSSs were designed, tailored to their usage either indoors or outdoors (see Figure S1 in the Supplementary Materials). The NO_2 and $PM_{2.5}$ sensors were from the company Alphasense (Great Notley, UK), model B43F and OPC-R1, respectively. Additionally, a temperature and relative humidity (RH) sensor model HYT221 from the company IST (Ebnat-Kappel, Switzerland) was included in both sensor systems. The indoor AQSSs were positioned in the participants' living room, chosen based on patients' feedback indicating it as their primary location for daily activities. Patients 1, 2, 3, 6, and 7 each had a single outdoor AQSS placed by their house. For patient 4, two AQSSs were placed outside on opposite sides of the house. One AQSS was positioned on the street-facing side of the house, while the other faced towards the garden. From here on, to differentiate between the two outdoor AQSSs for patient 4 they are referred to as either "street" or "garden". Outdoor data for patient 2 were lost due to a heavy storm. Patient 5 did not provide consent for an outdoor AQSS. The patients were unable to view the sensor data during the measurement period to avoid prompting changes in their everyday routines.

The air quality monitoring station operated by the University of Stuttgart at Hauptstätter Street was used for co-location. This station represents a traffic hotspot and is located in the city centre. The approximate locations of the patients' homes and the outdoor monitoring station are shown in Figure S1 in the Supplementary Materials.

2.3. Data Collection

A summary of the data collected (or excluded) for each patient is shown in Table S3. At the start of the experiment, each participant's home was mapped using observations from an in-person visit, and together with the patient, an environmental survey about the house, the daily routines, the building, and the area was conducted. Examples of information from the environmental survey include the type of landscape around the home, the distance from the nearest busy road, the type of stove, and whether there are additional residents in the home. Selected results from these environmental surveys are shown in Table S4 for all participants.

Patients were instructed to complete a logbook, recording hourly information across three distinct categories: patient location, activities at home, and room environment. The environmental category specifically included details such as window status (open, closed, or tilted), air conditioning use, and the presence of a functioning fireplace. However, none of the patients reported having either air conditioning or a fireplace in operation. Therefore, the analysis presented in this manuscript focuses solely on the window status within the room environment category. An example logbook is provided in Figure S2.

The category "activities at home", also called activity log, included ten common indoor activities. As activities were tracked each hour, sometimes multiple activities were recorded during the same hour. There were also some hours for which no data were provided. In the data analysis, several default values were selected to account for instances where a patient did not report any activities. The default values were set to "home", "unknown", and "window closed" accordingly. The three grouping categories, their possible values, and the default value are given in Table 1. Patients who had a garden or balcony (2, 3, 4, 5, 6, and 7) were asked to log when they used it during the day. For patients who had additional residents in their homes (3, 5, 6, 7), the activities of the additional residents were not recorded.

Participants were also asked to complete a daily health survey about the presence and severity of respiratory symptoms. Another key method to monitor asthma and COPD patient's health is to check lung function. For that purpose, each patient was given a peak flow meter, shown how to use it properly, and asked to self-test their peak expiratory flow (PEF) each evening. PEF is the rate at which air exits the lungs in one quick and forceful exhale and is given in units of L min^{-1}.

Table 1. Groups of activities tracked by patients and assigned default values.

Category	Possible Values	Default Value
Patient location	Home Not home Garden or balcony	Home
Window status	Window closed Window tilted Window open	Window closed
Activity	Sleeping Exercising Reading Computer TV or radio Cooking Eating Visitor Cleaning	Unknown

2.4. Data Analysis

2.4.1. Overview

As shown in Figure 1, source data from AQSSs, our outdoor monitoring station in Hauptstätter Street, health surveys, and logbooks were merged into a single comprehensive dataset, categorised by date and time, patient ID, and whether the data were collected indoors or outdoors. For the data gathered from participants' homes, the percentage of completeness after data cleaning was calculated and is presented in Table S5. None of the NO_2 sensors achieved 100% data completeness due to the exclusion of the warm-up periods.

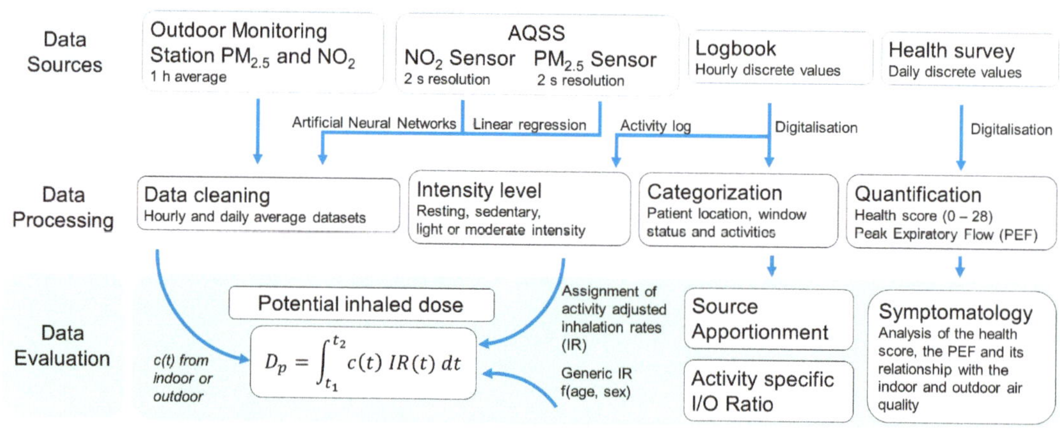

Figure 1. Schema of the data analysis.

The raw sensor data were corrected using artificial neural networks (ANN) and linear regression for NO_2 and $PM_{2.5}$, respectively, followed by conversion into hourly and daily averages. Logbook information was digitalised and categorised into the following groups: patient location, window status, and activities. This information was used in the I/O ratio and the source apportionment studies as well as for the calculation of the activity-specific IR according to the activity intensity levels.

Subsequently, the health score was calculated using the health surveys and, together with the PEF and the AQSS data, evaluated in a qualitative symptomatology study. Finally, the NO_2 and $PM_{2.5}$ potential doses (D_p) were calculated using pollutant concentrations

($c(t)$) from the outdoor monitoring station or the indoor AQSS and the activity-specific IR or the generic IR. The data analysis was carried out in the R software (v 4.2.2). The following sections provide a detailed description of the data processing and evaluation.

2.4.2. Sensor Data Quality Assurance

The calibration and evaluation of the sensors took place before deployment. A separate paper focusing on the calibration and performance evaluation of the NO_2 and $PM_{2.5}$ sensor units will be published soon. Data for $PM_{2.5}$ concentrations were taken from the raw sensor data every two seconds. First, data from times with any sensor overflow errors were removed. Next, the univariate linear regression shown in Equation (1) was applied to the $PM_{2.5}$ data before hourly averages were taken.

$$c_{corr} = m\ c_{raw} + b \quad (1)$$

Here, c_{corr} is the corrected pollutant concentration in μg m^{-3}, m is the correction coefficient, c_{raw} is the sensor's raw reading in μg m^{-3}, and b is the correction constant or offset in μg m^{-3}. These correction factors were found by co-locating reference-grade instruments and the indoor and outdoor AQSSs in indoor and outdoor conditions, respectively. Table S6 lists the values of the correction parameters for each sensor. The outdoor $PM_{2.5}$ sensor had a dryer at its inlet to avoid the effect of hygroscopic growth of particles at high RH. A detailed evaluation of the thermal dryer is described in Chacón-Mateos et al. [40].

The corrections of NO_2 sensor data were achieved using machine learning (ML) methods. Several ML models were investigated to determine which model best corrected the data for variations caused by changes in temperature and RH. Models applied during the study were support vector regressor (SVR), random forest regressor (RFR), and ANN together with multiple linear regression (MLR). Each model was run on both indoor and outdoor data. The predictions of the ML models were compared with the NO_2 concentrations measured by diffusion tubes placed in participants' homes alongside the indoor and outdoor AQSSs. ANN results were most comparable to the passive sampling data. Moreover, ANN was deemed to be the most robust, and capable of handling the influence of the RH and temperature on the sensor signal. The NO_2 sensor data were first averaged in 10 min intervals before being processed using the ANN model.

2.4.3. Activity Specific I/O Ratio

To investigate if AQSSs can be used to determine which indoor activities contribute most to indoor pollution and the effect of outdoor air and ventilation on indoor air, the I/O ratio was calculated for the nine logged activities. For hours during which participants logged multiple activities, each listed activity was considered individually rather than treating the multiple activities as a group. By splitting groups of activities, the calculated hourly concentration is attributed to each of the activities happening in that hour. This method is used to approximate the contribution of each activity to the total pollutant concentration but may lead to estimation errors. Take as an example, two activities logged in an hour, one which generates substantial pollution (cooking) and the other which does not generate any (sleeping). By assigning the hourly pollutant concentration average to the high-emission activity (cooking), its overall contribution may be underestimated. The reverse is true for a low-emission activity (sleeping), in that its contribution to the hour may be overestimated using this method.

To determine the I/O ratio associated with each activity, only indoor activities were included, and only if they occurred for multiple patients in the dataset. The categories "Not Home" and "Garden or Balcony" were therefore excluded from this analysis. For patient 4, who had two outdoor sensors, the data from the street-side sensor were used. As the indoor $PM_{2.5}$ concentration of patient 7 was, on average, 12 times higher than all other participants, the I/O ratio for $PM_{2.5}$ was evaluated separately. The outdoor $PM_{2.5}$ sensor in the house of patient 6 stopped working properly from 19 April 2020 until the end of the deployment on 30 April 2020, and therefore, only the first 15 days were used for the data analysis. Patients 2 and 5 were excluded from these results due to the lack of outdoor sensor data.

2.4.4. Source Apportionment

The activities recorded in the logbooks, combined with the sensor data, were used to determine the variation in indoor air quality due to participant actions. The goal was to determine which activities generated the highest pollution concentrations for NO_2 and $PM_{2.5}$ and whether they could be tracked using stationary AQSSs. For this analysis, activities were included as they were reported in the logbook, with multiple activities occurring during the same hour. Activities are only included if they occurred for a minimum of 30 h across all patients to focus only on the most common combinations of activities.

2.4.5. Symptomatology

To assess the feasibility of using AQSSs to investigate the relationship between symptom severity and air quality, a health questionnaire was developed based on the Asthma Control Test (ACT) and COPD Assessment Test (CAT). Symptoms included in the questionnaire were the following: feelings of tightness in their chest, dyspnea or shortness of breath, cough, sputum or coughing up mucus, wheezing, impairment of their daily life, and how much they used their rescue inhaler during the day. A sample of the health survey is shown in Figure S3 in the Supplementary Materials.

A health score was calculated based on the results of the daily questionnaire regarding the presence and severity of symptoms. Each of the possible answers to the questions in the health survey was rated from zero to four. The health score was calculated as the sum of these individual points for each single day. The maximum score possible was 28. A high health score means that the patient's symptoms were more severe. A score of zero would mean no symptoms were present on the day.

The results of the daily PEF measurements were analysed without further processing in L min^{-1}. The PEF device was able to measure airflows between 50 and 800 L min^{-1}. The health scores and the daily PEF measurements were examined individually for each patient.

2.4.6. Exposure Assessment

In this section, the calculations of the potential inhaled dose are explained. First, each activity in the logbook was assigned to an intensity level, based on how strenuous the activity may be. These estimates of intensity are taken from Chapter 6 of the U.S. EPA's Exposure Factors Handbook [41]. There are five intensity levels used here, as defined by the EPA. Resting corresponds to sleeping or napping, sedentary describes mainly static activities such as watching television or reading, light includes cooking or other standing activities, and moderate is described as walking, easy cycling, or climbing stairs [41]. This classification is used to predict the IR of a patient for the hour. The IR and the intensity level assigned for each activity according to age and sex are shown in Table S7. For hours where no activities were recorded, the generic IR was used. No IR or dose was calculated for the periods when patients were not home as the patient would not be in the same area of the sensor, making the calculations inaccurate for their true exposure.

Given that multiple activities with varying intensities were sometimes recorded within the same hour, we conducted an analysis to assess the influence of IR variability. We calculated the minimum, maximum, and mean IR and used them to estimate the potential inhaled dose indoors. The mean was estimated under the assumption that each recorded activity accounted for equal fractions of the hour. For example, if both "Cooking" and "Sleeping" were marked for 8:00 am, it was assumed that the person cooked and slept for 30 min each. To calculate the hourly minimum (or maximum) IR values, the lowest (or highest) intensity activities were used. The results were compared against the potential dose calculated using a generic IR. For the pollutant concentration, the indoor sensor data were used. Hours for which no specific activity-adjusted IR existed, for example, "Visitor", were assigned the generic IR. Only days when 85% of activity data were completed were included.

After determining the hourly IR values, the potential inhaled dose (D_p), which is the amount of pollutant that is inhaled by an individual [41], can be calculated as shown in Equation (2) [42].

$$D_p = \int_{t_1}^{t_2} c(t)\ IR(t)\ dt \tag{2}$$

where c is the indoor or outdoor pollutant concentration, IR is the inhalation rate, and t is the time. Doses are often presented as dose rates, i.e., the amount of dose per unit time (e.g., μg hour^{-1}) [42].

Krause [43] proposes three methods for calculating the potential dose. The first is to use a generic IR together with the pollutant data from an outdoor monitoring station. The second is to use the generic IR and pollutant concentrations from a portable AQSS, which is deployed in the same microenvironment as the individual. The third method is to calculate the potential dose from the activity adjusted IR and the personal AQSS pollutant data. In our work, an additional fourth method is used, which is to calculate the dose using the activity-adjusted IR with the outdoor monitoring station data. This fourth method is used to give a complete comparison of exposure estimation variability using local indoor or ambient outdoor data. Additionally, the comparison of the two IR (generic or activity-adjusted) calculations using AQSS data may give insight into the importance of using the activity-adjusted IR. If the calculated doses were similar, it would show that the use of an adjusted IR is unnecessary. In summary, the four methods used to calculate potential inhaled dose are as follows:

(A) Generic IR + outdoor monitoring station data
(B) Activity-adjusted IR + outdoor monitoring station data
(C) Generic IR + indoor AQSS data
(D) Activity-adjusted IR + indoor AQSS data

3. Results

3.1. Relationship between Indoor and Outdoor Air Quality

3.1.1. General Comparison

To make a comparison of the indoor and outdoor air quality measured by the AQSSs, the hourly concentrations of $PM_{2.5}$ and NO_2 for the entire measurement period are plotted in Figure 2 for each patient. The central box of the boxplot contains 50% of the data (interquartile range, IQR) and the inner line corresponds to the median. The whiskers extend to the smallest and largest observations within 1.5 times the IQR from the quartiles, with outliers plotted as individual dots. Outdoor data for patient 4 are taken from the street-facing AQSS. Patients 2 and 5 did not have any outdoor data to report.

The outdoor $PM_{2.5}$ concentrations were higher than the indoor concentrations for patients 1, 3, and 4. Median $PM_{2.5}$ concentrations in the home of patient 6 were similar for both indoor and outdoor microenvironments, with more occasional peaks observed indoors. All patients, excluding patient 7, had roughly similar indoor $PM_{2.5}$ concentrations among them, with median measurements falling below 10 μg m^{-3}. Patient 7 had much higher concentrations of $PM_{2.5}$ inside than outside, with the median concentration above 40 μg m^{-3}. From the logbook and the environmental questionnaire, it is known that patient 7 often lit multiple scented candles in the house and blew them out without proper ventilation, which is most probably the reason for the high $PM_{2.5}$ concentrations.

The comparison of indoor and outdoor NO_2 is shown in Figure 2b. Some slightly negative values were measured for NO_2 (patients 2, 4, and 6) which is an artefact of the ANN models used to process the data. We should also take into consideration that it is especially difficult to measure low NO_2 concentrations precisely with the tested electrochemical sensors as the relative expanded uncertainty is considerably larger than for high concentrations [44]. The average concentrations together with the relative expanded uncertainties (REU) of the NO_2 and $PM_{2.5}$ measurements are presented in Table S8.

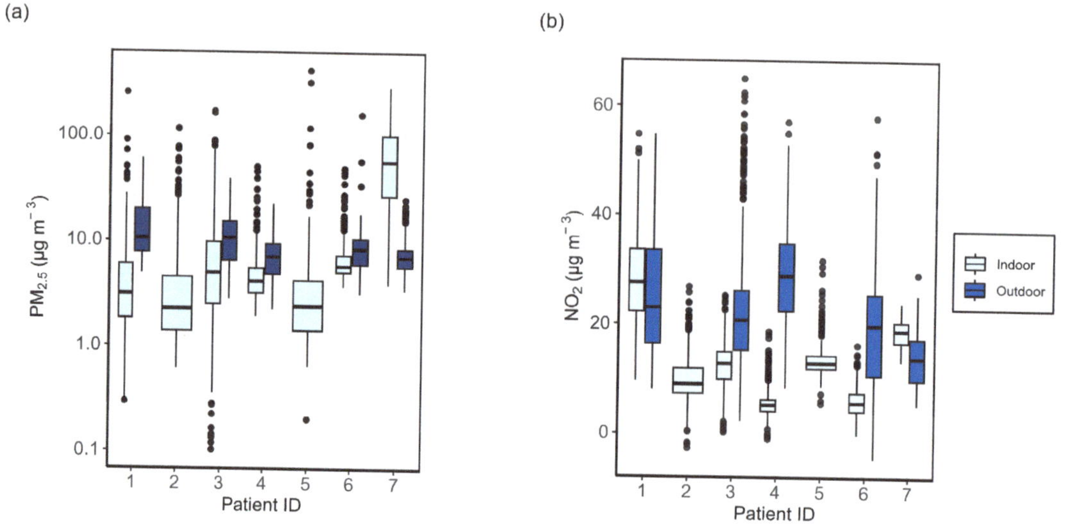

Figure 2. Hourly concentrations of (**a**) $PM_{2.5}$ and (**b**) NO_2 for each patient over the entire measurement period. Note that the *y*-axis in panel (**a**) is on a logarithmic scale and that, in panel (**b**), the unexpected high NO_2 concentrations in the house of patient 1 may be due to an overestimation caused by the ANN correction.

For NO_2, the outdoor concentrations were higher than indoors for patients 3, 4, and 6. Patients 1 and 7 showed the opposite trend. For patient 1, indoor NO_2 concentrations were roughly 30 μg m^{-3}. This is more than twice as high as the indoor NO_2 measurements of most other patients. As NO_2 passive samples were not collected for patient 1, it is challenging to determine whether the elevated indoor NO_2 concentrations accurately reflect true conditions or are a result of the ANN model used for calibration. Previous analysis of ML and MLR correction models does suggest an overestimation of NO_2 concentration by the ANN model for patient 7 [45].

For patient 4, two outdoor AQSSs were installed to determine if there were measurable differences in air quality on opposite sides of the house. Figure 3 shows the $PM_{2.5}$ and NO_2 concentrations measured indoors, as well as outside on the garden and street sides of the house. Both $PM_{2.5}$ and NO_2 concentrations are lower indoors than outdoors. In general, $PM_{2.5}$ concentrations tend to be more variable indoors compared with outdoors, while the opposite trend is observed for NO_2, with more consistent levels indoors and greater variability outdoors.

The street-facing AQSS measured higher $PM_{2.5}$ concentrations than the unit placed in the garden. There was a larger spread in garden-side NO_2 measurements, with a median of 35 μg m^{-3}, slightly higher than for the street-facing side (29 μg m^{-3}). The REU for the garden-side NO_2 sensor at the median was ±13 μg m^{-3} and was greater than the REU of the street-side NO_2 sensor (±7 μg m^{-3}). It is possible that the overestimations of the NO_2 concentrations seen at the garden side were due to the ANN calibration. The NO_2 passive samples measured an average of 24 μg m^{-3} and 23 μg m^{-3} for the street and the garden side, respectively, during the same period. This may indicate that the street sensor slightly overestimated the NO_2 concentrations, whereas the sensor placed in the garden overestimated the NO_2 concentrations by approximately 12 μg m^{-3}.

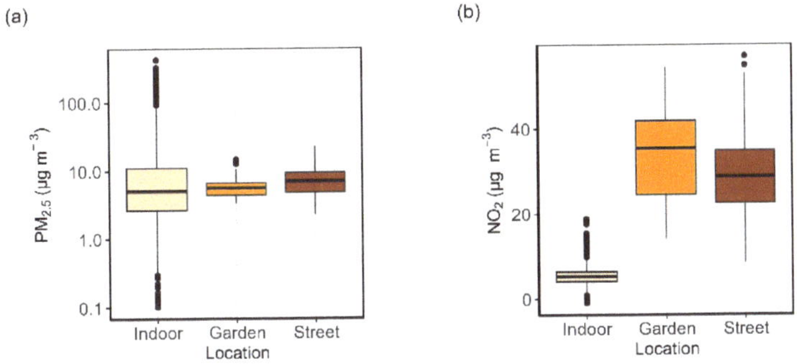

Figure 3. Hourly concentrations of (**a**) $PM_{2.5}$ and (**b**) NO_2 for patient 4 at indoor, garden, and street locations. Note that the *y*-axis in panel (**a**) is on a logarithmic scale.

The percentage of time that each patient spent at home, the status of the windows in the living room, and the time contribution of home activities are visualised in Figure 4. All patients spent the majority of the time at home, 83% on average, which is consistent with the results of previous studies [24]. Patient 2 spent the least amount of time at home (76%) while patient 4 spent the most time at home (93%). This means that the home environment is a crucial part of understanding personal exposure. The percentage of time spent at home in our study is slightly higher than the statistics compiled by Klepeis et al. [10], which reported that most people are at home for roughly 70% of the day. This difference could be due to COVID-19 related restrictions happening from the middle of March 2020 and lasting until the end of April 2020. The COVID-19 restrictions affected patients 3, 4, and 6.

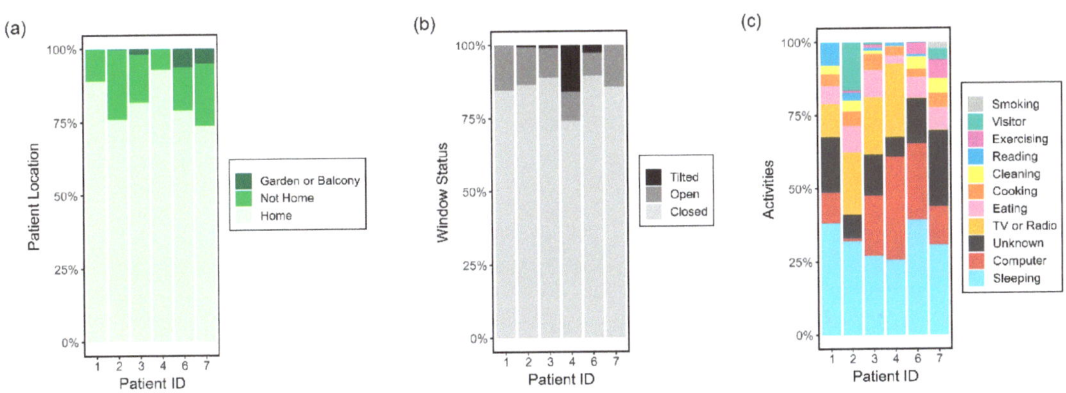

Figure 4. Percentage of patients' time grouped by (**a**) patient location, (**b**) window status, and (**c**) activities at home.

All patients had the windows closed for the majority of the time, 85% on average, with only four patients ever reporting to have tilted the windows, as opposed to fully opening them. Given that the measurements took place in winter and spring, it is expected that the windows would be closed most of the time. The activities that patients spent the most time on were "Sleeping", "Computer", and "TV or Radio". "Unknown" also accounted for a significant amount of each patient's time.

3.1.2. Activity Specific I/O Ratio

The mean I/O ratios associated with each activity, under various ventilation conditions (window closed, open, or tilted), are presented in Figure 5 for four of the participants (for patients 2 and 5, outdoor AQSS data were not available). As the $PM_{2.5}$ data collected in the house of patient 7 were extremely high due to the scented candles, the results of the activity specific $PM_{2.5}$ I/O ratio of patient 7 are presented in Figure S4.

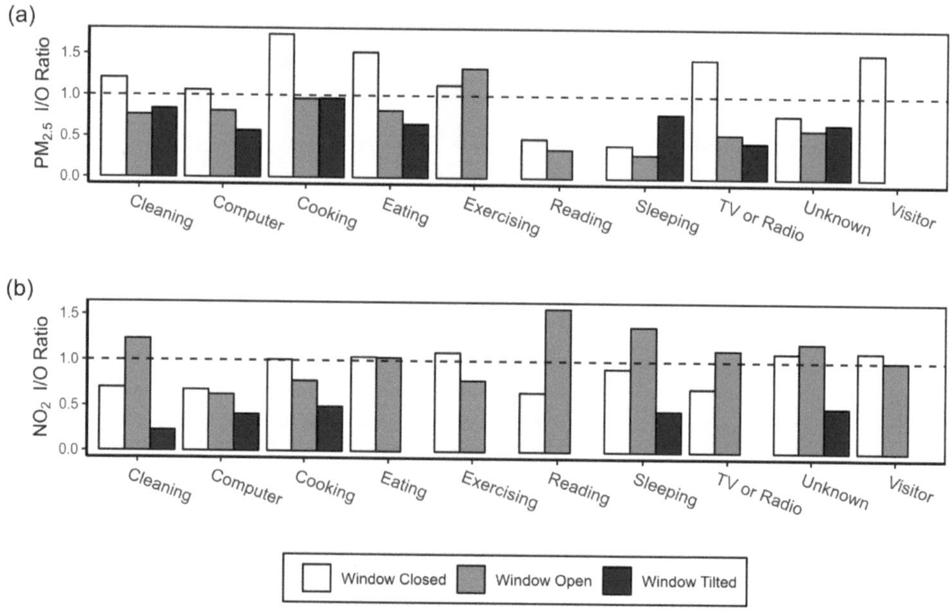

Figure 5. I/O ratios for (**a**) $PM_{2.5}$ and (**b**) NO_2 associated with individual activities, grouped by window status.

For all activities, $PM_{2.5}$ had the largest I/O ratio when the windows were closed, indicating that there are significant sources of $PM_{2.5}$ indoors. Another factor that may influence the results is the measurement period. The measurement campaign was conducted during the colder months when ventilation is typically minimised to reduce energy costs. Without adequate ventilation, the generated $PM_{2.5}$ can stay several hours in the air. The activities "Cleaning", "Cooking", "Eating", "TV or Radio", "Computer", "Exercising", and "Visitor" were found to be significant indoor sources of $PM_{2.5}$ when the windows were closed. Only "Sleeping" and "Reading" had an I/O ratio of less than one for all window statuses, indicating that the $PM_{2.5}$ concentration outdoors was higher than indoors. A fact observed in Figure 5 is that opening or tilting windows effectively reduces indoor $PM_{2.5}$ concentrations during activities such as "Cleaning", "Computer", "Cooking", "Eating", and "TV or Radio". The only activity that showed the opposite trend was sleeping. This is because indoor $PM_{2.5}$ concentrations were lower than outdoor concentrations during sleeping; therefore, keeping the window tilted for the whole night increased the I/O ratio. A detailed analysis of the hourly indoor and outdoor $PM_{2.5}$ and NO_2 concentrations together with the activities and the window status can be seen in the Supplementary Materials, Figures S5–S26. The results are presented in one-week segments, to be able to better associate pollutant peaks with specific activities and window status.

The I/O ratios for NO_2 were lower than one during most activities, suggesting there were fewer sources of NO_2 indoors. When the windows were closed, the only activities with an I/O ratio slightly greater than one were "Cooking", "Eating", "Exercising", "Unknown", and "Visitor". Cooking is known to be a source of NO_2, especially from the use of gas stoves or ovens [46]. Eating and cooking were frequently marked at the same hour. Moreover, the concentrations measured during "Eating" may have been created in the previous "Cooking" activity and what we measured was the "Post-cooking", i.e., the transportation of pollutants from the kitchen to the living room where the AQSS was located. For 50% of the activities, i.e., "Cleaning, "Reading", "Sleeping", "Unknown", and "TV or Radio", the I/O ratios for NO_2 were higher than one when windows were open. This is in accordance with findings from Stamp et al. [47], who recorded increased I/O ratios for NO_2 when windows were open. Other activities such as "Computer", "Cooking", and "Exercising" showed higher NO_2 I/O ratios when the windows were closed or tilted as compared with when they were open.

3.1.3. PM Advisory Study

From 21 January to 26 January 2020, a PM advisory (*Feinstaubalarm*) was active in Stuttgart during which sensors were deployed at the home of patient 1. These alerts are issued in conjunction with the German Weather Service (DWD) when certain conditions are met, including PM_{10} levels exceeding 30 μg m^{-3}, the absence of rain, and low wind speeds. Figure 6 shows the hourly $PM_{2.5}$ concentrations from the indoor and outdoor AQSSs during the event and the two preceding days. It can be observed that the outdoor $PM_{2.5}$ concentration reached its maximum (60 μg m^{-3}) on the night of 23 January and that the average $PM_{2.5}$ concentrations measured indoors increased gradually throughout the entire period of the PM alert.

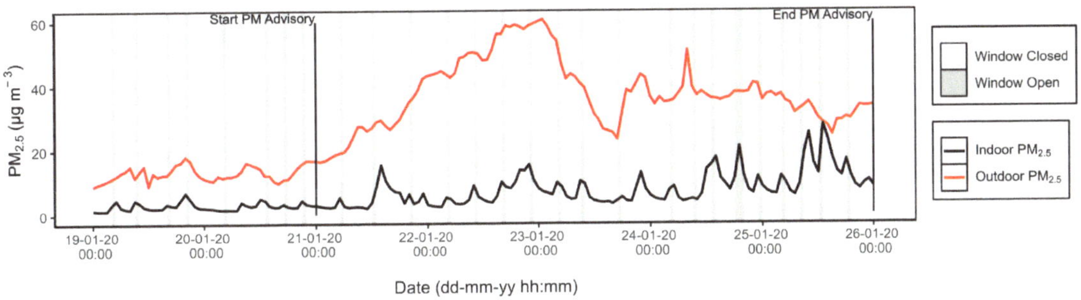

Figure 6. Indoor and outdoor $PM_{2.5}$ concentrations measured by the AQSSs deployed at the home of patient 1 during the PM alert, along with the corresponding window status. Periods when the window was open are indicated by a grey background.

The I/O ratio was analysed alongside window status and temperature to assess the impact of ventilation on indoor air quality under poor outdoor air quality conditions, as presented in Figure 7. A marked difference was observed in the change of the $PM_{2.5}$ I/O ratio when the windows were open. The peaks represent increases in indoor $PM_{2.5}$ concentrations as polluted outdoor air enters through the open window. In the six days preceding the PM alert, the average I/O ratio increased by 16% when the windows were open. However, during the PM alert period, the I/O ratio rose by 49% when the windows were open, compared with when they were closed. This suggests a significant impact of window opening on indoor air quality during periods of elevated outdoor pollution. Temperature drops on the 20th and 23rd, which were not associated with open windows, coincided with peaks in the I/O ratio. This confirms that the windows were open during those hours even though the patient did not register it.

Moreover, Figure 7 evidences that the $PM_{2.5}$ I/O ratio during nighttime hours progressively increased each day throughout the PM advisory period. This trend can be attributed to the infiltration of outdoor air into the house, which can occur through structural imperfections in window seals or doorways [48]. This phenomenon was observed in the home of patient 1, who reported in the environmental survey that the windows were 40 years old and inadequately sealed. Although the windows were kept open 2% less frequently during the PM alert period compared with the preceding week (16% vs. 18% of the time), indoor $PM_{2.5}$ concentrations increased by 85% during the alert period.

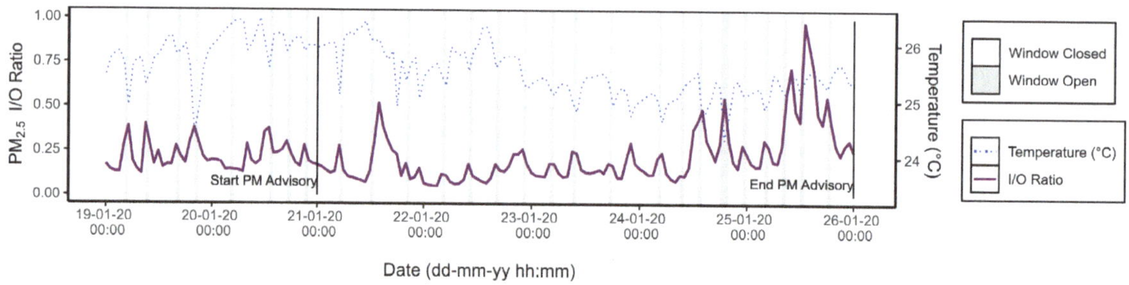

Figure 7. Time series of the mean hourly I/O ratio for $PM_{2.5}$ and indoor temperature during the PM alert, combined with window status.

The daily mean of the I/O ratio during the PM advisory and the two preceding days is shown in Figure 8. The daily indoor and outdoor $PM_{2.5}$ concentrations can be seen in Figure S27. There was an increase in the I/O ratio from 22 January to 25 January before it decreased again. The trend of increasing daily I/O ratio, together with the progressive rise in indoor $PM_{2.5}$ concentrations, may prove the infiltration of $PM_{2.5}$ in the home. The outdoor $PM_{2.5}$ concentrations began to decrease gradually on the 23rd, but the I/O ratio did not decrease until the 26th.

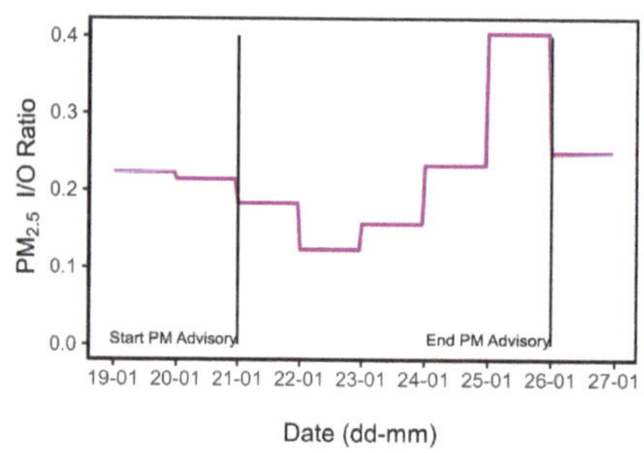

Figure 8. Daily $PM_{2.5}$ I/O ratio before, during, and after the PM alert.

3.2. Source Apportionment

The results of the $PM_{2.5}$ and NO_2 concentrations associated with each activity are shown in Figure 9. Here, activity and pollutant data were grouped across multiple patients. The individual results of all patients, except for patient 5 due to lack of activity data, are presented in the Supplementary Materials in Figures S28 and S29.

The highest NO_2 values across all activities were measured in the house of patient 1. As explained in Section 3.1.1, the ANN model used for the calibration of the sensor may have overestimated the concentrations in this particular case. The highest NO_2 concentrations corresponded to the category "Reading, TV or Radio". This unexpected result was significantly influenced by the generally higher concentrations measured in the house of patient 1. Other activities that were associated with high NO_2 concentrations were "Cooking", "Eating", and "Cleaning". Even though most patients had electric stoves, it is also possible for elevated NO_2 concentrations to occur if the oven is used [46]. Patient 3 was the only participant who used a gas stove, yet they did not exhibit significantly higher NO_2 concentrations compared with the other patients. NO_2 is a reactive gas and "deposition and reactions may occur during transport" which may lead to a decrease in concentration over a short distance [46]. This may lead to variations in NO_2 concentration throughout the home.

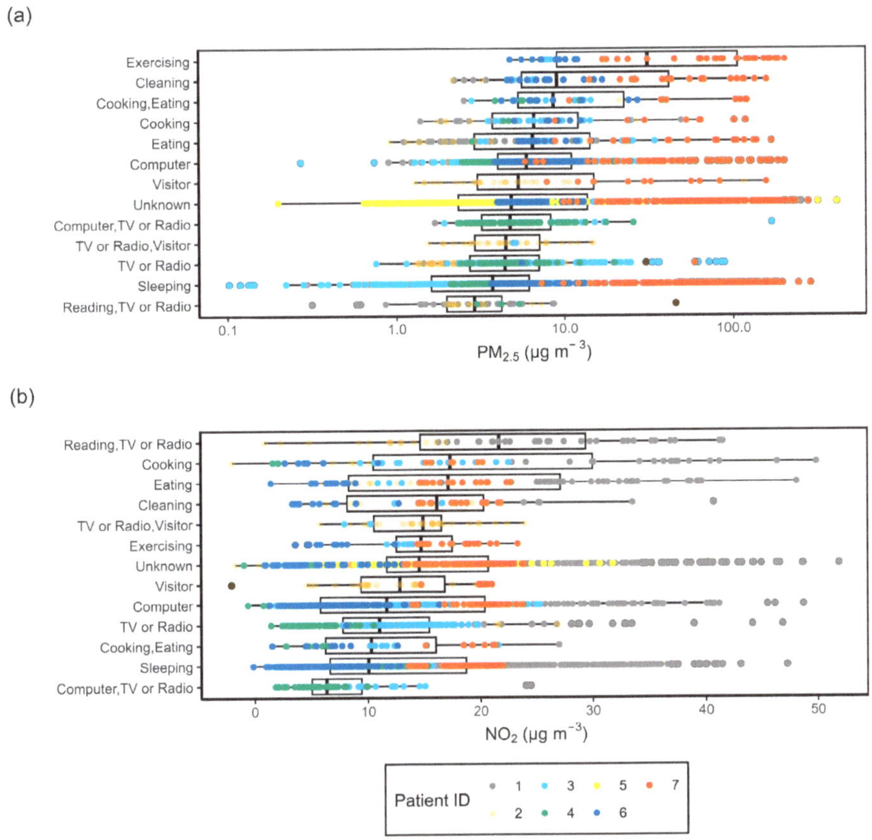

Figure 9. Indoor (**a**) $PM_{2.5}$ and (**b**) NO_2 concentrations associated with an activity or group of activities; patient values represented by colour-coded points. Note that the *x*-axis in panel (**a**) is on a logarithmic scale.

As for $PM_{2.5}$, "Cooking, Eating" showed higher concentrations than "Cooking". Several possible factors may explain the higher concentrations measured during "Cooking, Eating". Since most patients had closed kitchens, it is possible that emissions were contained within the kitchen until cooking was finished. At that point, the participant would have opened the door to move to another area of the house, allowing the kitchen pollution to reach the living room where the sensor was located. Another possible explanation is, again, the discrete nature of the logbook. It is possible that cooking occurred at the end of one hour, with the majority of any heating steps occurring at the beginning of the next hour and eating occurring at the end. In those cases, the activity group "Cooking, Eating" would contain the times when the most pollutants were generated. For both $PM_{2.5}$ and NO_2, high concentrations were measured during the activity "Eating", which may be explained by the same reasons as "Cooking, Eating".

As expected, most sedentary or resting activities were not associated with high pollutant concentrations. Concentrations of both pollutants were low during "Sleeping", "Computer", and "TV or Radio". During sedentary activities, PM settles, as it is not agitated by the physical movement of the participants. These hours were shown to have the lowest $PM_{2.5}$ concentrations, which is supported by other studies [43,49].

The activities recorded by Krause [43] differ from those logged in our work due to the use of portable sensors, which allowed Krause the study of different mobility patterns (car, walk, cycle, train, etc.). In their study, all home activities were grouped into one single category, except for sleeping. Though it was not possible to parse out the concentrations during different home activities, their observations during sleep were in good agreement with the results of this study. In both cases, the median $PM_{2.5}$ concentrations fell below 5 μg m^{-3}. The results for NO_2 varied slightly: in Krause's study, the median was less than 5 μg m^{-3}, compared with approximately 10 μg m^{-3} shown in Figure 9.

In Figure 10, the time series of the $PM_{2.5}$ concentration measured in the house of patient 2 is plotted to observe the temporal variation in indoor air quality with respect to the recorded activities. Results for all patients including the window status are shown in the Supplementary Materials, Figures S5–S26.

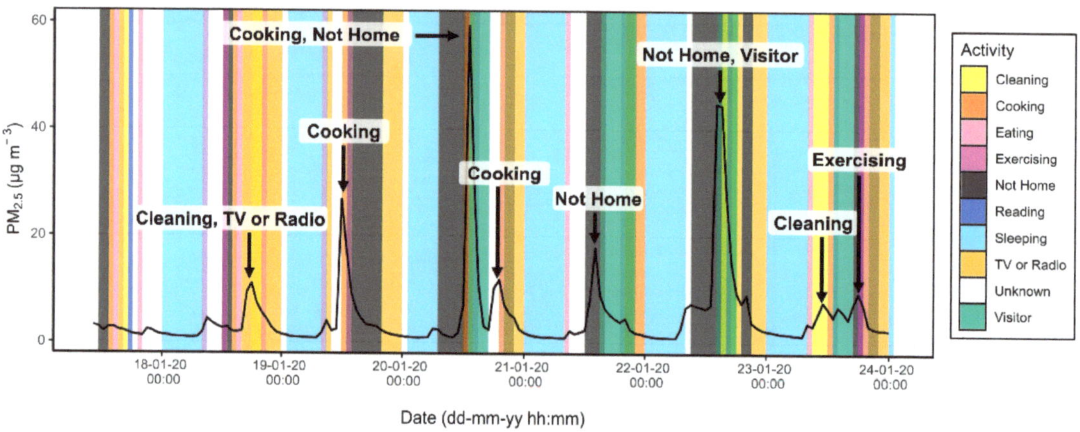

Figure 10. Hourly indoor $PM_{2.5}$ concentration for patient 2 from 18 January to 24 January 2020, combined with recorded activities.

The time series clearly shows that the $PM_{2.5}$ concentrations reached a minimum during sleeping times (light blue) and increased again in the morning as other actions started to occur. Some peaks, which are marked with arrows, occurred when cleaning or cooking were logged. Hours, when multiple activities were logged, are shown by overlapping colours as seen on 23 January. Other a priori unexpected peaks were also recorded, such as on 21 January, when a $PM_{2.5}$ peak occurred while the patient had marked "Not Home". This could be explained due to either a logging error by the patient or the presence of another person in the house.

3.3. Symptomatology

In this section, the data of the health score and the PEF measurements are combined with the air quality data measured by the AQSSs. The results of the health scores and PEF measurements are given in Table 2. Patients 1 and 3, who were diagnosed with COPD, had higher minimum, mean, and maximum health scores than the other patients, who had asthma. This metric shows that they did have more difficulty breathing in their day-to-day lives. Patients 4, 6, and 7 all reported days when they experienced no symptoms (health score = 0). Patient 5 did not complete the health survey, and patient 2 developed bronchitis during the second week of measurements, necessitating antibiotic treatment. Consequently, their health data have been excluded from the analysis. Missing values in PEF measurements were caused by a delay in the delivery of the peak flow meters.

Table 2. Mean, minimum, and maximum health score and PEF values for all participants.

Patient ID	Health Score (0–28)			PEF (L min^{-1})		
	Minimum	Mean	Maximum	Minimum	Mean	Maximum
1	10	14.5	20	-	-	-
2	-	-	-	-	-	-
3	6	11.1	18	500	540	580
4	0	1.6	4	370	400	430
5	-	-	-	-	-	-
6	0	1.0	5	800	800	800
7	0	1.2	4	290	324	370

The qualitative relationship between symptom severity and pollutant concentrations can be observed by plotting the pollutant concentrations and health symptoms together throughout the measurement period for each patient. The results for patients 1 and 3 are shown in Figures 11 and 12, respectively. Events or changes in the severity of symptoms are highlighted with blue rectangles. The results for the remaining patients are presented in the Supplementary Materials (Figures S30–S32).

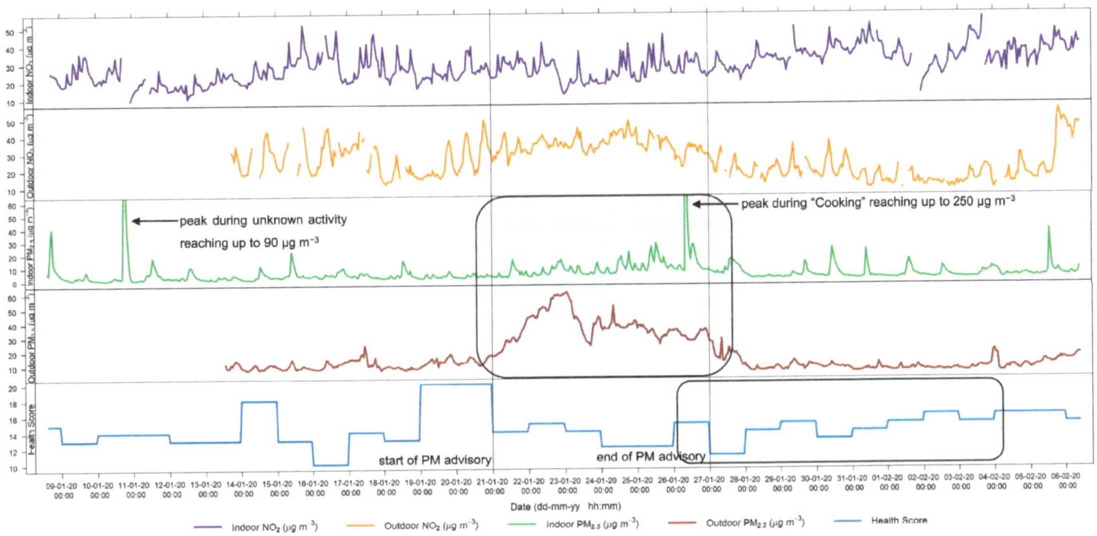

Figure 11. Hourly indoor and outdoor $PM_{2.5}$ and NO_2 concentrations and self-reported daily health score data from patient 1.

As discussed in the previous section, there was a PM advisory raised from 21 January to 26 January during the measurements in the house of patient 1. During this time, outdoor $PM_{2.5}$ concentrations increased gradually, and indoor concentrations progressively rose. There were constant changes in the health scores during the month, with the worst scores occurring on the 19th and 20th. During the week following the PM advisory (28 January–5 February), the health score also gradually increased.

Patient 3, shown in Figure 12, showed several indoor peaks of $PM_{2.5}$ during the measurements, as well as three weeks with increased outdoor concentrations. PEF measurements varied from day to day but consistently decreased whenever the health score increased, indicating that the health score is an accurate method for describing actual patient health. However, this correlation between health score and peak flow varied between the participants. The health score of patient 3 exhibited day-to-day variability. Generally, an increase in the health score was observed some days after a rise in outdoor $PM_{2.5}$. For instance, following the days with elevated PM concentrations (27–28 March), the increase in the health score on 29 March was also accompanied by a decrease in the PEF. A longer study would be necessary to determine if the observed pattern persists over longer periods.

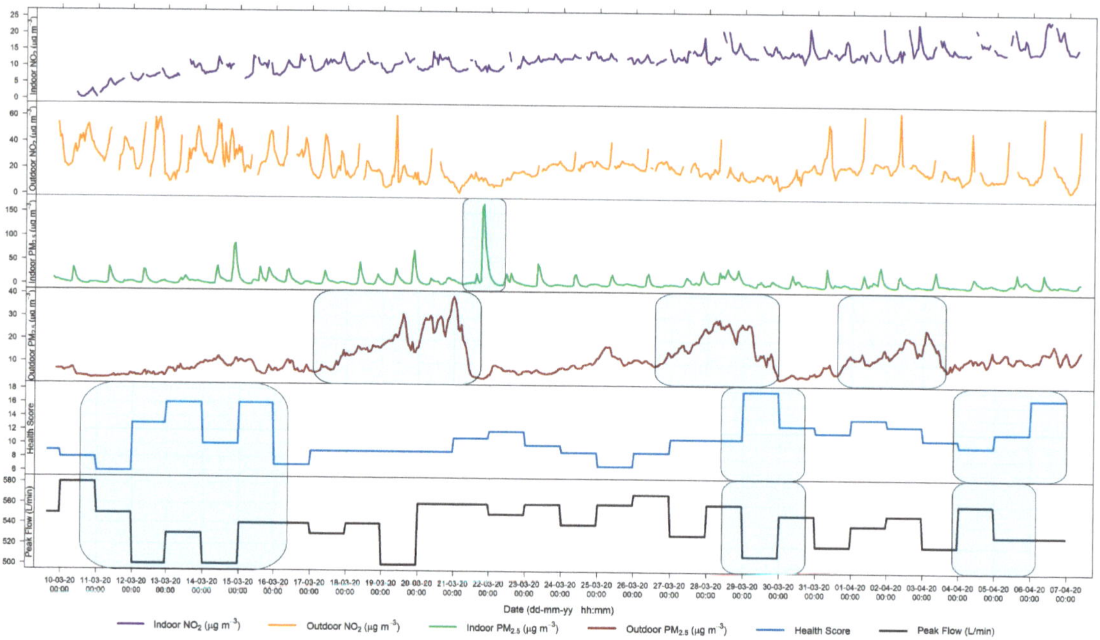

Figure 12. Hourly measured pollutant concentration and self-reported daily health score and PEF data from patient 3.

3.4. Exposure Assessment

The analysis of personal exposure was carried out in three parts. First, the influence of the estimated activity-adjusted IR on the potential inhaled dose was analysed. Second, the exposure misclassification was examined by comparing data from the outdoor monitoring station or generic IR with indoor data from AQSSs and activity-adjusted IR. Third, the activities with the greatest impact on personal exposure were identified.

3.4.1. Analysis of the Variability in the Inhalation Rate and Its Effect on the Potential Inhaled Dose

The statistical results of the average, minimum, and maximum daily potential inhaled doses (D_p) of $PM_{2.5}$ and NO_2, calculated using the activity-adjusted IR, as well as the daily doses calculated using the generic IR, are shown in Figure 13. There is little difference among the maximum, mean, or minimum of the D_p calculated using activity-adjusted IR. This may suggest that hours containing multiple activities do not significantly impact the total daily potential dose. It is also true that, for most hours during which two activities were performed, the activities had the same intensity level, resulting in the same IR. This can be seen in the activities listed in Figure 9 where only the most common activities were included. The most common pairs of activities were "Computer, TV or Radio", "Reading, TV or Radio", and "Cooking, Eating". All activities in these groups are classified as sedentary activities, except cooking which is classified as light intensity. As there is no significant difference between the maximum, mean, and minimum, the mean activity-adjusted IR was used for the calculation of the potential inhaled NO_2 and $PM_{2.5}$ doses. Moreover, it can also be seen in Figure 13 that the generic IR overestimates the potential inhaled doses compared with the activity-adjusted IR.

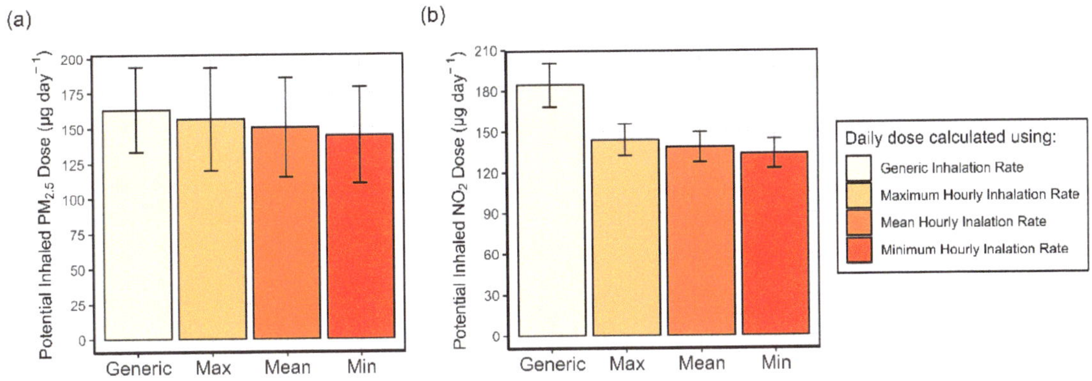

Figure 13. Comparison of daily mean potential inhaled dose of (**a**) $PM_{2.5}$ and (**b**) NO_2, calculated using the generic IR and the maximum, the mean, and the minimum activity adjusted IR. Error bars indicate the standard deviation in daily dose estimations.

A comparison of the daily doses for individual patients is shown in Figure 14. Patient 7 experiences up to 5 times higher daily doses of $PM_{2.5}$, due to the habit of lighting scented candles. Patients 1, 3, 4, and 6 show the same pattern as in Figure 13, i.e., generic IR produces the highest dose estimates. However, patients 2 and 7 show the opposite. This is because the calculated daily dose is affected by the individual's activity level. If an individual is more active at home, their activity adjusted IR will be higher than the generic rate for most of the hours. These differences are shown in Table 3, which lists the mean hourly activity adjusted and generic IR of each patient during active indoor hours, i.e., excluding sleeping hours. Patients 2 and 7 show a higher mean hourly activity adjusted IR compared with the generic IR. In these cases, the exposure will be underestimated when using a generic IR. For all the other cases, the generic IR is overestimated which triggers the overestimation of the daily doses seen in Figure 14. Based on these results, the use of a generic IR for the time spent indoors is not recommended. Further studies including more participants may see more differences in exposure between people with different lifestyles and levels of activity.

Table 3. Average of the generic and activity adjusted hourly IR for individual patients (excluding sleeping hours).

Patient ID	Hourly Mean IR (L min^{-1})	
	Activity Adjusted	**Generic**
1	7.0	9
2	9.2	6.8
3	8.3	9.9
4	5.3	8.5
6	9.5	12.1
7	9.5	8.5

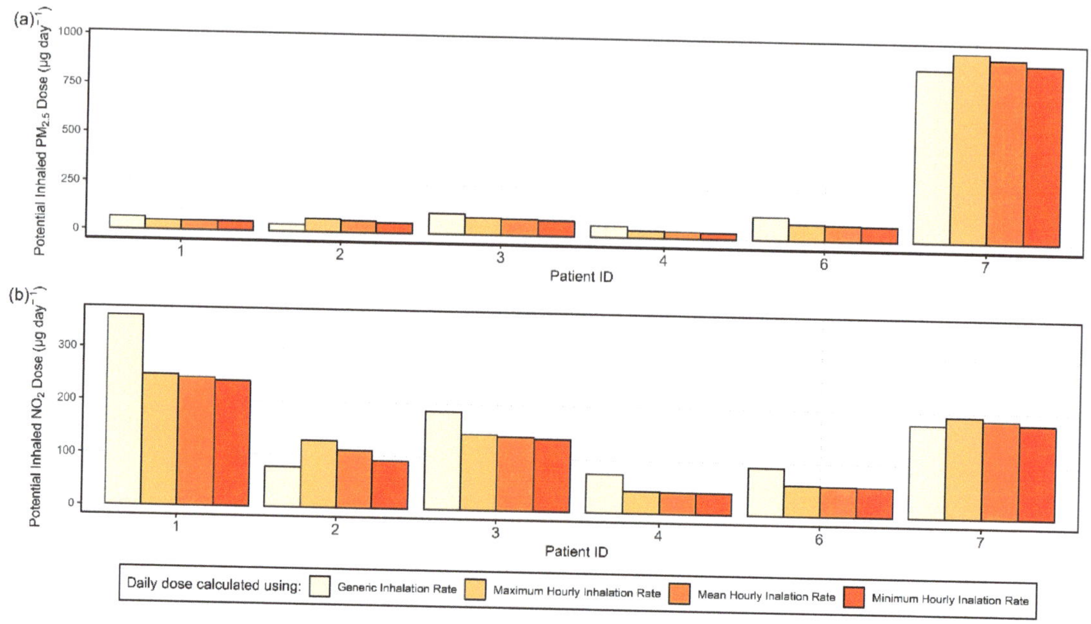

Figure 14. Daily potential dose of (**a**) $PM_{2.5}$ and (**b**) NO_2, calculated using generic, maximum, mean, and minimum IR, for individual patients.

3.4.2. Exposure Misclassification

The result of the daily mean potential NO_2 and $PM_{2.5}$ doses calculated using four different methods (see Section 2.4.6) across all patients is shown in Figure 15. Two trends emerge across the four methods. First, calculating daily potential doses using the outdoor monitoring station data, as in methods A and B, yields significantly different results compared with using indoor data from stationary AQSSs. Second, the use of the generic IR (methods A and C) overestimates the daily dose compared with the activity-adjusted IR.

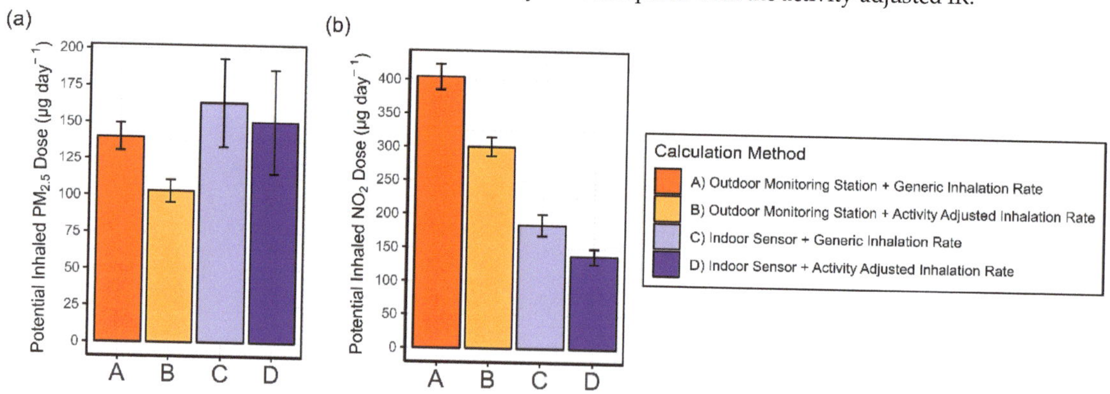

Figure 15. Comparisons of four potential inhaled dose calculation methods for (**a**) $PM_{2.5}$ and (**b**) NO_2. Error bars indicate standard deviation.

From Figure 15a, we can also derive that the indoor environment and the individual's habits have a strong impact on personal exposure. Patient 7 had a median indoor concentration of 56 $\mu g\ m^{-3}$, compared with 7 $\mu g\ m^{-3}$ at the Hauptstätter Street outdoor monitoring station, due to the continuous use of scented candles and improper ventilation.

Thus, the daily potential $PM_{2.5}$ doses calculated including the indoor $PM_{2.5}$ sensor data of patient 7 in the average are much higher than when using the data from the outdoor monitoring station. This indicates that the use of outdoor $PM_{2.5}$ data may result in either over- or underestimation, depending on the individual's habits. For NO_2, the trends shown in Figure 15b indicate a clear overestimation of the potential dose when using outdoor data from the monitoring station and generic IR compared with using indoor data and activity-adjusted IR. These findings may extend to other indoor environments with similar conditions and cooking habits.

The results of the potential dose have been compared with the outcomes from Krause [43] in the AIRLESS project with portable sensors. There, participants were split between those living in an urban area, Beijing, and a peri-urban area, Pinggu. From the comparison of the outcomes, it can be concluded that both studies agree on the fact that using the outdoor pollutant concentration from outdoor monitoring stations may trigger exposure misclassification.

Krause estimated the exposure in Beijing while at home using the activity adjusted IR and portable AQSSs to be roughly 400 and 300 $\mu g\ day^{-1}$ for $PM_{2.5}$ and NO_2, respectively. Those outcomes are higher than the estimations of 150 $\mu g\ day^{-1}$ for $PM_{2.5}$ and 140 $\mu g\ day^{-1}$ for NO_2 using stationary indoor AQSS data and activity-adjusted IR, as shown in Figure 15. The higher potential doses calculated by Krause for AIRLESS study participants may be due to two reasons. Firstly, Beijing has overall higher concentrations of both $PM_{2.5}$ and NO_2. Median indoor concentrations of $PM_{2.5}$ were nearly five times higher in Beijing compared with Stuttgart (25 versus 5 $\mu g\ m^{-3}$). For NO_2, median home concentrations were roughly 15 $\mu g\ m^{-3}$ in the AIRLESS project and 12 $\mu g\ m^{-3}$ averaged across participants in this study. The second reason may lie in Krause's use of a generic IR for the times when participants were at home. As we have seen in Figure 15, the use of a generic IR overestimates the results of potential $PM_{2.5}$ and NO_2 doses. In our study, we have proven that "home" is a very complex microenvironment where multiple activities with different intensity levels can occur. Considering that, in both studies, patients are at home for more than 80% of the time, using a generic IR for the entire time spent indoors may also be a source of exposure misclassification.

3.4.3. Activity-Specific Potential Inhaled Dose

To identify the indoor activities contributing most to personal exposure, the potential dose for $PM_{2.5}$ and NO_2 was calculated for each hour and grouped by the activities occurring during that hour, as shown in Figure 16. It is clear from Figure 16 that the highest potential doses were recorded during the most strenuous activities. "Exercising" was the activity with the highest intensity from those recorded on the activity log and is associated with the highest hourly doses of both $PM_{2.5}$ and NO_2. For $PM_{2.5}$, the median dose during "Exercising" (about 16 $\mu g\ hour^{-1}$) is more than twice as much as the next highest potential dose which occurs during "Cleaning" (about 6 $\mu g\ hour^{-1}$).

Calculating the potential dose using an activity-adjusted IR changes the weight that each activity carries for personal exposure as compared with simply using a generic IR. When looking at exposure based on potential dose, more strenuous activities such as exercising contribute more than sedentary ones. Activities such as cooking or cleaning, which are both known to generate pollutants and fall into the "light" or "moderate" intensity level, have both the highest concentrations and potential doses. Given that different indoor activities contribute variably to personal exposure based on their associated concentrations and/or intensity levels, it is important not to overlook these variations by relying on a generic IR for the time at home.

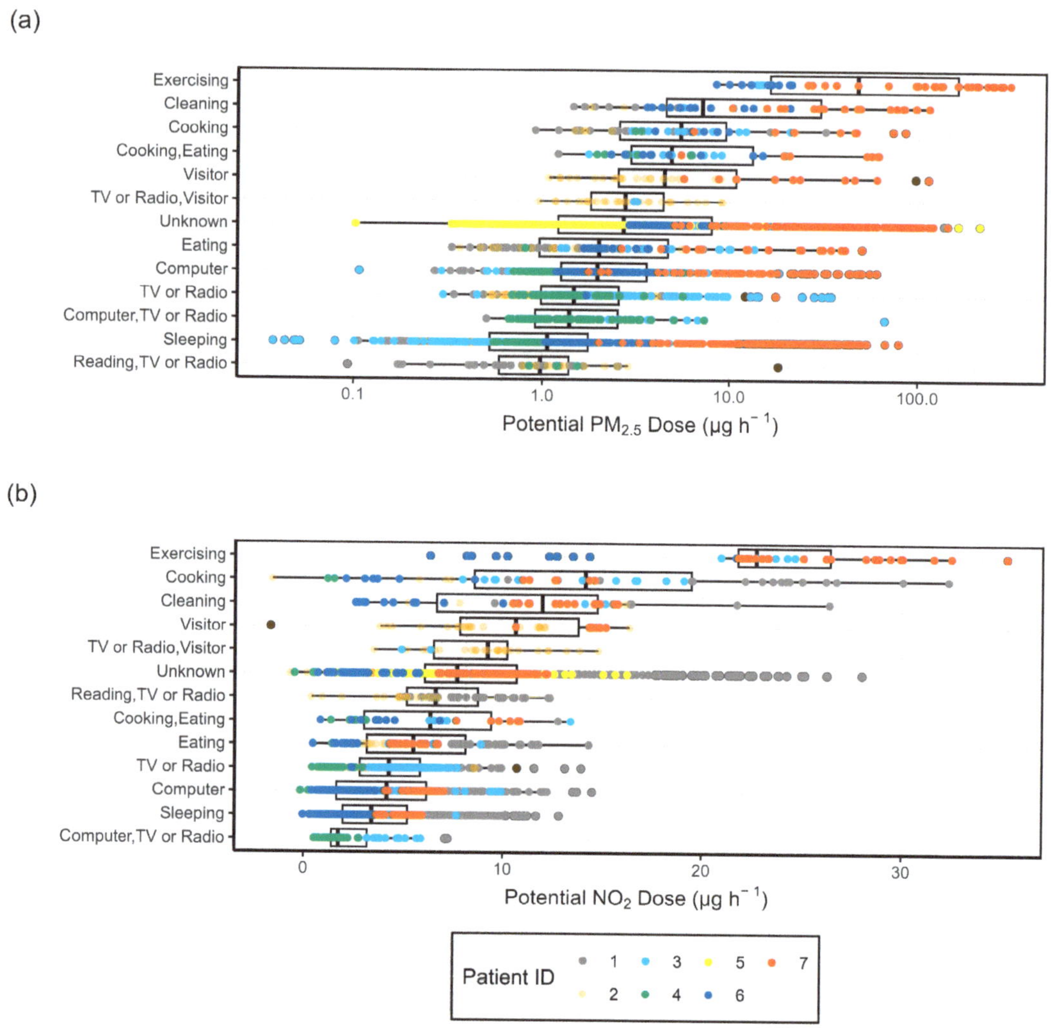

Figure 16. (**a**) $PM_{2.5}$ and (**b**) NO_2 potential inhaled dose associated with each unique group of activities. Patient values represented by colour-coded points. Note that the *x*-axis in (**a**) is on a logarithmic scale.

4. Discussion

4.1. Air Quality Sensors

4.1.1. Data Loss and Data Quality

The percentage of completeness of the sensor data was between 92% and 100% for the $PM_{2.5}$ sensors and between 76 and 99% for the NO_2 sensors. Data losses could have been reduced by adding Wi-Fi or LTE modules to the sensor systems. In that way, misfunctions and problems could have been identified sooner. The exclusion of data from the warm-up period for the NO_2 sensors caused the loss of the first hours in each of the sensor deployments, from 4 up to 24 h.

The primary challenge regarding the uncertainty of the sensor data lies in determining the suitability of the calibration parameters obtained two weeks prior to the campaign for use during the measurements at homes lacking reference instruments, particularly for indoor microenvironments. The transfer of calibration parameters becomes complex when relocating sensors from their original co-located positions, and ensuring accurate performance in the new location cannot be guaranteed [50]. The use of passive samples for NO_2 is a simple tool to obtain a reference value.

During the study, we detected inconsistencies in some of the calibrated NO_2 sensor data, for instance, the higher NO_2 concentration in the garden side compared with the street side during the measurements at the home of patient 4. Moreover, the indoor NO_2 concentrations at the home of patient 1 were most probably overestimated too. However, without a reference value, it is not possible to validate these results. Future studies should explore effective strategies for managing calibration transfer.

Lastly, some negative values were still present in the NO_2 datasets after calibration (patients 2, 4, and 6). Measuring low concentration levels with the current NO_2 electrochemical sensors implies a higher uncertainty of the measurements. However, we have seen that concentration peaks are well detected. Considering this, air quality sensors have more potential in those regions or indoor microenvironments where high concentrations are expected.

4.1.2. Use of Stationary Sensors

The AQSSs were positioned in the living room, chosen based on participants' reported highest occupancy, to ensure monitoring in the area most frequented by the patients. There is a possibility of underestimating exposure during activities occurring outside of the living room. Questions were raised, especially regarding cooking, which is known to raise pollutant concentrations. This may be particularly important for patients who have closed kitchens. In the future, multiple sensors could be placed in multiple areas of a single home to determine the difference across several rooms. Consideration should be given to deploying sensors in additional indoor environments, such as workplace offices, where patients routinely spend significant time. This broader sensor placement strategy can provide a more comprehensive understanding of personal exposure patterns.

The stationary deployment of AQSSs raises concerns about the accuracy of personal exposure estimates, particularly when patients are not in the sensor's vicinity, such as when away from home. However, the convenience and minimal disruption afforded by stationary AQSSs in the home, as opposed to carrying portable sensors, are advantageous, especially for long-term studies.

The results of this study, using sensor systems at fixed locations, were compared with those of a similar study from Krause [43] using portable sensors. During the study with portable sensors, it was found that even though the highest dose per minute was observed during commuting, the time in transit accounts for less than 10% of total time, with the home environment emerging as the predominant contributor to the total dose [43]. This suggests that our methodology of using stationary sensors for indoor and outdoor microenvironments may be sufficient to estimate exposure in long-term studies but may lack the sensitivity required to investigate short-term health effects resulting from pollution peaks.

Although it was shown that the exposure estimates calculated using data from indoor stationary sensors and portable sensors were similar due to the predominant contribution of the home environment, there remain scenarios where stationary sensor data may not suffice for determining exposure. In our study, the patients spent an average of 83% of their time at home. Individuals who spend more time outside their homes may necessitate the use of portable sensors or multiple stationary sensors strategically placed in environments where they spend the most time. This consideration is particularly crucial in regions with high pollution levels where outdoor exposure may significantly impact overall exposure levels.

4.2. Nature of Participant-Reported Data

Assessing the veracity of self-reported data from participants poses a significant challenge. In certain instances, recorded parameters may be correlated with proxy variables that facilitate the validation of reported activities. For example, sudden temperature changes may be an indicator of the opening of a window. However, most other activities recorded by patients are more difficult to verify.

Accurately determining individuals' actions without invasive monitoring presents a complex challenge. However, invasive monitoring may not always be feasible due to participant reluctance or study demands. Participants may be unwilling to continuously wear sensors or record data for extended periods. If study requirements are overly demanding, there is a risk of non-compliance. Providing monetary compensation to volunteers is an option, but it does not guarantee data quality.

While there is no feasible method to ensure complete cooperation and accuracy in recording data, good communication with participants, education about the study's objectives, proper training, and regular checks can enhance the likelihood of complete and accurate data collection.

4.3. Activity Specific I/O Ratio

The I/O ratios of activities, when grouped by window status, were affected by the discrete nature of activity reporting. Patients could only record an event as occurring for a full hour. Especially in winter, it is unlikely that any patient kept the windows open for an entire hour. Additionally, there could be a lag between certain activities and the measurement of the pollutant, resulting in peak concentrations being recorded in the following hour and causing the peak to be misassigned. Diapouli et al. [51] found that in their study of the I/O ratio of PM mass and number concentration in residential buildings during various activities, cooking caused the highest I/O ratio, which is consistent with the results for $PM_{2.5}$ presented in Figure 5. Together with "Cooking", hours when "Eating" occurred had high I/O ratios as well. "Eating" often took place in the same hour as, or immediately following, cooking. Most patients had closed kitchens, which could result in pollutants not reaching the sensors in the living room until cooking was finished and the patient had moved to another room to eat. These results include all instances of "Eating", whether it was the only activity during the hour or not. If "Eating" occurred simultaneously with "Cooking", the I/O ratios may be overestimated.

4.4. Source Apportionment

The results of the source apportionment showed that the data of the stationary $PM_{2.5}$ and NO_2 sensors together with the logbook are able to accurately attribute peaks in concentration to indoor activities occurring in the same hour. For this analysis, the activities that were recorded together for the same hour were considered as a unique group. In essence, "Cooking" is treated as a separate activity from "Cooking, Eating". When it was distributed to patients, 13 single activities were listed on the activity log. After completion of the pilot study, 113 unique groups of activities were recorded among all patients. This emphasizes the difficulty of discretizing the contribution of any single activity to personal exposure. It is not possible with this method to isolate emissions from a single activity during hours when multiple activities were logged simultaneously. Still, from the methods used here, general conclusions can be made about which indoor activities generate the highest pollutant concentrations.

4.5. Symptomatology

There is uncertainty in health data because symptoms vary greatly among individuals. Many factors can trigger asthma symptoms, including weather, exposure to allergens or other irritants, activity level, and strong emotions [52]. These triggers may build on each other to further exacerbate symptoms. This complexity makes it challenging to discern which symptoms are directly related to changes in air quality.

Moreover, the differences in how individuals perceive their symptoms make comparisons between multiple participants more complex. To overcome that, the health questionnaire was designed specifically to avoid the subjectivity of the patients, with answers aiming to quantify the symptom severity (e.g., coughed once briefly, coughed briefly several times, coughed almost every hour, etc.). Overall, the trends in how symptoms change become more important in drawing conclusions about larger populations. An increase in symptom severity is a strong indicator of how one's health is affected.

The use of other quantitative measurements can have disadvantages when participants complete measurements themselves. Even a simple lung function test may pose compliance difficulties in longer-term studies. Firstly, the training of patients by qualified personnel is crucial for correct data collection. Secondly, the effort a single patient puts in can vary from day to day, impacting the reliability of the results. It is known that patient motivation drops with time. Jiang et al. [53] found that in a two-week study of asthma symptom tracking, patient compliance had already dropped by the second week. Another study on asthma patient compliance with long-term PEF measurements found that after 6 months, only 50% of participants still recorded accurate PEF values [54]. Addressing these challenges is crucial for ensuring the reliability and validity of the data collected over extended periods.

Despite the limitations of patient-recorded health data, there are several advantages of using a symptom log. First, it is one of the least invasive methods for obtaining health data, as patients do not need to visit a medical office or make appointments. Secondly, the only type of lung function testing undertaken was the PEF, which is simple and easy to do at home. The health data collected in this study was independent of the day of the week, as patients reported daily. Additionally, unlike studies using hospital admission data, symptoms were recorded across all levels of severity. This enabled the identification of the onset of symptom aggravation, rather than solely noting when symptoms reached a specific severity threshold. Future studies with a large number of participants could investigate the lag effect between changes in symptoms, including the evaluation of specific symptoms, and changes in pollutant concentrations. Such studies could examine correlations starting from the same day and extending up to a week later, as identified to be the longest delayed response time [55].

In summary, although this pilot study cannot determine the precise influence and lag between peaks in pollutant concentration and subsequent increases in health scores, the presented results demonstrate the potential use of AQSSs for environmental epidemiology at fixed locations. Stationary sensors collect data from indoors and the surroundings effortlessly for the patient, which makes them an appropriate technique for long-term epidemiological studies. The continuous indoor and outdoor monitoring gives a better understanding of the quality of the air each individual breathes.

Moreover, the health score system developed for this study consistently shows plausible information on symptom intensity with fluctuations from day to day, so that changes in the state of health or symptom burden are easy to understand and can be analysed using the point scores. Even with varying disease severity, there are sufficient intra-individual fluctuations to document day-dependent deviations in symptom burden and relate them to the air quality data. The PEF measurement is a useful addition as an objective parameter.

4.6. Exposure Assessment

One of the current limitations of determining personal exposure is the variability in individual behaviour and activities. In this study, the activity log was used to determine the patient's activity level. However, there are some challenges when using a method such as the activity log. Partly, this is due to the previously mentioned limitations of using participant self-recorded data. However, some difficulties occur when participants must select actions for a discrete amount of time. For instance, there were many cases during the study where participants had marked multiple activities in one hour. This is logical, given that not all activities last a full hour, or start exactly on the hour. This practice introduces uncertainty, particularly when activities vary in intensity, potentially affecting the accuracy of exposure estimates. In this study, the mean IR was used, assuming equal time allocation for each recorded activity within an hour. The results of the variability analysis indicated that this assumption did not significantly alter the estimated exposure outcomes.

Another potential source of uncertainty is the methodology to calculate the potential inhaled dose. In this study, each activity was considered separately, and the entire amount of pollutant measured during an hour was attributed to each activity performed in that hour. This approach may have led to the under- or overestimation of the potential dose associated with each activity. For instance, if a "sedentary" activity such as sleeping occurred in the same hour as a "light" activity with high emissions like cooking, the dose attributed to the sedentary activity would be overestimated due to the contribution from the more active one.

The variation in the IR, which depends on overall individual fitness or health, may contribute to uncertainty in personal exposure estimates. IR values predicted by the EPA are typically calculated for healthy populations [56]. In this study, no adjustments were made to account for the fact that participants had respiratory problems. However, evidence suggests that this may not significantly affect accuracy. Corlin et al. [56] studied the IR in a population with a high percentage of individuals with respiratory or cardiac health conditions and found that EPA estimates remained accurate. They included adjustments for weight in their calculations, which were not possible in our study. Given the limited number of studies addressing this concern, it should be considered in future research.

Another limitation of using the EPA-estimated IR is the somewhat loose definition of activity intensity. For instance, if a patient recorded "Exercising", it was assumed to be light exercise indoors and labelled as "moderate intensity". However, the actual intensity could vary, either higher or lower, without confirmation of the exact nature of the exercise. For future studies, recording and including heart rate data could enhance the accuracy of determining an individual's activity level and improve IR calculations. Additionally, investigating the duration of elevated IR during more intense activities would provide valuable insights.

Finally, the comparison of the results of the potential inhaled dose with the AIRLESS study has shown that (I) both studies agree on pointing out outdoor air quality data from outdoor monitoring stations as a possible source of exposure misclassification, and (II) the results may be overestimated when assuming a generic IR for the time spent at home. It would be worth considering the possibility of assuming a generic IR for the time participants spend outdoors rather than for the time they spend indoors. As nearly 80% of the exposure occurs in the home environment, the contribution of travel methods (car/bus, train, walk, cycle, motorcycle) represents a small percentage of the total dose. This approach could reduce the exposure misclassification caused by the use of generic IR for the time spent at home while minimizing participant effort. In regions where people spend most of their time indoors, it may not be necessary to use portable air quality sensors to estimate personal exposure. This potential simplification could streamline both data collection and analysis processes, given the inherent complexities associated with deploying portable sensors, both from a participant engagement and data analysis perspective. For those cases, the methodology used in our pilot study, combined with multiple sensors placed in other indoor environments used by participants (e.g., work office), may be sufficient. Overall, our results underscore the importance of both indoor measurements and activity adjusted IR for accurately calculating personal exposure.

On the whole, even though our study was limited by the sample size (seven participants), it demonstrates the significant value of AQSSs in acquiring indoor data for exposure assessment. The widespread use of (calibrated) AQSSs will potentially help to reduce the bias in evaluating gender differences in mortality due to air pollution as indoor measurements reflect more accurately the quality of air to which women and girls are frequently exposed [57]. Furthermore, the granular data collected by AQSSs could drive targeted interventions, enabling policymakers to address specific sources of indoor air pollution, which is not regulated currently, thereby enhancing public health protection and potentially reducing long-term healthcare costs.

5. Conclusions

In this work, a pilot project to study the feasibility of using stationary air quality sensors for $PM_{2.5}$ and NO_2 in epidemiological research was conducted. It was found that the calibrated AQSSs to be used indoors and outdoors at fixed locations were able to run for the month of the study with practically no issues in performance. The results of the exposure assessment showed that using either generic IR or outdoor monitoring station data leads to exposure misclassification. In this study, individuals spent an average of 83% of their time at home. That implies that the use of stationary AQSSs is sufficient for tracking the majority of one's personal exposure. Future studies could scale up the methodology used here and, using multiple stationary AQSSs, overcome the limitations of this study, as we only measured the air quality in the living room.

A detailed analysis of the indoor and outdoor data measured by the AQSSs in the home of patient 1 revealed a leakage through the window sealing, demonstrating that indoor air quality is influenced not only by routines and behaviours but also by ventilation and building characteristics. Additionally, the source apportionment and activity specific I/O ratio results showed that data from stationary AQSSs, when combined with information from a logbook, can accurately identify and attribute concentration peaks to specific indoor activities. This approach also allows for the evaluation of the influence of outdoor air and ventilation patterns on indoor air quality.

The activity specific concentration and potential dose were calculated and compared with the results of the AIRLESS study. It was shown that there are significant differences in the weight an indoor activity has on personal exposure depending on its intensity. The calculation of the IR, taking into account the information of the activity log, reduces the uncertainty compared with the use of a generic IR in the home environment.

The results of this study showed consistent indications of symptom intensity fluctuations from day to day, making changes in health status or symptom burden easy to understand and analyse using point scores. Despite varying disease severity, sufficient intra-individual fluctuations were documented to relate day-dependent deviations in symptom burden to air quality data. The examination methods used (health survey, peak flow meter) are therefore considered to be valid for studying the effects of air pollution on vulnerable patient groups with chronic respiratory diseases or asthma. To exclude other influences or confounding variables, a parallel recording of activities and events in the logbook was essential. Overall, this study emphasizes the importance of measuring air quality indoors and tracking activity data for studying personal exposure and how air quality sensors may potentially find good use in the field of environmental epidemiology.

Supplementary Materials: The following supporting information can be downloaded at: https://www.mdpi.com/article/10.3390/s24175767/s1, Table S1: Dates of AQSS installation, checks, and collection for each patient; Table S2: Participant demographics; Figure S1: (a) Map of Stuttgart showing locations of the outdoor air quality monitoring station (blue circle) and participants' homes where sensors were deployed (yellow diamonds), (b) outdoor AQSS and (c) indoor AQSS; Table S3: Data collected for each participant; Table S4: Selected results of the environmental questionnaire filled out by patients; Figure S2: Logbook; Table S5: Data completeness after cleaning steps, presented as a percentage; Table S6: Correction parameters for each $PM_{2.5}$ sensor; Figure S3: Health questionnaire; Table S7: Inhalation rates for each intensity level, categorised by associated activities, age and gender; Table S8: Median NO_2 and $PM_{2.5}$ concentrations and uncertainties associated with each AQSS; Figure S4: I/O ratios for $PM_{2.5}$ associated with individual activities, grouped by window status for patient 7; Figure S5: Time series of pollutant concentration combined with logged activities and window status, for patient 1, week 1; Figure S6: Time series of pollutant concentration combined with logged activities and window status, for patient 1, week 2; Figure S7: Time series of pollutant concentration combined with logged activities and window status, for patient 1, week 3; Figure S8: Time series of pollutant concentration combined with logged activities and window status, for patient 1, week 4; Figure S9: Time series of pollutant concentration combined with logged activities and window status, for patient 2, week 1; Figure S10: Time series of pollutant concentration combined with logged activities and window status, for patient 2, week 3; Figure S11: Time series of pollutant concentration combined with logged activities and window status, for patient 2, week 4; Figure S12: Time series of pollutant concentration combined with logged activities and window status, for patient 3, week 1; Figure S13: Time series of pollutant concentration combined with logged activities and window status, for patient 3, week 2; Figure S14: Time series of pollutant concentration combined with logged activities and window status, for patient 3, week 3; Figure S15: Time series of pollutant concentration combined with logged activities and window status, for patient 3, week 4; Figure S16: Time series of pollutant concentration combined with logged activities and window status, for patient 4, week 1; Figure S17: Time series of pollutant concentration combined with logged activities and window status, for patient 4, week 2; Figure S18: Time series of pollutant concentration combined with logged activities and window status, for patient 4, week 3; Figure S19: Time series of pollutant concentration combined with logged activities and window status, for patient 6, week 1; Figure S20: Time series of pollutant concentration combined with logged activities and window status, for patient 6, week 2; Figure S21: Time series of pollutant concentration combined with logged activities and window status, for patient 6, week 3: Note that the outdoor sensor stopped working on 19 April 2020 at 3:00 am; Figure S22: Time series of pollutant concentration combined with logged activities and window status, for patient 6, week 4; Figure S23: Time series of pollutant concentration combined with logged activities and window status, for patient 7, week 1; Figure S24: Time series of pollutant concentration combined with logged activities and window status, for patient 7, week 2; Figure S25: Time series of pollutant concentration combined with logged activities and window status, for patient 7, week 3; Figure S26: Time series of pollutant concentration combined with logged activities and window status, for patient 7, week 4; Figure S27: Daily indoor and outdoor $PM_{2.5}$ concentrations during PM alert (from 21 January to 26 January 2020) during the measurement campaign in the home of patient 1; Figure S28: Individual activity specific $PM_{2.5}$ (left) and NO_2 (right) concentrations, for patients 6 and 7: Note that the $PM_{2.5}$ concentrations are on a logarithmic scale; Figure S29: Individual activity specific $PM_{2.5}$ (left) and NO_2 (right) concentrations, for patients 1, 2, and 3 and 4: Note that the $PM_{2.5}$ concentrations are on a logarithmic scale; Figure S30: Hourly indoor and outdoor (garden and street) $PM_{2.5}$ and NO_2 concentrations, self reported health score and PEF for patient 4; Figure S31: Hourly indoor and outdoor $PM_{2.5}$ and NO_2 concentrations, self-reported health score and PEF for patient 6: Note that the outdoor sensor stopped working on 19 April 2020 at 3:00 am; Figure S32: Hourly indoor and outdoor $PM_{2.5}$ and NO_2 concentrations, self-reported health score and PEF for patient 7.

Author Contributions: Conceptualization, M.C.-M., U.V. and C.W.; data curation, E.R.; formal analysis, M.C.-M.; funding acquisition, U.V. and C.W.; investigation, M.C.-M.; methodology, M.C.-M. and U.L.; project administration, U.V.; resources, F.H.; software, E.R.; supervision, U.V. and C.W.; validation, M.C.-M.; visualization, M.C.-M. and E.R.; writing—original draft, M.C.-M. and E.R.; writing—review and editing, U.V. All authors have read and agreed to the published version of the manuscript.

Funding: This work has received funding from the Ministry for Social Affairs and Integration Baden-Württemberg (grant no. AZ 53-5425.1/5).

Institutional Review Board Statement: This study was conducted according to the guidelines of the Declaration of Helsinki and approved by the Ethics Committee of the Baden-Württemberg Medical Association on 8 November 2019 under the reference number: F-2019-105.

Informed Consent Statement: Informed consent was obtained from all subjects involved in this study.

Data Availability Statement: The raw data supporting the conclusions of this article will be made available by the authors upon request.

Acknowledgments: Grecia Carolina Solís-Castillo, Bernd Laquai, and Ioannis Chourdakis are highly acknowledged for their valuable contributions during this feasibility study. Ángela Blanco is acknowledged for helping to digitalise the activity logs. Finally, the volunteers are greatly appreciated for making this study possible.

Conflicts of Interest: The authors declare that they have no known competing financial interests or personal relationships that could have appeared to influence the work reported in this paper.

References

1. Orellano, P.; Reynoso, J.; Quaranta, N.; Bardach, A.; Ciapponi, A. Short-term exposure to particulate matter (PM10 and $PM_{2.5}$), nitrogen dioxide (NO_2), and ozone (O_3) and all-cause and cause-specific mortality: Systematic review and meta-analysis. *Environ. Int.* **2020**, *142*, 105876. [CrossRef] [PubMed]
2. Thurston, G.D.; Kipen, H.; Annesi-Maesano, I.; Balmes, J.; Brook, R.D.; Cromar, K.; de Matteis, S.; Forastiere, F.; Forsberg, B.; Frampton, M.W.; et al. A joint ERS/ATS policy statement: What constitutes an adverse health effect of air pollution? An analytical framework. *Eur. Respir. J.* **2017**, *49*, 1600419. [CrossRef] [PubMed]
3. United Nations. The 17 Goals. Available online: https://sdgs.un.org/goals (accessed on 18 December 2023).
4. Fuller, R.; Landrigan, P.J.; Balakrishnan, K.; Bathan, G.; Bose-O'Reilly, S.; Brauer, M.; Caravanos, J.; Chiles, T.; Cohen, A.; Corra, L.; et al. Pollution and health: A progress update. *Lancet Planet. Health* **2022**, *6*, e535–e547. [CrossRef] [PubMed]
5. WHO. *WHO Global Air Quality Guidelines. Particulate Matter ($PM_{2.5}$ and PM10), Ozone, Nitrogen Dioxide, Sulfur Dioxide and Carbon Monoxide*; World Health Organization: Geneva, Switzerland, 2021; ISBN 978-92-4-003422-8.
6. Tonne, C. A call for epidemiology where the air pollution is. *Lancet Planet. Health* **2017**, *1*, e355–e356. [CrossRef] [PubMed]
7. Vilcassim, R.; Thurston, G.D. Gaps and future directions in research on health effects of air pollution. *EBioMedicine* **2023**, *93*, 104668. [CrossRef]
8. Martin, R.V.; Brauer, M.; van Donkelaar, A.; Shaddick, G.; Narain, U.; Dey, S. No one knows which city has the highest concentration of fine particulate matter. *Atmos. Environ. X* **2019**, *3*, 100040. [CrossRef]
9. Kornartit, C.; Sokhi, R.S.; Burton, M.A.; Ravindra, K. Activity pattern and personal exposure to nitrogen dioxide in indoor and outdoor microenvironments. *Environ. Int.* **2010**, *36*, 36–45. [CrossRef]
10. Klepeis, N.E.; Nelson, W.C.; Ott, W.R.; Robinson, J.P.; Tsang, A.M.; Switzer, P.; Behar, J.V.; Hern, S.C.; Engelmann, W.H. The National Human Activity Pattern Survey (NHAPS): A resource for assessing exposure to environmental pollutants. *J. Expo. Anal. Environ. Epidemiol.* **2001**, *11*, 231–252. [CrossRef]
11. Bulot, F.M.J.; Russell, H.S.; Rezaei, M.; Johnson, M.S.; Ossont, S.J.J.; Morris, A.K.R.; Basford, P.J.; Easton, N.H.C.; Foster, G.L.; Loxham, M.; et al. Laboratory Comparison of Low-Cost Particulate Matter Sensors to Measure Transient Events of Pollution. *Sensors* **2020**, *20*, 2219. [CrossRef]
12. Rajagopalan, S.; Brook, R.D. The indoor-outdoor air-pollution continuum and the burden of cardiovascular disease: An opportunity for improving global health. *Glob. Heart* **2012**, *7*, 207–213. [CrossRef]
13. Landrigan, P.J.; Fuller, R.; Acosta, N.J.R.; Adeyi, O.; Arnold, R.; Basu, N.N.; Baldé, A.B.; Bertollini, R.; Bose-O'Reilly, S.; Boufford, J.I.; et al. The Lancet Commission on pollution and health. *Lancet* **2018**, *391*, 462–512. [CrossRef]
14. Gordon, S.B.; Bruce, N.G.; Grigg, J.; Hibberd, P.L.; Kurmi, O.P.; Lam, K.-B.H.; Mortimer, K.; Asante, K.P.; Balakrishnan, K.; Balmes, J.; et al. Respiratory risks from household air pollution in low and middle income countries. *Lancet Respir. Med.* **2014**, *2*, 823–860. [CrossRef] [PubMed]
15. Okello, G.; Nantanda, R.; Awokola, B.; Thondoo, M.; Okure, D.; Tatah, L.; Bainomugisha, E.; Oni, T. Air quality management strategies in Africa: A scoping review of the content, context, co-benefits and unintended consequences. *Environ. Int.* **2023**, *171*, 107709. [CrossRef] [PubMed]
16. Hasenkopf, C.; Sharma, N.; Kazi, F.; Mukerjee, P.; Greenstone, M. *The Case for Closing Global Air Quality Data Gaps with Local Actors: A Golden Opportunity for the Philanthropic Community*; EPIC: Chicago, IL, USA, 2023.
17. Garland, R.M.; Altieri, K.E.; Dawidowski, L.; Gallardo, L.; Mbandi, A.; Rojas, N.Y.; Touré, N.d.E. Opinion: Strengthening research in the Global South—Atmospheric science opportunities in South America and Africa. *Atmos. Chem. Phys.* **2024**, *24*, 5757–5764. [CrossRef]
18. Tonne, C.; Basagaña, X.; Chaix, B.; Huynen, M.; Hystad, P.; Nawrot, T.S.; Slama, R.; Vermeulen, R.; Weuve, J.; Nieuwenhuijsen, M. New frontiers for environmental epidemiology in a changing world. *Environ. Int.* **2017**, *104*, 155–162. [CrossRef]
19. Delgado-Saborit, J.M. Use of real-time sensors to characterise human exposures to combustion related pollutants. *J. Environ. Monit.* **2012**, *14*, 1824–1837. [CrossRef]
20. Chatzidiakou, L.; Krause, A.; Popoola, O.A.M.; Di Antonio, A.; Kellaway, M.; Han, Y.; Squires, F.A.; Wang, T.; Zhang, H.; Wang, Q.; et al. Characterising low-cost sensors in highly portable platforms to quantify personal exposure in diverse environments. *Atmos. Meas. Tech.* **2019**, *12*, 4643–4657. [CrossRef]
21. Tékouabou, S.C.K.; Chenal, J.; Azmi, R.; Diop, E.B.; Toulni, H.; Nsegbe, A.d.P. Towards air quality particulate-matter monitoring using low-cost sensor data and visual exploration techniques: Case study of Kisumu, Kenya. *Procedia Comput. Sci.* **2022**, *215*, 963–972. [CrossRef]
22. Bousiotis, D.; Alconcel, L.-N.S.; Beddows, D.C.S.; Harrison, R.M.; Pope, F.D. Monitoring and apportioning sources of indoor air quality using low-cost particulate matter sensors. *Environ. Int.* **2023**, *174*, 107907. [CrossRef]
23. Bertrand, L.; Dawkins, L.; Jayaratne, R.; Morawska, L. How to choose healthier urban biking routes: CO as a proxy of traffic pollution. *Heliyon* **2020**, *6*, e04195. [CrossRef]
24. Chatzidiakou, L.; Krause, A.; Han, Y.; Chen, W.; Yan, L.; Popoola, O.A.M.; Kellaway, M.; Wu, Y.; Liu, J.; Hu, M.; et al. Using low-cost sensor technologies and advanced computational methods to improve dose estimations in health panel studies: Results of the AIRLESS project. *J. Expo. Sci. Environ. Epidemiol.* **2020**, *30*, 981–989. [CrossRef] [PubMed]

25. Jerrett, M.; Donaire-Gonzalez, D.; Popoola, O.; Jones, R.; Cohen, R.C.; Almanza, E.; de Nazelle, A.; Mead, I.; Carrasco-Turigas, G.; Cole-Hunter, T.; et al. Validating novel air pollution sensors to improve exposure estimates for epidemiological analyses and citizen science. *Environ. Res.* **2017**, *158*, 286–294. [CrossRef]
26. Mazaheri, M.; Clifford, S.; Yeganeh, B.; Viana, M.; Rizza, V.; Flament, R.; Buonanno, G.; Morawska, L. Investigations into factors affecting personal exposure to particles in urban microenvironments using low-cost sensors. *Environ. Int.* **2018**, *120*, 496–504. [CrossRef]
27. Xie, Q.; Dai, Y.; Zhu, X.; Hui, F.; Fu, X.; Zhang, Q. High contribution from outdoor air to personal exposure and potential inhaled dose of $PM_{2.5}$ for indoor-active university students. *Environ. Res.* **2022**, *215*, 114225. [CrossRef]
28. Amegah, A.K.; Dakuu, G.; Mudu, P.; Jaakkola, J.J.K. Particulate matter pollution at traffic hotspots of Accra, Ghana: Levels, exposure experiences of street traders, and associated respiratory and cardiovascular symptoms. *J. Expo. Sci. Environ. Epidemiol.* **2022**, *32*, 333–342. [CrossRef]
29. Steinle, S.; Reis, S.; Sabel, C.E.; Semple, S.; Twigg, M.M.; Braban, C.F.; Leeson, S.R.; Heal, M.R.; Harrison, D.; Lin, C.; et al. Personal exposure monitoring of $PM_{2.5}$ in indoor and outdoor microenvironments. *Sci. Total Environ.* **2015**, *508*, 383–394. [CrossRef] [PubMed]
30. Song, J. Towards space-time modelling of $PM_{2.5}$ inhalation volume with ST-exposure. *Sci. Total Environ.* **2024**, *948*, 174888. [CrossRef] [PubMed]
31. Lu, Y. Beyond air pollution at home: Assessment of personal exposure to $PM_{2.5}$ using activity-based travel demand model and low-cost air sensor network data. *Environ. Res.* **2021**, *201*, 111549. [CrossRef]
32. Che, W.; Frey, H.C.; Fung, J.C.H.; Ning, Z.; Qu, H.; Lo, H.K.; Chen, L.; Wong, T.-W.; Wong, M.K.M.; Lee, O.C.W.; et al. PRAISE-HK: A personalized real-time air quality informatics system for citizen participation in exposure and health risk management. *Sustain. Cities Soc.* **2020**, *54*, 101986. [CrossRef]
33. Chatzidiakou, L.; Krause, A.; Kellaway, M.; Han, Y.; Li, Y.; Martin, E.; Kelly, F.J.; Zhu, T.; Barratt, B.; Jones, R.L. Automated classification of time-activity-location patterns for improved estimation of personal exposure to air pollution. *Environ. Health* **2022**, *21*, 125. [CrossRef]
34. Chapizanis, D.; Karakitsios, S.; Gotti, A.; Sarigiannis, D.A. Assessing personal exposure using Agent Based Modelling informed by sensors technology. *Environ. Res.* **2021**, *192*, 110141. [CrossRef]
35. Amegah, A.K. Proliferation of low-cost sensors. What prospects for air pollution epidemiologic research in Sub-Saharan Africa? *Environ. Pollut.* **2018**, *241*, 1132–1137. [CrossRef] [PubMed]
36. Li, J.; Hauryliuk, A.; Malings, C.; Eilenberg, S.R.; Subramanian, R.; Presto, A.A. Characterizing the Aging of Alphasense NO_2 Sensors in Long-Term Field Deployments. *ACS Sens.* **2021**, *6*, 2952–2959. [CrossRef]
37. Sesé, L.; Gille, T.; Pau, G.; Dessimond, B.; Uzunhan, Y.; Bouvry, D.; Hervé, A.; Didier, M.; Kort, F.; Freynet, O.; et al. Low-cost air quality portable sensors and their potential use in respiratory health. *Int. J. Tuberc. Lung Dis.* **2023**, *27*, 803–809. [CrossRef]
38. Samad, A.; Kiseleva, O.; Holst, C.C.; Wegener, R.; Kossmann, M.; Meusel, G.; Fiehn, A.; Erbertseder, T.; Becker, R.; Roiger, A.; et al. Meteorological and air quality measurements in a city region with complex terrain: Influence of meteorological phenomena on urban climate. *Meteorol. Z. Contrib. Atmos. Sci.* **2023**, *32*, 293–315. [CrossRef]
39. Schwitalla, T.; Bauer, H.-S.; Warrach-Sagi, K.; Bönisch, T.; Wulfmeyer, V. Turbulence-permitting air pollution simulation for the Stuttgart metropolitan area. *Atmos. Chem. Phys.* **2021**, *21*, 4575–4597. [CrossRef]
40. Chacón-Mateos, M.; Laquai, B.; Vogt, U.; Stubenrauch, C. Evaluation of a low-cost dryer for a low-cost optical particle counter. *Atmos. Meas. Tech.* **2022**, *15*, 7395–7410. [CrossRef]
41. U.S. Environmental Protection Agency. *Exposure Factors Handbook: 2011 Edition*; EPA/600/R-09/052F; U.S. Environmental Protection Agency: Washington, DC, USA, 2011. Available online: https://www.epa.gov/expobox/about-exposure-factors-handbook (accessed on 23 December 2023).
42. U.S. Environmental Protection Agency. *Guidelines for Exposure Assessment*; U.S. Environmental Protection Agency: Washington, DC, USA, 1992; Volume 57, pp. 22888–22938.
43. Krause, A. *Using Novel Portable Air Quality Monitors to Improve Personal Exposure and Dose Estimations for Health Studies*; Apollo-University of Cambridge Repository: Cambridge, UK, 2021.
44. Chacón-Mateos, M.; Vogt, U.; Laquai, B.; García-Salamero, H.; Witt, C.; Liebers, U.; Heimann, F. *Evaluation of Air Quality Sensors for Environmental Epidemiology*; EGU General Assembly: Vienna, Austria, 2023.
45. García Salamero, H. Evaluation of Long-Term Measurements of NO_2 Sensors for Epidemiological Studies. Master's Thesis, University of Stuttgart, Stuttgart, Germany, 2021.
46. Fortmann, R.; Kariher, P.; Clayton, R. *Indoor Air Quality: Residential Cooking Exposures: Final Report*; ARCADIS Geraghty & Miller, Inc.: Sacramento, CA, USA, 2001.
47. Stamp, S.; Burman, E.; Chatzidiakou, L.; Cooper, E.; Wang, Y.; Mumovic, D. A critical evaluation of the dynamic nature of indoor-outdoor air quality ratios. *Atmos. Environ.* **2022**, *273*, 118955. [CrossRef]
48. Stasiulaitiene, I.; Krugly, E.; Prasauskas, T.; Ciuzas, D.; Kliucininkas, L.; Kauneliene, V.; Martuzevicius, D. Infiltration of outdoor combustion-generated pollutants to indoors due to various ventilation regimes: A case of a single-family energy efficient building. *Build. Environ.* **2019**, *157*, 235–241. [CrossRef]
49. Faria, T.; Martins, V.; Correia, C.; Canha, N.; Diapouli, E.; Manousakas, M.; Eleftheriadis, K.; Almeida, S.M. Children's exposure and dose assessment to particulate matter in Lisbon. *Build. Environ.* **2020**, *171*, 106666. [CrossRef]

50. deSouza, P.; Kahn, R.; Stockman, T.; Obermann, W.; Crawford, B.; Wang, A.; Crooks, J.; Li, J.; Kinney, P. Calibrating networks of low-cost air quality sensors. *Atmos. Meas. Tech.* **2022**, *15*, 6309–6328. [CrossRef]
51. Diapouli, E.; Eleftheriadis, K.; Karanasiou, A.A.; Vratolis, S.; Hermansen, O.; Colbeck, I.; Lazaridis, M. Indoor and Outdoor Particle Number and Mass Concentrations in Athens. Sources, Sinks and Variability of Aerosol Parameters. *Aerosol Air Qual. Res.* **2011**, *11*, 632–642. [CrossRef]
52. Vernon, M.K.; Bell, J.A.; Wiklund, I.; Dale, P.; Chapman, K.R. Asthma Control and Asthma Triggers. *J. Asthma Allergy Educ.* **2013**, *4*, 155–164. [CrossRef]
53. Jiang, H.; Han, J.; Zhu, Z.; Xu, W.; Zheng, J.; Zhu, Y. Patient compliance with assessing and monitoring of asthma. *J. Asthma* **2009**, *46*, 1027–1031. [CrossRef]
54. Côté, J.; Cartier, A.; Malo, J.L.; Rouleau, M.; Boulet, L.P. Compliance with peak expiratory flow monitoring in home management of asthma. *Chest* **1998**, *113*, 968–972. [CrossRef]
55. Dąbrowiecki, P.; Chciałowski, A.; Dąbrowiecka, A.; Badyda, A. Ambient Air Pollution and Risk of Admission Due to Asthma in the Three Largest Urban Agglomerations in Poland: A Time-Stratified, Case-Crossover Study. *Int. J. Environ. Res. Public Health* **2022**, *19*, 5988. [CrossRef]
56. Corlin, L.; Woodin, M.; Amaravadi, H.; Henderson, N.; Brugge, D.; Durant, J.L.; Gute, D.M. A field study to estimate inhalation rates for use in a particle inhalation rate exposure metric. *Sci. Total Environ.* **2019**, *696*, 133919. [CrossRef]
57. Véliz, K.D.; Alcantara-Zapata, D.E.; Chomalí, L.; Vargas, J. Gender-differentiated impact of $PM_{2.5}$ exposure on respiratory and cardiovascular mortality: A review. *Air Qual. Atmos. Health* **2024**, *17*, 1565–1586. [CrossRef]

MDPI AG
Grosspeteranlage 5
4052 Basel
Switzerland
Tel.: +41 61 683 77 34

Sensors Editorial Office
E-mail: sensors@mdpi.com
www.mdpi.com/journal/sensors

www.ingramcontent.com/pod-product-compliance
Lightning Source LLC
Chambersburg PA
CBHW041427210926
51850CB00013B/3374
* 9 7 8 3 7 2 5 8 4 2 4 1 4 *